T4-ALD-403

AN INTRODUCTION TO DISCRETE MATHEMATICS AND ITS APPLICATIONS

Kenneth Kalmanson
MONTCLAIR STATE COLLEGE
UPPER MONTCLAIR, NEW JERSEY

ADDISON-WESLEY PUBLISHING COMPANY

READING, MASSACHUSETTS • MENLO PARK, CALIFORNIA
DON MILLS, ONTARIO • WOKINGHAM, ENGLAND • AMSTERDAM
SYDNEY • SINGAPORE • TOKYO • MEXICO CITY • BOGOTÁ
SANTIAGO • SAN JUAN

Sponsoring Editor: Jeffrey Pepper
Project Editor: Marion E. Howe
Packaging Service: Cobb/Dunlop Publisher Services, Inc.
Text Designer: Rita Naughton
Cover Designer: Slide Graphics
Manufacturing Supervisor: Ann DeLacey

Library of Congress Cataloging in Publication Data

Kalmanson, Kenneth.
An introduction to discrete mathematics and its applications.

Bibliography: p.
Includes index.
1. Electronic data processing—Mathematics.
I. Title.
QA76.9.M35K35 1986 510 84-28459
ISBN 0-201-14947-8

ABCDEFGHIJK-VB-89876

PREFACE

This book was written while the author was directing a Sloan Foundation supported curriculum improvement project at Montclair State College on freshman and sophomore level mathematics. The book is designed for a one or two semester introductory course in discrete mathematics. The author and his colleagues have successfully used this book in manuscript form with both freshmen classes (as part of the Sloan Foundation supported program) as well as with more briskly paced sophomore level courses.

The only formal prerequisite for the first seven chapters of this text is two years of high school algebra. For Chapter 8, some experience with adding and multiplying formal series and finding roots of polynomials is useful but not essential. Calculus is not needed. Although algorithms are discussed at length starting with Section 1.3 and pervade the book as a whole, no computer programming is required. Any additional skills that are needed, such as matrix multiplication, are developed within the book.

Chapter 1 lays a foundation for the remainder of book by reviewing several key facts about numbers and sets while introducing many new ideas and techniques. For example, scientific notation is reviewed at the same time that we discuss normalized exponential form. Similarly, twos-complement subtraction is introduced right after we have reviewed simple binary arithmetic, and a review of simple clock arithmetic (which is often, but not always, discussed in elementary school) leads to an elementary discussion of the beautiful Chinese remainder theorem, an application that computer scientists find particularly useful. In fact, many such applications are given throughout the text.

Mathematical reasoning, set operations, and the algebraic laws that relate them to one another are of central interest in Chapter 2. Construction of a binary adder in Section 2.6 will quickly convince budding computer scientists of the merits of studying such abstract material. (Mathematics majors presumably do not need any such convincing here!) In a one-semester course one might continue with a brief discussion of Boolean algebras and minimization of logical circuits by plunging right into the middle of Section 7.5. (At that point we give a definition of a Boolean algebra as a set that is closed under certain operations and such that the operations obey certain laws.) A longer course (three quarters or two semesters in length) would take up these questions somewhat later in the general algebraic context of Chapter 7, where Boolean algebras are treated as lattices.

If one chooses to follow the "linear" order of the text, the question of how computers do addition via logical circuits forms a natural transition to Chapter 3 with its discussions of more sophisticated methods of counting, also known as "combinatorics."

Chapter 3 begins with basic material on permutations and combinations, but it contains additional topics—permutations with repetitions, inclusion/exclusion, probability, and recursion—that could enrich even a one-semester course. Sections 3.3–3.6 are independent of one another, but if they are taken in the order presented, a loose structure will become apparent. I highly recommend that the sections on recurrence relations and analysis of algorithms (3.6 and 3.7) be included in a first course on discrete mathematics. (Section 3.8 may be considered optional. Similarly, Chapter 8 contains a further discussion of recurrence relations, as well as an introduction to generating functions.) The next four chapters require no more in the way of combinatorics than what is contained in Sections 3.1 and 3.2. The only exception to this rule is a part of Section 5.4, which requires Section 3.7.

Chapters 4, 5, and 6 comprise the "geometric" or "topological" portion of this book. Multigraphs are discussed before digraphs because the multigraph has a simpler structure. Moreover, properties of digraphs can be deduced from those of multigraphs. The material on multigraphs and digraphs is developed around various models and algorithms to increase student interest and skill in problem solving. Some theorems are proved, but the emphasis is heavily on the side of developing intuition rather than on theorem proving.

The first five sections of Chapter 7 develop Boolean algebras as particular kinds of partial orders and lattices. As pointed out earlier, however, one can begin a discussion of Section 7.5 directly after Section 2.6 by starting with our "alternate definition" of a Boolean algebra. When time permits, however, the more leisurely, and algebraically comprehensive, treatment is preferred. Switching circuits are given in Section 7.6 as still another useful example of a Boolean algebra. The chapter closes with a discussion of groups and their application to coding theory.

If there is a common thread that runs through the various topics discussed under the heading of "discrete mathematics," it is the concept of an algorithm. Students are introduced to our plain language structured form of algorithms in Section 1.3 by means of the problem of computing sums. Algorithms are found, thereafter, in most sections of the text and in many of the exercises.

Algorithms should not be confused with computer programs, and, as pointed out earlier, computer programming is not required for use of this text. Computer programs are nice vehicles with which to illustrate algorithms, however, especially if one has the hardware on which to run them. It is for this reason that I have included an appendix of computer programs with a self-contained student study guide. The programs are written in BASIC so that they can be run on most readily available personal computers with little or (in the case of the Commodore 64K) no modification. Students are encouraged to write their own programs in the accompanying exercises.

Instructors who adopt this text for classroom use can get a free disk containing the programs that appear in the appendix from Addison-Wesley.

Instructors who like to include some formal computer programming in their discrete mathematics courses will find a set of computer programming exercises at the end of each chapter also. These can be assigned in any suitable language or, if one prefers, ignored.

Each section has its own set of exercises, and no exercise set requires computer programming. Each set is graded in difficulty with the more difficult exercises well toward the rear of the set.

Each set of exercises is preceded by a "completion review" of definitions, formulas, and other key facts discussed in the section. Rather than asking the student merely to read a list of such facts, the student is encouraged to actively recall them with a "warm up" type quiz. Answers to the completion reviews are provided immediately following each set of questions, whereas the answers to the exercise sets proper, about half of which are given, appear at the back of the book. There are over 1350 exercises in all!

One can give a variety of one-semester (or quarter) courses with this book. I have found the following *one-semester sequence*, for example, very successful: Chapter 1 (all except, perhaps 1.8); Chapter 2 (all); Sections 7.5 and 7.6; Sections 3.1, 3.2, 3.6, and 3.7; Sections 4.1, 4.2, 4.5,4.6, 5.1, 5.2, and 6.1.

A *more-advanced one-semester course* in combinatorics, multigraphs, and digraphs might begin by discussing algorithms (Section 1.3) and then proceeding with Chapters 3, 4, 5, 6, and 8.

In a *one-year course*, one could do all or most of Chapters 1–3 and Sections 4.1 and 4.2 during the first semester, followed by the remainder of the book, with a few omissions, perhaps, in the second semester.

In a *three-quarters sequence*, one could take Chapters 1 and 2, plus Section 3.1 and 3.2 during the first quarter; continuing through to Section 6.1 in the second quarter; and finishing most of the material remaining in the last quarter.

The following diagram shows the overall structure of this book, where a dotted line indicates a weak dependency and a solid line shows a strong dependency:

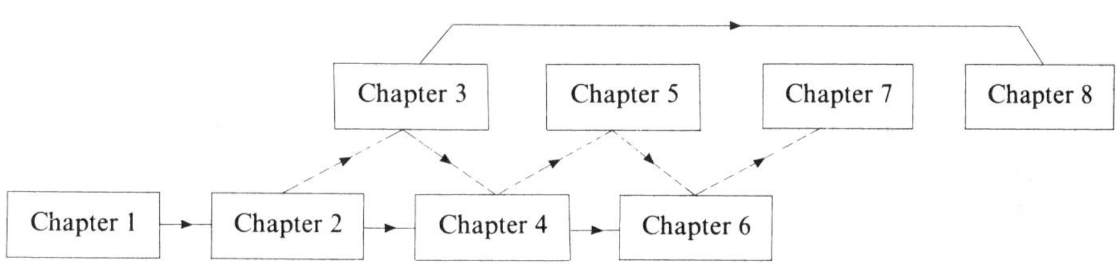

Furthermore, any of the following sections can be omitted within chapters without loss of continuity: 3.3, 3.4, 3.5, 4.3, 4.4, 5.2, 6.2, 6.3, 7.6, 7.8, 8.1, and 8.2.

As a final note, the end of an example is indicated by an open square, and the end of a proof is indicated by a solid square.

REVIEWERS

I am grateful for the comments and suggestions provided by the following people.

Ron Sandstrom (Fort Hays State University)
Harold Frederickson (Gould Electronics, El Monte, California)
Barbara Smith-Thomas (University of North Carolina at Greensboro)
Dana Richards (University of Virginia)
Keith Yale (University of Montana)
Stephen Hedetniemi (Clemson University)
Roxy Peck (California Polytechnic State University)

Donald Thompson (Pepperdine University)
Kevin T. Phelps (Georgia Institute of Technology)
Gerald Grossman (Oakland University, Michigan)
Robert Crawford (Western Kentucky University)

ACKNOWLEDGEMENTS

If it were possible, I would like to list the names of all of those whose suggested improvements and whose assistance in catching errors have helped me to make this a better book. In particular, my thanks to my editors Jeffrey Pepper at Addison Wesley and Rachel Hockett at Cobb/Dunlop; my colleagues at Monclair State College William Parzynski, Frank Servidio, James Stoddard, and Walter Westphal, who helped in class testing preliminary versions of the text; my wife Judith, and my son, Andrew, who helped with the typing and proofreading; my son, Matthew, who, in addition, supplied an early version of the program PRIMETEST; and my students, whose suggestions and encouragement were indispensable to this project. Finally, I would like to thank the Alfred E. Sloan Foundation for their grant to Montclair State College supporting our discrete mathematics project.

Kenneth Kalmanson

Upper Montclair, New Jersey

CONTENTS

Chapter 6 DIRECTED GRAPHS AND NETWORKS 269

Chapter 7 APPLIED MODERN ALGEBRA 315

Chapter 8 FURTHER TOPICS IN COUNTING AND RECURSION 385

Appendix PROGRAMS IN BASIC AND EXERCISES 421

CHAPTER 

SETS, NUMBERS, AND ALGORITHMS

The advent of the electronic computer has enabled us to keep track of and process data on a scale never before possible. The mathematics involved in the processing of data begins with arranging objects into categories or "sets," labeling them, ordering them, and counting them. Each of these, in turn, depends upon a clear understanding of the integers and the other real numbers. To achieve this understanding is the first goal of the present chapter.

The second major goal of this introductory chapter is to become accustomed to thinking in terms of step-by-step computational procedures, whenever possible. Although we will not, for the most part, discuss computer programs in the main body of this text, careful conception of such procedures is an important, and perhaps *the* most important, part of computer programming.

1.1 DEFINING SETS AND SUBSETS

A **set** is a well-defined collection of objects called its **members** or **elements**. Two sets A and B will be called **equal** if they have exactly the same members, or if neither has any members. In either case we will write "$A = B$." Otherwise we will write "$A \neq B$." Capital letters will be used to denote sets. We will generally reserve lowercase letters such as x, y, a, and b for the members of sets. Furthermore we will write "$x \in S$" when we mean that x is a member of S and "$y \notin S$" when y is not a member of S.

Pictures such as Fig. 1.1 are commonly used to illustrate set theoretic concepts. In these **Venn diagrams**, an oval represents a set. A dot within, or outside of, the oval represents an element that is a member of the set in the case of x, or is not a member of the set as in the case of y.

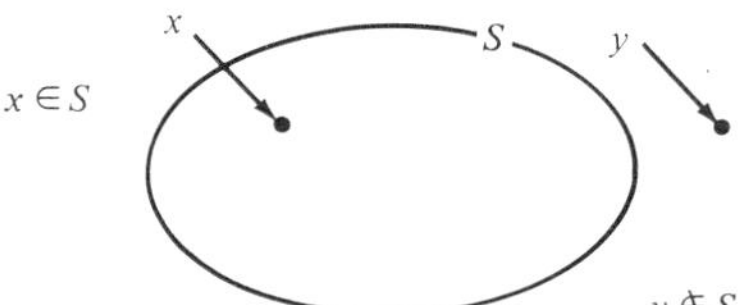

Figure 1.1 Venn diagrams show which elements are members of a set S and which are not.

To say that a set is **well defined** means that for any object x, either x belongs to the set or it does not belong to the set. Such a determination is most easily accomplished if all the elements of the sets are exhibited for your inspection. In this **listing**

method of defining a set, the elements of the set are displayed in some order, separated by commas, and the list is enclosed in braces.

Example 1

a) $T=\{0, 1, 2, 3, 4, 5, 6, 7, 8, 9\}$ is the set of whole numbers from 0 through 9, and $6 \in T$. The symbols themselves are called the **decimal digits**, and the first two, the "0" and the "1," are called the **binary digits**.

b) $C=\{$READ, INPUT, PRINT, GOTO$\}$ is a set of commands in the computer programming language BASIC (an acronym for *B*eginners' *A*ll-purpose *S*ymbolic *I*nstruction *C*ode) and PRINT $\in C$.

c) $P=\{$Apple, Atari, IBM, Commodore$\}$ is a set of companies that produce and sell microcomputers (also called "personal computers"). Observe that Osborne $\notin P$; even though Osborne manufactures microcomputers, it is not listed among the elements of this set.

d) $S=\{\{0\}, \{1\}, \{2\}, \{3\}, \{4\}, \{5\}, \{6\}, \{7\}, \{8\}, \{9\}\}$ is a set whose elements are sets themselves. (Such a set is sometimes called a *class* of sets.) Observe: $1 \notin S$, but $\{1\} \in S$, since the number 1 and the set containing this number are different objects. □

Take care not to confuse a list with a set. Unlike in a list, the order in which the elements of a set appear within the braces is unimportant and duplicates are ignored. Thus we have Example 2.

Example 2

$$\{r, a, t\} = \{t, a, r\} = \{t, a, r, a\}$$ □

Sometimes we define a set by listing only a portion of its members, relying upon the pattern of the elements in the list to show that, eventually, every element is accounted for. This is especially true for certain sets of numbers.

Example 3

a) The **set of positive integers from 1 to *n*** is written

$$\{1, 2, 3, \ldots, n\}.$$

The dots indicate that elements of the set may have been omitted from the list. The pattern of the first three elements listed implies a rule whereby we can supply the missing elements: (1) 1 is the first element; (2) add 1 to each element to get the next element; (3) when you reach n, stop. (n is the last element.)

b) The entire set of **positive integers** is given by

$$N = \{1, 2, 3, \ldots\}.$$

In this case the list has no last member; only steps (1) and (2) of Example 1(a) are relevant.

c) The **null set** or **empty set** is the unique set that has no elements whatever. (See Exercise 1.1-13.) It is denoted by the symbol "$\varnothing$" or by an empty pair of braces: $\{\ \}$. □

We say that **set *A* is a subset of set *B*** if every element of A is an element of B, or equivalently, if every element excluded from B is also excluded from A. This is illustrated in Fig. 1.2. We write $\boldsymbol{A \subset B}$ if A is a subset of B.

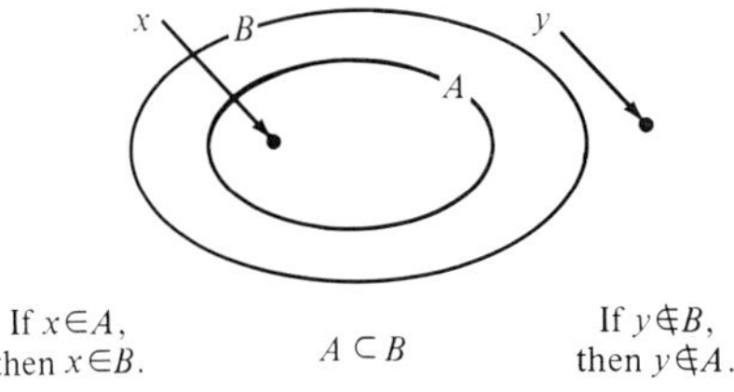

Figure 1.2 Set A is a subset of set B. The two equivalent definitions of a "$A \subset B$" are illustrated.

Example 4

a) The empty set $\varnothing$ is a subset of every other set. (See Exercise 1.1-14.)

b) Every set is a subset of itself.

c) If $A = \{1, 2, 3, 4\}$ and $B = \{1, 2, 3, 4, 5\}$, then $A \subset B$.

d) The set of positive integers is a subset of the set

$$\{0, 1, 2, 3, 4, \ldots\}.$$ □

The set of all subsets of a given set S is called the **power set of *S*** and is denoted by ***P*(*S*)**. Thus $S \in P(S)$ for any set S. The other elements of $P(S)$, if any, are called the **proper subsets of *S***.

Example 5 If $S = \{0, 1\}$, then $P(S) = \{\varnothing, \{1\}, \{0\}, \{0, 1\}\}$. □

If S is a well-defined set, then so is $P(S)$. Moreover, if S is well defined, then we can describe its subsets by means of **rules of inclusion** that tell us the distinguishing characteristics of the elements in the subset.

Example 6

a) $\{x \in N: x > 3\} = \{4, 5, 6, 7, \ldots\}$. The colon is read "such that." Hence we have the set of "all positive integers x such that x is greater than 3."

b) $\{x \in N: x < 0\} = \varnothing$, since there are no positive integers that are less than zero.

c) Let S be the set of full-time matriculated students in your school. Since the words "full-time matriculated" usually have a very definite meaning, we can accept S as a well-defined set. What is the "parent set"? We are not told, and in this case it is not really necessary to know, since it could be any one of several well-defined sets.

d) The set of "nice" students at your school is not well defined, however, unless one defines the word "nice" in an unambiguous way.

e) The set of "all sets that are not elements of themselves" is not well defined. Such a set S either would have to be an element of itself or not, and each of these logical possibilities leads to its exact opposite! If $S \notin S$, then S is not an element of itself. By definition of S, however, this implies that $S \in S$! But if $S \in S$, then S *is* an element of itself. We call such a contradiction ($S \in S$ and $S \notin S$) a **logical paradox**. (See Exercise 1.1-20 for the closely related "barber paradox.") □

Finite, Infinite, and Discrete Sets

The set of all points on an ordinary line is taken to be a well-defined set even though we cannot list its elements. (See Exercise 1.5-17.) We will assume that one can use any line as a **number line**, which we can think of as a sort of ruler extending without bound in each direction. To each point on the line there corresponds a number, called a **real number**, as shown in Fig. 1.3, in a way that reflects the usual order and distance relationships between real numbers.

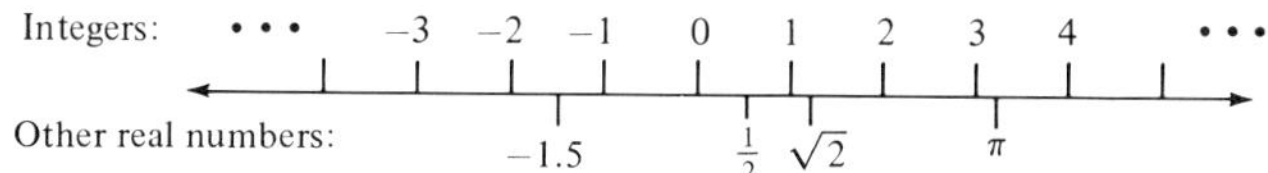

Figure 1.3 Integers and some other real numbers.

The numbers indicated above the number line in Fig. 1.3 form the set of **integers**, which we shall denote by Z. This set has no smallest or largest element, as opposed to the set of positive integers $N = \{1, 2, 3, \ldots\}$, which has the smallest element 1. Thus any listing of the set of integers such as

$$Z = \{0, 1, -1, 2, -2, 3, -3, \ldots\}$$

must exhibit the integers out of their natural order.

Our intuitive notions about ordering, labeling, counting, and simple arithmetic are formed by our experience with the set of positive integers. Subtraction eventually gives us 0 and the negative integers (in reverse order), whereas division and other operations make us aware of still larger sets of numbers—"larger" in the sense of set inclusion, of course. But at the same time, our ideas about ordering and so on become more sophisticated, and the way in which we use even the positive integers becomes more abstract and thereby more useful.

Let us reconsider how we order elements in a list. There must be a "first" element, a "second," and so forth, which leads to the idea of labeling these elements with the positive integers, $\{1, 2, \ldots\}$. This labeling is often indicated with subscripted letters of the form x_i, which is read "x sub i," and where i is usually taken to be a positive integer. Hence x_i is the ith term taken from a list, or **sequence**, having a "first term" x_1, a "second term" x_2, and so on. Sometimes we even begin with a "zeroth term" x_0, which is taken before the first term. The set of all these objects may be denoted by $\{x_1, x_2, x_3, \ldots\}$, or, if we are considering only a **finite sequence** having a last term x_n, by the set $\{x_1, x_2, x_3, \ldots, x_n\}$. Once again we caution the reader not to confuse the set, which is defined only by what the elements x_i are, with the list, which tells us, in addition, the order in which the elements must appear.

Example 7 The sequences 1, 2, 1, 2, ... and 2, 1, 2, 1, ... are different, even though the sets $\{1, 2, 1, 2, \ldots\}$ and $\{2, 1, 2, 1, \ldots\}$ are both equal to the set $\{1, 2\}$, and hence are equal to each other. □

Definition 1.1 We say that (a) a nonempty set S is **finite**, and that (b) **S has n elements**, indicated by writing (c) $|S| = n$, if all the elements of S can be written in a finite sequence $s_1, s_2, \ldots, s_n$ such that $s_i = s_j$ only if $i = j$. Moreover, the null set is also a finite set and $|\varnothing| = 0$. (d) All other sets are called **infinite**.

The set given in Example 7 is, of course, finite. Furthermore we are assuming, in effect, that the positive integers from 1 to n (Example 3a) form a finite set.

Notice that a sequence need not be finite, even when its terms form a finite set, since the terms may repeat the same numbers again and again.

Here are two further examples of finite sets.

Example 8

a) The set L of letters of the English alphabet is finite, since we can write $s_1 = a$, $s_2 = b$, $s_3 = c, \ldots, s_{26} = z$, where $s_i = s_j$ only if $i = j$. Hence $|L| = 26$.

b) The set of BASIC language instructions in the following simple computer program is finite. In referring to each instruction by its line number (which tells the order of execution of the commands), we have $s_1 = 100$, $s_2 = 110$, $s_3 = 120$, and $s_4 = 130$.

```
100  LET A=1
110  LET B=3
120  PRINT A+B
130  PRINT A-B
```

We see that we have a set of four distinct instructions (which result in the printing of the numbers 4 and -2). □

Definition 1.2 A set of real numbers S is **discrete** if for every x in S there is a positive distance d_x such that $|x-y|>d_x$ for every y in S. Here $|x-y|$ denotes absolute value.

The intuitive meaning of Definition 1.2 is that one can draw an interval about each number x of a discrete set that separates it from all other numbers in that set. One cannot do this, for example, with the number 0 in the set

$$T=\{1, 1/2, 1/3, \ldots, 1/n, \ldots, 0\}.$$

Hence T is not a discrete set.

However, it is not difficult to see that finite sets S of real numbers are discrete, since, if we are given x in S, then we may take d_x to be one half the minimum of the distances between x and all the other elements in S. Moreover, subsets of discrete sets are themselves discrete. And although the set Z of integers is an infinite set, it is also discrete. (Take $d_x=1/2$ for every integer x.) Hence all subsets of Z are discrete sets. We discuss some of these subsets in the following section.

Completion Review 1.1

Complete each of the following.

1. A well-defined collection of objects is called a ________________.
2. The set having no elements is called the ________________ or ________________ and is denoted ________________.
3. We write "$x \in S$" to mean that ________________; "$x \notin S$" means that ________________.
4. The listing method would give "________________" as the set whose elements are a, b, and c.
5. We write "$A \subset B$" to mean that ________________, while "$A=B$" means that the sets A and B have ________________. We say that "A is a proper subset of B" if ________________.
6. A rule of inclusion defines a set by telling ________________.

7. A statement that logically implies its own negation is known as a

________________.

8. The symbol "$P(S)$" denotes the ________________ of the set S. Its elements are

________________ of set S.

9. The set of positive integers is the set whose elements form the sequence

________________.

10. A set S is called "discrete" if for every x in S there is a ________________ such that for each y in S we have ________________.

11. A set is called "finite" if ________________. Otherwise we say that the set is

________________.

12. We write "$|S| = n$" if the set S has ________________.

Answers: **1.** set. **2.** empty set; null set; $\varnothing$. **3.** x belongs to S; x does not belong to S. **4.** $\{a, b, c\}$. **5.** A is a subset of B—every element of A is in B; have the same elements; an element of B is not in A, but $A \subset B$. **6.** a property of its elements that distinguish it from those not in the set. **7.** logical paradox. **8.** power set; the subsets. **9.** $1, 2, 3, \ldots, n, \ldots$. **10.** $d_x > 0$; $|x - y| < d_x$. **11.** it has exactly n elements; infinite. **12.** exactly n elements.

Exercises 1.1

For Exercises 1 to 8, which of the following are well-defined sets? Upon what do you base your answer in each case?

1. The large cities in the United States.
2. The cities in the United States that are state capitals.
3. The fast computer programs.
4. The negative integers.
5. $\{x \in N: 1 \leqslant x \leqslant 10\}$.
6. $\{x \in N: 1 < x < 2\}$.
7. The mathematics majors at your school.
8. The sets that are not members of themselves.

For Exercises 9 to 12 list the elements in each set—all of them if the set is finite.

9. $\{x: x \in N \text{ and } -3.5 < x < 5\}$.
10. $\{x: x^2 + 5x + 6 = 0\}$.
11. $\{1, 2, 3, \ldots, 20\}$.
12. $\{x: x \in N \text{ and } x \text{ is even}\}$.
13. Explain why, on the basis of set equality, there can be only one empty set.
14. Why is the empty set a subset of every other set? Base your answer on our "negative definition" of the term "subset."

15. Give a Venn diagram illustrating the following rule: If $A \subset B$ and $B \subset C$, then $A \subset C$.

16. Show that the rule in the last exercise is true—without using a picture. Just use the definition of set inclusion.

17. Which of the following pairs of sets are equal?

a) $\{a, b, c, b, c, a\}$ and $\{a, b, c\}$.

b) N and $\{2, 4, 6, \ldots\}$.

c) $\{\varnothing, \{a\}, \{b\}, \{a, b\}\}$ and $P(\{a, b\})$.

18. List the elements in $P(\{0, 1, 2\})$.

19. Explain why the following rule is true on the basis of the definitions of the terms involved: If $A \subset B$ and $B \subset A$, then $A = B$. Also, if $A = B$, then it follows that $A \subset B$ and $B \subset A$.

20. (The barber paradox) "In a certain town of clean-shaven men, one man is the only barber. The barber shaves every man in town who does not shave himself." Why is this a paradox? (*Hint*: Who shaves the barber?)

21. If the terms in the sequence 1, 2, 3, 1, 2, 3, 1, 2, 3, . . . are labeled x_i, where $i = 1, 2, 3, \ldots$, find **(a)** x_5; **(b)** x_{10}; **(c)** x_{100}.

22. Is the set $\{1, 2, 3, 1, 2, 3, 1, 2, 3, \ldots\}$ finite or infinite? Explain.

23. Why must every subset of a discrete set of numbers be discrete?

24. Find $|P(S)|$, where $S = \{0, 1, 2\}$.

25. Suppose that m and n are two members of a set of integers S. What is the greatest number of elements of S that might lie between m and n on the number line?

1.2 SETS AND FUNCTIONS

The idea of a correspondence between two sets is, as we have seen in our discussion on sequences, a very fundamental one. Indeed we could hardly do any labeling or counting without it. Although one can rely upon intuitive notions about sequences, or lists, of objects, as we did in Section 1.1 to get us started in our thinking about such things, not all sets of objects can be listed. The set of real numbers is one such set. (See Exercise 1.5-17.)

We can make the idea of correspondence more precise, and more general, if we base it upon sets whose elements are **ordered pairs**, that is, objects of the form (a, b), where $(a, b) = (c, d)$, if and only if $a = c$ and $b = d$. The **first elements** of each ordered pair are a and c, and the **second elements** are b and d.

Definition 1.3 The **Cartesian product** of two sets A and B is denoted by "$A \times B$," where

$$A \times B = \{(a, b): \quad a \in A \text{ and } b \in B\}.$$

Example 1 Let $A=\{0, 1\}$ and $B=\{r, s, t\}$. Then

a) $A \times B=\{(0, r), (0, s), (0, t), (1, r), (1, s), (1, t)\}$.

b) $B \times A=\{(r, 0), (r, 1), (s, 0), (s, 1), (t, 0), (t, 1)\}$.

c) $A \times A=\{(0, 0), (0, 1), (1, 0), (1, 1)\}$. □

Definition 1.4 Given two nonempty sets A and B, then we write (a) $\boldsymbol{f: A \rightarrow B}$, and we say that f is a (b) **function** from A into B, if f is a subset of $A \times B$, such that:

i) no two ordered pairs in f have the same first element, and

ii) for every $a \in A$, there is some $b \in B$, such that (a, b) is in f. The set A is called the (c) **domain** of f, and the set B is called the (d) **codomain** of f. The (e) **range** of f is the set

$$\{y \in B: \quad (x, y) \in f, \text{ for some } x \text{ in } A\}.$$

We say that y is the (f) **image** of x if $(x, y) \in f$, and in that case we write (g) $\boldsymbol{y=f(x)}$ (read "y equals f of x").

The idea of a function should not be entirely unfamiliar to you, but our present definitions may take some getting used to. Here are three examples to get you started.

Example 2 Let $A=\{0, 1\}$ and $B=\{r, s, t\}$ as in Example 1.

a) We may define the function $f: A \rightarrow B$ as the set

$$\{(0, r), (1, t)\},$$

in which r is the image of 0 and t is the image of 1. We may write

$$f(0)=r \quad \text{and} \quad f(1)=t.$$

The range of this function is the set $\{r, t\}$, and the domain of f is the set A. The codomain of f is B.

b) We may define the function $g: B \rightarrow A$ as the set

$$\{(r, 0), (s, 1), (t, 1)\},$$

where 0 is the image of r, and 1 is the image of both s and t. Hence

$$g(r)=0, g(s)=1, \quad \text{and } g(t)=1.$$

Notice that no two first elements of the function are equal, but the second elements of two different ordered pairs are allowed to be the same. Also notice that the domain of g is the set B, while A is the codomain and range.

c) We can define the function $h\colon A \to A$ from A into A as the set

$$\{(0, 1),\ (1, 0)\}.$$

Hence

$$h(0) = 1 \quad \text{and} \quad h(1) = 0.$$

The range, domain, and codomain of the function h are all A. □

Another way to think of functions is as correspondences between the elements in two sets. The arrows in diagrams such as those in Fig. 1.4 show how each element corresponds to its image. The functions are those of Example 2.

As we said previously, functions can help us give a more precise definition of some things we have already discussed. Here are some examples.

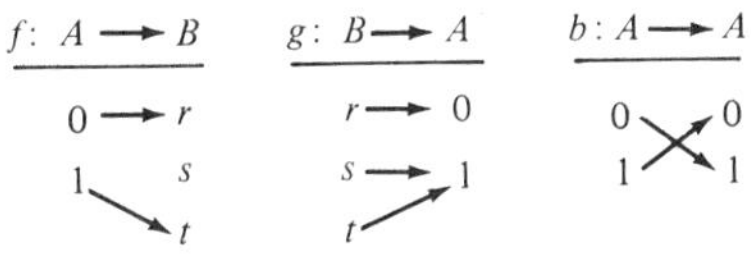

$f\colon A \to B$: $0 \to r$, $1 \to t$ (s has no preimage)

$g\colon B \to A$: $r \to 0$, $s \to 1$, $t \to 1$

$h\colon A \to A$: $0 \to 1$, $1 \to 0$

Figure 1.4 Function f is one-to-one and function g is onto, but only function h is a one-to-one correspondence from its domain onto its codomain.

Example 3

a) A sequence $s_1, s_2, s_3, \ldots$ is actually a function whose domain is the set $N = \{1, 2, 3, \ldots\}$ of positive integers. The ordered pairs in this function are of the form (n, s_n), where the second element is the term of the sequence. In other words, each n corresponds to, or labels, the term s_n exactly as we said before.

b) A finite sequence is a function whose domain is the set of the first n positive integers $\{1, 2, 3, \ldots, n\}$. As a more specific example, the finite sequence 1492, 1776, 1812, 2001 is actually the function $\{(1, 1492), (2, 1776), (3, 1812), (4, 2001)\}$. Notice that we could rearrange the ordered pairs in this set and still know exactly what the original sequence was. □

To sharpen our ideas about counting, the following definition is useful.

Definition 1.5 Let $f: A \to B$. We say that (a) ***f* maps *A* onto *B*** if B is the range of f. We say that (b) ***f* maps *A* one-to-one into *B*** if whenever $f(x_1)=f(x_2)$, then it follows that $x_1=x_2$. Functions that have both these properties are called (c) **one-to-one correspondences from *A* onto *B***.

Example 4 Referring to Fig. 1.4, we see that:

a) The function f does not map A onto B, since s is not the image of any element in A. But f does map A one-to-one ("1–1") into B, since each element in A has a different image in B.

b) The function g maps B onto A, because A is the range of g. That is, each element in A is the image of some element in B. However, g does not map B one-to-one into A, because $g(s)=g(t)=1$, and yet $s \neq t$.

c) The function h is a one-to-one correspondence from A onto itself, since it maps A both one-to-one and onto A. Observe that neither f nor g was a one-to-one correspondence from its domain into its codomain. □

Definition 1.6 If $f: A \to B$ is a one-to-one correspondence from A onto B, we define (a) the set $f^{-1} = \{(b, a) \in B \times A: (a, b) \in f\}$, and (b) we call f^{-1} the **inverse** of the function f. (Hence $f^{-1}(b)=a$ only if $f(a)=b$ and f^{-1} exists.)

The inverse of a one-to-one correspondence from A onto B, then, is a function from B onto A. (See Exercise 1.2-19a.) In fact it is a one-to-one correspondence from B onto A. (See Exercise 1.2-19b.) This is what allows us to speak of a "one-to-one correspondence *between* sets A and B." And knowing that there is a one-to-one correspondence between two sets is the same as knowing that they have the same number of elements (see Exercise 1.2-19c), at least if we are discussing finite sets.

We seem to have come right back to where we began at the end of the last section. But in fact we have, in addition to gaining increased precision, obtained a new way of looking at functions, which are also called "mappings," "correspondences," "transformations," and "operators" in various mathematical contexts.

Have we lost anything? Not at all. Perhaps you are used to seeing functions defined by statements such as

$$y=x^2 \quad \text{or} \quad f(x)=x^2,$$

for example, where we often assume that the domain is as large as possible when it has not been defined. Well we could just write

$$f=\{(x, y): x \text{ and } y \text{ are real numbers and } y=x^2\},$$

and we have restated our function more precisely. Of course,

$$g = \{(x, y): x \text{ and } y \text{ are integers and } y = x^2\}$$

is not the same function, since these are not the same sets.

What we have called "functions" you have known in many cases as the "graphs" of functions, which are plotted on an $x-y$ Cartesian coordinate system, as in Fig. 1.5. The figure shows f (or its "graph") as a continuous curve and g as a set of dots or "points" on this curve, and where the x and y "axes" are just real number lines, as in Section 1.1.

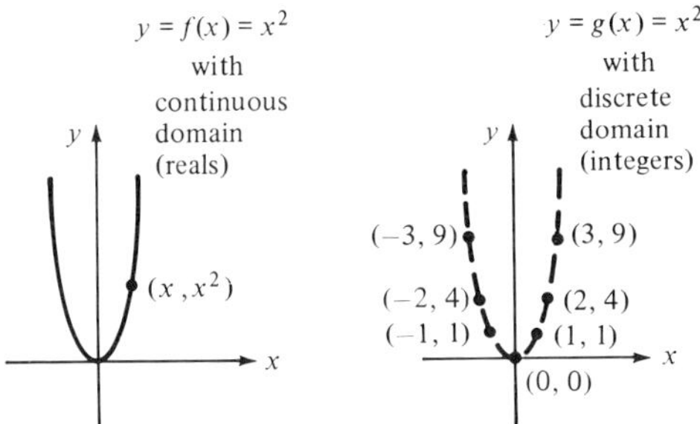

Figure 1.5 The domain of f is continuous, while the domain of g is discrete.

One further comment is in order before we end this section. Instead of the usual "$y=f(x)$," we will often use the notation

"Set $y \leftarrow f(x)$"

in our later work with computational methods. What this will mean, in general, is that you are to *substitute the value x into the function, or rule, f and assign the resulting value to y*. You will often know the value of x because of a previous statement of the form

"Set $x \leftarrow 1$,"

which just tells you to assign the value 1 to the variable x.

Example 5 Consider the following four statements. The numbers preceding each statement tell us the order in which they are read.

1. Set $x \leftarrow 2$.
2. Set $y \leftarrow x^2$.

3. Set $x \leftarrow y^3$.

4. Set $x \leftarrow x+1$.

Statement 1 just assigns the value 2 to x. Then in statement 2 we see that y is evaluated by means of the function $f(x)=x^2$. Since x is 2, *we assign the value* $f(2)=2^2=4$ *to the variable* y *in step* 2.

In statement 3 x is evaluated by means of the function $x=g(y)=y^3$, and the present value of y is 4. Hence x *is assigned the value* $g(4)=4^3=64$ *in step number 3.*

Finally x is evaluated in statement 4 by means of the function $h(x)=x+1$. Therefore we evaluate this function using the value assigned in step 3, and x *is now assigned its final value* $h(64)=64+1=65$.

Notice that the last step did *not* say that "$x=x+1$," which is clearly impossible. It merely told us how to make the last assignment to the variable x. □

Completion Review 1.2

Complete each of the following.

1. We say that the ordered pairs (a, b) and (c, d) are equal if and only if

________________.

2. The Cartesian product of sets A and B is written ________________ and consists of

________________.

3. A function $f: A \to B$ is a subset of ________________ such that no two (a, b) in f have ________________ and for every a in A there is an element b in B such that

________________.

4. If $f: A \to B$, then A is called the ________________ of f and B is called the ________________ of f. The range of f is the set ________________. We say that y is the image of x if ________________, and in that case we write ________________.

5. We say that a function f maps A onto B if ________________.

6. We say that a function f maps A one-to-one into B if ________________.

7. A function is a one-to-one correspondence of A onto B if f is ________________.

8. The inverse of a one-to-one correspondence f is denoted ________________ and is equal to ________________.

9. The notation "Set $y \leftarrow f(x)$" means that we substitute ________________ and assign the resulting value to the variable ________________.

10. The final values of x and y after the following sequence of instructions are ________________ and ________________: Set $x \leftarrow 2$; set $y \leftarrow x^2$; set $x \leftarrow y^3$; set $x \leftarrow x+1$.

Answers: **1.** $a=c, b=d$. **2.** $A \times B$; $\{(a, b): a \in A \text{ and } b \in B\}$. **3.** $A \times B$; the same first element; $(a, b) \in f$. **4.** domain; codomain; $\{y \in B: (x, y) \in f \text{ for some } x \text{ in } A\}$; $(x, y) \in f$; $y=f(x)$. **5.** B is the range of f. **6.** $f(c)=f(d)$ always implies $c=d$. **7.** one-to-one and onto B. **8.** f^{-1}; $\{(b, a) \text{ in } B \times A: (a, b) \in f\}$. **9.** x in $f(x)$; y. **10.** $x=65$; $y=4$.

Exercises 1.2

Let $A=\{a, b, c\}$ and $B=\{d, e, f, g\}$. Then in Exercises 1 to 9:

1. List the elements in $A \times B$.
2. List the elements in $B \times A$.
3. List the elements in $A \times A$.
4. List the elements in $B \times B$.
5. If $|A|=n$ and $|B|=m$, how many elements are in $A \times B$? Explain.
6. What are the range, domain, and codomain for the following function from A into B: $\{(a, e), (b, g), (c, d)\}$?
7. Why is each of the following *not* a function from A into B?
   a) $\{(a, d), (b, f), (c, h)\}$.
   b) $\{(a, d), (b, f), (c, g), (a, e)\}$.
   c) $\{(e, a), (g, b), (d, c)\}$.
8. Draw a diagram like Fig. 1.4 showing the correspondence given by the function in Exercise 6. Is that function **(a)** 1–1; **(b)** onto B; **(c)** a 1–1 correspondence between A and B? Explain.
9. What do we mean when we say that the sequence 2, 4, 6, 8, ... is a function? What is the domain of this function? Its range?
10. Repeat Exercise 9 for the finite sequence 2, 4, 6, 8.
11. Repeat Exercise 9 for the sequence $0, 1, -1, 2, -2, 3, -3, \ldots$.
12. Are functions in Exercises 9 to 11 one-to-one correspondences between the domain and range of each function? What does this seem to say about the sizes of these sets?
13. Find the inverse of the function $g: A \rightarrow B$ defined by $\{(1, 2), (2, 3), (4, 5)\}$ where A is the set of first and B is the set of second elements respectively.
14. Suppose, in Definition 1.6, the inverse of a function, we apply the definition to the function in Exercise 6—incorrectly, of course. What do we obtain? Is the result a function from B into A?
15. Write the definition of the "function" $y=x^3$ as a set of ordered pairs.
   a) What is the graph of this function?
   b) Does this function have an inverse? Explain.
16. Follow the directions in Exercise 15 for $y=x^4$.

17. What are the final values of x and y if we follow the sequence of the following instructions?

1. Set $x \leftarrow 3$.
2. Set $y \leftarrow 2x + 5$.
3. Set $x \leftarrow y - 4$.
4. Set $x \leftarrow x - 1$.

18. What are the final values of x and y if we follow the sequence of the following instructions?

1. Set $x \leftarrow 5$.
2. Set $x \leftarrow 3$.
3. Set $y \leftarrow 1$.
4. Set $y \leftarrow x^2$.

19. Let f be a one-to-one correspondence from A onto B.

a) Explain why f^{-1} must be a function from B into A.

b) Explain why f^{-1} must be a one-to-one correspondence onto A.

c) If two nonempty sets A and B are of size n, explain how one can find a one-to-one correspondence from A onto B. [*Hint:* Use the set $\{1, 2, \ldots, n\}$ as an intermediary, and draw a picture if necessary.]

20. The **identity function** $I: A \to A$ from a set A into itself is defined by $I(x) = x$ for all x in A. What is this function as a set of ordered pairs?

1.3 SUMS AND ALGORITHMS

The widespread availability of computers has made it possible for many people to do very complex arithmetic problems with a few taps on a keyboard. This in no way has made the need to understand arithmetic obsolete. On the contrary, it simply requires that people become more familiar with the conventional notation of algebra so that they can tap the full potential of their machines.

In this section we study some convenient notation and computational ideas associated with sums. For example, suppose that a function f is evaluated on the first n positive integers, giving us $f(1)$, $f(2), \ldots,$ and $f(n)$. Then the sum of these values might be denoted by $S(n)$. This sum is also a function, and $S(n)$ is given explicitly by the formula

$$S(n) = f(1) + f(2) + \cdots + f(n).$$

In actual computations addition is done two numbers at a time. (We say it is a "binary" operation.) To compute sums such as $S(n)$ in the usual way, therefore, we might use the following step-by-step procedure.

We "begin," in a sense, with a sum of zero, and we can formalize this by writing $S(0) \leftarrow 0$. We then add $f(1)$ to $S(0)$, obtaining $S(1) \leftarrow S(0) + f(1)$, or $S(1) = f(1)$. Addition of $f(2)$ to our previous sum can be written $S(2) \leftarrow S(1) + f(2)$, or $S(2) = f(1) + f(2)$. Similarly we assign to $S(3)$ the sum $S(2) + f(3) = (f(1) + f(2)) + f(3)$; and to $S(4)$ we assign the value $S(3) + f(4) = ((f(1) + f(2)) + f(3)) + f(4)$. Finally we know that the process has come to an end when we have assigned $S(n)$ the value $S(n-1) + f(n)$.

The procedure we have used to evaluate $S(n)$ can be written more succinctly in the following form.

Algorithm 1.1 ***Summation Procedure*** Given the function f and the positive integer n, to compute the sum $S(n) = f(1) + f(2) + \cdots + f(n)$,

1. [Initialize $S(0)$.] Set $S(0) \leftarrow 0$.
2. [Sum loop.] Repeat step 3 for $k = 1, 2, \ldots, n$.
   3. [Update $S(k)$.] Set $S(k) \leftarrow S(k-1) + f(k)$. (Recall that this means that $f(k)$ is added to the previous sum and the result is assigned to the new sum. Moreover, if $k < n$, we increase k by one and repeat this step. Otherwise we proceed with step 4.)
4. [Done.] Output $S(n)$ and stop.
5. End of summation procedure.

It is a good idea to test one's understanding of any computational procedure by trying to apply it to a specific example.

Example 1 Let us suppose that our function has the form $f(x) = x^2$ and that $n = 3$. These are known as **inputs**. In step 1 we say that we are giving the **initial value** of the sum $S(k)$, that is, the value of the sum when $k = 0$. Step 2 specifies that step 3 will be repeated with the values $k = 1$, 2, and 3 taken in succession. Hence step 3 is called a **loop**. Notice how indenting this step makes it stand out.

When $k = 1$ we have

$$S(1) \leftarrow S(0) + f(1) = 0 + 1^2 = 1.$$

When $k = 2$ we have

$$S(2) \leftarrow S(1) + f(2) = 1 + 2^2 = 5.$$

Finally, when $k = 3$ we have

$$S(3) \leftarrow S(2) + f(3) = 5 + 3^2 = 14.$$

Now we have satisfied step 2 and so the loop is complete. Step 4 requires only that we give the "answer," or **output**, which is 14, and that we stop. The last statement, step 5, is simply a signal that there are no further steps. □

Many of the computational procedures we present from now on will be in the form of our summation procedure. Some of the features you will notice are the following. Each step is preceded by a bracketed phrase explaining its purpose. Some steps will also be followed by a longer explanation in parentheses for better understanding. Even more pertinent are the following.

1. The steps are executed in the order they are numbered, unless otherwise noted (as in steps 3 and 4).
2. Inputs may be given even before step 1, but only finitely many inputs will be allowed, in general.
3. The steps themselves, aside from bracketed or parenthesized remarks, are well defined as a calculation, assignment, comparison, transfer to another step, or combination of these, for example.
4. There must be some output when the procedure ends.

Definition 1.7 A computational procedure will be called an **algorithm**, if, in addition to satisfying the foregoing four criteria, (5) it terminates after finitely many steps, and (6) each step can be completed in a finite amount of time.

Hence our computational procedure, Algorithm 1.1, given the inputs $f(x) = x^2$ and $n = 3$, may be called (after the Persian textbook writer Abu Ja'far Mohammed ibn Musa al-Khowarizmi, who lived in the ninth century A.D.) an algorithm, since the finiteness criteria have clearly been satisfied.

Sigma Notation

Sums of the form

$$f(1) + f(2) + f(3) + \cdots + f(n)$$

are often written using the greek letter sigma $\sum$, like this:

$$\sum_{k=1}^{n} f(k).$$

We define this "sigma notation" more generally as in Definition 1.8.

Definition 1.8 Given the function $f(k)$ and the integers r and s, such that $r \leqslant s$, we define (a) the **sigma notation**

$$\sum_{k=r}^{s} f(k) = f(r) + f(r+1) + f(r+2) + \cdots + f(s-1) + f(s).$$

The integers r and s are called (b) the **lower and upper limits of summation**, respectively, and the variable k is called (c) the **index of summation**.

Example 2

a) $\sum_{k=1}^{6} k = 1+2+3+4+5+6 = 21$. Observe that $f(k) = k$ in this example.

b) $\sum_{j=2}^{5} j^2 = 2^2 + 3^2 + 4^2 + 5^2 = 54$. Notice that the name of the index does not have to be k or in terms of k and that the first value of the index need not be 1.

c) $\sum_{k=1}^{n} k = 1 + 2 + \cdots + n$ is a variable sum. Writing the terms in reverse order:

$\sum_{k=1}^{n} k = n + (n-1) + \cdots + 1$, and adding vertically, we obtain

$$2\sum_{k=1}^{n} k = (n+1) + (n+1) + \cdots + (n+1) = n(n+1).$$

Then, after we divide by 2, we can write our sum in **closed form**, that is, an explicit function of the variable n:

$$\boxed{\sum_{k=1}^{n} k = \frac{n(n+1)}{2}}$$

d) The sum of the first n terms $a_0, a_0r, a_0r^2, \ldots, a_0r^{n-1}$ of a **geometric progression** (or sequence) having a first term a_0 and common ratio r can be written

$$\sum_{k=0}^{n-1} a_0r^k = a_0 + a_0r + a_0r^2 + \cdots + a_0r^{n-1}.$$

Hence

$$r\sum_{k=0}^{n-1} a_0r^k = a_0r + a_0r^2 + a_0r^3 + \cdots + a_0r^n, \text{ by multiplication.}$$

Subtracting the second line from the first and canceling repeated terms on the right-hand side, we obtain

$(1-r)\sum_{k=0}^{n-1} a_0r^k = a_0 - a_0r^n$, and, upon division by $1-r$,

$$\boxed{\sum_{k=0}^{n-1} a_0r^k = \frac{a_0 - a_0r^n}{1-r} \quad (r \neq 1)}$$

is the closed form. □

It is often desirable to obtain a closed form for a sum, as we did in Examples 2(c) and 2(d). It is obviously easier to write

$$\sum_{k=1}^{1,000,000} k = \frac{1,000,000(1,000,001)}{2} = 500,000,500,000$$

than to sum up one million numbers, even with a computer.

The following rules, whose proof we defer until after we have studied "mathematical induction" in Section 2.1, are useful in obtaining closed forms and otherwise manipulating sums.

Theorem 1.1 ***Summation Rules*** Given two functions f and g and a constant A:

a) $\displaystyle\sum_{k=m}^{n} f(k) + \sum_{k=m}^{n} g(k) = \sum_{k=m}^{n} (f(k) + g(k)).$

b) $\displaystyle\sum_{k=m}^{n} A \cdot f(k) = A \sum_{k=m}^{n} f(k).$

c) $\displaystyle\sum_{k=m}^{n} A = (n-m+1) \cdot A.$

Example 3 Suppose we have to evaluate the sum $\sum_{k=1}^{1000} (2k+3)$. Let us instead consider the corresponding sum $\sum_{k=1}^{n} (2k+3)$ with a variable upper limit n. Then we obtain the closed form as follows:

$$\sum_{k=1}^{n} (2k+3) = \sum_{k=1}^{n} 2k + \sum_{k=1}^{n} 3 \qquad \text{by Theorem 1.1(a).}$$

$$= 2\sum_{k=1}^{n} k + 3n \qquad \text{by Theorem 1.1(b) and (c).}$$

$$= 2\,\frac{n(n+1)}{2} + 3n \qquad \text{by Example 2(c).}$$

$$= n^2 + 4n \qquad \text{for all positive integers } n.$$

Hence

$$\sum_{k=1}^{1000} (2k+3) = (1000)^2 + 4(1000) = 1{,}004{,}000.$$

□

Polynomial Evaluation

Evaluation of a function of the form

$$p(x) = a_0x^n + a_1x^{n-1} + a_2x^{n-2} + \cdots + a_{n-1}x + a_n,$$

that is, a **polynomial**, is usually done by (1) substituting the required value of x, (2) finding its first through nth powers, (3) multiplying by the coefficients a_k, and (4) adding the products a_kx^{n-k}. Finding the powers of x, especially by means of built-in programs for y^x, often introduces serious errors. For this reason the required sums may be evaluated in the following "nested form," which we illustrate here for fourth-degree polynomials.

Definition 1.9 The **nested** form of the polynomial

$$p(x) = a_0x^4 + a_1x^3 + a_2x^2 + a_3x + a_4$$

is

$$p(x) = (((a_0x + a_1)x + a_2)x + a_3)x + a_4.$$

For example, the polynomial $p(x) = 5x^4 + 2x^3 - 6x^2 + x - 15$ would be written in nested form as

$$p(x) = (((5x + 2)x - 6)x + 1)x - 15.$$

If x were then assigned a value, we could evaluate $p(x)$ as follows:

$$5$$

$$5x + 2$$

$$(5x + 2)x - 6$$

$$((5x + 2)x - 6)x + 1$$

$$(((5x + 2)x - 6)x + 1)x - 15$$

The nested form of a polynomial actually requires fewer operations than the original to evaluate. (See Exercise 1.3-22.) Moreover, we can use the following algorithm to evaluate a fourth-degree polynomial without first writing out its nested form. All we need are the coefficients a_0 through a_4 and x.

Algorithm 1.2 To evaluate the polynomial $p(x) = a_0x^4 + a_1x^3 + a_2x^2 + a_3x + a_4$, given a_0, a_1, a_2, a_3, a_4, and x,

1. [Initialize P.] Set $P \leftarrow a_0$.
2. [Nesting loop.] Repeat step 3 for $k = 1, 2, 3$, and 4.
   3. [Multiply and add.] Set $P \leftarrow P \cdot x + a_k$. (We follow the nesting pattern of Definition 1.9. If $k < 4$, then we increase k by 1 and repeat this step. Otherwise we continue with step 4.)
4. [Output.] Output P and stop.
5. End of Algorithm 1.2.

After the value of $p(x)$ is initialized at a_0, steps 2 through 4 construct the value of P in the following pattern.

$$a_0$$

$$a_0x + a_1$$

$$(a_0x + a_1)x + a_2$$

$$((a_0x + a_1)x + a_2)x + a_3$$

$$(((a_0x + a_1)x + a_2)x + a_3)x + a_4: \textit{ Answer}$$

Example 4 We will illustrate our algorithm by evaluating

$$p(x) = (((3x + 2)x + 4)x - 5)x + 1 \quad \text{when } x = -3.$$

	3	2	4	−5	+1
(Add to first row.)		−9	21	−75	240
(Multiply by −3 these 1st and 2nd row sums.)	3	−7	25	−80	241 = $p(-3)$.

When the calculation is done by hand in the form we have given, it is often called **synthetic division** because *the numbers in the third row are the coefficients one* *obtains upon dividing $p(x)$ by $x-(-3)$.* That is, one can check by long division that

$$\frac{p(x)}{x-(-3)} = 3x^3 - 7x^2 + 25x - 80 + \frac{241}{x-(-3)}. \qquad \square$$

The reader should find little difficulty in modifying the algorithm we have illustrated for polynomials of degree other than 4. (See Exercise 1.3-20.) We should mention that this method of evaluating polynomials in nested form is often called **Horner's method**.

Completion Review 1.3

Complete each of the following.

1. An algorithm is a ________________ that requires only ________________ many steps, each of which requires only a ________________ amount of time to execute.
2. In the sigma notation $\sum_{k=r}^{s} f(k)$, k is called the ________________, while r and s are called the ________________.
3. A closed form of the sum $\sum_{k=1}^{n} k$ of the first n positive integers is ________________.
4. A closed form of the sum $\sum_{k=1}^{n-1} ar^k$ of a geometric progression is ________________.
5. A polynomial is a function that has the form ________________.
6. The nested form of $3x^2 + 4x + 5$ is ________________.
7. A method of evaluating polynomials that makes use of their nested form is called ________________ or ________________.
8. Three rules that may be used to simplify sums are ________________, ________________, and ________________.

Answers: **1.** computational procedure; finitely; finite. **2.** index of summation; limits of summation. **3.** $n(n+1)/2$. **4.** $(a-ar^n)/(1-r)$. **5.** $a^0x^n + a_1x^{n-1} + \cdots + a_{n-1}x + a_n$. **6.** $(3x+4)x+5$. **7.** synthetic division; Horner's method. **8.** (See Theorem 1.1.)

Exercises 1.3

1. Modify Algorithm 1.1 so that it will compute the sum of the terms $f(2), f(3), \ldots, f(n-1)$.
2. Modify Algorithm 1.1 so that it will compute the sum of the terms $f(2), f(4), f(6), \ldots, f(2n)$.

3. What would happen in Algorithm 1.1 if the value of n were set at 3.5? Could we then call our procedure an algorithm?

4. What would happen if $f(x)=\sqrt{x}$ were the function we "inputed" in Algorithm 1.1? On what does your answer depend?

5. Write the sums in Exercises 1 and 2 in sigma notation.

For each of Exercises 6 to 17, write out the first three terms in the sum, write the closed form, if appropriate, and evaluate the sum.

6. $\sum_{k=1}^{100} k.$ **7.** $\sum_{k=10}^{100} k.$ **8.** $\sum_{k=0}^{100} k.$

9. $\sum_{j=1}^{n} (j+1).$ **10.** $\sum_{j=1}^{100} (j+1).$ **11.** $\sum_{k=1}^{100} (3k+2).$

12. $\sum_{k=0}^{n} 2^k.$ **13.** $\sum_{k=0}^{100} 2^k.$ **14.** $\sum_{k=0}^{n} 3\cdot(\tfrac{1}{2})^k.$

15. $\sum_{k=1}^{100} 2.$ **16.** $\sum_{k=10}^{100} 2.$ **17.** $\sum_{k=0}^{100} 0.$

18. Evaluate $\sum_{k=0}^{4} 2^k x^{4-k}$ if $x=3$.

19. Suppose that someone offers to pay you in one of two ways: **(a)** a dollar on the first day, two dollars the next, three dollars the next, and so on or **(b)** a dime on the first day, 20¢ the next, 40¢ the next, 80¢ the next, and so on, each way for 30 days. Which gives the larger sum?

20. Write each of the following polynomials in nested form: $p(x)=$
(a) $x^4+x^3+x^2+x+1$; **(b)** $2x^4-x+6$; **(c)** $2x^3-x^2+x^2-x+3$;
(d) x^5+2x.

21. How would you modify Algorithm 1.2 so that we can use it to evaluate Exercises 20(c) and (d)?

22. Compare the number of operations needed to evaluate a fourth-degree polynomial in **(a)** standard and **(b)** nested form. **(c)** Do (a) and (b) for an nth-degree polynomial.

23. The factor and remainder theorems say that if $p(x)$ is a polynomial, then
(a) when $p(x)$ is divided by $x-r$, we get

$$p(x)/(x-r)=\text{a polynomial}+p(r)/(x-r),$$

and **(b)** $p(x)$ is evenly divisible by $x-r$ if the only if $p(r)=0$. Based on the foregoing, divide $p(x)$ in Example 4 by $x+2$ and determine whether -2 is a *root* of $p(x)$, that is, if $p(-2)=0$.

1.4 INTEGERS AND ALGORITHMS

The set of integers Z has, in addition to the set of positive integers, many other interesting and important subsets. The next few examples will renew our acquaintance with some of them. We will also encounter a somewhat less familiar, but no less interesting, set of integers discovered in the thirteenth century by Leonardo of Pisa, who was also called "Fibonacci."

Example 1 a) The set of **whole numbers** $= \{x \in Z: x \geqslant 0\}$.

b) The set of **even numbers** $= \{x \in Z: x = 2n$ for all n in $Z\}$.

c) The set of **odd numbers** $= \{x \in Z: x = 2n + 1$ for all n in $Z\}$.

d) The set of **Fibonacci numbers** $= \{1, 2, 3, 5, 8, 13, \ldots\}$.

e) The set of **prime numbers** $= \{2, 3, 5, 7, 11, 13, \ldots\}$. □

The first three sets in Example 1 should be quite familiar to you. They have been given here using rules of inclusion that tell you the general form of the number. The last two sets have been given by the listing method, and the first six elements of each of these sets are shown.

All the sets in Example 1 are infinite subsets of Z. How can we decide whether a particular integer is a member of one of these sets? Except for the first six Fibonacci and prime numbers, we cannot do this by direct comparison. However, we now discuss two other methods that are open to us.

Example 2 Let us determine whether 123 is an odd number. If it is, according to the form given in Example 1(c), it must have the form $2n + 1$, for some integer n. Hence we must solve the equation

$$2n + 1 = 123$$

and see whether or not we arrive at a solution in which n is an integer. Subtracting 1 and dividing the result by 2 gives us

$$n = (123 - 1)/2 = 61.$$

Since 61 is an integer, we conclude that 123 is an odd number. □

To determine whether 123 is a Fibonacci number, however, requires a somewhat different procedure. We have not been given the general form of these numbers, but it is not difficult to make a good guess after some experimentation.

Example 3 After the first two Fibonacci numbers, 1 and 2, the pattern that stands out in Example 1(d) is this:

$$3 = 2 + 1$$

$$5 = 3 + 2$$

$$8 = 5 + 3$$

$$13 = 8 + 5$$

$$\cdots\cdots$$

$$a_n = a_{n-1} + a_{n-2}.$$

In other words, each number, after the second one in the sequence, is the sum of the two previous numbers. Hence we can now predict that the Fibonacci number that comes after 8 and 13 will be

$$21 = 13 + 8. \qquad \square$$

We say that the Fibonacci numbers are determined "recursively," that is, in terms of previous members of the sequence. Even though we cannot solve the equation, $a_n = a_{n-1} + a_{n-2}$, by techniques as simple as those in Example 3, we can give an algorithm to determine whether any number greater than 2 is a Fibonacci number.

Algorithm 1.3 To determine whether $x > 2$ is a Fibonacci number,

1. [Initialize a_{n-1}, a_{n-2}, n.] Set $a_1 \leftarrow 1$, $a_2 \leftarrow 2$, and $n \leftarrow 3$. (Recall that the symbol $\leftarrow$ means that an assignment is being made.)

2. [Begin a_n loop.] Repeat steps 3a through 3d until $a_n > x$.
   3a. [Calculate a_n.] Set $a_n \leftarrow a_{n-1} + a_{n-2}$. (The calculation of the Fibonacci number a_n is based upon the latest values of a_{n-1} and a_{n-2}.)
   3b. [$a_n = x$.] If $a_n = x$, then output "x is a Fibonacci number" and stop.
   3c. [Next n.] Set $n \leftarrow n + 1$. (This step can only be reached when the current a_n is less than x. Hence we now return to step 3a and calculate the next a_n.)

4. [$a_n > x$.] Output "x is not a Fibonacci number" and then stop. (All subsequent Fibonacci numbers must be larger than x.)

5. End of Algorithm 1.3.

Example 4 Let us take $x = 123$ as our initial input into Algorithm 1.3. Then after the assignments $a_1 \leftarrow 1$, $a_2 \leftarrow 2$, and $n \leftarrow 3$ are made, we find, with $n = 3$,

$$a_3 \leftarrow a_2 + a_1 = 2 + 1 = 3 < 123.$$

The conditions of steps 2 and 3b are not fulfilled, and so they give no output. That is, since $x = 123 > a_3 = 3$, we set $n \leftarrow 4$ and return to step 3(a). Continuing as before, we obtain

$$a_4 \leftarrow a_3 + a_2 = 3 + 2 = 5 < 123\text{: Setting } n \leftarrow 5 \text{ and returning to step 3(a),}$$

$$a_5 \leftarrow a_4 + a_3 = 5 + 3 = 8 < 123\text{: Setting } n \leftarrow 6 \text{ and returning to step 3(a),}$$

$$a_6 \leftarrow a_5 + a_4 = 8 + 5 = 13 < 123\text{: Setting } n \leftarrow 7 \text{ and returning to step 3(a),}$$

....

$$a_{10} \leftarrow a_9 + a_8 = 55 + 34 = 89 < 123\text{: Setting } n \leftarrow 11 \text{ and returning to step 3(a),}$$

$$a_{11} \leftarrow a_{10} + a_9 = 89 + 55 = 144 > 123.$$

At this point step 4 will give us the *output*: "123 is not a Fibonacci number." □

We now have two procedures: One is algorithmic and the other as in Example 2, may be described as algebraic. (The word "algebra," incidentally, comes from the title of al-Khowarizmi's textbook.) Can we use either to determine whether 123 belongs to the set of prime numbers? The list of primes given in Example 1(e) may remind you that prime numbers have something to do with arithmetic operations. In fact it is with respect to the arithmetic operation of division that we can give a precise definition of a prime number.

Definition 1.10 If a and b are integers and their quotient a/b is also an integer, then (a) we say that ***b* divides *a*** (or "a is divisible by b," or "b is a divisor of a"). (b) A **prime number** (or "prime") is a positive integer greater than 1 whose only positive divisors are itself and 1.

Example 5 The number 124 is not a prime, since it is clearly divisible by 2. In fact we can find all the positive divisors of 124 by writing

$$124 = (1)(2)(2)(31),$$

giving us the divisors 1, 2, 4, 62, and 124. Notice that the last three factors in (1)(2)(2)(31) are primes. We can check that 31 is a prime by dividing it in succession by the numbers 2, 3, 4, 5, . . . , 30. In no case is the quotient an integer, so 31 is prime. □

Algebraic methods for determining primality are beyond the scope of this book. Example 5, however, suggests a straightforward algorithmic test that can be used, in principle, to determine whether or not a positive integer is a prime number.

Algorithm 1.4 To determine if an integer $p > 2$ is a prime,

1. [Initialize divisor.] Set $D \leftarrow 2$.
2. [Begin division loop.] Repeat steps 3a to 3d while $D \leqslant (p/2) + 1$.
   - 3a. [Find quotient.] Set $q \leftarrow p/D$.
   - 3b. [Output not prime.] If q is an integer, then output "p is not a prime" and stop.
   - 3c. [Output prime.] If $D \geqslant p/2$, then output "p is a prime" and stop. (No larger D will divide p.)
   - 3d. [Next D.] Set $D \leftarrow D + 1$. (We reach this step when D is still less than $p/2$. Hence when we increase D by 1, the condition in step 2 will be satisfied. So we repeat steps 3a to 3d.)
4. End of Algorithm 1.4.

The reader is encouraged to use Algorithm 1.4 to determine whether 123 and 223 are prime numbers. (See Exercise 1.4-8.) Observe that step 3a calls for an exact calculation of the quotient p/D. For example, one might have to find 123/62. In its present form, p/D, it creates no problem. But if one tries to change 123/62 to a decimal fraction, the calculation could conceivably go on forever. Hence our procedure would not terminate, and, in effect, our "algorithm" would be no algorithm at all! And even if we were content to "chop off" our quotients at, for example, the 10,000,000th decimal place, we would still have difficulty determining primes greater than 20,000,000, since the remainder, $p - Dq$, of such a quotient might be less than 1/10,000,000 and yet not zero. (See Exercise 1.4-10.)

The problem of testing large numbers for primality is actually of great practical value in fields such as cryptanalysis, where coded account numbers are often formed by multiplying together very large primes. (See "The Mathematics of Public-Key Cryptography" by Martin Hellman, *Scientific American*, August 1979.) Very sophisticated computer-assisted methods have recently been developed to determine the primality of 100-digit numbers in minutes. (See "The Search for Prime Numbers" by Carl Pomerance, *Scientific American*, December 1982.)

For purposes of simple arithmetic, primes can be thought of as the building blocks of the integers, not only because of their applications in advanced technology, but also in day-to-day hand calculations. These both depend on the following fact.

Theorem 1.2 ***The Fundamental Theorem of Arithmetic*** Every positive integer greater than 1 can be written uniquely as a product of 1 and its prime divisors, except for the order of the factors.

Example 6 The products (1)(2)(5)(5)(3) and (2)(5)(3)(1)(5) are essentially the same way of factoring 150 into prime divisors and 1; the 5 appears twice, and the 2 and the 3 each appear once. The order of the factors is unimportant. (So is the 1, which we usually omit; it is just harder to state Theorem 1.2 without it.) □

One can easily determine whether two integers have any factors in common if one is given their *prime factorizations* (i.e., their factorizations into primes).

In particular, one can find the **greatest common divisor** of two positive integers a and b, which is defined as the largest positive integer dividing both a and b. Their greatest common divisor (gcd) can be found by taking the product of the distinct terms $p_i^{m_i}$, where p_i is a prime factor of a or of b, and m_i is the minimum of the number of factors p_i in a and the number of factors p_i in b. If, instead, we take the product of the factors $p_i^{M_i}$, where we let M_i be the maximum of the number of factors p_i in a and the number of factors p_i in b, then we obtain the **least common multiple** (lcm) of a and b. The lcm of a and b is clearly the smallest integer that is divided by both a and b. Here is an example illustrating these definitions.

Example 7 The gcd of 300 and 250 is 50, since $300=(2^2)(3)(5^2)$ and $250=(2^1)(5^3)$. Hence if we let $p_1=2$, $p_2=3$, and $p_3=5$, then $m_1=1$, $m_2=0$, and $m_3=2$. Therefore,

$$p_1^{m_1}p_2^{m_2}p_3^{m_3}=(2^1)(3^0)(5^2)=50.$$

Moreover, the lcm of 300 and 250 is 1500, since $M_1=2$, $M_2=1$, and $M_3=3$. Therefore

$$p_1^{M_1}p_2^{M_2}p_3^{M_3}=(2^2)(3^1)(5^3)=1500. \qquad \square$$

We close this section with the observation that Exercise 7 gives an alternate way to find the gcd of two positive integers, a way that does not require their prime factorization.

Completion Review 1.4

Complete each of the following.

1. The set of whole numbers equals________________.
2. The set of even numbers, given by a rule of inclusion, equals________________.

3. The set of odd numbers, given by a rule of inclusion, equals________________.

4. A positive integer p greater than 1 is a prime if p has no________________ other than itself and 1.

5. The Fibonacci numbers are given by the relation $a_n =$________________ and the initial conditions $a_0 = a_1 =$________________. We say that these numbers are defined ________________.

6. If a and b are integers, we say that a divides b whenever there is an integer c such that ________________.

7. The fundamental theorem of arithmetic says that every positive integer greater than 1 can be________________.

8. The largest positive integer dividing two integers a and b is known as their ________________.

9. The smallest positive integer than can be divided by two integers a and b is called their ________________.

Answers: **1.** $\{0, 1, 2, 3, \ldots\}$. **2.** $\{x: x = 2n$ for all n in Z$\}$. **3.** $\{x: x = 2n + 1$ for all n in Z$\}$. **4.** divisors. **5.** $a_{n-1} + a_{n-2}$; 1; recursively. **6.** $b = ac$. **7.** factored uniquely into prime factors. **8.** greatest common divisor. **9.** least common multiple.

Exercises 1.4

1. Why is each of the sets discussed in Example 1 discrete?
2. What is the difference between the set of whole numbers and the set of positive integers?
3. How could you demonstrate that the set of even numbers is infinite? the set of odd numbers? the Fibonacci numbers?
4. Show that 2456 is an even number by going through a demonstration as in Example 2.
5. Determine whether 12,345 is a Fibonacci number by using Algorithm 1.3.
6. **a)** Write an algorithm, modeled after Algorithm 1.3, that will determine if a number x belongs to the set

$$\{a_n: a_n = a_{n-1} + 2 \text{ for } n > 1, \text{ and } a_1 = 3\}.$$

   b) Calculate a_{10}.

7. **a)** Write an algorithm, modeled after Algorithm 1.3, that will determine if a number $x > 1$ belongs to the set

$$\{a_n: a_n = a_{n-1} + 2a_{n-2} \text{ if } n > 2, \text{ and } a_1 = a_2 = 1\}.$$

b) Calculate a_{10}.

8. Use Algorithm 1.4 to determine whether **a)** 123 and **b)** 223 are primes.
9. How can one improve step 2 in Algorithm 1.4?
10. Can a computer that is accurate to 16 places (digits) use Algorithm 1.4 to determine the primality of a 10-digit integer? Explain.
11. Give the prime factorization of each of the following.
    (a) 24; **(b)** 36; **(c)** 37; **(d)** 48; **(e)** 65; **(f)** 99.
12. What are the positive integral divisors of each of the numbers in Exercise 1.4–11?
13. The *Euclidean algorithm* is a method for finding the greatest common divisor of two positive integers. It goes as follows:

    Given two positive integers m and n, find their greatest common divisor (the largest integer dividing m and n) by,

    1. [Find remainder.] Divide m by n, and let r be the remainder.
    2. [Is remainder 0?] If $r=0$, output "n is the greatest common divisor." Then stop. Otherwise:
    3. [Interchange.] Set $m \leftarrow n$, $n \leftarrow r$, and go back to step 1.

    a) What guarantees that the algorithm will eventually stop?
    b) Could one run into a problem in step 1 if one uses a hand calculator? Explain.
    c) Use the Euclidean algorithm to calculate the greatest common divisor of 235 and 146.
14. The *sieve of Eratosthenes* (c. 200 B.C.) is a method for finding all primes less than or equal to a given positive integer n. Here is a rough version: "Remove 1 and all multiples of 2 other than 2 from the set of integers from 1 to n. Retain the first remaining integer, call it p, and remove all multiples of p. Then repeat the last step with the first remaining integer (if any) greater than p, making it the new p, until this step no longer can be repeated. All the numbers less than or equal to n that were not removed are primes."
    a) Use this procedure in its present form to find all primes $\leqslant 123$.
    b) Try to rewrite the method in good algorithmic form, as specified in Definition 1.3 (Section 1.3).
15. Suppose that a, b, and p are positive integers such that p divides both a and b. Prove that p divides $(a+b)$. (*Hint*: There must be positive integers r and s such that $a=rp$ and $b=sp$ . . . .)
16. Show that the set of primes is infinite. (*Hint*: If there were only the primes p_1, p_2, . . . , p_n such that $p_1 < p_2 < \cdots < p_n$, then one of these primes p_i divides $p_1 p_2 \ldots p_n + 1$. Why? This implies that the prime p_i divides 1. Why? But this is impossible . . . .)
17. Use the fundamental theorem of arithmetic to show that if a prime p divides n^2, then p must divide n.

1.5 RATIONAL AND REAL NUMBERS

A **rational number** is one that can be expressed in the form a/b, where a and b are integers, $b \neq 0$. Thus a rational number is a ratio of integers. Every integer is a rational number because, if n is an integer, then we may write $n = n/1$. Rational numbers do not have a unique representation as ratios. For example, $1/2 = 2/4 = 3/6$. On the other hand, the fundamental theorem of arithmetic, Theorem 1.2, permits us to refer to rational numbers in the following standard way.

Definition 1.11 A rational number $a/b \neq 0$ is **in lowest terms** if $|a|$ and $|b|$ are positive integers having no common prime factors.

Example 1 The fractions 3/4, 4/7, and 15/14 are in lowest terms, but the fractions 6/8, 12/21, and 30/28 are not in lowest terms, since the last three are equal to 3/4, 4/7, and 15/14 respectively. □

One can also express a common fraction in **decimal form**, that is, by thinking of the fraction a/b as an indicated division $a \div b$, where the quotient is expressed as a whole number and an expansion $0.a_1a_2a_3 \ldots$ in the decimal digits a_i called a **decimal fraction**.

Example 2 Some common decimal equivalents are

$$1/2 = 0.5,\ 3/4 = 0.75,\ 3/8 = 0.375,\ 3/2 = 1.5, \text{ and } 2/3 = 0.666 \ldots$$

where in the last number the 6's continue forever. This is also indicated by writing $0.\bar{6}$. It is an example of a **nonterminating decimal**. Observe that $2/3 \neq 0.666$, since

$$0.666 = 666/1000 < 2/3. \qquad □$$

Treating 2/3 as if it were equal to 0.666 is known as a **truncation error** since it refers to truncating (cutting off) the decimal expansion of 2/3 prematurely. Truncation errors can lead to serious consequences, as you may recall in our test for primality, Algorithm 1.4.

Every rational number has either a terminating or a nonterminating decimal form, where we will consider decimal forms ending in infinite sequence of zeros as terminating. Thus 1.10000 . . . is a terminating form since there are only zeros after the right-hand "1." However, *if the nonterminating form of a decimal expansion ends in a sequence of 9's, we can reexpress the decimal expansion as a terminating 1.*

Example 3 Consider the decimal 0.999 . . . . Multiplying this by 10 gives us 9.999 . . . . Subtracting the first number from the second gives us

$$9.999 \ldots - 0.999 \ldots = 9.$$

But if $10x - x = 9$, we must have $9x = 9$ and $x = 1$.

More generally, let us consider **repeating decimals**, that is, decimal digit expansions of the form $0.a_1a_2 \ldots a_k\overline{b_1b_2 \ldots b_r} =$

$$0.a_1a_2 \ldots a_k\overbrace{b_1b_2 \ldots b_r}\overbrace{b_1b_2 \ldots b_r}\overbrace{b_1b_2 \ldots b_r}\overbrace{b_1 \ldots},$$

where, after the first k digits, there is a sequence $b_1b_2 \ldots b_r$ that repeats itself ever after. For example, in the expansion

$$0.46\overbrace{23}\overbrace{23}\overbrace{23}\overbrace{23}\ldots,$$

the sequence "2, 3" is repeated forever. Hence $0.46\overline{23}$... is a repeating decimal. □

Theorem 1.3 Every rational number can be expressed as an integer plus a repeating decimal. Moreover, every repeating decimal is a rational number.

The following examples illustrate how one can make the changes between fractional and decimal forms that are implied by Theorem 1.3.

Example 4

a) To express 23/99 as a repeating decimal, divide 99 into 23. When the remainder of 32 appears a second time, we have a signal that the sequence "23" will repeat ever after.

```
      0.232 . . . .
  99 |23.0000 . . .
      19 8
     | 3 20  <-
     -2 97     \
       230       Remainder of 32: Repeats itself after the decimal point.
      -198     /
      | 320  <-
        etc.
```

b) To express $0.3232\ldots = r$ as a ratio of integers a/b, write $100r = 32.3232\ldots$; so $100r - r = 32$; and

$$r = 32/99.$$ □

In Example 4(a) we are using the usual method of "repeated subtraction and bringing down zeros" to divide 99 into 23. More generally, we can now see that if the division indicated by the fraction a/b does not terminate, then, after the decimal point, there can be at most $b - 1$ different nonzero remainders. These are the positive integers less than b. Hence as soon as one of these repeats itself, the entire preceding sequence of remainders must also repeat themselves since the step of "bringing down zeros" is repeated.

We can also generalize Example 4(b) as Algorithm 1.5.

Algorithm 1.5 To change $0.\overline{a_1 a_2 \ldots a_k} = r$ to a quotient, given digits $a_1, a_2, \ldots, a_k$:

1. [Input k.] Set $k \leftarrow$ the length of the sequence of repeated digits.
2. [Calculate d.] Set $d \leftarrow 10^k - 1$ (d is the required denominator).
3. [Calculate numerator.] Set $n \leftarrow$ the integer whose digits are $a_1, a_2, \ldots, a_k$ (that is, the integer written "$a_1 a_2 \ldots a_k$").
4. [Output r.] Output: r is the fraction "n/d." Then stop.
5. End of Algorithm 1.5.

To see why our algorithm works, we note that multiplying $r = 0.\overline{a_1 a_2 \ldots a_k}$ by 10^k gives us

$$a_1 a_2 \ldots a_k . \overline{a_1 a_2 \ldots a_k}.$$

Hence

$$10^k r - r = a_1 a_2 \ldots a_k.$$

And so

$$r = (a_1 a_2 \ldots a_k)/(10^k - 1) = n/d.$$

Irrational Numbers

Given Theorem 1.3, it is clear that a nonterminating, nonrepeating decimal is not a rational number. You can probably give a recipe to construct one of these so-called **irrational numbers**, but they are easier to find than that.

Theorem 1.4 The square root of 2 is an irrational number.

Proof: We will give a **proof by contradiction** of this fact, by assuming the contrary (that $\sqrt{2}$ is rational) and obtaining a contradiction based upon our assumption.

If $\sqrt{2}$ is rational, we can write $\sqrt{2} = a/b$, which is a quotient of integers in *lowest terms*. Squaring, we get

$$2 = a^2/b^2 \quad \text{or} \quad 2b^2 = a^2.$$

Therefore, 2 is a divisor of $a^2 = a \cdot a$, and so 2 must also divide the integer a. (See Exercise 1.4-17.) Hence we can write $a = 2n$ for some integer n. Therefore, substituting $2n$ in the foregoing equation gives us

$$2b^2 = (2n)^2 = 4n^2,$$

or

$$b^2 = 2n^2.$$

As before, 2 divides b^2 and so 2 divides b. But this gives us our contradiction, since 2 also divides a, and we were able to assume that a/b was in lowest terms.

We conclude that we *cannot* write $\sqrt{2} = a/b$, and that $\sqrt{2}$ is not rational. ■

We examine the structure of this kind of proof more closely in Section 2.5, along with other proof techniques.

Computation and Scientific Notation

Digital devices, from the simplest hand calculator to the most complicated digital computer, treat all numerical data as if they were rational numbers.* This is so because every digital device can store only a finite, and often fixed, number of digits in any numeral. The number of digits is called the **fixed precision** of the device. Some inexpensive hand calculators, for example, can only store seven or eight digits of a numeral; others can store as many as 12, even though they may only display seven or eight. Apart from the obvious difficulty that not all numbers are rational, even a rational number may have a very long or nonterminating decimal expansion. Hence we are sometimes forced to use imperfect representations, or "approximations," of our numerical data.

The standard decimal expansion of any real number is of the form

$$r = a_1 a_2 a_3 \ldots a_n . b_1 b_2 \ldots b_m \ldots$$

or

$$r = a_1 \cdot 10^{n-1} + a_2 \cdot 10^{n-2} + \cdots + a_n \cdot 10^0 + b_1 \cdot 10^{-1} + b_2 \cdot 10^{-2} + \cdots + b_m \cdot 10^{-m} + \cdots,$$

where the a_i's and b_i's are the decimal digits. The nonzero digit furthest to the left is called the **leading** (or **most** or **first**) **significant digit of r**. The digits to its right are called

*Some specialized programs, however, can accept data such as $\sqrt{2}$ and manipulate these symbolically.

successively the **second, third,** . . . , and ***n*th significant digits** of the number r. For example, the first, second, and third significant digits of the number 0.005798 are 5, 7, and 9, respectively, since the zeros here do not matter. But the first three significant digits of the number 7.06888 . . . are 7, 0, and 6, in that order, the same as for the number 7.06000 . . . . If we are allowed only three significant digits, is 7.06 an adequate representation of 7.0688 . . .? No, it is not, because 7.06888 . . . is closer to 7.07 than to 7.06. Whereas 7.06 is a truncated representation of 7.06888 . . . , we say that 7.07 is a "rounded-off" representation, and we write "7.06888 . . . $\doteq$ 7.07."

Definition 1.12 ***Rounding-off Rule*** To round off a decimal expansion of r to n significant digits, where the significant digits of r are $a_1, a_2, \ldots, a_n, a_{n+1}, \ldots$:

1. Add 1 to a_n if a_{n+1} is greater than or equal to 5. (If $a_n = 9$, we set $a_n \leftarrow 0$ and carry the 1 to a_{n-1}, as usual.)
2. Change the digits $a_{n+1}, a_{n+2}, \ldots$ to zeros.

Example 5 Rounding each of the following to three significant digits:

a) 7.0349 is rounded down to 7.03, and 7.0349 $\doteq$ 7.03.

b) 70.351 is rounded up to 70.4, and 70.351 $\doteq$ 70.4.

c) 0.70350 is rounded up to 0.704, and 0.70350 $\doteq$ 0.704.

d) 7095 is rounded up to 7100 and 7095 $\doteq$ 7100. □

Observe how each number in Example 5 was rounded to a different *decimal* place: (a) to the hundredths, (b) to the tenths, (c) to the thousandths, and (d) to the tens place. All have been rounded, however, to three significant digits. To make this a bit clearer in Example 5(d), though, we might have presented the answer as (710)(10), since the final zero of 7100 "has no significance."

Although proper rounding can help us avoid truncation errors, rounding is in itself no cure-all for the basic limitations of a digital calculating device. There are many additional ways, however, to help overcome the problem of fixed precision, but most of them belong in a course on numerical analysis. The two additional methods that we will mention concern standard ways of representing numbers of very large magnitude or that are very close to zero.

Definition 1.13 A nonzero number is said to be written in (a) **scientific notation** if it has the form $k \cdot 10^n$, where n is an integer and k is a number in decimal form whose magnitude is at least 1 but less than 10.

The number $k \cdot 10^n$ is in (b) **normalized exponential form** if the magnitude of k is at least 0.1 but less than 1.

Example 6 Writing our approximations from Example 5 first in scientific notation and then in normalized exponential form gives us Table 1.1.

Table 1.1 Scientific Notation—Normalized Exponential Form

	Number		Scientific notation		Normalized exponential form
a)	7.0349	$\doteq$	$7.03 \cdot 10^0$	=	$0.703 \cdot 10^1$
b)	70.351	$\doteq$	$7.04 \cdot 10^1$	=	$0.704 \cdot 10^2$
c)	0.70350	$\doteq$	$7.04 \cdot 10^{-1}$	=	$0.704 \cdot 10^0$
d)	7095	$\doteq$	$7.10 \cdot 10^3$	=	$0.710 \cdot 10^4$

□

Completion Review 1.5

Complete each of the following.

1. A rational number is a quotient of________________.
2. A rational number is in lowest terms if it is in the form a/b, where a and b ________________.
3. Every repeating decimal is a________________ number. Every nonterminating, nonrepeating decimal is an________________ number.
4. The positive square root of 2 is an example of a________________ number. We proved this by giving a proof by________________.
5. In the decimal expansion of a number, the nonzero digit furthest to the left is called the________________ of the number. Taking the latter as the first, the rth digit from the left is called the________________ of the number.
6. Rounding off the number 2/3 as 0.666 is called a________________ error. The correct way to round 2/3 to three significant figures is________________.
7. A number written in the form $k \cdot 10^n$, where n is an integer and the magnitude of k is at least 1 but less than 10, is said to be in________________ notation. If the magnitude of k is at least 0 but less than 1, then the number is said to be in________________.

Answers: **1.** integers, with a nonzero denominator. **2.** have no common prime factors. **3.** rational; irrational. **4.** irrational; contradiction. **5.** first (or leading or most) significant digit; rth significant digit. **6.** truncation; 0.667. **7.** scientific; normalized exponential form.

Exercises 1.5

1. Which of the following fractions are in lowest terms?
(a) 3/6; **(b)** 4/3; **(c)** 32/10; **(d)** 12/1; **(e)** 231/65.

2. Express the following fractions exactly in decimal form.
(a) 3/8; **(b)** 5/4; **(c)** 2/3; **(d)** 4/5; **(e)** 1/6; **(f)** 1/7.

3. Which of the decimals in Exercise 2 are terminating? Which ones are nonterminating?

4. Express each of the following decimals in the form m/n, where m and n are integers, $n \neq 0$: **(a)** 0.2323 . . . ; **(b)** 0.2323; **(c)** 0.2222 . . . ; **(d)** 0.2222; **(e)** 2.3232 . . . ; **(f)** −1.4444 . . . ; **(g)** 4.9999 . . . ; **(h)** $0.\overline{123}$; **(i)** $0.1\overline{23}$; **(j)** $0.12\overline{3}$.

5. Suggest a modification of Algorithm 1.5 so that it can be applied to Exercise 4, parts (e), (f), (g), (i), and (j). (We presume you have found a way to handle those numbers.)

6. Explain why every terminating decimal is a rational number.

7. Explain why the following decimal is an irrational number: 0.1010010001000010000010 . . . (the nth 1 and the $n+1$st 1 are separated by n zeros in general).

8. Give a proof by contradiction that $\sqrt{3}$ is irrational.

9. Is $\sqrt{2}+1$ rational or irrational? (*Hint*: Use the fact that $\sqrt{2}$ is known to be irrational and addition of fractions.)

10. What sort of error (truncation or rounding) is illustrated by the following incorrect statements? **(a)** $5/3 \doteq 1.666$; **(b)** $\pi = 3.141$; **(c)** $\sqrt{2} = 1.414$; **(d)** $0.9999999999999 = 1$.

11. What are the first and second significant digits in each of the following?
(a) 345; **(b)** 0.00354; **(c)** 3.500; **(d)** 25,000.

12. Round each of the numbers in Exercise 11 to the second significant digit and then express your answer in scientific notation.

13. Round each of the numbers in Exercise 11 to the second decimal place and then express your answer in scientific notation.

14. If we want to make a list that includes every positive rational number at least once, we can combine infinitely many lists, as shown below, by following the arrows. How could we include each positive rational number *exactly* once? (*Note*: m and n are positive integers.)

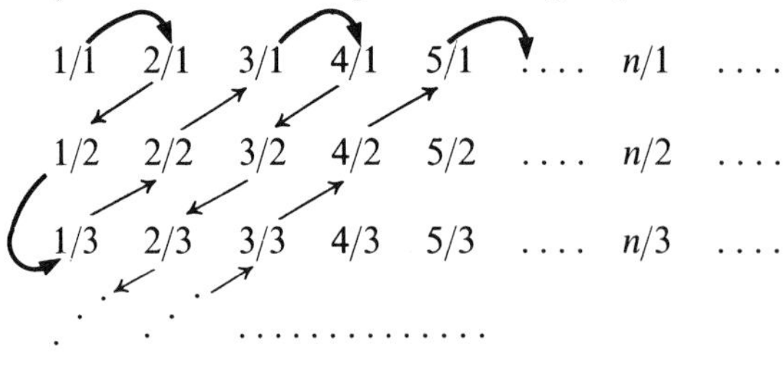

15. Use Exercise 14 to show that one can make a list containing every rational number exactly once. (Hence we say that the rational numbers are **countable**.)

16. In spite of the fact that we can make a list containing every rational number exactly once, the set of rational numbers is not a discrete set of numbers. Why not? (*Hint*: Review Definition 1.2, Section 1.1, and consider the difference between a set and a list.)

17. Verify the steps in the following proof, which shows that one cannot make a list containing all real numbers r, where $0 \leqslant r < 1$.

a) We can represent each r by a *unique* decimal digit expansion of the form $0.a_1a_2a_3\ldots$ if the a_i's are not all 9's from some point onward. (Why is it unique?)

b) *Given* any list of real numbers $a_1, a_2, a_3, \ldots$ of the form

$$a_1 = 0.a_{11}a_{12}a_{13}a_{14}\ldots\ldots,$$
$$a_2 = 0.a_{21}a_{22}a_{23}a_{24}\ldots\ldots,$$
$$a_3 = 0.a_{31}a_{32}a_{33}a_{34}\ldots\ldots,$$
$$\vdots \qquad \ldots\ldots\ldots\ldots\ldots\ldots$$

(*Note*: a_{ij} is the jth digit in the ith number on this list.)

the following real number is not on the list: $b = 0.b_1b_2b_3b_4\ldots b_n\ldots$, where

$$b_n = \begin{cases} 1 & \text{if } a_{nn} \neq 1. \\ 2 & \text{if } a_{nn} = 1. \end{cases}$$

(Check that $0 \leqslant b < 1$.)

18. Why does Exercise 17 imply that one cannot list every real number, not just those between 0 and 1? (Hence we say that the real numbers are **uncountable**.)

1.6 OTHER SYSTEMS OF NUMERATION

We find the decimal system of numeration in which we use the 10 decimal digits 0, 1, 2, . . . , 9 to represent all the real numbers very natural and convenient. After all, we do have a total of 10 fingers on our hands. In spite of this frequently made observation, people have used number systems with many other **bases**, or **radixes**, such as 5 (the number of fingers on one hand), 12, 20, and 60, when the need has arisen. In fact we still use 60 as a standard when measuring angles and time in minutes and seconds. This is not the same as a number system with base 60 though, because base 60 would require 60 different symbols. The base then is the number of distinct digits the system requires.

The **binary system** of numeration requires only the two digits 0 and 1 to represent all the real numbers. It lends itself nicely to electronic circuitry that has an off–on pattern. This, in turn, simplifies the storage of numerical data, in a way analogous to the sequence of light bulbs in Fig. 1.6. Each bulb gives 1 "bit" of information, namely, whether a particular digit is a 0 ("off") or a 1 ("on").

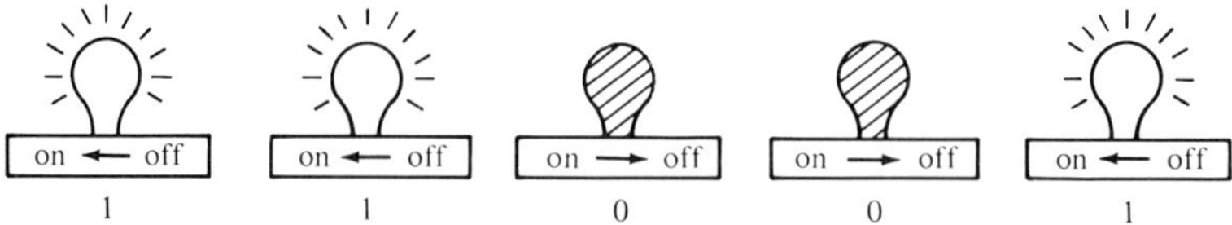

Figure 1.6 An on–off pattern of light bulbs represents the number 25 in binary.

Definition 1.14 A **binary numeral** is one that is in the form

$$a_n \cdot 2^n + a_{n-1} \cdot 2^{n-1} + \cdots + a_0 \cdot 2^0 + b_1 \cdot 2^{-1} + \cdots + b_m \cdot 2^{-m} + \cdots$$

where the a_i's and b_i's are 0's or 1's, and m and n are nonnegative integers.

For purposes of hand or machine computation, we can think of our binary numerals as terminating at, say, b_m. In that case we would write the binary numeral in Definition 1.14 as

$$(a_n a_{n-1} \ldots a_0 . b_1 b_2 \ldots b_m)_2.$$

Where there is no danger of confusion regarding the base,

$$a_n a_{n-1} \ldots a_0 . b_1 b_2 \ldots b_m$$

is what we write.

Example 1 a) To write the binary numeral $(11001)_2$ in decimal notation, we first observe that there is no decimal point (or "binary" point). Hence the digit in the first place at the right tells us the number of 2^0's. Adding all the indicated powers of 2 gives us the number 25.

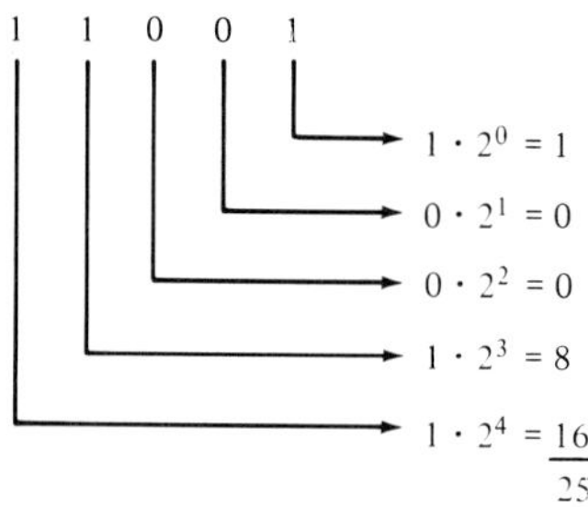

Figure 1.7 Binary to decimal.

b) Similarly the binary numeral $(1.01)_2 =$

$$1 \cdot 2^0 + 0 \cdot 2^{-1} + 1 \cdot 2^{-2} = (1.25)_{10}. \qquad \square$$

If you glance once more at Fig. 1.6, you will realize that the off–on pattern of the light bulbs gives an electronic representation of the number 25—in binary. Each digit in the binary numeral 11001 holds 1 *bit* of information. This is the basic unit of information that a computer can handle. (We should mention that modern computers use transistors that have conducting and nonconducting states to represent bits. We used light bulbs because they are easier to visualize.)

Octal and Hexadecimal Systems

It clearly takes more room, in general, to write a number in binary as opposed to decimal notation. Twenty-five, for example, requires the five-digit numeral (or **string**) 11001 in binary, compared with the two-digit string "25" in decimal notation. The considerable length of binary strings becomes especially cumbersome in computer printouts. One way to alleviate this problem is to break up binary strings into groups of four digits each, as shown in Fig. 1.8. Each string of four digits is a binary number itself. Reading these numerals from left to right in Fig. 1.8, we have

$$A = 10 \quad B = 11 \quad C = 12 \quad D = 13 \quad E = 14 \quad F = 15$$

as the six **hexadecimal digits** we will need beyond the decimal digits 0, 1, 2, 3, 4, 5, 6, 7, 8, and 9 to represent numbers in **base 16**.

1010	1011	1100	1101	1110	1111
A	B	C	D	E	F

Figure 1.8 Binary numerals and hexadecimal digits.

A careful definition of a **hexadecimal numeral** (one in base 16) can be modeled after our definition of a binary numeral. (See Exercise 1.6-16.) Suffice it to say, for now, that in hexadecimal notation, each digit represents a power of $16\ (= 2^4)$.

Example 2 a) $(\text{A29F})_{16} = 10 \cdot 16^3 + 2 \cdot 16^2 + 9 \cdot 16^1 + 15 \cdot 16^0 = 41{,}631$.

b) $(\underbrace{1010}_{A}\underbrace{1011}_{B}\underbrace{1100}_{C}\underbrace{1101}_{D}\underbrace{1110}_{E}\underbrace{1111}_{F})_2 = (ABCDEF)_{16} = 11{,}259{,}375$, since

$(ABCDEF)_{16} = 10 \cdot 16^5 + 11 \cdot 16^4 + 12 \cdot 16^3 + 13 \cdot 16^2 + 14 \cdot 16^1 + 15 \cdot 16^0$. $\square$

Our last example illustrates how much more compactly numbers can be written in hexadecimal as opposed to binary or even decimal notation. Computer scientists say that each of the six hexadecimal digits A, B, C, D, E, and F contains one-half "byte" of information, because 1 **byte** = 8 bits.* Moreover, it is particularly easy to convert binary numerals into hexadecimal. Just subdivide the binary numeral, starting from the binary point, into half-byte groups of four digits. Initial or terminal zeros may be added if necessary.

Example 3 $(101101.001)_2 = (0010\ 1101\ .\ 0010)_2 = (2D.2)_{16}$. □

Using a radix of 8 is another popular way to represent numbers in computer science, giving us the **octal** system of numeration. The octal system uses the familiar eight digits 0, 1, 2, 3, 4, 5, 6, and 7. Each position represents a power of 8.

Example 4

a) $(23 \times 5)_8 = 2 \cdot 8^3 + 3 \cdot 8^2 + 7 \cdot 8^1 + 5 \cdot 8^0 = 1277.$

b) $(101010)_2 = (101\ 010)_2 = (52)_8 = 5 \cdot 8^1 + 2 \cdot 8^0 = 42.$ □

Example 4(b) demonstrates the simple conversion of binary numerals into octal by grouping the binary digits in *strings of length* 3. It is even easier to go from octal or hexadecimal to binary, of course. One only need remember the binary string for each octal or hexadecimal digit.

Example 5

a) $(2501)_8 = (010\ 101\ 000\ 001)_2$.

b) $(2A01)_{16} = (0010\ 1010\ 0000\ 0001)_2$. □

From Decimal to Binary

The most *direct*, and laborious, *method* for converting a decimal numeral to binary is to subtract all powers of 2 that can be subtracted from the number until nothing remains. We begin with the highest power of 2 that is less than the given number.

Example 6 Consider the decimal number 11.875. To get nonnegative differences we subtract in succession

$$\left.\begin{array}{lll} 11.875 - 1 \cdot 2^3 &= 3.875 \longrightarrow & 1 \\ 3.875 - 0 \cdot 2^2 &= 3.875 \longrightarrow & 0 \\ 3.875 - 1 \cdot 2^1 &= 1.875 \longrightarrow & 1 \\ 1.875 - 1 \cdot 2^0 &= 0.875 \longrightarrow & 1 \\ 0.875 - 1 \cdot 2^{-1} &= 0.375 \longrightarrow & 1 \\ 0.375 - 1 \cdot 2^{-2} &= 0.125 \longrightarrow & 1 \\ 0.125 - 1 \cdot 2^{-3} &= 0.000 \longrightarrow & 1 \end{array}\right\} \text{binary digits.}$$

Hence $(11.875)_{10} = (1011.111)_2$. □

*More generally, a byte is the smallest amount of information that can be referenced within a computer realization of an algorithm. This amount is 8 bits on many, but not all, machines.

A method that is sometimes used for somewhat larger numbers involves handling the integer part and the fractional parts separately. *For the integer part of the decimal, we divide the integer repeatedly by 2*; the remainders after each division, written in reverse order, give the binary representation of the integer part. This is justified by noticing that, for example,

$$1 \cdot 2^3 + 0 \cdot 2^2 + 1 \cdot 2^1 + 1 \cdot 2^0 =$$

$$((1 \cdot 2 \ + 0)2 \ \ + 1)2 \ \ + 1.$$

The terms 1, 1, 0, and 1 will be the remainders, therefore, when we repeatedly divide 11 by 2.

To make the procedure a bit more definite, we can write it in the following algorithmic form.

Algorithm 1.6 To write a positive integer n as a binary numeral,

1. [Initialize i, m.] Set $i \leftarrow 1$, $m \leftarrow n$. (We prepare to find the last significant digit.)
2. [Binary digit loop.] Repeat steps 3a–3d until $q_i = 0$:
   - 3a. [Divide.] Divide m by 2.
   - 3b. [Find quotient.] Set $q_i \leftarrow$ the integral part of the quotient in step 3a.
   - 3c. [Find remainder.] Set $r_i \leftarrow$ the remainder in step 3a. ($m = 2q_i + r_i$.)
   - 3d. [$q_i > 0$?] If $q_i > 0$, then set $m \leftarrow q_i$ and set $i \leftarrow i + 1$. (The previous quotient will be divided if we return to step 3a. Otherwise we will continue with step 4.)
4. [$q_i = 0$.] Output $(r_i r_{i-1} \ldots r_1)_2$ and stop.
5. End of Algorithm 1.6.

Example 7 To change 37 to binary notation, we divide as follows:

$37 \div 2 = 18$, with a remainder of 1 ($q_1 = 18$; $r_1 = 1$),

$18 \div 2 = 9$, with a remainder of 0 ($q_2 = 9$; $r_2 = 0$),

$9 \div 2 = 4$, with a remainder of 1 ($q_3 = 4$; $r_3 = 1$),

$4 \div 2 = 2$, with a remainder of 0 ($q_4 = 2$; $r_4 = 0$),

$2 \div 2 = 1$, with a remainder of 0 ($q_5 = 1$; $r_5 = 0$),

$1 \div 2 = 0$, with a remainder of 1 ($q_6 = 0$; $r_6 = 1$).

Reading the remainders in reverse order, we have

$$(37)_{10} = (100101)_2. \qquad \square$$

Similarly, if we write the fractional part of Example 6 as

$$0.875 = \tfrac{1}{2}(1 + \tfrac{1}{2}(1 + \tfrac{1}{2} \cdot 1)),$$

we are led to a rule whereby we *repeatedly double the decimal fraction and subtract the integer part of the products, until we obtain a product of 1.* This, in effect, strips away the factors of $\frac{1}{2}$ in succession, from left to right, yielding the binary digits as the integer parts of each product in the correct order. (Hence $0.875 = (0.111)_2$.)

Example 8 To write 0.3125 as a fractional binary number of the form $(0.d_1 d_2 d_3 \ldots)_2$, we multiply as follows:

Integer part

$$2 \cdot 0.3125 = 0.625. \text{ Hence } d_1 = 0.$$

$$2 \cdot (0.625 - d_1) = 2 \cdot 0.625 = 1.25. \quad \text{Hence } d_2 = 1.$$

$$2 \cdot (1.25 - d_2) = 2 \cdot 0.25. = 0.5. \quad \text{Hence } d_3 = 0.$$

$$2 \cdot (0.5 - d_3) = 2 \cdot 0.5 = 1.0. \quad \text{Hence } d_4 = 1\text{: } \textit{Stop}, \text{ product} = 1.$$

Thus $(0.3125)_{10} = (0.0101)_2 = (0.d_1 d_2 d_3 d_4)_2$. For verification notice that

$$0.3125 = \tfrac{1}{2}(0 + \underbrace{\tfrac{1}{2}(1 + \underbrace{\tfrac{1}{2}(0 + \underbrace{\tfrac{1}{2} \cdot 1}_{0.5})}_{0.25})}_{0.625}) = (0.0101)_2. \qquad \square$$

The method illustrated in Example 8 also can be presented in algorithmic form. We prefer to leave this as an exercise for the reader. (See Exercise 1.6-12.)

Completion Review 1.6

Complete each of the following.

1. The number 10 is known as the ________________ of the decimal system of numeration, which uses the digits ________________.

2. The binary system of numeration uses a base of__________ and the digits __________.

3. The octal and hexadecimal systems use bases of__________ and __________ respectively.

4. The hexadecimal system uses the additional digits *A*, *B*, *C*, *D*, *E*, and *F*, which stand for the numbers__________ through__________ respectively.

5. To change an integer from base 10 to base 2 notation, we repeatedly divide by __________. The__________, written in reverse order, give the binary numeral.

6. To change a decimal fraction to binary numeration, we repeatedly__________. The__________, written in order, give the binary fraction.

Answers: **1.** base or radix; 0, 1, 2, 3, 4, 5, 6, 7, 8, 9. **2.** 2; 0 and 1. **3.** 8; 16. **4.** 10; 15. **5.** 2; remainders. **6.** multiply by 2; integer parts.

Exercises 1.6

1. Write out each of the following binary numerals as the sum of powers of 2. Then write the numbers in decimal notation. **(a)** 101; **(b)** 1000; **(c)** 1111; **(d)** 0001; **(e)** 1111111111.
2. Follow the directions of Exercise 1 for the following binary numerals. **(a)** 0.1; **(b)** 1.1; **(c)** 10.0011; **(d)** 1000.0001.
3. Why can the number 1 be written as the nonterminating binary numeral 0.111 . . . ? (*Hint*: Multiply by 2 and subtract the original number from the product.)
4. What binary numeral is indicated by the following illustration?

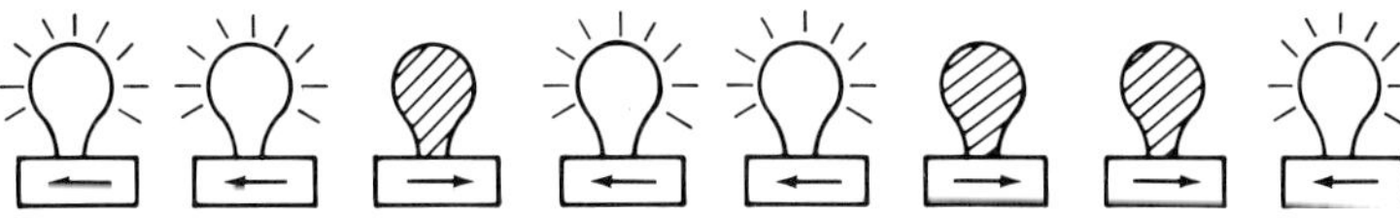

5. How many bits of information do each of the numbers in Exercise 1 carry?
6. Write each of the following hexadecimal numerals in binary and decimal form. **(a)** $(A123)_{16}$; **(b)** $(103F)_{16}$; **(c)** $(0.E)_{16}$.
7. Write each of the following octal numerals in binary and decimal form. **(a)** $(1230)_8$; **(b)** $(707)_8$; **(c)** $(0.7)_8$.
8. Break up the binary numeral 11010.011 into half-bytes and convert it to hexadecimal form.
9. Convert the following decimal numbers to binary using Algorithm 1.6: **(a)** 29; **(b)** 100; **(c)** 200.

10. Suggest a modification of Algorithm 1.6 for converting from decimal notation to **(a)** octal; **(b)** hexadecimal.

11. Use the algorithms you obtained in Exercise 10 to convert the decimal number 1234 to **(a)** octal; **(b)** hexadecimal.

12. Write out the method illustrated in Example 8 in algorithmic (step-by-step) form.

13. Convert to binary: **(a)** 0.59375; **(b)** 27.046875.

14. Write out a careful definition of **(a)** a hexadecimal number and **(b)** an octal number based on the form of Definition 1.14.

15. Explain why, in terms of the number of operations performed, the following way of evaluating Example 2(a) is more efficient than raising 16 to the required powers: $((10 \cdot 16 + 2) \cdot 16 + 9) \cdot 16 + 15$.

16. Use the way we illustrate in Exercise 15 to evaluate $10 \cdot 16^5 + 11 \cdot 16^4 + 12 \cdot 16^3 + 13 \cdot 16^2 + 14 \cdot 16^1 + 15 \cdot 16^0$.

17. Write an algorithm describing the method we suggest for the evaluation of six-digit hexadecimals in Exercises 15 and 16.

18. Modify the algorithm of Exercise 12 for hexadecimals, and apply your algorithm to the decimal numeral 0.875.

19. The following table gives the ASCII-8 code (*A*merican *S*tandard *C*ode for *I*nformation *I*nterchange—pronounced "as-key") for the 10 decimal digit characters 0–9 and the 26 "alphabetic" characters A–Z. These 36 characters are represented as either eight binary digits or as the numerically equivalent hexadecimal numeral. The first 4 bits in the binary representation give the numerical equivalent of the characters 0–9 or the relative position of the characters A–Z in the alphabet. The second 4 bits (or "zone" bits) tell whether the character is a digit or belongs to one of two groups of letters.

Fill in the 10 missing ASCII-8 binary and hexadecimal representations, (a)–(j).

Table ASCII-8

Char.	Zone	Numeric	Hex::Char.	Zone	Numeric	Hex::Char.	Zone	Numeric	Hex.
0	0101	0000	50::A	1010	0001	A1::P	1011	0000	B0
1	0101	0001	51::B	1010	0010	A2::Q	1011	0001	B1
2	0101	(a)	52::C	1010	0011	A3::R	1011	(b)	B2
3	0101	0011	(c)::D	1010	(d)	A4::S	1011	0011	B3
4	0101	(e)	54::E	1010	(g)	A5::T	1011	0100	(g)
5	0101	0100	55::F	1010	0110	A6::U	1011	0101	B5
6	0101	(h)	(i)::G	1010	0111	A7::V	1011	0110	B6
7	0101	0111	57::H	1010	1000	A8::W	1011	0111	(j)
8	0101	1000	58::I	1010	1001	A9::X	1011	1000	B8
9	0101	1001	59::J	1010	1010	AA::Y	1011	1001	B9
			K	1010	1011	AB::Z	1011	1011	BA
			L	1010	1100	AC			
			M	1010	1101	AD			
			N	1010	1110	AE			
			O	1010	1111	AF			

1.7 BINARY ARITHMETIC AND TWO'S COMPLEMENTS

The invention of the decimal system of numeration was a considerable advance over Roman numerals, which were not position oriented and did not have a zero. The binary system simplifies the arithmetic operations of addition, subtraction, multiplication, and division even further. And what is more, the binary system is well suited to machine calculation. In this section we give a brief description of the four common arithmetic operations using binary numbers. Later, in Section 2.6, we see how binary arithmetic can be carried out on an analog of machine circuitry.

Binary Addition and Subtraction

The facts one needs to know in order to do binary addition are very simply stated*:

$$0+0=0$$

$$0+1=1$$

$$1+0=1$$

$$1+1=10$$

is really all there is to it! We can also arrange the foregoing combinations of binary digits as shown in the addition Table 1.2.

Table 1.2 **Binary Addition**

+	0	1
0	0	1
1	1	10

Hence when we add two binary digits, a carryover to the next place can only occur in the two cases:

$$1+1=10 \quad \text{and} \quad 1+(1+1)=11.$$

Both of these cases are illustrated in the following example.

Example 1 To add the binary numbers (by which we mean "numbers written with binary numerals") 101.0101 and 1001.11, we first line up their binary points, adding terminal or initial zeros to the strings if desired. We then proceed as in decimal addition, adding pairs of corresponding digits, or triples, if there is a carryover from the preceding step. Carryovers are taken into account as shown in Fig. 1.9.

*Note: All numerals in this section are taken base 2.

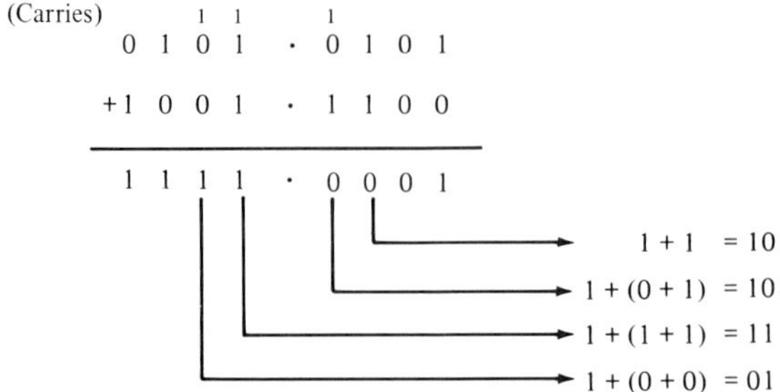

Figure 1.9 Binary addition. □

Subtraction of binary numbers can be done in the same way that we subtract decimal numbers, that is, by the method of "borrowing" whenever necessary from digits to the left of the pair of digits being considered. Just bear in mind that we are borrowing "2's," "4's," and so on instead of the usual "10's," "100's" "1000's," and other powers of 10.

Example 2 When we subtract 1 from 100, we cannot borrow initially from the 2's digit of 100. So we borrow from the 4's digit instead. This gives us 10 in the 2's digit. Next, borrowing 1 from the 2's digit leaves us with 01 in the 2's digit. The subtraction is completed as shown in Fig. 1.10.

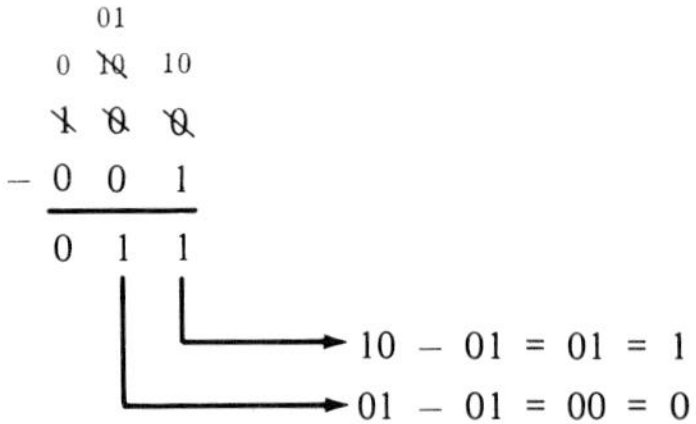

Figure 1.10 Binary subtraction. □

Subtraction of binary numbers is usually accomplished somewhat differently than shown here when it is done on digital calculators. In fact it is usually changed into an addition problem in a way that we describe at the end of this section.

Binary Multiplication and Division

Multiplication of binary numbers is reduced to a repeated addition problem just as it is with decimal numbers. That is, each digit of the first factor multiplies the second factor, and each of the resulting sums is shifted over one more place to the left than the last. However, working with binary numbers simplifies this process as a result of the following binary number multiplication facts.

$$0 \times 0 = 0$$

$$0 \times 1 = 0$$

$$1 \times 0 = 0$$

$$1 \times 1 = 1$$

These facts give us the following binary multiplication table.

Table 1.3 Binary Multiplication

×	0	1
0	0	0
1	0	1

We may use Table 1.3 just as we do an ordinary (decimal number) times table. But there is one important difference: In binary multiplication there is no carrying. Indeed in a problem of the form "(multiplier) times (multiplicand)," the multiplicand is copied whenever the multiplier digit is a 1, and it is not copied (that is, its digits are replaced by zeros) whenever the multiplier digit is a 0. Here are two illustrations.

Example 3 We may carry out the multiplications 11 × 11101 and 101 × 111 as follows:

```
            11101                    111
             × 11                  × 101
            -----                  -----
            11101                    111
           11101                    000
          -------                  111
Product:  1010111                 ------
                       Product:   100011
```

Figure 1.11 Binary multiplication. □

Just as multiplication is changed into repeated addition, division of one binary number by another becomes a repeated subtraction problem. We may proceed as in long division of decimal numbers.

Example 4 We can divide 11 into 1010111 as follows: Trial dividends are found for 11 by simple comparison. The first two digits in 1010111 give us the number 10, which is smaller than 11; so the first trial dividend is 101. Subtracting 11 from 101 leaves a remainder of 10. We bring down the next digit, a zero, and continue with the trial dividend 100 as shown in Fig. 1.12.

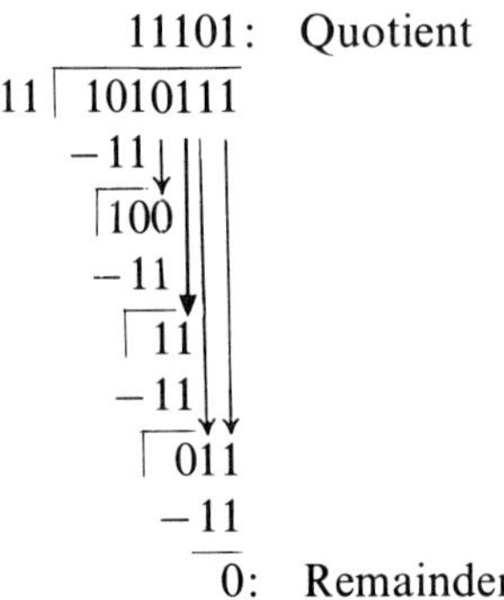

Figure 1.12 Binary division. □

Two's Complements and Subtraction

Subtraction of binary numbers can be reduced to an addition procedure by means of what is called the method of "two's complements." This is how subtraction is done on digital calculators.

We will assume that each of our binary numbers is a nonnegative integer that is represented by $n-1$ binary digits, using leading zeros if necessary. That is, our numbers will be less than 2^{n-1}. Moreover, we will add an additional leading zero for a "sign check," giving us n places in all.

Our method of subtraction will consist of finding the difference $x-y$ by means of the equation

$$x - y = x + (2^n - y) - 2^n.$$

While the foregoing equation seems to give us one more subtraction than $x-y$ did, the fact that we are using binary numbers will enable us to avoid the usual subtraction algorithm entirely.

Let us first consider the term $2^n - y$ in the equation. This term is called the **two's complement** of y. It can be found without subtraction by means of the following two steps.

1. *Change each digit of y to its opposite.* That is, ones become zeros and zeros become ones. This gives what is called the **one's complement** of y.

2. *Add 1 to the one's complement of y.*

The reason we obtain the two's complement, $2^n - y$, is that step 1 subtracts y from the number $2^n - 1$, which is represented by n ones, while step 2 completes the subtraction of y from $(2^n - 1) + 1 = 2^n$.

Example 5 Let us suppose that our numbers will be less than 2^5 $(n-1=5)$ and that they will have a zero in 2^5's place. Then Table 1.4 shows how to find the two's complements of 001101, 010010, and 000000.

Table 1.4 Finding One's and Two's Complements*

Number	001101	010010	000000
One's complement	110010	101101	111111
Add 1	+1	+1	+1
Two's complement	110011	101110	1000000

**Note*: We usually drop the leading one in the two's complement of zero.

□

Let us now consider the second subtraction in the equation, that is, where we subtract the 2^n in $x+(2^n-y)-2^n$. There are two cases.

Suppose that *x is greater than or equal to y*. The way that a machine detects the case $x \geqslant y$ is by noting a 0 in the 2^{n-1}'s place when we add x to the two's complement of y. It will be there because $0 \leqslant x-y<2^{n-1}$.

Adding the two's complement of y to x will give us a sum of $(x-y)+2^n$. This is at least as large as 2^n because $x-y \geqslant 0$. Therefore, this sum will have a 1 in the 2^n's place as its leading digit. Hence we can find $x-y$ in this case by simply ignoring the leading 1, since this subtracts 2^n from $(x-y)+2^n$ and yields $x-y$. In a machine that handles only n digits, this leading 1 would, in fact, never appear.

Example 6 To find the difference $x-y=010010-001101$, we find the two's complement of 001101, which, by Example 5 is 110010, and add it to x. Thus

$$010010+110011=1\overset{\checkmark}{0}00101.$$

The 0 (see the check mark) in the 2^5's place, that is, the second significant digit of 1000101, tells us to drop the leading 1. So our answer is 000101. (In a machine that handles only six digits, this leading 1 would never appear.)

Original	Sum with two's complement	Answer
010010 −001101 →	010010 + 110011	
	1000101	→ 000101

□

Now let us *suppose that x is less than y*, so that $x-y$ is negative. A digital machine could sense this case by the fact that when we add x to the two's complement of y we get a 1 in the 2^{n-1}'s place, since

$$(x-y)+2^n<2^n \quad \text{and} \quad x-y<2^{n-1}.$$

Hence

$$2^{n-1} < (x-y) + 2^n < 2^n.$$

We complete our calculation in this case, therefore, by taking the negative of the two's complement of $x + (2^n - y)$. The reason that this works is because we can write

$$x + (2^n - y) - 2^n = -(2^n - (x + (2^n - y))).$$

Example 7 Now let us find the difference $x - y = 001101 - 010010$. We first find the two's complement of y, which, by Example 5, is 101110, and add it to x. Thus

$$001101 + 101110 = \overset{\checkmark}{1}11011.$$

The 1 (see the check mark) in the 2^5's place tells us to take the two's complement and append a minus sign. Hence our answer is

$$-(000100 + 1) = -000101.$$

Table 1.5 **Subtraction with Two's Complements ($x - y$; $x < y$)**

	Original		Sum with two's complement		Answer
x:	001101		001101		
y:	−010010	→	+101110		
			111011	→	−000101

□

In summary, to find the difference $x - y$ of two binary numbers each of which is less than 2^n:

1. Find the two's complement of y (with respect to 2^n).
2. Add this two's complement to x. Then either:
   3a. Ignore the 2^n's digit (when the 2^{n-1}'s digit is a 0) and read off the rest of the sum in 2.

Or:

   3b. Take the two's complement of the sum in step 2 with a negative sign (when the 2^{n-1}'s digit is a 1).

Completion Review 1.7

Complete each of the following.

1. For binary addition the pertinent facts are $0+0=$____________,
$0+1=1+0=$____________, $1+1=$____________, and, for carrying,
$1+1+1=$____________.
2. The binary multiplication depends upon the facts $0\cdot 0=0\cdot 1=1\cdot 0=$____________,
and $1\cdot 1=$____________.
3. Binary subtraction may be done by borrowing, but it is more often accomplished on machines by means of the method called____________.
4. If y is a nonnegative number less than 2^n, then $2^n - y$ is known as the____________ of y.
5. To find the two's complement of y, we first find its____________ by changing each digit of y to its____________. Then we____________.
6. To find $x-y$ we may add the two's complement of____________ to ____________. If x is $\geqslant y$, we ignore the leading____________ in the answer and read off the rest of the number. If $x<y$, we take the____________.

Answers: **1.** 0; 1; 10; 11. **2.** 0; 1. **3.** two's complementation. **4.** two's complement (with respect to 2^n). **5.** one's complement; opposite; add 1. **6.** y; x; 1; two's complement of the sum with a negative sign.

Exercises 1.7

Do each of the indicated additions in Exercises 1 to 6.

1. $1101+1011$.
2. $1101+101$.
3. $111+11111$.
4. $10.01+101.1$.
5. $111.11+0.111$.
6. $1111+1.111$.

Do each of the indicated subtractions in Exercises 7 to 10 by means of the borrowing method.

7. $11011-10110$.
8. $10000-101$.
9. $10.11-1.001$.
10. $10000-0.01$.

Do each of the indicated multiplications in Exercises 11 to 14.

11. 111×10110.

12. 1001×10110.

13. 1.01×1101.

14. 1.101×1.01.

(*Note*: The number of places in the answer after the binary point is additive, as usual.)

Do each of the indicated divisions in Exercises 15 and 16.

15. $10110/10$.

16. $10110/110$.

Find the one's and two's complements of each binary number in Exercises 17 to 20 with respect to 2^7.

17. 101101.

18. 100000.

19. 1011.

20. 0.

21–24. Do each of Exercises 7 to 9 using the method of two's complements. (*Hint*: For Exercises 8 and 9 first multiply by an appropriate power of 2 to move the binary points.)

Do Exercises 25 and 26 by the method of two's complements.

25. $101 - 11111$.

26. $10.1 - 1.1111$.

1.8 MODULAR ARITHMETIC

Systems of arithmetic that deal with finite sets of numbers, as opposed to the infinite set of real numbers, are of great importance in applications of mathematics and computer science. These systems figure in the statistical design of experiments, error-correcting codes, cryptography, and strategies for reducing the amount of carrying in ordinary addition.

The odometer on most cars provides an immediate example of the use of a finite set of numbers for addition, since most odometers only range from 0 to 99,999 miles (neglecting tenths of a mile). Any mileage added after 99,000 has been reached starts again at "mile zero." We say that arithmetic on these odometers is being carried out "modulo 100,000" or "mod 100,000."

A finite arithmetic is even easier to visualize on a 12-hour clock face, where one o'clock can be thought of as one more than 12, two more than 11, or even 12 more than one. As examples of some consequences of addition on a 12-hour clock face, that is, of addition "mod 12," we have $13 \equiv 1 \pmod{12}$, $14 \equiv 2 \pmod{12}$, and $24 \equiv 12 \pmod{12}$, where the symbol "$\equiv$" is read "equivalent to." We observe that $13 - 1 = 12$, $14 - 2 = 12$, and $24 - 12 = 12$. We also have $27 \equiv 3 \pmod{12}$, since $27 - 3 = 24 = (2)(12)$. Hence "27" represents two full revolutions around the clock, starting at "3."

More generally, we state the following definition.

Definition 1.15 Let a, b, and n be integers, $n > 0$. Then we write

$$a \equiv b \pmod{n}$$

if there is an integer q such that $a - b = qn$. We will say that **a is congruent to b, modulo n.**

Example 1

a) $27 \equiv 15 \pmod{12}$, since $27 - 15 = (1)(12)$.

b) $27 \equiv 15 \pmod{6}$, since $27 - 15 = (2)(6)$.

c) $27 \equiv 15 \pmod{3}$, since $27 - 15 = (4)(3)$.

d) But 27 is not congruent to 15 (mod 5), because $27 - 15 = 12$ is not equal to $(q)(5)$ for any integer q. In other words, 12 is not divisable by 5. □

The connection between equivalence modulo n and divisability of integers is based upon the following theorem, a theorem that goes by the name of the **division algorithm.** We state this theorem without proof.

Theorem 1.5 If a and b are integers, b not equal to zero, then there exist unique integers q and r (called the "quotient" and "remainder," respectively) such that $0 \leqslant r < |b|$ and $a = bq + r$.

In other words, if we divide a by b, then there are a unique quotient q and a unique remainder r. Thus dividing $a = 32$ by $b = 5$ yields a quotient of $q = 6$ and a remainder $r = 2$. Moreover, $32 = (5)(6) + 2$, and $0 \leqslant 2 < |5|$.

Notice that a is divisable by b iff $r = 0$.

We can now state and prove the following.

Theorem 1.6 Two integers a and b have the same remainder on division by a positive integer n iff $a \equiv b \pmod{n}$.

Proof: By the division algorithm, we may assume that there are integers q_i and r_i such that

$$a = q_1 n + r_1, \quad \text{where } 0 \leqslant r_1 < |n|$$

and

$$b=q_2n+r_2, \quad \text{where } 0\leqslant r_2<|n|.$$

Suppose that a and b have the same remainder upon division by n, that is, that $r_1=r_2$. Then

$$a-b=q_1n-q_2n=(q_1-q_2)n.$$

Since q_1-q_2 is an integer, then $a\equiv b \pmod n$, by definition.

Conversely, suppose that $a\equiv b \pmod n$, that is, that there is an integer d such that $a-b=dn$. Then

$$(q_1n+r_1)-(q_2n+r_2)=(q_1-q_2)n+(r_1-r_2)=dn$$

and so n divides r_1-r_2. (See Exercise 1.7-24.) But in that case $0\leqslant|r_1-r_2|<|n|$ implies that $r_1-r_2=0$, that is, $r_1=r_2$, which completes the proof. ■

Thus 27 and 3 have the same remainder upon division by 12—namely, 3—and one therefore may write $27\equiv 3 \pmod{12}$.

Congruence arithmetic (mod n) can be carried out as in the following examples.

Example 2

a) $27+17=44=(12)(3)+8$. Hence $27+17\equiv 8 \pmod{12}$.

b) Similarly, $(27)(17)=459=(38)(12)+3$. Hence

$$(27)(17)\equiv 3 \pmod{12}.$$

c) Both Examples 2(a) and 2(b) can also be done by noting that $27\equiv 3 \pmod{12}$ and $17\equiv 5 \pmod{12}$. Hence

$$27+17\equiv 3+5\equiv 8 \pmod{12}$$

and

$$(27)(17)\equiv(3)(5)\equiv 15\equiv 3 \pmod{12},$$

as before. □

As Example 2(c) illustrates, one can do arithmetic modulo n entirely within the set of integers $\{0, 1, 2, \ldots, n-1\}$, that is, within the set of possible remainders upon division by n. (See Exercise 1.7-24.) Tables 1.6 and 1.7 indicate how this is done for addition and multiplication modulo 3 and modulo 4.

The arithmetic systems (sometimes called "residue systems") described by these tables are known as Z_3 and Z_4, respectively, that is, the **integers modulo 3** and the

Table 1.6 **Arithmetic Mod 3**

+	0	1	2
0	0	1	2
1	1	2	0
2	2	0	1

·	0	1	2
0	0	0	0
1	0	1	2
2	0	2	1

Table 1.7 **Arithmetic Mod 4**

+	0	1	2	3
0	0	1	2	3
1	1	2	3	0
2	2	3	0	1
3	3	0	1	2

·	0	1	2	3
0	0	0	0	0
1	0	1	2	3
2	0	2	0	2
3	0	3	2	1

integers modulo 4. More generally, for any positive integer n, the set $\{0, 1, 2, \ldots, n-1\}$, together with operations addition and multiplication modulo n, define the system of **integers modulo n, denoted Z_n**.

Recall that what we call subtraction is really a form of *addition*, since $x-y=x+(-y)$. It is not necessary to add negatives of the elements $0, 1, 2, \ldots, n-1$ to Z_n in order to carry out subtraction modulo n. Indeed Z_n already contains all of its "negatives," or, as we prefer to call them, **additive inverses**. The additive inverse of each element x in Z_n is defined to be that element y in Z_n such that $x+y=0$. For example, in Z_4, we identify 2 with "-2," since $2+2=0$ (mod 4). In other words, the additive inverse of 2 in Z_4 is 2. (What is the additive inverse of 2 in Z_3?) Hence to calculate $1-2$ in Z_4 we must instead write

$$1+(-2)\equiv 1+2\equiv 3 \pmod 4.$$

Similarly in Z_4 we have

$$3-2\equiv 3+(-2)\equiv 3+2\equiv 1 \pmod 4.$$

We have already seen an application of subtraction modulo n in the preceding section, where we considered subtraction by two's complements. The two's complement of y is just 2^n-y, for some integral power of y, and $2^n-y\equiv -y \pmod{2^n}$. Our method of finding the difference $x-y$ required us to calculate $x+(2^n-y)$, which is congruent to $x-y \pmod{2^n}$.

If addition is to be thought of as addition of additive inverses, then division (or at least division with zero remainders) should be thought of as multiplication by multiplicative inverses, where the **multiplicative inverse** of an element x is that element y such that $xy=1$. In general one can do such divisions in Z_n only under special conditions. To see why this should be so, consider what it means to divide the ordinary integer 3 by the ordinary integer 2. Instead of the quotient 3/2, we could equivalently consider the product of 3 with the multiplicative inverse of 2, namely, the rational number 1/2. In general the multiplicative inverse of an ordinary integer is not an integer. Similarly, numbers in Z_n do not generally have their multiplicative inverses in Z_n. For example, what we denote by the symbol "2" in Z_4 has no multiplicative inverse in Z_4. (Check the multiplication table for Z_4.) Hence one cannot do exact division by 2 in Z_4, in general. Similarly, construction of the multiplication table for Z_{10}, for example, would show that one cannot do exact division by either 2 or 5 in Z_{10}. Notice that in Z_3, however, we have $(2)(2) \equiv 1 \pmod 3$. And so exact division by 2 is always possible in Z_3, since 2 is its own multiplicative inverse in Z_3.

The Chinese Remainder Theorem

Many digital computers allow only a fixed maximum number of binary digits to represent an integer. In this mode, therefore, they can only represent integers up to a certain size, say n. Hence arithmetic with integers on such machines might be considered as being carried out modulo n, that is, in the residue system Z_n.

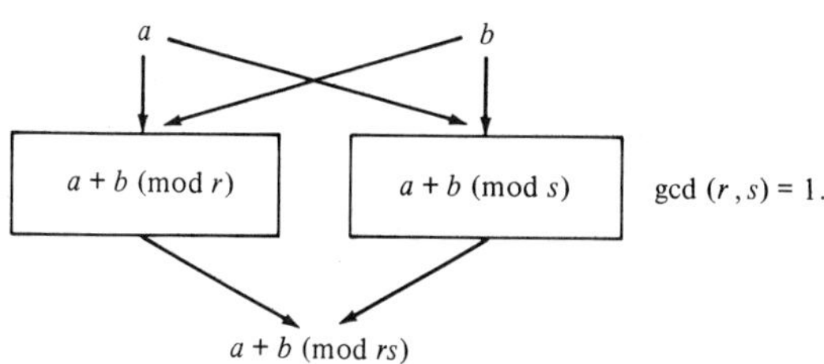

Figure 1.13 The Chinese remainder theorem is used to simplify modular arithmetic.

Now arithmetic in Z_n can often be done in terms of residue systems Z_r and Z_s, where $n = rs$ and where r and s are considerably smaller than n. Such an arrangement often allows us to reduce the number of "carries" in addition problems by adding numbers in Z_r and in Z_s having fewer digits than their counterparts in Z_n. This will, in turn, enable one to carry out the operations more quickly. We shall discuss a technique for accomplishing this feat with what is called the "Chinese remainder theorem."

First, however, recall that "$A \times B$" denotes the set of ordered pairs (a, b) of elements from sets A and B. Moreover, we will say that two integers are **relatively prime** if they have no prime factors in common (that is, their g c d $= 1$). For example, 3 and 4 are relatively prime, since their only prime factors are 3 and 2 respectively.

We may now state the **Chinese remainder theorem** as Theorem 1.7.

Theorem 1.7 Suppose that $m = rs$, where r and s are relatively prime positive integers, and let $Z_m^* = Z_r \times Z_s$. Then the mapping

$$f: Z_m \to Z_m^*$$

given by

$$f(x) = (x(\text{mod } r),\ x(\text{mod } s))$$

is a one-to-one correspondence. Moreover, for every x and y in Z_m we have

$$f(x+y) = (x(\text{mod } r),\ x(\text{mod } s)) \oplus (y(\text{mod } r),\ (y(\text{mod } s)),$$

and

$$f(xy) = (x(\text{mod } r),\ x(\text{mod } s)) \odot (y(\text{mod } r),\ y(\text{mod } s)),$$

where the operations $\oplus$ and $\odot$ indicate that addition and multiplication, respectively, are to be performed coordinatewise in Z_r and in Z_s.

What we mean when we say that operations are to be performed "coordinatewise" is that, for example, $(a, b) \oplus (c, d) = (a+c, b+d)$. If this is understood, then we can write the conclusions of Theorem 1.7 more simply as $f(x+y) = f(x) \oplus f(y)$ and $f(xy) = f(x) \odot f(y)$.

Let us illustrate Theorem 1.7 by showing how to do arithmetic in Z_{12} in terms of Z_3 and Z_4.

Example 3 We can factor $12 = (2)^2(3)$. Let $r = 4$ and $s = 3$. Then r and s have no common prime factors. The correspondence f, given in Theorem 1.7, tells us, for example, that

$$f(6) = (6(\text{mod } 4),\ 6(\text{mod } 3)) = (2, 0),$$

and

$$f(11) = (11(\text{mod } 4),\ 11(\text{mod } 3)) = (3, 2). \qquad \square$$

The entire correspondence between Z_{12} and Z_{12}^* is given explicitly in Table 1.8. The correspondence is obtained by listing the elements of Z_4 and of Z_3 over and over again in the first and second coordinates, respectively, of Z_{12}^*. For example, the list 0, 1, 2, 3 appears in the first coordinate three times in succession.

Table 1.8 The Mapping of Z_{12} into $Z_4 \times Z_3 = Z^*_{12}$

Z_{12}	$\rightarrow$	$Z_4 \times Z_3$
0		(0, 0)
1		(1, 1)
2		(2, 2)
3		(3, 0)
4		(0, 1)
5		(1, 2)
6		(2, 0)
7		(3, 1)
8		(0, 2)
9		(1, 0)
10		(2, 1)
11		(3, 2)

Example 4 Find the sum $6+11$ (mod 12) and the product (6)(11) (mod 12) using the Chinese remainder theorem and Table 1.8.

We may write

$$f(6+11) = (2, 0) \oplus (3, 2) = (2+3 (\text{mod } 4),\ 0+2 (\text{mod } 3)) = (1, 2).$$

But (1, 2) is the image of 5 in Z_{12}. Hence our theorem tells us that $6+11 \equiv 5$ (mod 12), which is, of course, easily verified. The point, however, is that *we can avoid addition of 6 and 11 modulo 12 and still get the correct sum.*

In a similar manner, we can determine the product of (6)(11) (mod 12) from

$$f((6)(11)) = (2, 0) \odot (3, 2) = (2 \cdot 3 (\text{mod } 4),\ 0 \cdot 2 (\text{mod } 3)) = (2, 0).$$

But, according to our table, (2, 0) is the image of 6. Thus $(6)(11) \equiv 6$ (mod 12). This agrees with the calculation $(6)(11) = 66 = (5)(12) + 6$, which again shows that $(6)(11) \equiv 6$ (mod 12). □

The advantage of doing arithmetic with the aid of Table 1.8 is that one adds and multiplies numbers that are considerably smaller than 12.

One can avoid having to construct the entire table of correspondences in order to find the inverse images of elements. Instead one can rely upon the observation that *if $f^{-1}(1, 0) = x$ and if $f^{-1}(0, 1) = y$, then $f^{-1}(a, b) = ax + by$ (modulo n).* For example, $f^{-1}(1, 0) = 9$ and $f^{-1}(0, 1) = 4$ implies that

$$f^{-1}(3, 2) = (3)(9) + (2)(4) = 27 + 8 \equiv 11 \ (\text{mod } 12).$$

An obvious disadvantage of this method, however, is that it requires the very arithmetic modulo 12 that we tried to avoid. Nevertheless there are methods for finding

$f^{-1}(a, b)$ that avoid this pitfall. (See Tremblay and Monahar, 33, pp. 352–359, for example.)

As a final note on this section, we point out that for integers n whose prime power factorization is $r_1 r_2 \ldots r_k$, where each r_i is a power of a distinct prime, one can map the elements of Z_n into "ordered k-tuples" of numbers from the k residue systems Z_{r_i}, and one does arithmetic coordinatewise in each of these residue systems. (See Exercise 1.8-27.)

Completion Review 1.8

Complete each of the following.

1. If a, b, and n are integers such that $a - b = qn$ for some integer q, then we say that a is __________ to b __________, and we write __________.
2. The division algorithm says that if a and b are any integers, $b \not\equiv 0$, then there exist unique integers q and r such that $a =$ __________, where r is such that __________.
3. Every element in Z_n has its additive inverse in Z_n. (*True or False*?)
4. Every nonzero element in Z_3 has its multiplicative inverse in Z_3. (*True or False*?) More generally, every nonzero element in Z_n has its multiplicative inverse in Z_n. (*True or False*?)
5. The __________ gives a one-to-one correspondence f between Z_{rs} and __________ that preserves sums and products, coordinatewise modulo r and modulo s, if r and s are __________ integers.
6. The theorem in Question 5 enables us to reduce the amount of __________ in arithmetic, thereby speeding up calculations.
7. If, in this theorem, $f^{-1}(1, 0) = x$ and $f^{-1}(0, 1) = y$, then $f^{-1}(a, b) =$ __________ (modulo n).

Answers: **1.** congruent; modulo n; $a \equiv b \pmod{n}$. **2.** $bq + r$; $0 \leqslant r < |b|$. **3.** True. **4.** True; False. **5.** Chinese remainder theorem; $Z_r \times Z_s$; relatively prime. **6.** carrying. **7.** $ax + by$.

Exercises 1.8

For Exercises 1 to 8 tell whether each statement is true or false, and explain why.

1. $3 \equiv 5 \pmod{2}$.
2. $33 \equiv 47 \pmod{2}$.

3. $33 \equiv 100{,}033 \pmod{100{,}000}$.
4. $3 \equiv 39 \pmod{12}$.
5. $0 \equiv 25 \pmod{12}$.
6. $12 \equiv 12 \pmod{12}$.
7. $3 \equiv 17 \pmod{5}$.
8. $27 \equiv 16 \pmod{7}$.

For Exercises 9 to 12 find the quotient q and remainder r when the integer a is divided by the nonzero integer b. Show that r and q satisfy the division algorithm.

9. $a=4$, $b=9$.
10. $a=9$, $b=4$.
11. $a=-13$, $b=28$.
12. $a=-13$, $b=-28$.
13. List the complete set of remainders when integers are divided by 9.
14. List the complete set of remainders when integers are divided by 16.
15. List the elements in Z_5, the set of integers modulo 5.
16. List the elements in Z_7, the set of integers modulo 7.
17. Give the addition and multiplication tables for Z_5.
18. Give the addition and multiplication tables for Z_7.
19. Solve each of the following equations in Z_5. **(a)** $3+3 \equiv x \pmod{5}$; **(b)** $(3)(3) \equiv x \pmod{5}$; **(c)** $3+x \equiv 1 \pmod{5}$; **(d)** $3x \equiv 1 \pmod{5}$.
20. Solve each of the following equations in Z_7. **(a)** $4+4 \equiv x \pmod{7}$; **(b)** $(4)(4) \equiv x \pmod{7}$; **(c)** $4+x \equiv 2 \pmod{7}$; **(d)** $4x \equiv 1 \pmod{7}$.
21. **(a)** Give a table showing the one-to-one correspondence of Z_{10} onto $Z_2 \times Z_5$ guaranteed by the Chinese remainder theorem. Then use your table to find **(b)** $8+9 \pmod{10}$ and **(c)** $(8)(9) \pmod{10}$.
22. **(a)** Give a table showing the one-to-one correspondence of Z_{20} onto $Z_4 \times Z_5$ guaranteed by the Chinese remainder theorem. Then use your table to find **(b)** $17+14 \pmod{20}$ and **(c)** $(17)(14) \pmod{20}$.
23. Show that if $a|b$ ("a divides b") and $a|c$, then $a|(b+c)$ and $a|(b-c)$.
24. Show that if $a \equiv b \pmod{n}$ and $c \equiv d \pmod{n}$, then $(a+c) \equiv (b+d) \pmod{n}$, and $ac \equiv bd \pmod{n}$.
25. **(a)** Prove that if, in the Chinese remainder theorem, $f^{-1}(1, 0)=x$ and $f^{-1}(0, 1)=y$, then $f^{-1}(a, b) \equiv ax+by \pmod{n}$. *Hint*: Find $f(ax+by)$.
(b) Apply part (a) to Exercises 22(b) and (c).
26. Give an algorithm for constructing the one-to-one correspondence from Z_n onto $Z_r \times Z_s$ guaranteed by the Chinese remainder theorem, where $n=rs$, and r and s have no prime factors in common.
27. **a)** Give a generalization of the Chinese remainder theorem for a positive integer $n=rst$, where r, s, and t are three positive integers having no prime factors in common. (*Hint*: Use ordered triples.)
b) Give the correspondence guaranteed by your generalization for $n=30=(2)(3)(5)$.

COMPUTER PROGRAMMING EXERCISES

These exercises are for students with knowledge of a high-level programming language. Both these students and especially those with little or no programming experience may find the appendix and Programs A1 to A7 of interest.

In Exercises 1 to 4 let each set A of n elements (real numbers) be represented by a one-dimensional array of length $n+1$ ending in a terminal zero. The elements are listed consecutively, and no set contains zero.

1.1. Write a program that determines whether a particular element is a member of a set A.

1.2. Write a program that counts the number of elements in a set A.

1.3. Given two sets, A and B, write a program that determines if A is a subset of B.

1.4. Write a program that determines the power set of a set A. That is, your program should list all subsets of A.

1.5. Write a program that determines if a two-dimensional array defining a set of ordered pairs of real numbers is a function and, in that case, outputs the function's domain and range.

1.6. Write a program that determines if a function defined as a two-dimensional array is one to one and gives the inverse of such a function from the range of the function into its domain.

1.7. Write a program that determines if a function defined as a two-dimensional array is onto a given set.

1.8. Write a program that calculates the sum $f(1)+f(2)+\cdots+f(n)$ for a particular function f and positive integer n by implementing Algorithm 1.1, Section 1.3.

1.9. Write a program that evaluates a given polynomial $p(x)$ for a particular value of x by using the nesting procedure of Algorithm 1.2, Section 1.3.

1.10. Write a program that determines if a particular number x is a Fibonacci number by implementing Algorithm 1.4, Section 1.4.

1.11. Write a program that determines if a particular integer x is a prime by implementing Algorithm 1.4, Section 1.4.

1.12. Write a program that lists all positive primes less than a given n by means of "the sieve of Eratosthenes." (See Exercise 1.4-14.)

1.13. Write a program that finds the greatest common divisor of two integers by means of the Euclidean algorithm. (See Exercise 1.4-13.)

1.14. Use Exercise 13 to write a program that converts a given fraction of integers (rational numbers) to lowest terms.

1.15. Write a program that converts a repeating decimal of the type $0.\overline{a_1a_2\,..\,a_n}$ into fractional form using Algorithm 1.5, Section 1.5.

1.16. Write a program that converts a positive integer in decimal notation into one in binary notation using Algorithm 1.6, Section 1.6.

1.17. Write a program that converts a binary numeral into **(a)** decimal notation; **(b)** octal notation; **(c)** hexadecimal notation.

1.18. Write a program that finds the two's complement of a binary numeral with respect to a given 2^n.

CHAPTER 2

SETS, LOGIC, AND COMPUTER ARITHMETIC

In this chapter we introduce further operations on sets and their formal symbolic manipulation. We also indicate how these operations might be handled on a digital machine.

A second major goal of this chapter is to make you more fully aware of the acceptable ways of demonstrating mathematical truth and formal principles of logic. Logic is not, however, only the concern of those who are interested in abstract argument. Indeed in the final section of this chapter we apply formal logic to construct the sort of "adding machine" that is a fundamental component of modern digital computers.

2.1 EXAMPLES, COUNTEREXAMPLES, AND MATHEMATICAL INDUCTION

In most mathematics one usually reasons from the general to the particular. This is known as **deductive reasoning**. That is, we rely on general principles, or theorems, to tell us what we want to know about particular objects.

Example 1 The angle sum of any triangle is 180 degrees. Figure 2.1 is a triangle, having $\measuredangle A$, $\measuredangle B$, and $\measuredangle C$. From this we deduce that the sum of the measures of $\measuredangle A + \measuredangle B + \measuredangle C = 180$ degrees. No measurements of these particular angles are needed once we know the general principle.

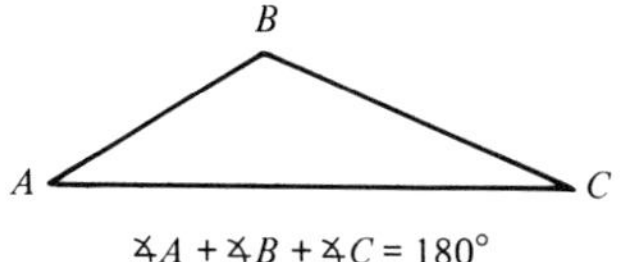

$\measuredangle A + \measuredangle B + \measuredangle C = 180°$

Figure 2.1 If the figure is a triangle, then we deduce that the sum of the measures of its angles is 180 degrees.

□

Examples are sometimes used to *disprove* an assertion. Disproving something is also a kind of proof. When we use an *example* to disprove an assertion we say that we have given a **proof by counterexample**.

Example 2 Suppose that someone tells you that "the product of every two odd numbers greater than 1 is always even." You can disprove this assertion with the counterexample

$$(3)(5) = 15,$$

which is an odd number. Hence you have shown that the product of two odd numbers greater than 1 is *not always* even. (It is at least *sometimes* odd.) □

Does the example $(3)(5) = 15$ show that the product of two odd numbers is *always* odd? No! Nor will any further examples show this.

Words such as "some" and "all" are called **quantifiers**. Our assertion is not about "some" (an **existential** quantifier) products being odd, but rather that "all" (a **universal** quantifier) the products xy are odd for "every" (another universal quantifier) pair of odd numbers x and y. If one wants to prove an assertion containing a universal quantifier such as "all" or "every," one must give a deductive proof (unless the "universe" of all objects under discussion amounts to only a few objects, of course).

The following deductive proof is called **direct**, since we proceed directly from our input (of two arbitrary odd numbers in the present case), through a series of conclusions, the last one being what we wanted to prove. (We say more about this in Section 2.5.)

Example 3 To prove that the product of two odd numbers is always odd, we may recall that $2n+1$ is the general form of an odd number, for any integer n. Given another integer m, then $2m+1$ is another odd number. (They are the same odd number only if $n=m$.) Hence $(2n+1)(2m+1)$ is what the product of two odd numbers looks like, in general. However,

$$(2n+1)(2m+1) = 4mn + 2n + 2m + 1 = 2(2mn + n + m) + 1,$$

and the last expression is clearly an odd number, since $2mn + n + m$ is an integer for every m and n. This completes the proof. □

Inductive reasoning bases general principles upon particular examples. This is what one usually does in real-life situations and, with many refinements, in the natural and social sciences.

Example 4 No entering college students are 100 feet tall. How do I know? Well I've seen thousands, and none were even close! □

While arguments like those in Example 4 have a certain appeal, they are *not* acceptable mathematically. In mathematics general principles must be based upon deductive, not inductive, reasoning, with a few important exceptions.

One of the most important exceptions to this rule is the following observation about the positive integers, the set

$$N = \{1, 2, 3, 4, \ldots, n, \ldots\}.$$

Positive Integers Axiom Suppose you have any subset $M \subset N$ such that

1. $1 \in M$.
2. Whenever you know that a particular positive integer $k \in M$, then you can show that $k+1 \in M$ as well.

Then we must conclude that $M = N$.

Why is this observation true? Since $1 \in M$, by step 2, $1+1=2 \in M$; but then step 2 implies that $2+1=3 \in M$, and so $3+1=4 \in M$, and so forth. This may seem a perfect deductive argument at first, but all it really amounts to is a kind of definition of the positive integers. Nevertheless most people, even mathematicians, are willing to accept the conclusion given *axiomatically*, that is, on faith. If we do, we obtain as a bonus the following proof technique, which tells us how to verify statements $S(n)$ (read "S of n") about every positive integer n.

Theorem 2.1 ***Principle of Mathematical Induction*** Let $S(n)$ be a statement concerning every positive integer n. Suppose we can verify that (a) $S(1)$ is true. (The statement is true with respect to 1.) Suppose further that (b) whenever $S(k)$ is assumed true for some integer $k \geqslant 1$, it can be shown that $S(k+1)$ must also be true (k is chosen arbitrarily). Then we may conclude that $S(n)$ is true for all positive integers n.

Example 5 Let $S(n)$ be the statement, "The formula

$$2+4+6+\cdots+2n = n(n+1)$$

is true for every positive integer n." Then, for example, $S(1{,}000{,}000)$ says that the sum of the first 1,000,000 positive, even integers is

$$1{,}000{,}000(1{,}000{,}001) = 1{,}000{,}001{,}000{,}000.$$

One could verify this fact by translating the following algorithm, for instance, into a computer program and running it.

1. [Initialize n and SUM.] Set SUM←0.
2. [SUM loop.] Repeat steps 3 for $n = 2, 4, 6, \ldots, 2000000$.
    3. [Increase SUM] Set SUM←SUM + n.
4. [Done.] Output SUM, and stop.
5. End algorithm. □

Since 1,000,000 numbers must be added, however, this calculation would take a significant amount of time, even on a typical personal computer. Our alternative is to prove the formula $S(n)$ by induction.

Example 6 To prove that the formula $S(n)$, that is,

$$2+4+6+\cdots+2n=n(n+1),$$

holds for every positive integer n, by Theorem 2.1(a) we must first (a) prove that $S(1)$ is true. But $S(1)$ just says that

$$2=1(1+1)=2.$$

Hence we have verified the truth of $S(1)$. We proceed to step (b). Prove that $S(k+1)$ is true if $S(k)$ is true, for an arbitrary positive integer $k \geqslant 1$. The forms that these statements take is of the greatest importance. We are assuming for *some* $k \geqslant 1$ that

$$2+4+\cdots+2k=k(k+1). \qquad \text{[induction assumption]}$$

We will reduce the left-hand side of the formula $S(k+1)$ until, it is hoped, we obtain the right-hand side, making use of the "induction assumption."

$$\begin{aligned} 2+4+\cdots+2(k+1) &= \quad \text{(by writing the next to last addend, } 2k) \\ (2+4+\cdots+2k)+2(k+1) &= \quad \text{(by the induction assumption)} \\ k(k+1)+2(k+1) &= \quad \text{(by the distributive law)} \\ (k+2)(k+1) &= \quad \text{(by rearranging terms)} \\ (k+1)((k+1)+1). \end{aligned}$$

Since the last expression is what one gets on the right-hand side of $S(n)$ upon substitution of $k+1$ for n, *we have verified that the truth of $S(k+1)$ follows from the truth of $S(k)$.*

Conclusion: $S(n)$: $2+4+\cdots+2n=n(n+1)$ is a true statement for *every* positive integer n. □

A typical error in the "induction step," part (b), is to make your k too particular. For example, students will sometimes say "let us assume $S(3)$ and prove $S(4)$ follows." This sort of thing is not sufficient. The k of part (b) must be particular and, at the same time, completely arbitrary. Only in part (a) of the proof do we deal with a specific number, namely, 1.

The principle of mathematical induction is sometimes stated for the set of whole numbers, which includes both the positive integers and zero. In that case $S(0)$ is verified in step (a); and we substitute 0 for 1 in step (b) as well. Similarly *one can obtain a valid generalization by substituting an initial whole number n_I for the number 1 in* *steps (a) and (b). This enables us to prove statements S(n) valid for "all whole numbers greater than or equal to n_l."* (See Exercise 2.1-7.) Here is still another useful variation of the principle of mathematical induction (or just "induction"), which is in the spirit of what we have just described.

Theorem 2.2 ***Alternative Principle of Mathematical Induction*** Let $S(n)$ be a statement concerning every whole number n greater than or equal to an initial whole number n_1. Suppose that we can verify that both

a) $S(n_1)$ is true; and

b) whenever $S(k)$ is assumed true for every whole number k such that $n_1 \leqslant k < n_2$, where n_2 is an arbitrary whole number $> n_1$, then we can show that $S(n_2)$ is also true. Then we may conclude that $S(n)$ is true for all whole numbers n, $n \geqslant n_1$.

It can be shown that all three versions of induction that we have stated (two explicitly and one implicitly) are equivalent, in the sense that each one implies the other.

We will now use Theorem 2.2 to prove part of the fundamental theorem of arithmetic, the part that asserts that any whole number greater than 1 can be factored into primes. But first let us recall that a prime is an integer greater than 1 that has no positive integral divisors except itself and 1. All other integers greater than 1 are called *composite numbers*. These therefore can be written as a product of two integers, as in $n = rs$, where both r and s are greater than 1 and less than n. It is this observation that gives us a chance to use Theorem 2.2.

Example 7 Let $S(n)$ be the statement that any whole number greater than 1 can be factored into its prime divisors.

a) The first whole number to which the statement applies is $n_1 = 2$. But $S(2)$ is clearly true, because 2 is a prime and $2 = 2$.

b) Now we assume that $S(k)$ is true for all whole numbers k, $2 \leqslant k < n_2$, where n_2 is an arbitrary whole number.

If n_2 is a prime, then $S(n_2)$ is obviously true, as in part (a). Hence we will suppose that n_2 is a composite number. As we have already observed, if n_2 is composite, then it can be factored into a product of two whole numbers r and s, where

$$1 < r < n_2 \quad \text{and} \quad 1 < s < n_2.$$

By our induction assumption, both r and s can be factored into their prime divisors. But this also factors $n_2 = rs$ into prime divisors. Therefore, we have shown that $S(n_2)$ is true.

***Conclusion*:** $S(n)$ is true for all whole numbers ≥ 2. □

Let us emphasize once again that the "induction step," part (b) of any of the three induction principles, requires deductive reasoning. Students sometimes get the impression that one is merely assuming what one wants to prove in part (b). In our first version, however, the quantifier "some" should alert you that we are not assuming that $S(k)$ holds in all cases. And even in our alternate version, the quantifier "every" is restricted by the particular n_2 that was chosen first.

The method of mathematical induction can be loosely compared to a sort of "proof algorithm" into which we input the integer 1 or n_1 and output a proof. The part of this analogy that holds most strongly is this: One must try the method with particular examples really to understand it, and the more examples, the better!

Completion Review 2.1

Complete each of the following.

1. The practice of drawing conclusions from general principles by means of logic is known as ________________.

2. A specific example that shows that a general assertion is false is known as a ________________.

3. A "quantifier" is a word or phrase that tells us ________________ elements in a set to which a statement applies. Words such as "all" or "every" are called ________________ quantifiers, while phrases such as "there exist" are called ________________ quantifiers.

4. The "positive integers axiom" states that ________________ is a positive integer; moreover, whenever k is a positive integer, so is ________________.

5. The (first) principle of mathematical induction instructs us to prove a statement $S(n)$ about all ________________ true in two steps: First, show that ________________ is true. Second, (the induction step) assume that $S(k)$ is true for ________________, and use this to show ________________.

6. We can also use the principle of mathematical induction to show that $S(n)$ is true for ________________.

7. As an alternative induction step, we may use the assumption that $S(k)$ is true for all integers n, such that ________________ for a positive integer k in order to show that $S(k)$ is true.

8. An integer greater than 1 that is not a prime is called a ________________.

Answers: **1.** deductive reasoning. **2.** counterexample. **3.** how many; universal; existential. **4.** 1; $k+1$. **5.** positive integers; $S(1)$; some positive integer k; $S(k+1)$ is true. **6.** for all positive integers n such that $n \geqslant$ a positive integer n_1. **7.** $n_1 \leqslant n < k$. **8.** composite number.

Exercises 2.1

1. Show that each of the following statements is false by means of a counterexample.
  a) The sum of every pair of odd numbers is odd.
  b) The product of an even number and an odd number is odd.
  c) Every quadrilateral with four equal sides is a square.
  d) If x and y are integers with $x > y$, then $x^2 > y^2$.
  e) Every positive integer can be represented as a product of its prime divisors.
  f) $1^2 + 2^2 + \cdots + n^2 = (n(n+1)/2)^2$ for all positive integers n.

2. Give a direct proof of each of the following.
  a) The sum of two even numbers is always even.
  b) The product of two even numbers is always even.
  c) The product of an odd number and an even number is always even.
  d) The square of an odd number is always odd.

3. For each of the following statements $S(n)$, state $S(1)$ and $S(4)$.
  a) $1 + 3 + 5 + \cdots + 2n - 1 = n^2$.
  b) $1 + 2 + 3 + \cdots + n = n(n+1)/2$.
  c) The angle sum of any convex polygon having n sides is $(n-2) \cdot 180$ degrees.
  d) It is possible to make up postage of exactly n¢ by using combinations of only 2¢ and 5¢ postage stamps.

4. Which of the statements $S(1)$ and $S(4)$ in Exercise 3 are true?

5. Prove each of the formulas in Exercises 3(a) and (b) true for all $n \in N$ by induction. (*Hint*: Use Theorem 2.1.)

6. Prove that each of the following formulas holds for all $n \in N$.
  a) $1 + 4 + 7 + \cdots + (3n-2) = n(3n-1)/2$.
  b) $1 + 5 + 9 + \cdots + (4n-3) = n(2n-1)$.
  c) $1^2 + 2^2 + 3^2 + \cdots + n^2 = n(n+1)(2n+1)/6$.
  d) $1^3 + 2^3 + 3^3 + \cdots + n^3 = (1 + 2 + 3 + \cdots + n)^2$.
  e) $\dfrac{1}{(1)(3)} + \dfrac{1}{(3)(5)} + \dfrac{1}{(5)(7)} + \cdots + \dfrac{1}{(2n-1)(2n+1)} = \dfrac{n}{2n+1}$.

7. State the generalization of Theorem 2.1 that one obtains by just substituting n_1 for 1 and adding the qualifying phrase "$n \geqslant n_1$" where needed in Theorem 2.2.

8. Use Exercise 7 to prove Exercise 3(d) for $n \geqslant 4$.

9. **a)** Carefully state what each step does in the algorithm we gave in Example 5.
  b) Modify that algorithm so that it calculates the sum of the even integers from 26 to 2,500,000. What is this sum?

10. Give an algorithm, as in Example 5, that will calculate the product of the first 100 positive odd integers.

11. If one calculates the sum of the first one million positive even integers on some hand-held calculators (such as the TI-30) with the aid of the formula $2+4+\cdots+2n=n(n+1)$, one obtains an answer of 1×10^{12}, in scientific notation. Is this answer "good enough"?
12. Prove the summation rules of Theorem 1.1, Section 1.3, by mathematical induction.

2.2 SET OPERATIONS

In the course of discussing the set of real numbers and some of its subsets, we had to introduce some of the language of set theory. Therefore, the words "set," "subset," and "null set," and the symbols "$x \in A$" and "$B \subset A$" are somewhat familiar to you. You have also been introduced to certain sets of numbers and simple relationships between them. If $N=$ the set of positive integers, $W=$ the set of whole numbers, $Z=$ the set of integers, $Q=$ the set of rational numbers, and $R=$ the set of real numbers, then we have

$$N \subset W \subset Z \subset Q \subset R.$$

This can also be illustrated as in Fig. 2.2.

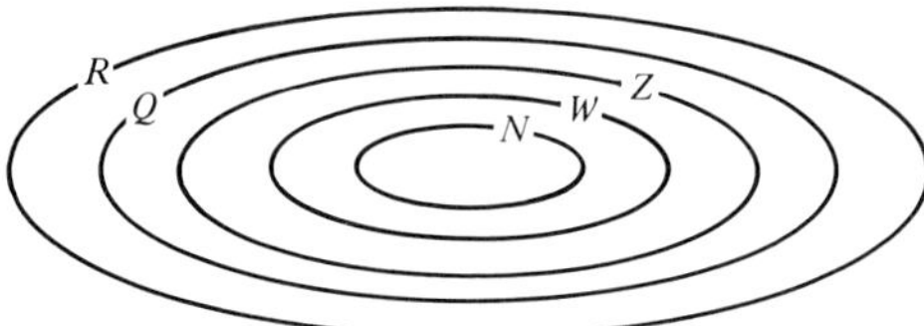

Figure 2.2 A Venn diagram showing the subset relationships between the real numbers, rational numbers, whole numbers, integers, and positive integers.

In Section 1.5 we learned that one has to "add" the irrational numbers to the rational numbers to get the entire set of real numbers. The correct way to describe this additive process is with the word "union." We say that the union of the rational numbers and the irrational numbers yields the set of real numbers. The union is but one example of an operation on *two* sets (hence a *binary* operation). Another commonly used set operation is that of "intersection."

Definition 2.1 The (a) **union** of two sets A and B is written "$A \cup B$," and $A \cup B = \{x: x \in A \text{ and/or } x \in B\}$. The (b) **intersection** of sets A and B is written "$A \cap B$," and $A \cap B = \{x: x \in A \text{ and } x \in B\}$.

If we let I = the set of irrational numbers, then

$$R = Q \cup I.$$

Moreover,

$$Q \cap I = \varnothing$$

since a real number cannot be both rational and irrational. When two sets intersect in the empty set, we say that they are **disjoint**. To take another example, if E denotes the even integers and D denotes the odd integers, then $E \cap D = \varnothing$, that is, the sets of even numbers and odd numbers are disjoint. Moreover, $E \cup D = Z$. We will say that E and D **partition** the set of integers, because they divide it into two disjoint sets whose union is Z. Similarly, Q and I partition the real numbers.

Partitioning the set under discussion (which is called the **universe of discourse** or **universal set**) can be very useful, especially if one is dealing with finite sets that one wishes to enumerate. Before proceeding further with this, let us generalize our definitions of union and intersection for more than two sets at a time.

Definition 2.2 Given a collection of n sets $A_1, A_2, \ldots, A_{n-1}, A_n$, we define the (a) **union** and (b) **intersection** of these sets, respectively, as

$$(A_1 \cup A_2 \cup \ldots \cup A_{n-1}) \cup A_n \quad \text{and} \quad (A_1 \cap A_2 \cap \ldots \cap A_{n-1}) \cap A_n.$$

Furthermore, we write $\bigcup_{i=1,2,\ldots,n} A_i$ and $\bigcap_{i=1,2,\ldots,n} A_i$ for the union and intersection, respectively. (Sometimes we write $\bigcup_{i=1}^{n} A_i$ and $\bigcap_{i=1}^{n} A_i$.)

These are, of course, recursive definitions (see Section 1.4) because they depend on the definitions of unions and intersections for $n-1$ sets. In the case of $n = 3$, for example,

$$(A_1 \cup A_2 \cup A_3) = \bigcup_{i=1,2,3} A_i = (A_1 \cup A_2) \cup A_3,$$

and

$$(A_1 \cap A_2 \cap A_3) = \bigcap_{i=1,2,3} A_i = (A_1 \cap A_2) \cap A_3.$$

One can see that the foregoing definitions of unions and intersection amount to the following.

$$\bigcup_{i=1,\ldots,n} A_i = \{x\colon x \in A_i \quad \text{for at least one } i = 1, 2, 3, \ldots, n\},$$

and

$$\bigcap_{i=1,\ldots,n} A_i = \{x : x \in A_i \quad \text{for every } i = 1, 2, 3, \ldots, n\}.$$

Definition 2.3 We may define the (a) **difference** $A - B$ of two sets by

$$A - B = \{x : x \in A \quad \text{and} \quad x \notin B\}.$$

This is sometimes called the "relative complement of B with respect to A." If U is the universe of discourse and B is a subset of U, then (b) the **complement of B**, written $\bar{B}$, is

$$\bar{B} = \{x : x \in U \quad \text{and} \quad x \notin B\}.$$

Sometimes the part about the universe is omitted in the definition of $\bar{B}$, but it is always implicitly understood. The universe, of course, varies with the discussion. Whatever U happens to be, however, we always have

$$\bar{U} = \varnothing \quad \text{and} \quad \bar{\varnothing} = U.$$

Example 1

a) $\{d, o, u, b, t\} \cup \{e, r\} = \{d, o, u, b, t, e, r\}$.

b) $\{d, o, u, b, t\} \cap \{e, r\} = \varnothing$.

c) $\{d, o, u, b, t, e, r\} \cap \{d, e, b, t\} = \{d, e, b, t\}$.

d) $\{d, o, u, b, t, e, r\} \cup \{d, e, b, t\} = \{d, o, u, b, t, e, r\}$.

e) $\{d, o, u, b, t, e, r\} - \{d, e, b, t\} = \{o, u, r\}$. □

We can visualize the operations of union, intersection, and complementation more readily by using Venn diagrams, as we did in Section 1.1. In Fig. 2.3(a) and (b) the sets A and B are represented by circular regions or ovals and the universal set U by the interior of a rectangle. The shaded regions represent $A \cup B$ and $A \cap B$ respectively.

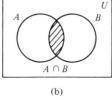

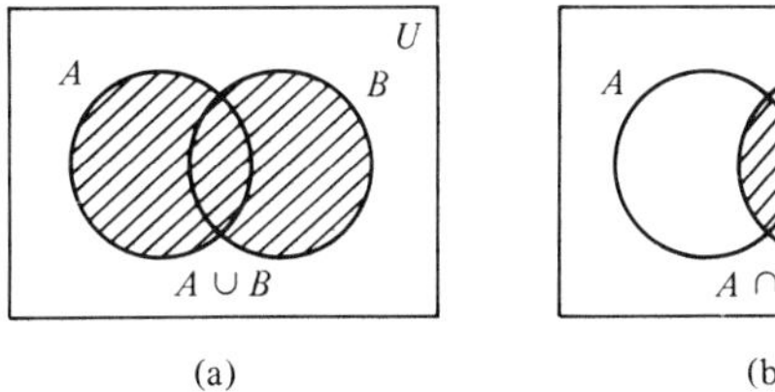

Figure 2.3 (a) A union B. (b) A intersect B.

Venn diagrams are useful in bringing out some of the more obvious properties of sets, such as in Fig. 2.3.

Theorem 2.3 If A and B are any two sets, then

1. $(A \cap B) \subset A \subset (A \cup B)$, and
2. $(A \cap B) \subset B \subset (A \cup B)$.

Moreover, if $A \subset B$, then

3. $A \cap B = A$, and
4. $A \cup B = B$. (See Exercise 2.2-6.)

In Figs. 2.4(a) and (b) we have Venn diagrams illustrating the differences between the sets $A - B$ and $\bar{B}$.

Venn diagrams involving three or more sets A, B, and C are a bit more complicated. Two examples are given in Figs. 2.5(a) and (b). The diagram for $(A \cap B) \cap C$ is drawn by first locating $A \cap B$, and then finding the part of the latter that overlaps C. Similarly, for $(A \cup B) \cap \bar{C}$, we find the part of $A \cup B$ that does *not* overlap C.

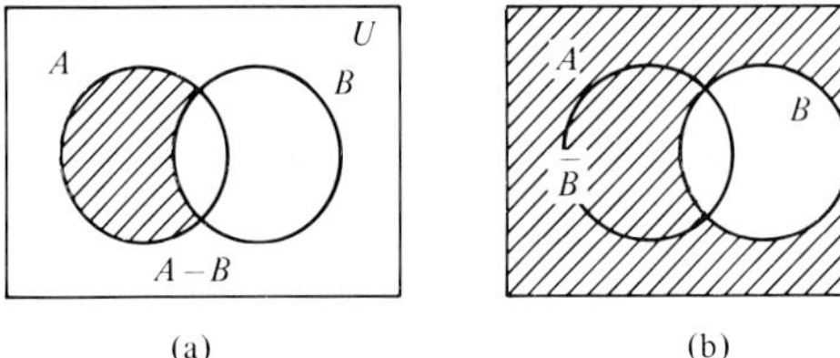

Figure 2.4 (a) The complement of B relative to a set A. (b) The complement of B (relative to the universal set).

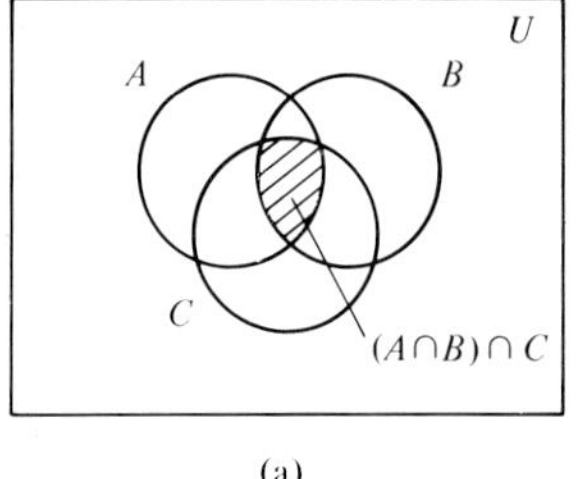

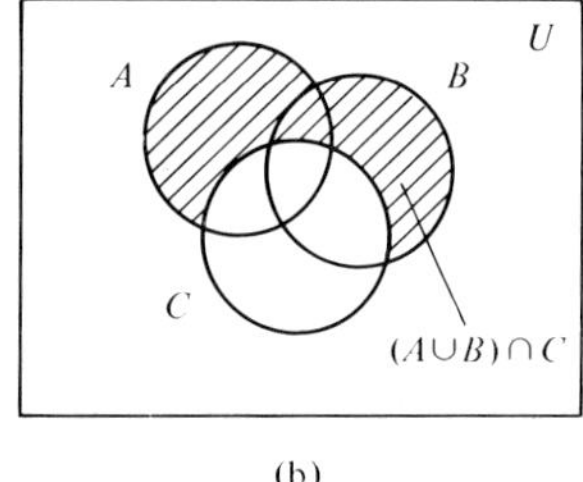

(a) (b)

Figure 2.5 Venn diagrams illustrating operations on three sets, A, B, and C.

Subsets and Binary Notation

Suppose that we have a set $U = \{a_1, a_2, \ldots, a_n\}$, each of whose elements a_i has been given a unique label $i = 1, 2, \ldots, n$. Then it is possible to label all the subsets of U so that one can tell which elements are in a subset merely by looking at the label of the subset.

Theorem 2.4 ***Subset Rule*** The subset S of $U = \{a_1, a_2, \ldots, a_n\}$ is given the subscript (or label) $i_1 i_2 \ldots i_n$, where $i_j = 1$ or 0 as the element a_j belongs or does not belong, respectively, to the set S.

Example 2 Let $U = \{a_1, a_2, a_3, a_4\}$. Then $n = 4$. The subset $\{a_2, a_4\}$ of U is denoted "S_{0101}." Notice that only the second and fourth digits of the subscript are 1's. Similarly, $S_{1001} = \{a_1, a_4\}$, $S_{0000} = \varnothing$, and $S_{1111} = U$. The length of subscripts can be shortened by changing the binary subscript to decimal. For example, $S_{1001} = S_9$ and $S_{0101} = S_5$. One can change back to binary when necessary. ☐

Recall that the digit 1 in the $n+1$st place of a binary number stands for $1 \cdot 2^n$. Hence there are 2^n nonnegative binary numbers having n digits or less. This in turn implies Theorem 2.5.

Theorem 2.5 If $|U| = n$, then U has 2^n subsets.

Using the ***n*-bit binary sequence notation** of Theorem 2.4, a computer (or a human being) need only compare the subscripts of two subsets of U to determine if they are equal. (See Exercise 2.2-9.)

Many programming languages have logical operations, sometimes called "AND," "OR," and "NOT," which can facilitate taking intersections, unions, and complements of sets, respectively, if we are making use of this binary notation for subsets. The AND (or its analog in the computer language of choice) compares two binary numbers b_1 and b_2 one digit at a time. It yields a binary number that has 1 in the ith place only when b_1 and b_2 both have 1 in the ith place. The OR returns a 1 in the ith place when either of the numbers b_1 or b_2 has the digit 1 in the ith place. Finally, the NOT operates on a single binary number at a time (so it is called a *unary* operation). It returns the digit 1 for each digit 0, and the digit 0 for each digit 1 in the binary number.

Theorem 2.6 Given two subsets S_r and S_p of a finite set U that have been labeled with binary numerals according to Theorem 2.4. To find $S_r \cap S_p$, $S_r \cup S_p$, and $\bar{S}_r$:

Set $t \leftarrow (r \text{ AND } p)$; $u \leftarrow (r \text{ OR } p)$; $c \leftarrow (\text{NOT } r)$.

Then

$$S_r \cap S_p = S_t;\ S_r \cup S_p = S_u;\ \bar{S}_r = S_c.$$

Example 3 Let $U = \{a_1, a_2, a_3, a_4, a_5\}$. Given S_{10100} and S_{11010}, we find 10100 AND 11010 = 10000, 10100 OR 11010 = 11110, and NOT 10100 = 01011. Hence

$$S_{10100} \cap S_{11010} = S_{10000}, \quad S_{10100} \cup S_{11010} = S_{11110}, \quad \text{and} \quad \bar{S}_{10100} = S_{01011}.$$

□

Partitions and Counting

As we mentioned earlier in this section, a **partition** of a set A is a collection of subsets A_i of A whose union equals all of A and which are **pairwise disjoint**, that is, such that no two of the subsets overlap. In this case it is obvious that we can add up the number of elements in a finite set A using the following rule.

Theorem 2.7 Let $A_1, A_2, \ldots, A_n$ be n finite sets such that $A_i \cap A_j = \varnothing$ unless $i = j$. Then

$$\left| \bigcup_{i=1,\ldots,n} A_i \right| = |A_1| + |A_2| + \cdots + |A_n|.$$

Proof: We proceed by mathematical induction on n. If $n = 1$, there is nothing to prove. If $n = 2$, on the other hand, then the sum $|A_1| + |A_2|$ clearly counts each element of the union $A_1 \cup A_2$ once and only once whenever $A_1 \cap A_2 = \varnothing$.

Next suppose that our theorem is true for some $n = k > 1$, and that the A_i, $i = 1, 2, \ldots, k+1$, are finite sets such that $A_i \cap A_j = \varnothing$ unless $i = j$. We will leave it for the reader to show that in this case we must have $(A_1 \cap A_2 \cap \ldots \cap A_k) \cap A_{k+1} = \varnothing$. (See Exercise 2.2-14.) But, by our induction hypothesis, we have

$$|A_1 \cup A_2 \cup \ldots \cup A_k| = |A_1| + |A_2| + \cdots + |A_k|$$

for the k mutually disjoint sets $A_1, A_2, \ldots, A_k$. Therefore, applying the case $n = 2$ to the set $(A_1 \cup A_2 \cup \ldots \cup A_k) \cup A_{k+1} = \bigcup_{i=1,\ldots,k+1} A_i$, we obtain

$$\left|\bigcup_{i=1,2,\ldots,k+1} A_i\right| = |(A_1 \cup A_2 \cup \ldots \cup A_k) \cup A_{k+1}|$$

$$= |A_1 \cup A_2 \cup \ldots \cup A_k| + |A_{k+1}|$$

$$= (|A_1| + |A_2| + \cdots + |A_k|) + |A_{k+1}|.$$

This proves our theorem for the case $n = k+1$ and completes the proof by mathematical induction. ■

The next example shows how we may use Theorem 2.7.

Example 4 A survey of the 80 computer science majors who graduated from DPU in 1982 revealed the following. By the end of their junior year, 40 had completed "Data Structures I"; 24 had completed "Assembler I"; 28 had completed "File Processing." Furthermore, among those who had completed "Data Structures I," eight had completed "Assembler I," 11 had completed "File Processing," and five students had completed all three courses. Finally, the survey showed that 12 of those who had completed "Assembler I" had also completed "File Processing."

If we let

$U =$ the set of the 80 computer science majors,

$A =$ the set of those who completed "Data Structures I,"

$B =$ the set of those who completed "Assembler I,"

$C =$ the set of those who completed "File Processing,"

then

$$|A| = 40, \quad |B| = 24, \quad |C| = 28, \quad |A \cap B| = 8,$$

$$|A \cap C| = 11, \quad |B \cap C| = 12, \quad \text{and} \quad |A \cap B \cap C| = 5.$$

The sets A, B, and C do *not* partition the set U of all computer science students. For one thing, $|U| = 80$, but $|A| + |B| + |C| = 40 + 24 + 28 = 92$. Moreover, $A \cap B \neq \varnothing$, since $|A \cap B| \neq 0$. To partition U correctly we can make a Venn diagram, as in Fig. 2.6 in which U has been partitioned into eight sets $S_1, S_2, \ldots, S_8$. These eight sets can be described in terms of unions, intersections, and complements of A, B, and C, as for example, with $A \cap B \cap C = S_1$ and $(\overline{A \cup B \cup C}) = S_8$. Now we have been told that $|S_1| = 5$, but how many are in S_8, that is, how many students had not finished any of the three courses by the end of their junior year?

Referring to Figs. 2.6 and 2.7, we begin by replacing S_1 with $|S_1| = 5$ in the Venn diagram. Then in subtracting from the pairwise intersections, we obtain

$$|S_2| = 8 - 5 = 3, \quad |S_3| = 11 - 5 = 6, \quad \text{and} \quad |S_4| = 12 - 5 = 7.$$

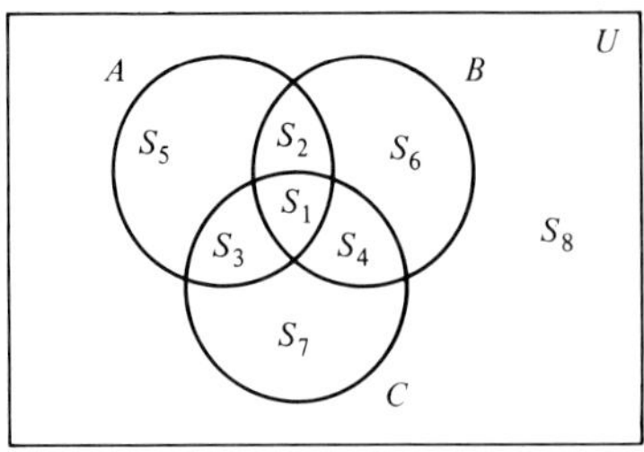

Figure 2.6 Partitioning the universe U into eight disjoint subsets, or cells.

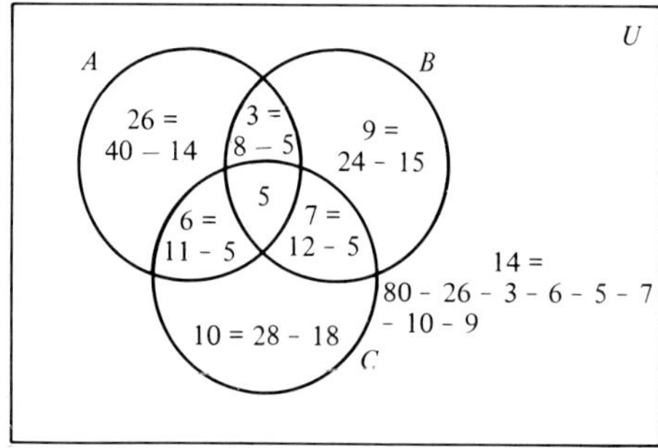

Figure 2.7 Deducing the number of elements in each partition cell from information about the number of elements in unions and intersections of sets A, B, and C.

The parts that A, B, and C do not have in common are also found by subtraction:

$$|S_5| = 40 - (5+3+6) = 26, \quad |S_6| = 24 - (5+3+7) = 9,$$
$$|S_7| = 28 - (5+6+7) = 10.$$

Since $|U| = 80$, we know that $|S_8| = 14 (= 80 - 26 - 3 - 6 - 5 - 7 - 10 - 9)$. Hence there were 14 computer science majors who had not taken any of the courses "Data Structures I," "Assembler I," or "File Processing" by the end of their junior year. □

Completion Review 2.2

Complete each of the following

1. The set of all elements in set A or in set B is called ________________ and is denoted by ________________.

2. The set of all elements in both set A and set B is called ________ and is denoted by ________.
3. The set $A - B$ is the set of all elements in ________. The complement of set B (with respect to the universal set U) is the set of all elements ________ and is denoted by ________.
4. If $A \cap B = \varnothing$, we say that the sets A and B are ________.
5. A partition of the set U is a collection of sets A_i whose ________ equals U and which are mutually ________.
6. In n-bit binary sequence notation, if element a_i is in set S, then the ith subscript of S is a ________. Otherwise the ith subscript of S is a ________.
7. An operation on two elements at a time is called a ________ operation. An operation on one element at a time is called a ________ operation.
8. S_{0011} AND S_{1010} = ________.
9. S_{0011} OR S_{1010} = ________.
10. NOT S_{0011} = ________.

Answers: **1.** A union B; $A \cup B$. **2.** A intersect B; $A \cap B$. **3.** A and not in B; U and not in B; $\bar{B}$. **4.** disjoint. **5.** union; disjoint. **6.** 1; 0. **7.** binary; unary. **8.** S_{0010}. **9.** S_{1011}. **10.** S_{1100}.

Exercises 2.2

1. Let U = {1, 2, 3, 4, 5, 6, 7, 8, 9, 10}; $A = \{4, 5, 7, 8, 9\}$; and $B = \{3, 5, 9, 10\}$. List the elements in each of the following.
(a) $\bar{A}$; **(b)** $\bar{B}$; **(c)** $A \cap B$; **(d)** $A \cup B$; **(e)** $\overline{A \cap B}$; **(f)** $\overline{A \cup B}$; **(g)** $\bar{A} \cap \bar{B}$; **(h)** $\bar{A} \cup \bar{B}$; **(i)** $\bar{\varnothing}$; **(j)** $A - B$; **(k)** $B - A$; **(l)** $\bar{U}$.

2. Make six Venn diagrams like the one shown here. Then shade in the region corresponding to: **(a)** $\bar{A} \cap B$; **(b)** $A \cap \bar{B}$; **(c)** $\overline{A \cup B}$; **(d)** $\overline{A \cap B}$; **(e)** $\bar{A} \cap \bar{B}$; **(f)** $\bar{A} \cup \bar{B}$.

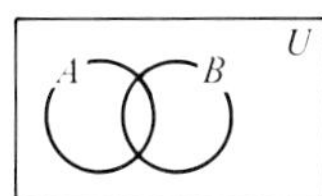

3. If we define $A \oplus B = (A - B) \cup (B - A)$, called the **symmetric difference** of A and B, find $A \oplus B$ for A and B of Exercise 1.

4. Indicate the symmetric difference in a Venn diagram as in Exercise 2.

5. Given subsets A, B, and C of a universal set U, draw a Venn diagram for the following sets, assuming that they all intersect one another. **(a)** $(A \cup B) \cup C$;

(b) $(A \cup B) \cap C$; **(c)** $(A \cap B) \cup C$; **(d)** $(A \cap C) \cup (B \cap C)$;
(e) $(A \cup C) \cap (B \cup C)$; **(f)** $(A \cap \bar{B}) \cap C$; **(g)** $\bar{A} \cap (B \cap C)$;
(h) $(\bar{A} \cup \bar{B}) \cup \bar{C}$.

6. Use a Venn diagram to convince yourself of the truth of Theorem 2.3. Does your diagram amount to a proof?

7. How many subsets does a set of five elements contain?

8. How many elements are in $P(U)$, if $|U| = 10$?

9. How would you designate the subsets A and B of set U in Exercise 1 using binary subscript notation?

10. Use Theorem 2.6 to find $A \cup B$, $A \cap B$, and $\bar{A}$ for the subsets in Exercise 1.

11. A study of the reading habits of 250 college sophomores revealed that 158 read *Time* magazine, 139 read *Playboy* magazine, and 100 read both *Time* and *Playboy*.
a) How many sophomores read either *Time* or *Playboy*?
b) How many sophomores read exactly one of these magazines?

12. In Exercise 11 it was also found that 53 students read *Esquire* magazine, 25 read both *Playboy* and *Esquire*, but none read both *Time* and *Esquire*.
a) How many read all three magazines?
b) How many read none of these magazines?

13. A genetics experiment required the collecting of data on three characteristics of a number of pea plants: height (tall versus short), color (green versus black), and smoothness (smooth versus wrinkled). It was found that 74 were short; 60 were black; 64 were short and wrinkled; 48 were short and black; 78 were wrinkled; 40 were black, wrinkled, and short; 42 were black and wrinkled; and eight were smooth, green, and tall. **(a)** How many were black, tall, and wrinkled? **(b)** How many were short, smooth, and green? **(c)** How many plants were surveyed in all?

14. Prove by mathematical induction that if $A_i \cap A_j = \varnothing$ for $i, j = 1, 2, \ldots, n$, $i \neq j$, then $A_1 \cap A_2 \cap \ldots \cap A_n = \varnothing$.

15. Prove that $|A \cup B| = |A| + |B| - |A \cap B|$.

16. Prove that $|A \oplus B| = |A| + |B| - 2|A \cap B|$.

2.3 THE ALGEBRA OF SET OPERATIONS

The arithmetic operations of addition and multiplication obey various algebraic laws that we have become accustomed to taking for granted. Among these are the so-called "associative laws"

$$(a+b)+c = a+(b+c) \quad \text{and} \quad (ab)c = a(bc),$$

which permit us to write "$a+b+c$" and "abc" without any fear of ambiguity. The associative laws show that the parentheses, which tell which operations are carried out first, really have no effect on the final sum or product.

Somewhat less obvious is the "distributive law"

$$a(b+c)=ab+ac,$$

which shows us how to mix the operations of addition and multiplication. On the left-hand side we first add and then multiply; on the right, however, one first takes the products ab and ac and then adds. Notice that although "multiplication is distributive over addition," it is not true that "addition is distributive over multiplication." The following examples illustrate what this means.

Example 1 a) To illustrate the principle that $a(b+c)=ab+ac$, for all real a, b, and c:

$$2\cdot(3+4)=2\cdot 7=14 \quad \text{and} \quad 2\cdot 3+2\cdot 4=6+8=14.$$

b) To show that $a+(bc)\neq(a+b)\cdot(a+c)$ for some real a, b, c:

$$2+(3\cdot 4)=2+12=14 \quad \text{but} \quad (2+3)\cdot(2+4)=5\cdot 6=30.$$

Observe that part (a) is *not* a proof of the distributive law. (The key word is "all.") But part (b) *does* give a proof that inequality holds for *some* a, b, and c ($a=2$, $b=3$, and $c=4$). □

The set operations that we discussed in the previous section are less familiar than the arithmetic operations. Rules governing set unions, intersections, and complementation are sometimes, but not always, analogous to those for addition, multiplication, and the taking of negatives. We now discuss how to discover and establish valid laws of set operations.

Venn diagrams can be used as an aid to intuition, compensating, to some extent, for our lack of extensive experience with set operations.

Example 2 Our definition for the intersection $A\cap B\cap C$ of more than two sets at a time was given in Section 2.2 as

$$(A\cap B)\cap C.$$

To decide whether this should always give the same result as

$$A\cap(B\cap C),$$

where the $B\cap C$ intersection is taken first, we can make Venn diagrams as in Figs. 2.8(a) and (b). These were each formed by intersecting just two sets at a time, as the different patterns of shading indicate. The regions where the different shadings overlap are the same in both (a) and (b). It is the region in common to all three sets, as seen in (c).

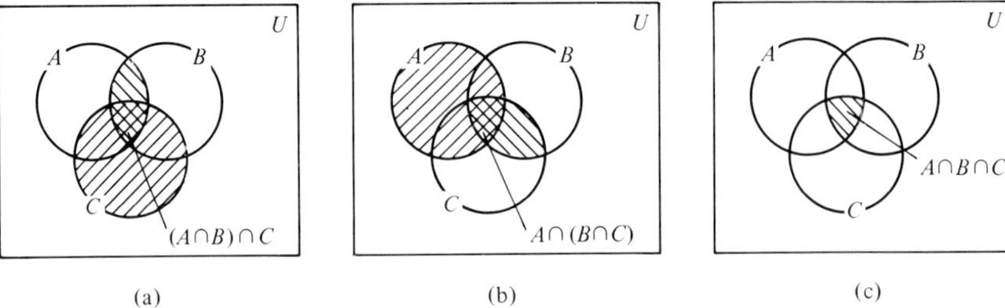

Figures 2.8 Verifying the associative law $(A \cap B) \cap C = A \cap (B \cap C)$ with Venn diagrams. □

One can give a proof of the *associative law for set intersections*

$$(A \cap B) \cap C = A \cap (B \cap C)$$

for all sets A, B, and C, based entirely upon the definition of the intersection of two sets. [Although the Venn diagram technique may seem very convincing, it is not entirely general. See Exercises 2.3-18(a) and (b).] Such proofs of set equality are often based upon the observation that $A = B$ *if and only if* $A \subset B$ *and* $B \subset A$, that is, if every element in A is an element of B, and conversely. In fact we are going to leave the proof of this associative law to the reader. We illustrate the technique, however, by using it to prove one of the "distributive laws" for sets. (Unlike arithmetic, set operations have *two* distributive laws.)

Example 3 To prove the *distributive law* for sets

$$A \cap (B \cup C) = (A \cap B) \cup (A \cap C),$$

we first show that $A \cap (B \cup C) \subset (A \cap B) \cup (A \cap C)$. Let us suppose that $x \in A \cap (B \cup C)$. By definition of intersection, $x \in A$ and $x \in B \cup C$. From the latter, $x \in B$ or $x \in C$. If $x \in B$ is true, then $x \in A \cap B$; if $x \in C$, then $x \in A \cap C$. Since one or the other must be true, we conclude that $x \in (A \cap B) \cup (A \cap C)$. Hence $A \cap (B \cup C) \subset (A \cap B) \cup (A \cap C)$.

Conversely, suppose that x is any element in the set $(A \cap B) \cup (A \cap C)$. Then $x \in A \cap B$ or $x \in A \cap C$, by definition of set unions. If the first is true, then $x \in A$ and $x \in B$; if the second, then $x \in A$ and $x \in C$. Therefore, in both cases $x \in A$, and either $x \in B$ or $x \in C$. Hence $x \in A \cap (B \cup C)$. This shows that $(A \cap B) \cup (A \cap C) \subset A \cap (B \cap C)$, as required, and completes the proof. □

Illustrating the distributive law that we proved in Example 3 with Venn diagrams requires that we represent $A \cap (B \cup C)$ and $(A \cap B) \cup (A \cap C)$ in separate diagrams and compare the results. (See Fig. 2.9.)

There are several key properties of the operations of set unions, intersections, and complementation that can be proved directly as in Example 3, and illustrated with Venn diagrams as in Fig. 2.9. We list them in Theorem 2.8 and leave the verification of their correctness for the reader to check in the exercises.

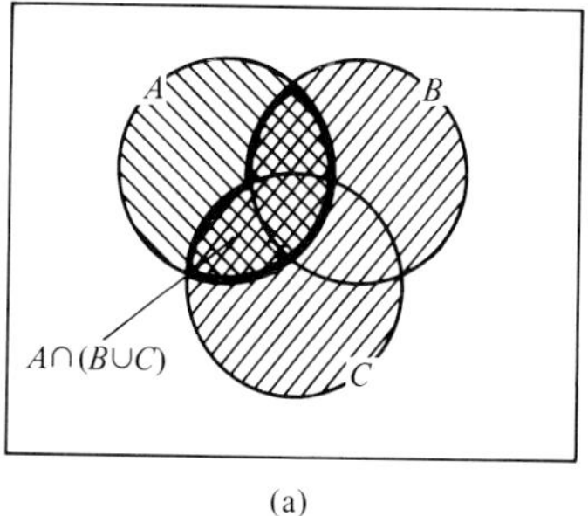

(a)

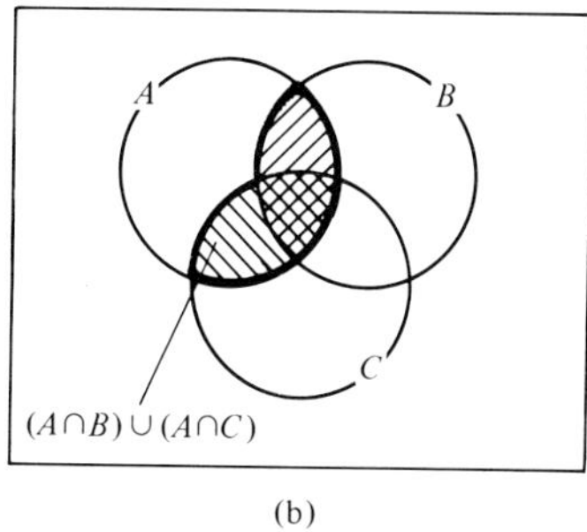

(b)

Figure 2.9 Verifying the distribute law $A \cap (B \cup C) = (A \cap B) \cup (A \cap C)$ with Venn diagrams.

Theorem 2.8 ***Laws of Set Operations*** Let A, B, and C be any subsets of some universe U. Then each of the following holds.

1. Commutative

1. (a) $A \cup B = B \cup A$. 1. (b) $A \cap B = B \cap A$.

2. Associative

2. (a) $A \cup (B \cup C) = (A \cup B) \cup C$. 2. (b) $A \cap (B \cap C) = (A \cap B) \cap C$.

3. Distributive

3. (a) $A \cup (B \cap C) = (A \cup B) \cap (A \cup C)$. 3. (b) $A \cap (B \cup C) = (A \cap B) \cup (A \cap C)$.

4. Identity (U)

4. (a) $A \cup U = U$. 4. (b) $A \cap U = A$.

5. Identity ($\varnothing$)

5. (a) $A \cup \varnothing = A$. 5. (b) $A \cap \varnothing = \varnothing$.

6. Idempotence

6. (a) $A \cup A = A$. 6. (b) $A \cap A = A$.

7. 8. Complements

7. (a) $A \cup \bar{A} = U$. 7. (b) $A \cap \bar{A} = \varnothing$.

8. (a) $\bar{U} = \varnothing$. 8. (b) $\bar{\varnothing} = U$.

9. De Morgan's

9. (a) $(\overline{A \cup B}) = \bar{A} \cap \bar{B}$. 9. (b) $(\overline{A \cap B}) = \bar{A} \cup \bar{B}$.

10. Involution

$(\overline{\bar{A}}) = A$.

If we make the (arbitrary) association of union, intersection, and complementation with ordinary addition, multiplication, and the taking of negatives of real numbers, we see many analogs of arithmetic laws in Theorem 2.8. The associative and commutative laws are the outstanding examples; but the "involution" law, number 10, says something like "the negative of the negative is the original." This holds for real numbers too! Notice, however, that there are some laws that have no obvious analog, such as the laws named for the logician Augustus De Morgan. Moreover, there are two distinct distributive laws for sets, as opposed to the one law we discussed for numbers. (See Example 1 again.) The system of algebra for sets is, therefore, not identical with the ordinary algebra of real numbers. It is called a "Boolean algebra," for the mathematician George Boole, who studied general algebraic systems with very similar laws, laws that exhibit a great deal of symmetry.

Definition 2.4 Let S be a statement of set equality. Then the **dual of S**, denoted by S^*, is obtained by replacing each of the symbols $\cup$, $\cap$, U, and $\varnothing$ in S with the symbols $\cap$, $\cup$, $\varnothing$, and U respectively.

Example 4 If we let S be any statement on the left in Theorem 2.8, and we follow the replacement rule in Definition 2.4, we obtain the statement on the right. Thus

$$\overline{(A \cap B)} = \bar{A} \cup \bar{B} \quad \text{is the dual of} \quad \overline{(A \cup B)} = \bar{A} \cap \bar{B}.$$

To take another example, $\bar{\bar{A}} = A$ is the dual of itself because there is nothing to replace. □

Suppose that we have deduced a statement S by means of a chain of set equalities based upon the rules in Theorem 2.8. One could then take very similar steps to establish the dual S^*. This is illustrated in the following example.

Example 5 We will simplify the expressions $\overline{(A \cup \bar{B})}$ and $\overline{(A \cap \bar{B})}$ simultaneously using corresponding laws of set operations in Theorem 2.8.

$$\begin{aligned} \overline{(A \cup \bar{B})} &= \bar{A} \cap \bar{\bar{B}} && \text{by 9(a)} & \qquad \overline{(A \cap \bar{B})} &= \bar{A} \cup \bar{\bar{B}} && \text{by 9(b)} \\ &= \bar{A} \cap B && \text{by 10.} & &= \bar{A} \cup B && \text{by 10.} \end{aligned}$$

□

Since we can always do such simplifications "in parallel," we have the following rule.

Theorem 2.9 ***Principle of Duality*** Any general statement S of the identity of two sets is true if and only if the dual statement S^* is true as well.

Example 6 Let S be the statement $(A \cap \varnothing) \cup (A \cap U) = A$. Then the dual of this statement, S^*, is $(A \cup U) \cap (A \cup \varnothing) = A$. If we can prove that S is a true statement, then the truth of S^* will follow automatically, according to Theorem 2.9. Thus

$$\begin{aligned}(A \cap \varnothing) \cup (A \cap U) &= A \cap (\varnothing \cup U) && \text{by distributive law 3(b)}\\ &= A \cap U && \text{by identity law 4(a)}\\ &= A && \text{by identity law 4(b).}\end{aligned}$$

Having proved statement S, the truth of statement S^* is guaranteed by Theorem 2.9. □

Completion Review 2.3

Complete each of the following.

1. The associative law for set unions says __________; the associative law for set intersections says __________.
2. The distributive law of union over intersection says __________; the distributive law of intersection over union says __________.
3. The commutative law of set union says __________; the commutative law of set intersection says __________.
4. The identity laws say that __________, __________, __________, and __________.
5. The idempotence laws say that __________ and that __________.
6. The laws of complements say that __________, __________, __________, and __________.
7. De Morgan's laws say that __________ and that __________.
8. The involution law says that __________.
9. The dual of a statement S about the identity of two sets is formed by replacing each of the symbols U, $\varnothing$, $\cup$, and $\cap$ by __________, __________, __________, and __________ respectively.
10. The dual of S is denoted by __________. The __________ says that S and its dual are both true or both false.

Answers: **1.** $A \cup (B \cup C) = (A \cup B) \cup C$; $A \cap (B \cap C) = (A \cap B) \cap C$. **2.** $A \cup (B \cap C) = (A \cup B) \cap (A \cup C)$; $A \cap (B \cup C) = (A \cap B) \cup (A \cap C)$. **3.** $A \cup B = B \cup A$; $A \cap B = B \cap A$. **4.** $A \cup U = U$; $A \cap U = A$; $A \cap \varnothing = \varnothing$; $A \cup \varnothing = A$. **5.** $A \cup A = A$; $A \cap A = A$. **6.** $A \cup \bar{A} = U$; $A \cap \bar{A} = \varnothing$; $\bar{U} = \varnothing$; $\bar{\varnothing} = U$. **7.** $\overline{(A \cup B)} = \bar{A} \cap \bar{B}$; $\overline{(A \cap B)} = \bar{A} \cup \bar{B}$. **8.** $\overline{(\bar{A})} = A$. **9.** $\varnothing$; U; $\cap$; $\cup$. **10.** S^*; principle of duality.

Exercises 2.3

Exercises 1 to 4 refer to operations with real numbers. Give an example to prove each statement.

1. Subtraction is not commutative $(a - b \neq b - a)$.
2. Division is not commutative $(a/b \neq b/a)$.
3. Subtraction is not associative $(a - (b - c) \neq (a - b) - c)$.
4. Subtraction is not distributive over addition $(a - (b + c) \neq (a - b) + (a - c))$.
5. Let $A = \{1, 4, 5\}$, $B = \{1, 3, 4\}$, and $U = \{1, 2, 3, 4, 5\}$. Illustrate each part of Theorem 2.8 using these sets whenever possible.

For Exercises 6 to 8, give two Venn diagrams each supporting

6. The associative law, 2(a).
7. The distributive law, 3(a).
8. De Morgan's laws, 9(a) and 9(b).

Use *definitions of the symbols* to prove the laws referred to in Exercises 9 to 13.

9. The commutative laws, 1(a) and 1(b).
10. The identity laws, 5(a) and 5(b).
11. The idempotence laws, 6(a) and 6(b).
12. The laws of complements, 7 and 8.
13. The involution law, 10.

The following are a little more involved. Prove, in Exercises 14 to 16,

14. The distributive law, 3(a).
15. The associative laws, 2(a) and (b).
16. De Morgan's laws, 9(a) and 9(b).
17. Write the duals of the following statements. **(a)** $\overline{(\bar{A} \cap \bar{B})} = (A \cup B)$; **(b)** $(A \cap U) \cup (B \cap A) = A$; **(c)** $(\bar{A} \cap B) \cup (A \cap B) = B$.
18. (a)–(c) Prove each assertion of Exercise 17.
19. (a)–(c) Prove the dual of each assertion of Exercise 17.

2.4 TRUTH SETS AND TRUTH TABLES

One could not do mathematics or write computer programs without logically correct arguments and operations. What we usually mean by the word "logical" is "carefully reasoned" or "without introducing false suppositions." However, logic is actually the study of **propositions**, that is, statements that can be described as either true or false, but not both.

Example 1 The following are propositions.

a) Today is pay day.

b) This course is called "Discrete Math Structures."

c) Shirley is a computer science major.

d) The algorithm will terminate in exactly 10 steps.

The following are not propositions.

e) What time is it?

f) Turn on the computer terminal. □

Notice that Example 1(e) and (f) are not declarative sentences. Example 1(e) is a question, while part (f) is a command. Hence it is not appropriate to describe these as being true or false.

The truth or falsehood of each of our other examples, 1(a) to 1(d), depends upon the particular subject or circumstances. Some algorithms will end in 10 steps; others will not. Some Shirleys may well be computer science majors; others may be majoring in classical civilizations. We call such statements **hypothetical**, while statements whose truth does not depend on the particular circumstances are called **categorical**.

Example 2

a) $x+1=3$ is a hypothetical statement. It is true when $x=2$. It is false for all other values of x.

b) $x+x=2x$ is a categorical statement, since its truth does not depend on the value of x. In fact it is true for all real numbers x.

c) $x=x+1$ is also categorical, being false for all values of x. □

Propositions that are always true are called **tautologies**. Thus Example 2(b) is a tautology. Definitions are generally tautologies as well. For example, "a quadruped is a four-legged creature" is always true because the word "quadruped" can mean nothing but "four-legged creature." The mere form of the statement is what makes it a tautology. Likewise, the form of Example 2(c) ensures that it will always be false. We call such statements **contradictions**.

We will denote propositions with lowercase letters such as p or q, especially when these are simple declarative statements. We will say that their **truth values** are T (for "true") or F (for "false").

Simple statements can be combined to form more complicated statements of the forms "p and q," "p or q," and "if p then q," for example. From the point of view of logic these forms are very different in the ways that their truth values are obtained from the truth values of their components p and q. Let us look at some examples of compound statements of these forms.

Example 3 Let us suppose that p denotes the statement, "My hand-held calculator won't go on," and let q denote the proposition, "The batteries need replacing." Then the form "p and q" is "My hand-held calculator won't go on and the batteries need replacing"; the form "p or q" is "My hand-held calculator won't go on or the batteries need replacing"; the form "if p then q" is "If my hand-held calculator won't go on, then the batteries need replacing." As we shall see, the logical meanings of these statements are quite different. □

In addition to analyzing whether or not an argument is correct, studying the truth values of various parts of a computer program can help determine if it will operate as expected. Computer languages, such as BASIC, FORTRAN, COBOL, and PL1, have commands that can be used to test whether or not a given condition has been satisfied. That is, the computer can assign truth values to a statement and then, depending upon whether the statement is true or false, go on to execute another command.

Since we are not assuming knowledge of any particular programming language, let us illustrate the preceding with a simple algorithm.

Example 4 Looking back at Algorithm 1.6 and Example 7 in Section 1.6, we recall that the problem was to change 37 into binary notation. Step 3 of the algorithm was the one whose truth value needed to be checked at each stage of the loop: Was the remainder 0 or not? If you will glance at the calculations in Example 7, you will see that the first five times this statement was encountered, its truth value was F, and consequently another division had to take place. On the sixth time, the quotient q_i was zero. So the truth value was T, and the algorithm terminated. □

We shall henceforth confine our discussion to how truth values of simple statements affect the truth values of the compound statements to which they belong. We are often assisted by examining the **truth sets** of these propositions, that is, the set of circumstances under which the proposition is true.

In Fig. 2.10 we have let P denote the set of circumstances under which the proposition p is true. Then the set $\bar{P}$, the complement of P, will stand for the circumstances in which p is false. Indeed we can think of the set $\bar{P}$ as the truth set of what is called *the negation of* p or *not-p* and denoted $\neg p$.

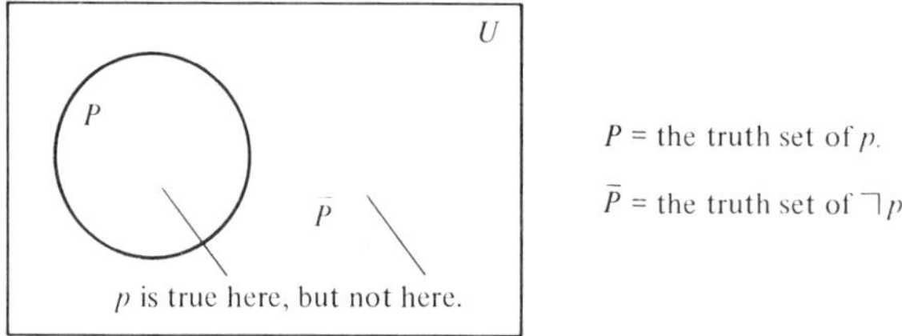

Figure 2.10 The truth set of a proposition contains the elements of the universe for which the proposition is true.

Forming the negation of a proposition is even simpler than the previously mentioned ways of getting new propositions from old ones. So it is fitting that we give its **truth table** first. This is a table that shows how the truth values of p affect those of $\neg p$. There are only two values of p to consider:

Table 2.1 Truth Table of a Proposition and Its Negation

p	$\neg p$
T	F
F	T

Thus $\neg p$ is false when p is true, and $\neg p$ is true when p is false. For example, if it is true that "my hand-held calculator won't go on," then it is false that "it is not the case that my hand-held calculator won't go on." That is, the calculator *will* go on. Or, for example, if it is false that "the batteries need replacing," then it is true that "it is not the case that the batteries need replacing." That is, they do *not* need replacing. We urge you to try to avoid awkward constructions that the negation of a statement often seems to require. Double negatives are particularly awkward, but they are also very easy to avoid in the sense that p can be substituted for $\neg(\neg p)$, as described in Definition 2.5.

> **Definition 2.5** Two propositions p and q are **logically equivalent** (or just "equivalent") if they have exactly the same truth values in every circumstance. (Hence they have the same truth set.) In that case we will write $p = q$.

Clearly any proposition is logically equivalent to itself. To see that p and $\neg(\neg p)$ are equivalent, we need only examine the truth set of $\neg(\neg p)$ in Table 2.2.

Table 2.2 Truth Tables of $\neg p$ and $\neg(\neg p)$

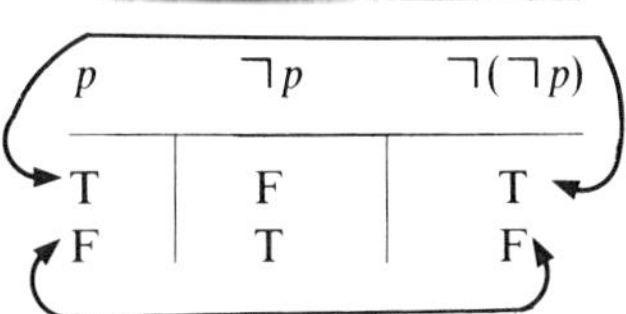

p	$\neg p$	$\neg(\neg p)$
T	F	T
F	T	F

The effect of the negation symbol "$\neg$" is to change a truth value from T to F or from F to T. This is how the third column of Table 2.2 was obtained. For our purposes it will be only the truth values of a proposition that really matter. This is why logically equivalent statements can be substituted for one another: it is their truth values in which we are interested!

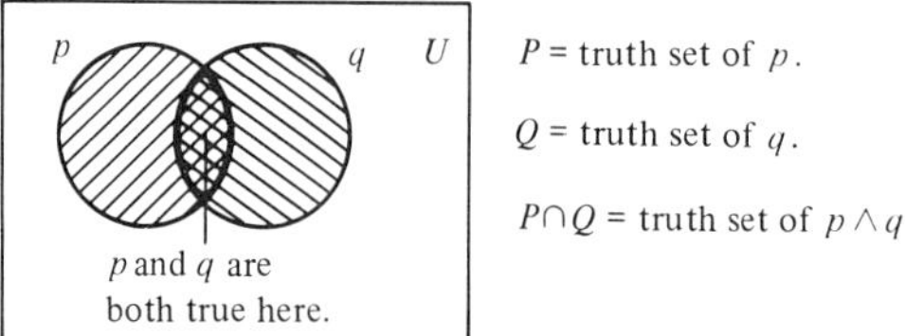

Figure 2.11 The truth set of $p \wedge q$ consists of elements of U for which p and q are both true.

Now suppose that we have two distinct propositions p and q with their truth sets as in Fig. 2.11. Let us again consider the statements p: My hand-held calculator will not go on; and q: The batteries need replacing. The statement, "My hand-held calculator won't go on *and* the batteries need replacing," is known as the **conjunction of p and q**, and is written as "$p \wedge q$."

The truth set of $p \wedge q$ is $P \cap Q$, the set where both statements are true simultaneously. Hence $p \wedge q$ is only true when both p and q have the truth values T, as in Table 2.3.

Table 2.3 Truth Table of $p \wedge q$

p	q	$p \wedge q$
T	T	T
T	F	F
F	T	F
F	F	F

Contrast $p \wedge q$ with the statement, "My hand-held calculator won't go on *or* my batteries need replacing." The word "or" in mathematics is always taken to mean "and/or," as, for example, when an element belongs to the set $P \cup Q$. Such an element belongs to the set P and/or the set Q. We will think of the union $P \cup Q$ as the truth set of "p or q," written $p \vee q$ and called the **disjunction of p and q**. Its truth table is given in Table 2.4. It is clear that $p \wedge q$ and $p \vee q$ are not logically equivalent. (However, $p \vee q$ and $q \vee p$ are equivalent. So are $p \wedge q$ and $q \wedge p$.)

Table 2.4 Truth Table of $p \vee q$

p	q	$p \vee q$
T	T	T
T	F	T
F	T	T
F	F	F

We can combine propositions using more than one logical symbol (or "connective") at a time, as we see in Example 5.

Example 5 In Table 2.5(a) and (b) we have constructed the truth tables for $(\neg p) \vee q$ and $\neg(p \wedge (\neg q))$ respectively. Observe that we must combine the truth values of p and q in each of the possible four ways TT, TF, FT, and FF, where the first value is that of p and the second that of q. We gradually build up the required compound statement, using the algebraic order indicated by the parentheses as our guide.

Table 2.5

p	q	$\neg p$	$(\neg p) \vee q$
T	T	F	T
T	F	F	F
F	T	T	T
F	F	T	T

(a)

p	q	$\neg q$	$p \wedge (\neg q)$	$\neg(p \wedge (\neg q))$
T	T	F	F	T
T	F	T	T	F
F	T	F	F	T
F	F	T	F	T

(b)

We can now see that the propositions $(\neg p) \vee q$ and $\neg(p \wedge (\neg q))$ are logically equivalent, since they have the same truth values for corresponding pairs of values of p and q. Therefore, we may write $(\neg p) \vee (q) = \neg(p \wedge (\neg q))$. □

This might be a good place to point out that all tautologies are logically equivalent to one another, since they all have the one truth value T. Similarly, all contradictions are logically equivalent to one another. Hence any tautology or contradiction may be denoted by the letters $\mathscr{T}$ or $\mathscr{F}$, respectively.

Example 6 We need only two lines in the truth tables of $p \wedge \mathscr{T}$ and $p \vee \mathscr{F}$.

Table 2.6

p	$\mathscr{T}$	$p \wedge \mathscr{T}$
T	T	T
F	T	F

(a)

p	$\mathscr{F}$	$p \vee \mathscr{F}$
T	F	T
F	F	F

(b)

Hence

$$p \wedge \mathscr{T} = p \quad \text{and} \quad p \vee \mathscr{F} = p.$$

□

Example 7 For our final example we will give the truth table of the compound of three simple statements, p, q, and r. Notice that the truth values of these statements must be combined in eight different ways to obtain the truth table for $r \wedge (p \vee q)$.

Table 2.7 A Truth Table of the Compound of Three Simple Statements Generally Requires Eight Rows

p	q	r	$p \vee q$	$r \wedge (p \vee q)$
T	T	T	T	T
T	T	F	T	F
T	F	T	T	T
T	F	F	T	F
F	T	T	T	T
F	T	F	T	F
F	F	T	F	F
F	F	F	F	F

□

Completion Review 2.4

Complete each of the following.

1. A proposition is a statement that is either ________________ or ________________.
2. A categorical proposition has only one ________________.
3. A ________________ proposition's truth values depend upon the particular circumstances.
4. A proposition that is always true is called a ________________. One that is always false is called a ________________.
5. The two truth values are ________________ and ________________.
6. The set of all circumstances under which a statement is true is called the ________________ of the statement.
7. A proposition that has the opposite truth values of p is called the ________________ of p and is denoted ________________.
8. A proposition that is only true when both p and q are true is called their ________________. It is denoted by ________________.
9. A proposition that is only false when both p and q are false is called their ________________. It is denoted by ________________.

10. Two propositions that have exactly the same truth values under the same circumstances are said to be ________________.

Answers: **1.** true; false. **2.** truth value. **3.** hypothetical. **4.** tautology; contradiction. **5.** T; F. **6.** truth set. **7.** negation; $\neg p$. **8.** conjunction; $p \wedge q$. **9.** disjunction; $p \vee q$. **10.** logically equivalent.

Exercises 2.4

1. Which of the following statements is a proposition? A tautology? A contradiction?
  a) Oh, what a beautiful morning!
  b) If the sun is shining, then the sun is shining.
  c) The statement is true, or I'm a monkey's uncle.
  d) It was the worst of times, and it was the best of times.

2. Let p denote the statement, "The course is enjoyable"; let q denote the statement, "The presentation is stimulating"; and let r denote the statement, "The material is significant." Write each of the following in symbolic form.
  a) The material is significant and the presentation is stimulating, but the course is not enjoyable. ("but" = "and").
  b) It is not the case that both the course is enjoyable and, at the same time, the presentation is not stimulating.

3. Referring to Exercise 2, write each of the following in good standard English.
  (a) $p \vee q$; **(b)** $(\neg p) \wedge (\neg q)$; **(c)** $\neg(p \vee (\neg q))$; **(d)** $r \wedge (p \vee q)$; **(e)** $(r \wedge p) \vee (r \wedge q)$.

4. Construct truth tables for parts (b) to (e) in Exercise 3.

5. Construct truth tables for each of the following. **(a)** $\mathscr{T} \wedge \mathscr{F}$; **(b)** $\mathscr{T} \vee p$; **(c)** $\mathscr{F} \wedge p$; **(d)** $\mathscr{T} \vee \mathscr{F}$.

6. Draw the Venn diagrams of the truth sets in parts (b) to (e) of Exercise 3, as well as for part (f) $(\neg p) \vee q$.

7. Show that $\neg(p \wedge q)$ and $(\neg p) \vee (\neg q)$ are logically equivalent.

8. Show that $(p \vee q) \vee (\neg p)$ is a tautology.

9. Show that $(p \wedge q) \wedge (\neg p)$ is a contradiction.

10. How many rows does one need for a truth table dealing with a compound statement composed of the simple statements $p_1, p_2, p_3, \ldots, p_n$? Why?

11. Show that the following two statements are logically equivalent.
  a) I will go to the races either if there is no examination tomorrow or if there is an examination tomorrow and Peacemaker is running.
  b) I will go to the races tomorrow either if there is no examination tomorrow or if Peacemaker is running.
  (*Hint*: Disregard the "if.")

12. How many times in the course of the following algorithm is the truth value of step 3 "T"?
  1. Set $m \leftarrow 120$; $n \leftarrow 96$.
  2. Divide m by n and let r be the remainder.
  3. If $r = 0$, the algorithm terminates. Output: "n is the gcd of 96 and 120." Stop.

4. Otherwise set $m \leftarrow n$, $n \leftarrow r$, and go back to step 2.
(*Note*: The Euclidean algorithm given here calculates the greatest common divisor of m and n.)

2.5 LAWS OF LOGIC AND RULES OF REASONING

Statements of the form "p implies q" are the ones most people think of in connection with the word "logic." Roughly translated, such statements mean that the truth of p requires the truth of q, or if p is true, then q is true.

Example 1 Suppose that we would like to check the validity of the statement, "The fact that (p:) my calculator turns on implies that (q:) its batteries are good." In other words, can we believe that "p implies q"? It seems reasonable. But we should ask ourselves what it would take for us *not* to believe it as a generally valid test of good batteries. We can prove the test is false only if the calculator goes on with bad batteries! Thus our belief in the statement "p implies q" will be shaken only when p is true but q is false. □

Looking back at Example 5, Section 2.4, we see that both of the logically equivalent propositions $\neg(p \wedge (\neg q))$ and $(\neg p) \vee q$ are false only when p is true but q is false. We can, in fact, read the statement $\neg(p \wedge (\neg q))$ as, "It is not the case that p holds and q does not." This is what we really mean when we say **p implies q**. Other ways of saying this are that p is a **sufficient condition** for q, or q is a **necessary condition** for p, or merely, **if p then q**, all of which are written

$$p \rightarrow q.$$

□

Example 2 Let p be the statement, "My hand-held calculator turns on," and let q denote the statement, "Its batteries are good," as before. Then we can read "$p \rightarrow q$" in the following ways:

a) If my hand-held calculator turns on, then (I conclude that) its batteries are good.

b) A sufficient condition (to decide) that its batteries are good is that my hand-held calculator turns on.

c) A necessary condition for my hand-held calculator to turn on is that its batteries are good. □

We ask you to observe what for some will be completely obvious: One must make small modifications in the way the original propositions were stated in order to connect them in a grammatically acceptable sentence.

The truth values of $p \rightarrow q$ correspond to those of $(\neg p) \vee q$ or, equivalently, to those of $\neg(p \wedge (\neg q))$. Consequently the truth sets of these three propositions are the same.

Table 2.8

p	q	$p \to q$	$(\neg p) \vee q$
T	T	T	T
T	F	F	F
F	T	T	T
F	F	T	T

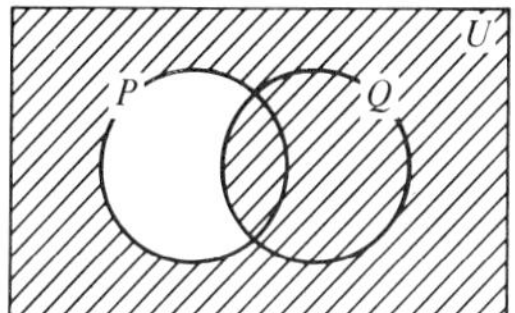

Figure 2.12 The truth set of $p \to q$ is $P' \cup Q$.

If p is both a **necessary and sufficient condition** for q, it is clear that we can write $(p \to q) \wedge (q \to p)$. This is abbreviated by writing $p \leftrightarrow q$, which can also be read "p if and only if q," or "p iff q." Thus the truth table for $p \leftrightarrow q$ should be the same as that for $(p \to q) \wedge (q \to p)$; that is, these statements are logically equivalent by definition. Thus Table 2.9(b) actually defines Table 2.9(a).

Table 2.9

p	q	$p \leftrightarrow q$
T	T	T
T	F	F
F	T	F
F	F	T

(a)

p	q	$p \to q$	$q \to p$	$(p \to q) \wedge (q \to p)$
T	T	T	T	T
T	F	F	T	F
F	T	T	F	F
F	F	T	T	T

(b)

Definitions always give necessary and sufficient conditions for a thing to be true. Indeed a definition is always logically equivalent to the thing it is describing. Thus "a rose is a rose," if this is seriously meant to be a definition, is of the form "$r \leftrightarrow r$." It is logically correct, in any case, even if it does not tell you how to recognize a rose when you see one. For example, we defined "$p \leftrightarrow q$" to be "$(p \to q) \wedge (q \to p)$." Hence the proposition $(p \leftrightarrow q) \leftrightarrow ((p \to q) \wedge (q \to p))$ should always be true, independent of the truth values of p and q: And it is!

Example 3 Table 2.10 shows that $(p \leftrightarrow q) \leftrightarrow ((p \to q) \wedge (q \to p))$ is always true, or, in other words, that it is a tautology. The truth values from the third and fourth columns are obtained from the Tables 2.9(a) and (b).

Table 2.10

p	q	$p \leftrightarrow q$	$((p \rightarrow q) \wedge (q \rightarrow p))$	$(p \leftrightarrow q) \leftrightarrow ((p \rightarrow q) \wedge (q \rightarrow p))$
T	T	T	T	T
T	F	F	F	T
F	T	F	F	T
F	F	T	T	T

□

Rules of Reasoning

Some of the most common errors in reasoning involve confusing the statement $p \rightarrow q$ with its **converse** $q \rightarrow p$.

Example 4 A student, when asked to define a "square," answered that "a square is a quadrilateral with four equal sides." We can put this answer into the form $p \rightarrow q$ by saying, "If S is a quadrilateral with four equal sides, then S is a square." This is incorrect, of course, since the four angles need not be equal just because the sides are equal. The student probably meant the converse, $q \rightarrow p$, which says, "If S is a square, then S is a quadrilateral with four equal sides." This is true, but it is not enough to define a square. Hence not only does our example show that *$p \rightarrow q$ and $q \rightarrow p$ are not equivalent* (which one can tell just by looking at their truth tables), but it also illustrates a point that bears repeating: *A definition must give necessary and sufficient conditions for what is being defined.* Equal sides are necessary, but not sufficient, to define a square, for example. □

The necessity and sufficiency in a definition constitute one of several rules of reasoning that are commonly employed in mathematical discussions. You will notice however, that all of those we discuss in the following are based on the idea of logical equivalence.

Definition 2.6(a) ***Modus ponens*** is the principle that states if

1. a rule $p \rightarrow q$ (called (b) the **major premise**) is true,
2. and p (called (c) the **minor premise** or (d) **hypothesis**) is also true, then
3. q (called (e) the **conclusion**) must follow.

Modus ponens can be stated as the tautology (See Exercise 2.5-12.)

$$((p \rightarrow q) \wedge p) \rightarrow q.$$

Example 5 Modus ponens arguments are sometimes diagrammed as in the following example.

Major premise ($p \to q$): If my calculator turns on, then its batteries are good.
Minor premise (p) : My calculator turns on.

Conclusion (q) : Therefore, its batteries are good. □

Another well-established rule of reasoning is that an implication $p \to q$ is logically equivalent to its **contrapositive**,

$$(\neg q) \to (\neg p),$$

the statement formed by substituting the negations of the conclusion and the hypothesis for p and q respectively. (See Exercise 2.5-10.)

Example 6 The statement,

a) "If my calculator turns on, then its batteries are good"
is equivalent to its contrapositive,
"If its batteries are not good, then my calculator will not turn on."

Likewise, the statement,

b) "If $\sqrt{2}$ is a rational number, then $\sqrt{2}$ can be written as a quotient of integers, a/b, in lowest terms"
is equivalent to its contrapositive,
"If $\sqrt{2}$ cannot be written as a quotient of integers, a/b, in lowest terms, then $\sqrt{2}$ is not a rational number." □

You will recall that our proof of the irrationality of $\sqrt{2}$ actually used the first implication in Example 6(b). Our argument can be diagrammed as follows:

($p \to q$): If $\sqrt{2}$ is a rational number, then $\sqrt{2}$ can be written as a quotient of integers, a/b, in lowest terms.
$\neg q$: $\sqrt{2}$ cannot be written as a quotient of integers in lowest terms.

$\neg p$: Therefore, $\sqrt{2}$ is not a rational number.

The validity of our argument can be seen by substituting $(\neg q) \to (\neg p)$ for $p \to q$ in the major premise. This gives us:

$(\neg q) \to (\neg p)$: If $\sqrt{2}$ cannot be written as a quotient of integers, a/b, in lowest terms, then $\sqrt{2}$ is not a rational number.
$\neg q$: $\sqrt{2}$ cannot be written as a quotient of integers, a/b, in lowest terms.

$\neg p$: Therefore, $\sqrt{2}$ is not a rational number.

Since $[((\neg q) \to (\neg p)) \wedge (\neg q)] \to \neg p$ is a modus ponens argument and, therefore, a tautology, we see that our first line of reasoning was also a valid, albeit indirect, way of using the contrapositive to make a correct argument. This manner of using the contrapositive is sometimes called "indirect proof." It may be stated more formally as Definition 2.7.

Definition 2.7 The tautology $((p \to q) \wedge (\neg q)) \to (\neg p)$ is known as the principle of **modus tollens**. (See Exercise 2.5-13.)

Simplification of Propositions

As we indicated earlier, many logical equivalents can be obtained indirectly by comparing the truth sets of propositions. Thus our laws of set operations in Section 2.3 immediately yield Theorem 2.10's equivalences for propositions.

Theorem 2.10 ***Laws of Logical Connectives*** Let p, q, and r be propositions. Then the following equivalences hold.

1. Commutative

1. (a) $p \vee q = q \vee p$ 1. (b) $p \wedge q = q \wedge p$

2. Associative

2. (a) $p \vee (q \vee r) = (p \vee q) \vee r$ 2. (b) $p \wedge (q \wedge r) = (p \wedge q) \wedge r$

3. Distributive

3. (a) $p \vee (q \wedge r) = (p \vee q) \wedge (p \vee r)$ 3. (b) $p \wedge (q \vee r) = (p \wedge q) \vee (p \wedge r)$

4. Identity ($\mathcal{T}$: always true)

4. (a) $p \vee \mathcal{T} = \mathcal{T}$ 4. (b) $p \wedge \mathcal{T} = p$

5. Identity ($\mathcal{F}$: always false)

5. (a) $p \vee \mathcal{F} = p$ 5. (b) $p \wedge \mathcal{F} = \mathcal{F}$

6. Idempotence

6. (a) $p \vee p = p$ 6. (b) $p \wedge p = p$

7, 8. Negations

7. (a) $p \vee (\neg p) = \mathcal{T}$ 7. (b) $p \wedge (\neg p) = \mathcal{F}$
8. (a) $\neg \mathcal{T} = \mathcal{F}$ 8. (b) $\neg \mathcal{F} = \mathcal{T}$

9. De Morgan's Laws

9. (a) $\neg(p \vee q) = (\neg p) \wedge (\neg q)$ 9. (b) $\neg(p \wedge q) = (\neg p) \vee (\neg q)$

10. Involution

$\neg(\neg p) = p$

One can also verify these equivalences by making truth tables, as in Section 2.4. Also following Section 2.4, one can use these equivalences to verify additional equivalences, or just to simplify needlessly complicated propositions.

Example 7 Suppose that the board of trustees of a certain firm adopts the following resolution: "The new computer system will be purchased in the event that the old system is not operational or if the old system is operational and the new system is more efficient." Let p denote the statement, "The old system is operational." Let q denote the statement, "The new system is more efficient." Then the new computer system will be purchased if $(\neg p) \vee (p \wedge q)$ is a true statement. But

$$
\begin{aligned}
(\neg p) \vee (p \wedge q) &= ((\neg p) \vee p) \wedge ((\neg p) \vee q) && \text{by Theorem 2.10 (3a)} \\
&= (p \vee (\neg p)) \wedge ((\neg p) \vee q) && \text{by Theorem 2.10 (1a)} \\
&= \mathcal{T} \wedge ((\neg p) \vee q) && \text{by Theorem 2.10 (7a)} \\
&= ((\neg p) \vee q) \wedge \mathcal{T} && \text{by Theorem 2.10 (1b)} \\
&= ((\neg p) \vee q) && \text{by Theorem 2.10 (4b).}
\end{aligned}
$$

Hence the board could have simplified its resolution to read, "The new computer system will be purchased if the old system is not operational or the new system is more efficient." □

Completion Review 2.5

Complete each of the following.

1. The symbolic expression of "p implies q" or "if p then q" is ________________. In this case we say that p is a ________________ condition for q and that q is a ________________ condition for p.
2. In the truth table for "if p then q" one obtains an F only when ________________.
3. The symbolic expression for "p if and only if q" is ________________. The truth table for this proposition gives T when ________________.
4. The converse of "p implies q" is ________________. The contrapositive is ________________. The statement "p implies q" is logically equivalent to its ________________.
5. In a "p implies q" statement, p is called the ________________ and q is called the ________________.

6. Modus ponens has the logical form ____________________. Modus ponens is the logical basis of ____________________ proof.

7. Modus tollens has the logical form ____________________. Modus tollens is the logical basis of ____________________ proof or proof by ____________________.

8. Modus ponens and modus tollens are both ____________________.

9. De Morgan's laws for logical connectives are ____________________ and ____________________.

10. The laws of logical connectives parallel those for ____________________, and they imply that the principle of ____________________ holds for propositions.

Answers: **1.** $p \to q$; sufficient; necessary. **2.** p is true and q is false. **3.** $p \leftrightarrow q$; p and q have the same truth values. **4.** $q \to p$; $\neg q \to \neg p$; contrapositive. **5.** hypothesis; conclusion. **6.** $[(p \to q) \wedge p] \to q$; direct. **7.** $[(p \to q) \wedge \neg q] \to \neg p$; indirect; contradiction. **8.** tautologies. **9.** $\neg(p \wedge q) = \neg p \vee \neg q$; $\neg(p \vee q) = \neg p \wedge \neg q$. **10.** set operations; duality.

Exercises 2.5

1. Let p denote the statement, "The course is enjoyable"; let q denote the statement, "The material is significant"; and let r denote the statement, "The presentation is stimulating." Write each of the following in symbolic form.
   a) If the presentation is stimulating, then the course is enjoyable.
   b) If the course is not enjoyable, then either the material is not significant or the presentation is not stimulating.
2. Write the converse of the statement in Exercise 1(a).
3. Write the contrapositive of the statement in Exercise 1(a).
4. Referring to Exercise 1, write each of the following in standard English.
   (a) $r \to p$; **(b)** $(p \wedge q) \to r$; **(c)** $p \to (q \vee r)$; **(d)** the converse of $r \to p$; **(e)** the contrapositive of $r \to p$.
5. Let p denote the statement, "The professor is delighted"; let q denote the statement, "The class performed admirably"; and let r denote the statement, "We will have a party." Write each of the following in symbolic form.
   a) If the class performed admirably, then the professor is delighted.
   b) We will have a party if and only if the professor is delighted or the class performed admirably.
6. Write the converse of Exercise 5(a).
7. Write the contrapositive of Exercise 5(a).
8. Write truth tables for each of the following: **(a)** $(\neg p) \to q$; **(b)** $(\neg p) \leftrightarrow (\neg q)$; **(c)** $p \to (p \vee q)$; **(d)** $p \leftrightarrow p$; **(e)** $p \leftrightarrow (\neg p)$.
9. Which of the propositions in Exercise 8 are tautologies? Which ones are contradictions?

10. Prove that the statement $p \to q$ is logically equivalent to its contrapositive $(\neg q) \to (\neg p)$.

11. Show that the converse $q \to p$ of a statement $p \to q$ is logically equivalent to the statement's **inverse** $(\neg p) \to (\neg q)$.

12. Show that the principle of modus ponens $((p \to q) \wedge p) \to q$ is a valid form of reasoning by showing that it is a tautology.

13. Show that the principle of modus tollens $((p \to q) \wedge (\neg q)) \to (\neg p)$ is a valid form of reasoning by showing that it is a tautology.

14. What kind of error is being made in the following definitions?

a) "A square is a quadrilateral having four equal angles."

b) "A man is a featherless biped."

15. Present the following statement as a modus-ponens-type argument: "My computer program ran perfectly because I completely debugged it."

16. Show that $(\neg p) \vee (p \wedge q) = (\neg p) \vee q$ using the laws of logical connectives, Theorem 2.10.

17. The following instructions were given to a maintenance worker.

a) The automatic monitoring system will be in operation if and only if there is a large payroll in the office or there is nobody in the office.

b) The electric power should be turned on if it is not the case that nobody is in the office and the automatic monitoring system is not in operation. So the worker decided simply to leave the power on. Was she right? [*Hint*: Make a substitution in (b) for (a).]

18. Explain why any logical statement that is written with the logical symbols $\wedge$, $\vee$, $\neg$, $\to$, and $\leftrightarrow$ can be written with the symbols $\vee$ and $\neg$ alone. (*Hint*: Consider De Morgan's laws.)

19. Write propositions equivalent to the following that use only the negation and disjunction symbols. **(a)** $p \wedge q$; **(b)** $\neg(p \vee q) \to (p \wedge q)$; **(c)** $p \leftrightarrow q$.

2.6 LOGIC GATES AND COMPUTER ARITHMETIC

The popular conception of a computer is that of a complicated adding machine, a kind of superabacus. A computer can do arithmetic at dazzling speed, of course. But it would be closer to the truth if one thought of a computer as a logic machine, because the way a computer does addition is by means of the logical rules we developed in the previous sections. The computer (by which we mean a modern digital computer) has simple electronic circuits that are able to receive input corresponding to "true" and "false" in terms of voltages from one or more sources. Then, depending upon the kind of circuitry, a voltage is emitted, which again corresponds to either "true" or "false."

Let us now be more specific without getting into the actual construction of the different kinds of circuits. Recall that when we write numbers in binary notation we use only the digits 0 and 1, as in Table 2.11.

We will use the digit "1" to represent both "high voltage" and "T" (for "true") and the digit "0" for both "low voltage" and "F" (for "false"). Let us also remember that

Table 2.11 Binary and Decimal Numeration

Binary									Decimal		
2^7	2^6	2^5	2^4	2^3	2^2	2^1	2^0		10^2	10^1	10^0
1	0	1	1	0	0	1	1	=	1	7	9
128+	0+	32+	16+	0+	0+	2+	1	=	100+	70+	9

addition in binary is done one binary digit, or bit, at a time. We can transfer information about the bits in terms of high or low voltage to and from devices called **logic gates**. We will discuss the three types of logic gates, corresponding to the logical connectives "not," "and," and "or." Not surprisingly they are called the **NOT-gate** (or **inverter**), the **AND-gate**, and the **OR-gate**. The symbols used for these devices are given below in Fig. 2.13. (Gates called NAND and NOR are used in practice. See Exercises 2.6-23 and 2.6-24.)

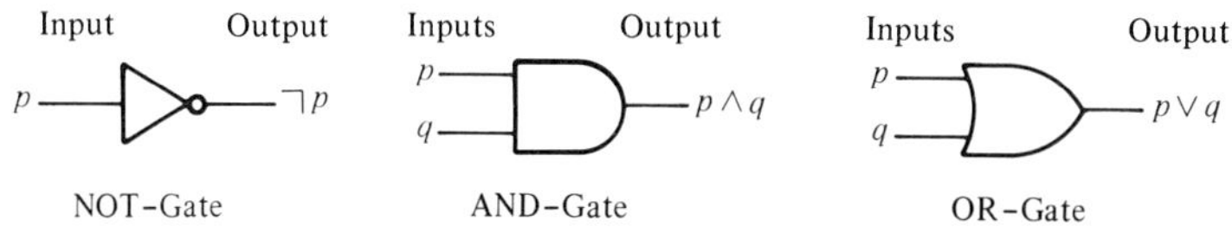

Figure 2.13 The three logic gate symbols that are used to construct logical circuits.

Example 1 We can now show how each of the three logic gates treats its high- or low-voltage input, which you will recall is represented by a 1 or a 0 respectively. We simply construct tables just like the truth tables we studied in the previous sections.

Table 2.12 (a) NOT-Gate, (b) AND-Gate, and (c) OR-Gate

(a)

p	$\neg p$
1	0
0	1

(b)

p	q	$p \wedge q$
1	1	1
1	0	0
0	1	0
0	0	0

(c)

p	q	$p \vee q$
1	1	1
1	0	1
0	1	1
0	0	0

□

We can construct more complicated circuits by **concatenating** logic gates, that is, by making the output of one gate the input of another gate according to the rules of Table 2.12.

Example 2 A logic gate corresponding to the proposition $p \rightarrow q$ can be constructed if we use its logical equivalent $(\neg p) \vee q$. Hence we need to concatenate a NOT-gate with an

Figure 2.14 A "p implies q"-gate.

OR-gate, as in Fig. 2.14. It operates according to Table 2.13. We see that the only way to get a low-voltage output is with a high-voltage input from p and a low-voltage input from q.

Table 2.13

p	q	$p \to q$
1	1	1
1	0	0
0	1	1
0	0	1

□

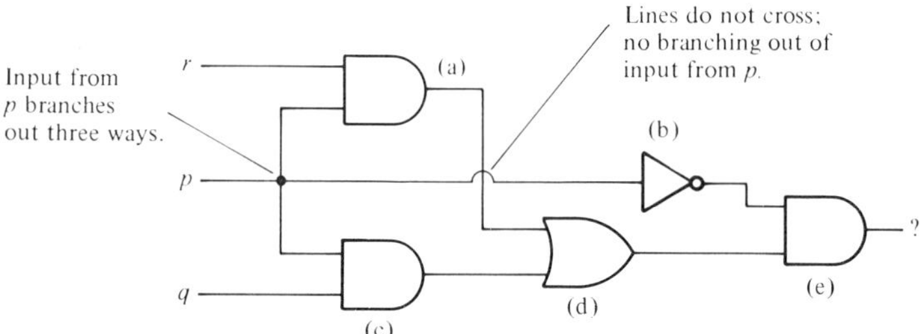

Figure 2.15 Logic gates are concatenated to form more complex logical circuits. Dots indicate branching out of input; loops indicate that the input "wires" do not actually cross and that there is no branching.

We can also do this kind of problem in reverse, asking for the symbolic representation of a given concatenation of logic gates, as in Fig. 2.15.

We observe that the input from p goes three places: (a) into an AND-gate with r; (b) into a NOT-gate; and (c) into an AND-gate with q. The resulting outputs $(p \wedge r)$ and $(p \wedge q)$ are inputs for an OR-gate (d). The output $(p \wedge r) \vee (p \wedge q)$ at (d) and the output $(\neg p)$ at (b) are input for the AND-gate at (e). Therefore, the circuit has the logical representation

$$[(p \wedge q) \vee (p \wedge r)] \wedge (\neg p).$$

One of the useful applications of logical representation of these circuits is that if the representation can be simplified, then so can the circuit, often at a great saving in space and money.

Example 3 To see how to simplify the second circuit in Example 2, we can write the following sequence of equivalences:

$$((p \wedge q) \vee (p \wedge r)) \wedge (\neg p) = \quad \text{by Theorem 2.10 (3a)}$$
$$(p \wedge (q \vee r)) \wedge (\neg p) \quad = \quad \text{by Theorem 2.10 (2b and 1b)}$$
$$(p \wedge (\neg p)) \wedge (q \vee r) \quad = \quad \text{by Theorem 2.10 (7b)}$$
$$0 \wedge (q \vee r) \quad = 0 \quad \text{by Theorem 2.10 (5b),}$$

where we have used the laws of logical connectives with 0 for $\mathscr{F}$. But our result implies that the only output of this circuit will be low voltage. (The circuit is a "contradiction.") Hence any contradiction, such as $p \wedge (\neg p)$, can be substituted for it.

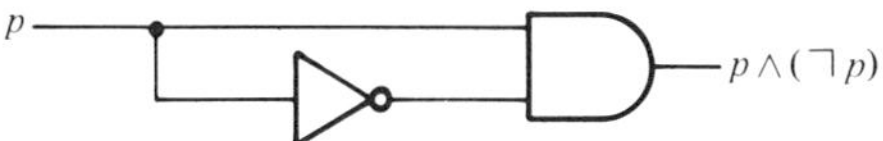

Figure 2.16 A "contradiction" circuit. The only output is *F*.

We can discard the inputs from q and r altogether. □

It is possible, although not necessarily desirable, to achieve another kind of simplification. We can actually construct any kind of circuit that makes use of AND-, OR-, and NOT-gates using only two or fewer kinds of logic gates. (Also see Exercises 2.6-23 and 26.6-24.)

Example 4 The circuit in Fig. 2.17 has the same output as an OR-gate, and yet it is constructed by concatenating only NOT-gates and an AND-gate.

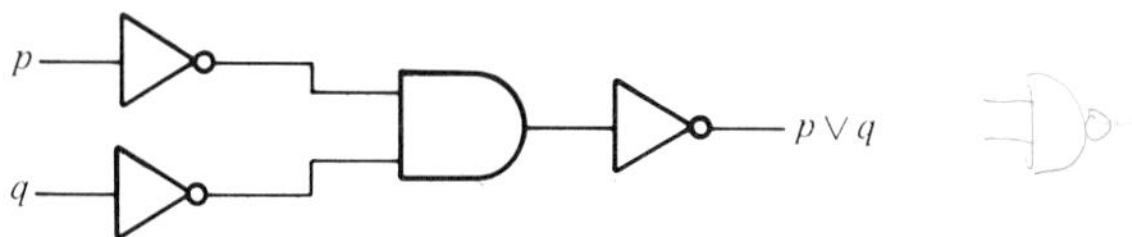

Figure 2.17 This circuit is equivalent to an OR-gate.

To see that we get the output as claimed, observe that the circuit may be written

$$\neg((\neg p) \wedge (\neg q)) = (\neg(\neg p)) \vee (\neg(\neg q)) \quad \text{by Theorem 2.10 (9b)}$$
$$= p \vee q \quad \text{by Theorem 2.10 (10).}$$

□

Computer Addition: A Binary Adder

Now that we have some experience with logic gates, let us get back to our problem of constructing an adding device that uses the logic gates with inputs and outputs of high and low voltages. To add two binary numbers, we must be able to carry on binary arithmetic at each place. Moreover, as in Table 2.14, we must know what to do when

a) the binary digit in each of the addends is "0" or "1,"

b) a "0" or a "1" has been "carried over" from the addition in the last place, and

c) a "0" or a "1" is to be "carried over" into the next place.

Table 2.14

(Carried:	1 1 1 0 0)	
Add x:	$(0\ 1\ 0\ 1\ 1)_2 =$	11
To y:	$(0\ 1\ 1\ 1\ 0)_2 =$	$+14$
Get sum:	$(1\ 1\ 0\ 0\ 1)_2 =$	25

Hence we will need input from three sources (the carried digit and the digits from x and y), and two outputs (the new carried digit and the sum digit). We can construct the circuits that give each of the two outputs separately, one for the carried digit and one for the sum digit.

Example 5 The first thing we need to do is analyze what digit is carried when combinations of 0's and 1's are added from the bits for the numbers x and y, in addition to that which is carried over, which we will call c, from the last addition. We summarize the results in Table 2.15.

Table 2.15

x	y	c	Carry bit
1	1	1	1
1	1	0	1
1	0	1	1
1	0	0	0
0	1	1	1
0	1	0	0
0	0	1	0
0	0	0	0

It can be verified that Table 2.15 is the truth table of $((x \vee c) \wedge y) \vee (x \wedge c)$. Hence to construct a device that will deliver the correct output to the carry bit, we need only follow the design of the circuit in Fig. 2.18.

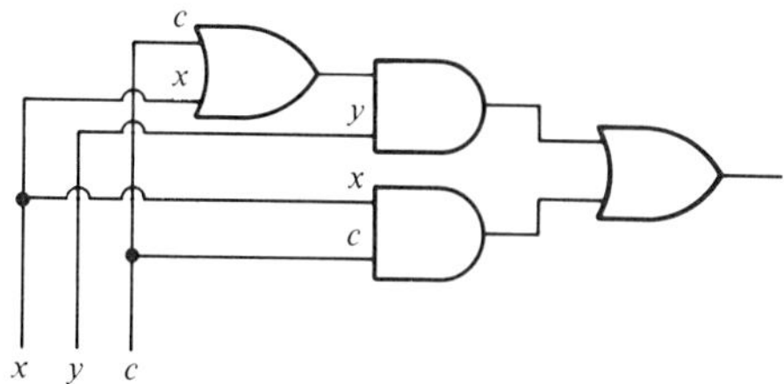

Figure 2.18 Determining the carry bit. □

Example 6 Now we need to analyze what digit is displayed in the sum when the inputs from x, y, and c are combined. These results are displayed in Table 2.16.

Table 2.16

x	y	c	Sum bit
1	1	1	1
1	1	0	0
1	0	1	0
1	0	0	1
0	1	1	0
0	1	0	1
0	0	1	1
0	0	0	0

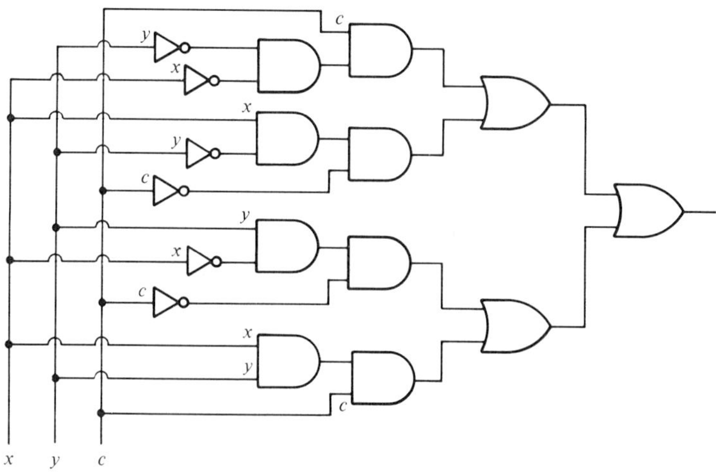

Figure 2.19 Determining the sum bit.

Here is the symbolic form of a circuit that will deliver the correct output to the sum bit:

$$\{[((\neg x) \wedge (\neg y)) \wedge c] \vee [(x \wedge (\neg y)) \wedge (\neg c)]\} \vee \{[((\neg x) \wedge (y)) \wedge (\neg c)] \vee [(x \wedge y) \wedge c]\}.$$

Although it looks somewhat intimidating at first glance, it just says that the way to get a 1 is with either three 1's or with one 1 and two 0's. The corresponding circuit is shown in Fig. 2.19. □

The Algebra of Circuits

When we combine the circuits for determining the carry and sum bits, we obtain what is called a **full adder**, as in Fig. 2.20.

We may define circuits to be equal when they have the same output for corresponding inputs. We may also define special circuits $\mathscr{T}$ and $\mathscr{F}$ whose only outputs are 1 and 0, respectively, regardless of the input. Given these definitions, we obtain algebraic laws for combining logical circuits. These laws are none other than those given in Theorem 2.10, the laws of logical connectives, which enable us to study logical circuits in the framework of an algebraic system called a "Boolean algebra." We discuss these algebraic systems in Sections 7.5 and 7.6. (Those who wish to know more about Boolean algebras without reading Sections 7.1 through 7.4 should begin with Definition 2.15.)

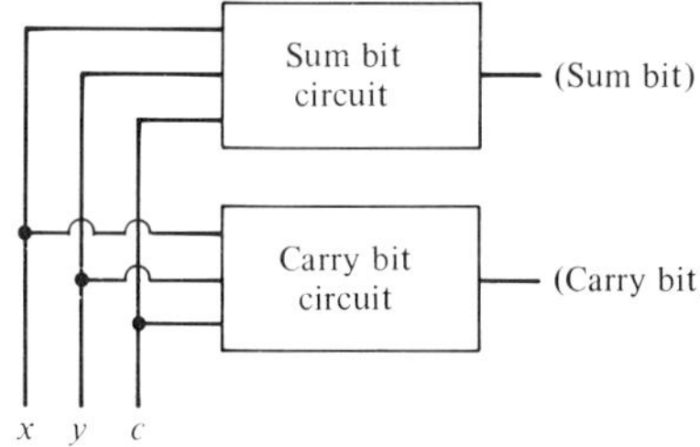

Figure 2.20 A full adder requires the integration of the sum and carry bit circuits.

It is a common practice, which we follow in this book, to discuss questions such as "what is the simplest form of a given logical circuit?" in the context of Boolean algebras. Indeed some authors refer to logical circuits entirely in terms of the notation of these algebraic systems. Hence the reader should be prepared for the following notational variations when reading about logical circuits: the symbols +, *, and ′ instead of the symbols $\vee$, $\wedge$, and $\neg$ that we have been using for disjunctions, conjunctions, and negations respectively.

Completion Review 2.6

Complete each of the following.

1. A logic gate that has input from p and is equivalent to $\neg p$ is called a ________________ gate or an ________________.
2. A logic gate that has inputs from p and q and is equivalent to $p \vee q$ is called an ________________ gate.
3. A logic gate that has inputs from p and q and is equivalent to $p \wedge q$ is called an ________________ gate.
4. The process of making the output of one logic gate an input for another logic gate is called ________________.
5. Two logical circuits are equivalent if they have ________________ for all possible inputs.
6. A logical circuit that can add the corresponding bits of two binary numbers and a carry from the addition of a preceding pair is called a ________________.

Answers: **1.** NOT; inverter. **2.** OR. **3.** AND. **4.** concatenation. **5.** the same (or corresponding) output. **6.** full adder.

Exercises 2.6

For Exercises 1 to 14 give a circuit diagram corresponding to the given logical expression, using the logic gates OR, AND, and NOT.

1. $(\neg(p \vee q)) \vee (p \vee q)$.
2. $(p \wedge (\neg q)) \wedge ((\neg p) \wedge q)$.
3. $(p \vee q) \wedge (p \vee r)$.
4. $(p \wedge q) \vee (p \wedge r)$.
5. $(\neg(p \vee q)) \wedge (p \vee q)$.
6. $(\neg(p \wedge q)) \vee (p \wedge q)$.
7. $(p \wedge q) \rightarrow (r \vee s)$.
8. $(p \vee q) \rightarrow (r \wedge s)$.
9. $(p \rightarrow q) \rightarrow r$.
10. $r \rightarrow (p \rightarrow q)$.
11. $(\neg(\neg(\neg p)))$.
12. $(p \wedge (p \rightarrow q)) \rightarrow q$.
13. $((\neg q) \wedge (p \rightarrow q)) \rightarrow (\neg p)$.
14. $(r \wedge (p \leftrightarrow q)) \rightarrow s$.

For Exercises 15 to 20 write a logical statement corresponding to the given circuit and then simplify both if possible.

15.

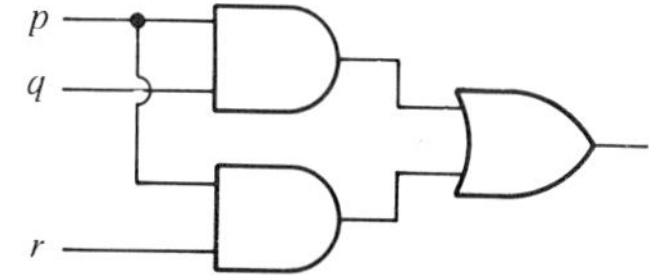

16.

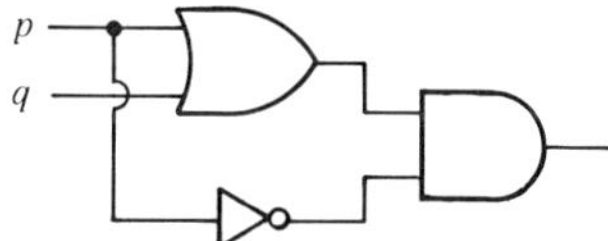

17. **18.** **19.** **20.**

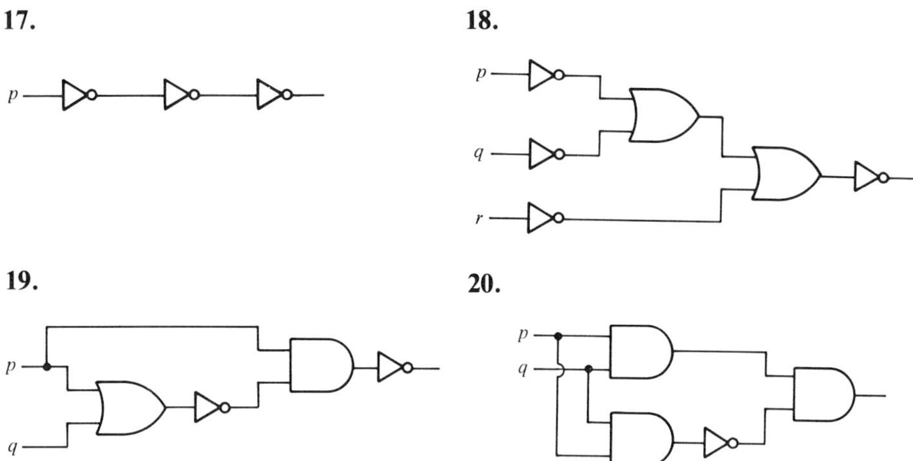

21. Show how an AND-gate can be replaced by interconnected OR-gates and NOT-gates.

22. Show how an OR-gate can be replaced by interconnected AND-gates and NOT-gates.

23. A NAND-gate has $\neg(p \wedge q)$ as its output when the inputs are p and q. Its symbolic expression is given as follows:

p, q — NAND — $\neg(p \wedge q)$

Show how one can replace OR-gates, AND-gates, and NOT-gates entirely by representing each of these only in terms of NAND-gates.

24. A NOR-gate has $\neg(p \vee q)$ as its output when the inputs are p and q. Its symbolic expression is given as follows:

p, q — NOR — $\neg(p \vee q)$

Show how one can replace OR-gates, AND-gates, and NOT-gates entirely by representing each of these only in terms of NOR-gates.

25. Design a circuit that will give the outputs required in the following table. The three sources of input are a, b, and c. [*Hint*: Form a conjunction such as $(a \wedge b \wedge \neg c)$ wherever there is an output of 1, and then take the disjunction of all your conjunctions.] Check your answer.

a:	1	1	1	1	0	0	0	0
b:	1	1	0	0	1	1	0	0
c:	1	0	1	0	1	0	1	0
Output:	0	1	0	0	0	1	0	1

26. Design a circuit that will give the following required outputs, as in Exercise 25.

a:	1	1	1	1	0	0	0	0
b:	1	1	0	0	1	1	0	0
c:	1	0	1	0	1	0	1	0
Output:	1	1	1	0	1	0	1	1

27. To design a circuit that will serve as a 1's complementer, we need to analyze what we want such a circuit to do. Basically there will be a control bit for the sign, which we will denote by c, and an input bit (digit) denoted by p. We want to obtain the following table, which just says "change the input iff $c = 1$":

c	p	Output
1	1	0
1	0	1
0	1	1
0	0	0

Check that the following circuit gives us the desired output and write it in logical form.

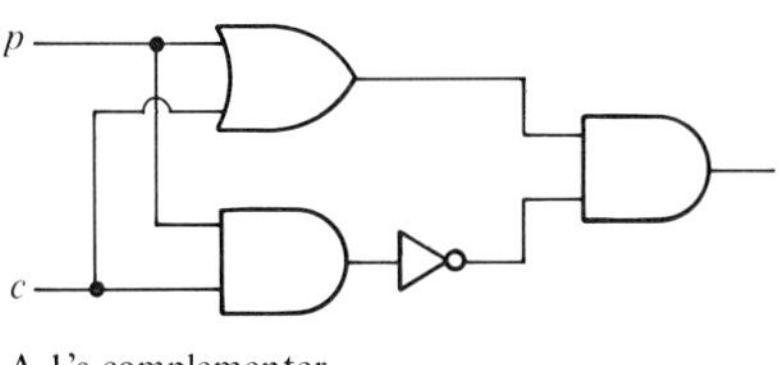

A 1's complementer

COMPUTER PROGRAMMING EXERCISES

(Also see the appendix and computer programs A8 to A9.)

2.1. Write programs that will verify the formulas given in Exercise 2.1-6 for large values of n, and run each program with $n = 1000$.

2.2. Write a program that will calculate the sum $(1)(2) + (2)(3) + \cdots + (n)(n+1)$ and output the values of such sums for $n = 1$ to 20. Use your output to guess at a formula for the sum, and then test your formula for $n = 20$. If your test seems to bear out your formula, try to prove that your formula is true for all positive integers n by mathematical induction.

2.3. Suppose that sets A having n elements are represented by one-dimensional arrays of length $n + 1$, including a terminal zero, and that zero is not an element of any set. Given sets A and B, write programs that will find $A \cup B$, $A \cap B$, and $A - B$.

2.4. Write a computer program that represents subsets of a given universe U with binary numbers according to Theorem 2.4, Section 2.2. Have your program find $A \cup B$, A-complement, $A \cap B$, and $A - B$ according to Theorem 2.6.

2.5. Given a finite set U and three of its subsets A, B, and C, write a program that will find the number of elements in each of the eight subsets partitioning U, as in Fig. 2.6, Section 2.2.

In the following exercises use the conventions mentioned at the end of Section 2.6, namely, $0 \leftrightarrow \mathrm{F}$, $1 \leftrightarrow \mathrm{T}$, $\neg \leftrightarrow {}'$, to write programs evaluating expressions such as $p \vee (q \wedge (\neg r))$, or $p + (q \cdot (r'))$, for input triples of zeros and ones such as (0, 1, 1). Use a language that evaluates these so-called "Boolean expressions" directly.

2.6. Write a program that prints the truth table of a logical statement.

2.7. Write a program that determines if two logical statements are equivalent.

2.8. Write a program that determines if $p \rightarrow q$ or if $q \rightarrow p$.

CHAPTER 3

COUNTING

3.1 THE MULTIPLICATION PRINCIPLE AND PERMUTATIONS enumeration, tree diagrams, multiplication rule, $n!$, permutations of n distinct objects taken r at a time, $P(n, r) = n!/(n-r)!$

3.2 COMBINATIONS AND BINOMIAL COEFFICIENTS combinations of n distinct objects taken r at a time, $C(n, r) = n!/[r!(n-r)!]$, binomial theorem, binomial coefficients, Pascal's triangle, $C(n+1, r) = C(n, r) + C(n, r-1)$

3.3 REPETITIONS AND PARTITIONS multisets, permutations with repetitions, $n!/[(r_1!)(r_2!) \ldots (r_k!)]$, ordered partitions, $C(n+r-1, n)$

3.4 INCLUSION–EXCLUSION generalizations of $|\bar{A} \cap \bar{B}| = |U| - |A| - |B| + |A \cap B|$ for n subsets of U, derangements

3.5 APPLICATIONS OF COUNTING FROM PROBABILITY TO PIGEONHOLES sample spaces, basic outcomes, events, probability distributions, $\Pr(E)$, addition rules, pigeonhole principle

3.6 RECURRENCE MODELS recursion formulas, difference equations, recurrence models, initial conditions, Fibonacci numbers, $a_n = a_{n-1} + a_{n-2}$, $s_k = s_{k-1} + f(k)$, $a_n = a_{n-1} + d$, $a_n = r \cdot a_{n-1}$

3.7 CLOSED FORMS AND ANALYSIS OF ALGORITHMS closed-form solutions of recurrence relations, $a_n = r \cdot s_{n-1} + a_0$, backward substitution, sorting and searching algorithms, files, records, keys, bubble sort, comparison step, Bose-Nelson method, divide-and-conquer techniques

3.8 DIVIDE-AND-CONQUER RELATIONS $a_n = ka_{n/k} + f(n)$, $F(n) = O(G(n))$, merge-sort

COMPUTER PROGRAMMING EXERCISES

In this chapter we are concerned with variations on the question "how many?" If you reply with the question "how many of what?," then you will soon understand why we shall devote this entire chapter to counting and what this chapter has to do with the previous ones.

The key to success in counting often has more to do with having the right description of what it is one wants to count than with the speed with which one can do arithmetic. For example, what do we mean when we ask for *the number of inputs into a circuit whose logical description is the statement*

$$(p \wedge q) \vee r?$$

If we are counting the number of *sources*, the answer is clearly *three*; p, q, and r. If we are counting the number of *inputs from each source*, the number is *two*: 0 or 1. The *sum of the inputs from the three sources* would then be *six*: two from p, two from q, and two from r. Our experience tells us, however, that the significant question is, "In how many ways can we combine the individual inputs from sources p, q, and r?" Recalling the tables of Section 2.6, we realize that we should be counting *sequences of zeros and ones of length three*, such as 0, 1, 0, corresponding to assignments of values to p, q, and r respectively. And, of course, there are *eight* of these.

Defining the things that we wish to count often involves an element of choice on our part, such as a selection of the values of p, q, and r. For this reason we could say that we are studying the "mathematics of choice," which is also known as "combinatorics," since it is usually necessary to combine several interrelated choices.

3.1 THE MULTIPLICATION PRINCIPLE AND PERMUTATIONS

If we have a clear idea of the things we want to count, then an effective, but sometimes slow, way to find their total number is to make a complete list of them, that is, **enumerate** them, making sure that each item appears on the list exactly once. The following example indicates a systematic way that this can occasionally be done, as well as a fundamental principle that can be justified by the procedure.

Example 1 Suppose that we want to assemble a home computer system consisting of a monitor (or TV screen), a microprocessor, and a printer. If we can choose any one of three types of monitors, two types of microprocessors, and two types of printers, how many different systems can we assemble?

A **tree diagram**, such as Fig. 3.1, can be used to show what the choices are at each stage. Notice how the choices "branch out" to produce a complete list of distinct combinations of choices, which, in this example, are the 12 distinct home computer systems.

The answer could have been obtained without enumeration had we simply

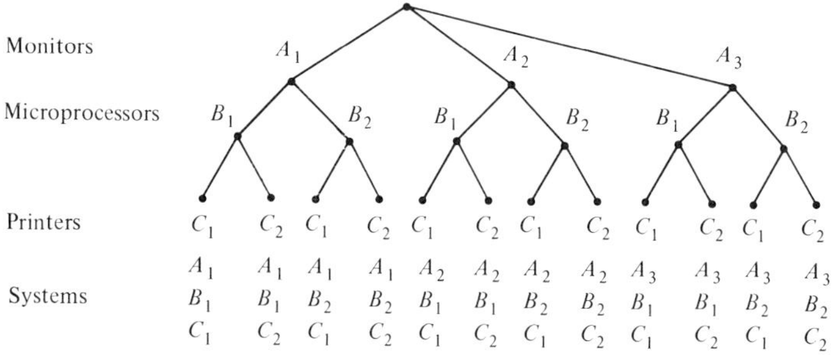

Figure 3.1 The tree diagrams enumerates all 12 possible distinct home computer systems.

multiplied the number of choices available at each stage. Thus

$$(3)(2)(2) = 12. \quad \square$$

It is clear that we can generalize the multiplication idea in our last example to situations in which we have to make choices in n stages, where $n \geqslant 1$. This generalization may be stated as the following rule (Theorem 3.1).

Theorem 3.1 ***Multiplication Rule*** If there are m_1 ways to choose the first item, and if having made that choice, there are then m_2 ways to choose the second item, followed by m_3 ways to choose the third, . . . , and finally m_n ways to choose the nth item, then there is a total of

$$m_1 m_2 m_3 \ldots m_n$$

ways to choose all n items.

Proof: We prove Theorem 3.1 by mathematical induction on n.

If we only have $n = 1$ item to choose in m_1 ways, then there is nothing to multiply. Hence the rule is clearly true.

Suppose, therefore, that our rule is true for some $n = k \geqslant 1$, that we have already chosen the first k items, and that we can now choose the $k + 1$st item in m_{k+1} ways. By hypothesis the first k items could be chosen in a total of $p = m_1 m_2 \ldots m_k$ ways. For each of these p choices, however, we can choose the next item in m_{k+1} ways. Hence the total number of ways to choose the $k + 1$ items is found by adding the p summands $m_{k+1} + m_{k+1} + \cdots + m_{k+1}$, giving us a total of

$$p \cdot m_{k+1} = m_1 m_2 \ldots m_k m_{k+1}$$

ways. This proves Theorem 3.1 for $n = k + 1$ and completes the proof by mathematical induction. ∎

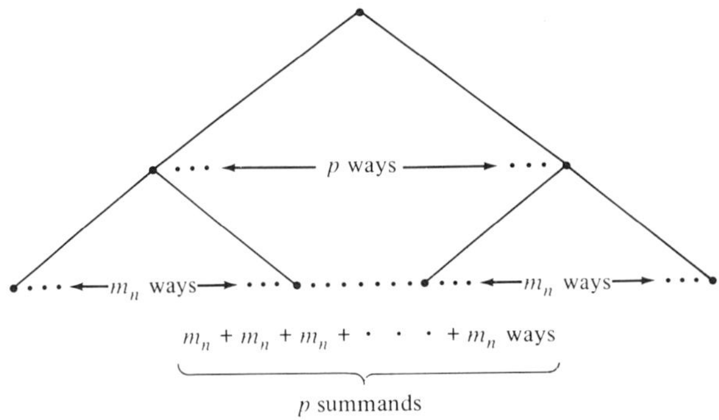

Figure 3.2 In this illustration of our proof of the multiplication rule, $p = m_1 m_2 \ldots m_{n-1}$ by our induction hypothesis.

An illustration of our proof of the multiplication rule is given in the tree diagram of Fig. 3.2. Notice how the "first" stage of the tree diagram includes all p ways of making the first k choices.

Example 2 How many nonnegative whole numbers can be expressed by binary numerals having six digits or less? The answer is 2^6, because each digit, including leading zeros, which we can later discard if necessary, can be chosen in one of two ways. That is, the first from the left is 0 or 1, the next is 0 or 1, and so on, giving

$$(2)(2)(2)(2)(2)(2) = 2^6 = 64$$

ways to choose all six digits. □

Example 3 How many subsets are there of a set having six distinct elements? The answer is again $2^6 = 64$, because we can make a subset of the given set by deciding, for each element of the set, whether or not it goes into the subset. Since there are two choices for each element, we have

$$(2)(2)(2)(2)(2)(2) = 2^6$$

subsets in all. □

Example 4 Doing so much thinking has made you hungry, and the local pizza parlor advertises "OVER 50 KINDS OF PIZZA!" But when you drive up, you see the small print, which lists only the choices "extra cheese, sausage, meatballs, anchovies, peppers, or pepperoni." Did the sign lie? No, because Tony, the chef, will be happy to include any of them on your pie, or leave it off for a total of

$2^6 = 64$ different kinds of pizza. □

Two of the preceding examples may not have seemed as if they required any "choice" on your part at first, but on closer inspection you see that each problem can be reformulated in just this way. You will see that this is true of the next example as well.

Example 5 A student has six questions on an examination that he can answer in any order he chooses. In how many different orders can he answer the questions? (No, the answer is *not* 2^6 this time.)

The student has six choices for the question he will answer first. After he chooses the first question, he will have only five choices for the second, four choices for the third, and so on, giving him a total of

$$(6)(5)(4)(3)(2)(1) = 720$$

orders in which he can answer the six questions. □

Multiplications such as the foregoing are common in combinatorics. For this reason it is convenient to introduce the notation "$n!$," which is read **"*n* factorial"** and where

$$n! = n(n-1)(n-2)\ldots(3)(2)(1)$$

if n is a positive integer, and $0! = 1$.

For example,

$$3! = (3)(2)(1) = 6, \quad 6! = (6)(5)(4)(3)(2)(1) = 720, \quad \text{and}$$
$$10! = (10)(9)(8)(7)(6)(5)(4)(3)(2)(1) = 3{,}628{,}800.$$

The following algorithm uses the observations that

$$n! = n\cdot(n-1)!, \quad \text{for } n \geqslant 1, \quad \text{and } 0! = 1$$

to calculate n factorial.

Algorithm 3.1 To calculate $n!$ for a nonnegative integer n,

1. [Initialize F.] Set $F \leftarrow 1$. ("F" stands for "factorial.")
2. [$0!$.] If $n = 0$, then output F and stop. Otherwise,
3. [F loop.] Repeat step 4 for $k = 1, 2, \ldots, n$.
   4. [Multiply.] Set $F \leftarrow F\cdot k$.
5. [Done.] Output F.
6. End of Algorithm 3.1.

Factorials have a way of growing very rapidly; hence calculations requiring division of factorials are best handled by cancellation. For example, $100! > 10^{100}$, but

$$100!/99! = (100 \cdot 99!)/99! = 100.$$

Failing to take advantage of cancellation in such calculations can present serious overload problems even for a high-speed computer. (See Exercise 3.1-18.)

Factorials typically arise in selection problems concerned with the *order* in which things appear or are chosen.

Definition 3.1 If r items are to be selected from among n distinct items and arranged in the order selected, then each ordered selection is called a **permutation of n items taken r at a time.**

Thus the order in which the student decided to answer his exam in Example 5 involved permutations of six distinct things taken six at a time. Here is another example.

Example 6 Suppose that we would like to assign an office to each of our three systems analysts, whom we shall call a_1, a_2, and a_3. If we have seven rooms available, there are

$$(7)(6)(5) = 210$$

different ways we can choose three of these rooms and assign them, in order, to analysts a_1, a_2, and a_3. Notice that the order in which the rooms are selected determines who gets which room. Moreover, after the third room is selected (out of the remaining four), there is no reason to make further selections. □

Observe that our calculation in Example 6 could also have been made as follows:

$$(7)(6)(5) = \frac{(7)(6)(5)(4)(3)(2)(1)}{(4)(3)(2)(1)} = \frac{7!}{4!} = \frac{7!}{(7-3)!}.$$

Theorem 3.2 ***Permutation Rule*** The number $P(n, r)$ of permutations of n distinct items taken r at a time is given by

$$P(n, r) = \frac{n!}{(n-r)!} = n(n-1)(n-2)(n-3)\ldots(n-r+1).$$

(*Note*: The second equality is valid if n and r are not zero.)

Example 7 Let us evaluate (a) $P(3, 3)$, (b) $P(3, 2)$, (c) $P(3, 1)$, and (d) $P(3, 0)$ and, at the same time, enumerate the corresponding permutations of objects from the set $\{x, y, z\}$.

a) $P(3, 3) = \frac{3!}{(3-3)!} = \frac{3!}{0!} = \frac{3!}{1} = (3)(2)(1) = 6$, corresponding to:

xyz, yxz, zxy, zyx, yxz, and xzy.

b) $P(3, 2) = \frac{3!}{(3-2)!} = \frac{3!}{1!} = (3)(2) = 6$, corresponding to

xy, xz, yx, yz, zx, and zy.

c) $P(3, 1) = \frac{3!}{(3-1)!} = \frac{3!}{2!} = 3$, corresponding to x, y, and z.

d) $P(3, 0) = \frac{3!}{(3-0)!} = 1$, corresponding to the one "way" of selecting and arranging none of the objects. □

Our next example shows how two permutations can be combined with the aid of our multiplication rule (Theorem 3.1).

Example 8 How many license plate "numbers" can be formed consisting of three distinct decimal digits, followed by three distinct letters? The first thing to notice is that the selections are being made from two different sets of objects. Hence this is not a simple permutations problem, although permutations are involved. Indeed the number of ways we can select the decimal digits is

$$P(10, 3) = (10)(9)(8) = 720;$$

the number of ways we can then select the three letters is

$$P(26, 3) = (26)(25)(24) = 15{,}600. \text{ (See Fig. 3.3.)}$$

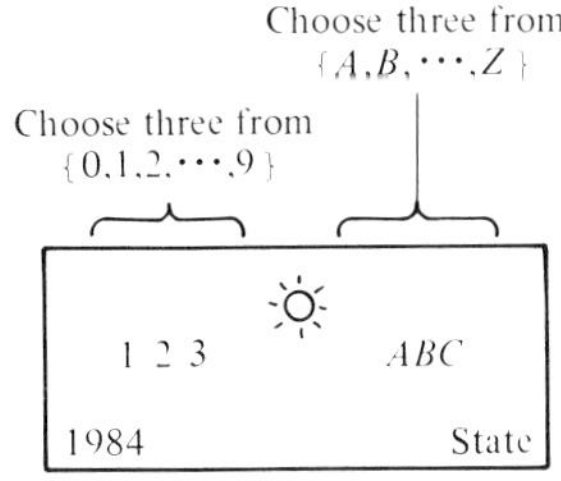

Figure 3.3 Determining the number of license plates having three digits and three letters.

The multiplication rule tells us that we can make a total of

$$P(10, 3) \cdot P(26, 3) = (720)(15{,}600) = 11{,}232{,}000$$

distinct license plates. □

The most important question to ask yourself in all of these problems is not "what do I multiply?" but rather "exactly what am I counting?" The next example shows another way in which a simple ordered selections problem can develop a little twist.

Example 9 A circular dial on a television screen will have the five letters P, Q, R, S, and T placed on it in some order. How many different orders are possible?

At first glance you might think that the answer is $P(5, 5) = 5!$ But this not so because, as we all know, dials are meant to be turned, as in Fig. 3.4. All of the arrangements shown on the dial are equivalent and, in fact, our choice of position for, say, the P is quite arbitrary. However, the first position adjacent to P's, in a clockwise direction, is now determined by wherever P is. There are $P(4, 4) = 4! = 24$ ways to place the remaining four letters in this direction and, therefore, 24 distinct ways in all to place the five letters around the dial. □

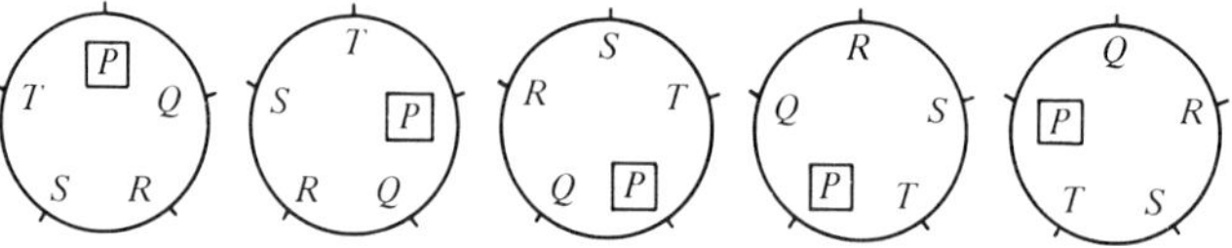

Figure 3.4 Five equivalent arrangements on a movable dial.

Completion Review 3.1

Complete each of the following.

1. To count a set of objects by listing them so that each object appears once and only once is called________________.

2. The________________ rule says that if there are n_1 ways to choose the first item, and then n_2 ways to choose the second, . . . , and finally n_k ways to choose the kth, then there are________________ ways to make all k choices in the given order.

3. A pictorial representation of the number of ways to make a sequence of choices is the ________________ diagram.

4. If $n \geqslant 1$, then $n! =$________________. Moreover, $O! =$________________.

5. If r items are to be selected from among n distinct items and arranged in the order selected, then each ordered selection is called a __________.

6. $P(n, r) =$ __________ $=$ __________ counts the number of __________.

Answers: **1.** enumeration. **2.** multiplication; $n_1 n_2 \ldots n_k$. **3.** tree. **4.** $n(n-1)(n-2)\ldots 1$; 1. **5.** permutation of n things taken r at a time. **6.** $n!/(n-r)!$; $n(n-1)\ldots(n-r+1)$; permutations of n things taken r at a time.

Exercises 3.1

1. Evaluate **(a)** $3!$; **(b)** $6!$; **(c)** $8!$; **(d)** $12!$; **(e)** $0!$.
2. Evaluate **(a)** $5!/4!$; **(b)** $8!/6!$; **(c)** $1/0!$; **(d)** $20!/19!$; **(e)** $20!/(20-2)!$; **(f)** $5!/((3!)(2!))$; **(g)** $10!/((5!)(5!))$.
3. Simplify **(a)** $n!/(n-1)!$; **(b)** $(n+1)!/n!$; **(c)** $n!/(n-2)!$.
4. Evaluate **(a)** $P(7, 3)$; **(b)** $P(7, 7)$; **(c)** $P(7, 0)$; **(d)** $P(17, 0)$; **(e)** $P(17, 1)$; **(f)** $P(17, 3)$; **(g)** $P(17, 17)$.
5. If a certain style of shoe comes in six sizes, three widths, and four colors, how many varieties of this shoe must the storekeeper have on hand to satisfy any customer who might want this shoe?
6. There are five roads connecting city A to city B, and four roads connecting cities B and C. How many ways can one travel from city A to city C using roads going through B?
7. If a certain species of pea plant has either a smooth or a wrinkled pea, either a tall or a short stem, and either a black or a green pea, how many varieties of this species are there?
8. Make tree diagrams illustrating Exercises 5, 6, and 7.
9. In how many ways can a judge give out first, second, and third prizes if there are three contestants?
10. How many four-letter "words" can be made from the letters of the word ACRIMONY if any four-letter sequence counts as a word, but no letter is used more than once?
11. Do Exercise 10 with the word AROUND.
12. In how many ways can three books on computer science, four books on biology, and five books on mathematics be arranged on a shelf if the books in the same category must stay together?
13. Find the number of ways in which five boys and five girls can be seated in a row if boys and girls must have alternate seats.
14. How many different ways can six people sit around a circular table if different ways are distinguished by **(a)** who sits next to whom; **(b)** who sits to the right and the left of each person?
15. How many ways can six keys be arranged on a key ring? (Careful! This is a three-dimensional problem.)

16. How many different license plates can be made having a sequence of three digits 0 through 9 followed by a sequence of three letters a through z, with no digits or letters repeated?
17. Use Algorithm 3.1 to compute 5!.
18. Try to compute 100!/99! on your hand calculator without first canceling 99 factors in the numerator and denominator. What is the result?
19. Write an algorithm to compute $P(n, r)$ that avoids division.
20. Prove that if $|S| = n$, then $|P(S)| = 2^n$.
21. Prove that if $|A| = m$ and $|B| = n$, then $|A \times B| = mn$.

3.2 COMBINATIONS AND BINOMIAL COEFFICIENTS

In the previous section we were concerned with the order in which our selections were made. There are times when the order is not important, as in the following example.

Example 1 Suppose that your professor of discrete structures wants to schedule three exams next week in such a way that there is not more than one exam per day. If an exam can be scheduled for any day, Monday through Friday, in how many different ways can he or she select exam days?

The problem is a bit ambiguous as stated, since it is not clear whether the order in which the exams are given is important. If the order is a factor in the problem, the answer is clearly $P(5, 3) = 5!/2! = 60$ ways. If the order is not important, only which days are selected, then the answer is

$$\frac{1}{3!}(5!/2!) = 10 \text{ ways}$$

since any subset of three days from the set of five can be arranged in 3! permutations. For example, {M, T, W} = {M, W, T} = {T, W, M} = {T, M, W} = {W, M, T} = {W, T, M}. □

Definition 3.2 A subset of r items taken from a set of n items is called a **combination of n things taken r at a time.**

Theorem 3.3 ***Combinations Rule*** The number $C(n, r)$ of combinations of n distinct items taken r at a time is given by

$$C(n, r) = \frac{n!}{r!(n-r)!} = \frac{n(n-1)(n-2)\ldots(n-r+1)}{r!}.$$

Proof: As in Example 1 each subset of r things can be permuted $r!$ ways (Theorem 3.2). Thus there are $r! \cdot C(n, r)$ ways of arranging a combination of n things taken r at a time. Hence $r! \cdot C(n, r) = P(n, r)$. Dividing by $r!$ we have

$$C(n, r) = (1/r!)(P(n, r)) = \frac{n!}{r!(n-r)!}$$

as claimed. ■

Example 2 In how many ways can we choose four of the 50 states in which to set up corporate headquarters?

Since the objects we are counting are subsets of size 4 taken from a set of size 50, the answer is

$$C(50, 4) = \frac{50!}{4!(50-4)!} = \frac{50 \cdot 49 \cdot 48 \cdot 47}{4 \cdot 3 \cdot 2 \cdot 1} = 230{,}300$$

different combinations of four states. □

It can easily be shown (see Exercise 3.2-3) that

$$C(n, 0) = C(n, n) = 1 \quad \text{and} \quad C(n, 1) = n$$

for all integers $n \geqslant 0$. Moreover, if we look at the last fraction in Example 2, we see that we can calculate $C(50, 4)$, for example, by alternately multiplying and dividing the numbers 50, 4, 49, 3, 48, 2, 47, and 1. These observations lead us to the following multiply–divide algorithm for calculating $C(n, r)$.

Algorithm 3.2 To compute $C(n, r)$ for nonnegative integers n and r by alternately multiplying and dividing,

1. [$C(n, n)$; $C(n, 0)$.] If $r = n$ or if $r = 0$, then output 1 and stop.
2. [$C(n, 1)$.] If $r = 1$, then output n and stop.
3. [Initialize M, D, C.] Set $M \leftarrow n$, $D \leftarrow r$, and $C \leftarrow n/r$. (M and D are numerator and divisor respectively.)
4. [Multiply–divide loop.] Repeat step 5 while $D > 1$.
    5. [Multiply–divide.] Set $M \leftarrow M - 1$, $D \leftarrow D - 1$, and then $C \leftarrow C \cdot M/D$. (If $D > 1$, then we repeat this step before going on to step 6.)
6. [Done.] Output C.
7. End of Algorithm 3.2.

The student should try to compute $C(100, 5)$, for example, by means of Algorithm 3.2 and a hand-held calculator. (See Exercise 3.2-5.)

The numbers $C(n, r)$ are sometimes written ${}_nC_r$, C_r^n, or $\binom{n}{r}$ and are referred to as **bionomial coefficients**. The reason for this name is their prominence in the formula of Theorem 3.4.

Theorem 3.4 ***Binomial Theorem*** $(x+y)^n =$

$$C(n, 0)\cdot x^n y^0 + C(n, 1)\cdot x^{n-1}y^1 + \cdots + C(n, r)\cdot x^{n-r}y^r + \cdots + C(n, n)\cdot x^0 y^n$$

for any nonnegative integer n.

Proof: To see why the binomial theorem is true, observe that the expansion of $(x+y)^n$ is found by taking the n-fold product:

$$(x+y)^n = \overbrace{(x+y)(x+y)\ldots(x+y)}^{n \text{ factors}}.$$

This product is calculated by use of the distributive law of multiplication, according to which the rth term is formed by choosing y's from r factors and x's from the remaining $n-r$ factors. This yields $x^{n-r}y^r$ in $C(n, r)$ ways, since there are $C(n, r)$ combinations of the n factors taken r at a time. So the rth term is $C(n, r)\cdot x^{n-r}y^n$ as the theorem claims. ■

Example 3 Let us expand the binomial $(x-y)^5$ by first noticing that $(x-y)^5 = (x+(-y))^5$. We will also need the following binomial coefficients.

$$C(5, 0) = 5!/[(0!)(5!)] = 1. \quad C(5, 3) = 5!/[(3!)(2!)] = 10.$$
$$C(5, 1) = 5!/[(1!)(4!)] = 5. \quad C(5, 4) = 5!/[(4!)(1!)] = 5.$$
$$C(5, 2) = 5!/[(2!)(3!)] = 10. \quad C(5, 5) = 5!/[(5!)(1!)] = 1.$$

Hence $(x-y)^5 = (x+(-y))^5 =$

$$1\cdot x^5(-y)^0 + 5\cdot x^4(-y)^1 + 10\cdot x^3(-y)^2 + 10\cdot x^2(-y)^3 + 5\cdot x^1(-y)^4 + 1\cdot x^0(-y)^5$$
$$= x^5 - 5x^4y + 10x^3y^2 - 10x^2y^3 + 5xy^4 - y^5.$$ □

The binomial coefficients can also be obtained using a pictorial scheme known as **Pascal's triangle**, given in Fig. 3.5.

Each successive row of Pascal's triangle is obtained from the numbers in the last row following the addition scheme in the diagram. Alternatively one can use the

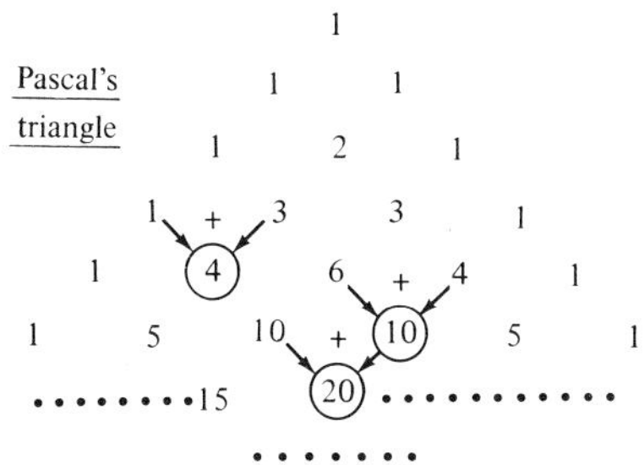

Coefficients of $(x+y)^0$
Coefficients of $(x+y)^1$
Coeffic)ents of $(x+y)^2$
Coefficients of $(x+y)^3$
Coefficients of $(x+y)^4$
Coefficients of $(x+y)^5$
Coefficients of $(x+y)^6$

Figure 3.5 Each successive row of Pascal's triangle is obtained from the last row by following the addition scheme in the diagram.

recursion formula

$$C(n+1, r) = C(n, r-1) + C(n, r),$$

which, with the additional conditions

$$C(n, n) = C(n, 0) = 1,$$

is merely a symbolic way of expressing what is in Pascal's pictorial scheme (Blaise Pascal, 1623–1662).

One can verify the recursion formula by expanding both sides (see Exercise 3.2-16), but let us see how these formulas are used.

Example 4 To calculate $C(4, 2)$ by means of the formula, we write

$$C(4, 2) = C(3, 2) + C(3, 1).$$

But $C(3, 1) = C(2, 1) + C(2, 0) = [C(1, 1) + C(1, 0)] + 1 = [1 + 1] + 1 = 3$. Moreover,

$$C(3, 2) = C(2, 2) + C(2, 1) = 1 + 2 = 3.$$

Hence

$$C(4, 2) = 3 + 3 = 6.$$ □

The method illustrated in Example 4 has the advantage of avoiding multiplications and divisions entirely. And, of course, if n is not too large, one can write out several lines of Pascal's triangle rather quickly. But now let us return to our main theme, which is that of counting.

One of the traditional ways of sharpening one's counting skills is to consider the events that can occur in well-known board or card games. There are several interesting counting problems concerned with card games, especially with the game known as poker.

In one version of poker, each player is dealt a "poker hand" of five cards drawn from a deck of 52 playing cards. The deck of cards is arranged in four suits: hearts, diamonds, spades, and clubs, each consisting of cards face-valued 1 (or ace) through 10, and a jack, queen, and king. There is a total of

$$C(52, 5) = 52!/((5!)(47!)) = 2{,}598{,}960$$

poker hands that a player can receive. Certain hands are distinguished from others, as in the next example.

Example 5 A "full house" is a poker hand consisting of three cards with one face value and two cards with a second face value. Let us calculate the number of such hands. This can be done in stages. We illustrate each stage with part of a decision tree (Fig. 3.6).

1. Choose the face value of the three of one kind in $C(13, 1) = 13$ ways.

2. Choose three suits for this face value in $C(4, 3) = 4$ ways.

3. Choose a different face value for the pair $C(12, 1) = 12$ ways.

4. Choose two suits for this face value $C(4, 2) = 6$ ways.

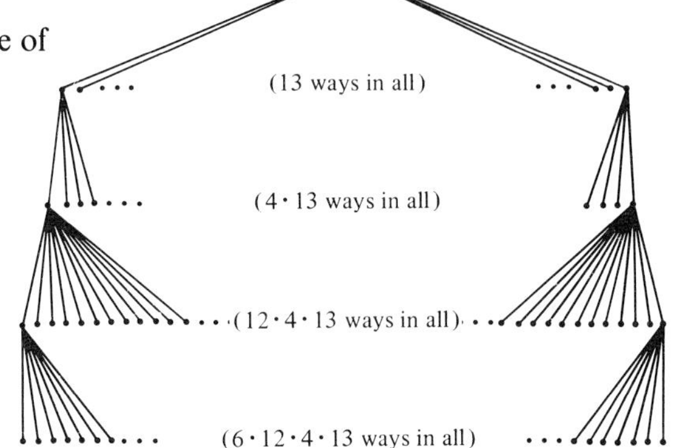

Figure 3.6 Determining the number of "full houses" with a tree diagram.

Combining these characteristics gives us, by the multiplication rule, a total of

$$(13)(4)(12)(6) = 3744$$

"full houses." □

Completion Review 3.2

Complete each of the following.

1. The symbol $C(n, r)$ stands for the number of________________ of n things taken r at a time.

2. A combination of n things taken r at a time is actually a ____________ of size r taken from a ____________.
3. The numbers $C(n, r)$ are known as the ____________ coefficients because $(x+y)^n =$ ____________.
4. We may evaluate $C(n, r)$ by means of the quotient ____________.
5. Pascal's triangle is a method for the evaluation of $C(n, r)$ by means of the recurrence formula $C(n, r) =$ ____________, where $C(n, 0) = C(n, n) = 0$____________ for all n.

Answers: **1.** combinations. **2.** subset; set of size n. **3.** binomial; $\sum_{r=0}^{n} C(n, r)x^{n-r}y^r$. **4.** $n!/[r!(n-r)!]$. **5.** $C(n-1, r) + C(n-1, r-1)$; 1.

Exercises 3.2

1. Use Theorem 3.3 to evaluate each of the following. **(a)** $C(6, 0)$; **(b)** $C(6, 1)$; **(c)** $C(6, 2)$; **(d)** $C(6, 3)$; **(e)** $C(6, 4)$; **(f)** $C(6, 5)$; **(g)** $C(6, 6)$.
2. Evaluate each of the following. **(a)** $C(8, 8)$; **(b)** $\binom{8}{5}$; **(c)** C_1^8; **(d)** ${}_8C_7$; **(e)** $C(8, 3)$.
3. Show that **(a)** $C(n, n) = C(n, 0) = 1$ and **(b)** $C(n, 1) = n$ for all nonnegative whole numbers n.
4. Show that $C(n, r) = C(n, n-r)$ for all nonnegative whole numbers n and r, $n \geqslant r$.
5. Evaluate $C(100, 5)$ using Algorithm 3.2 and a hand-held calculator. Write out the result of each step in the instructions.
6. Find $C(100, 95)$ using Exercises 4 and 5.
7. Find the number of combinations of seven objects taken four at a time.
8. Find the number of subsets of size 3 in a set of size 8.

9–14. Expand the following binomials using the binomial theorem.

9. $(x+y)^7$	**10.** $(p+q)^8$	**11.** $(a+b)^4$
12. $(x-y)^7$	**13.** $(x+2y)^6$	**14.** $(x^2+y)^4$

15. Write out sufficiently many lines of Pascal's triangle to expand the binomial $(x+y)^{10}$.
16. Verify the formula $C(n+1, r) = C(n, r) + C(n, r-1)$ by writing out both sides and combining fractions on the right-hand side.
17. Use the formula in Exercise 16 together with the conditions $C(n, n) = C(n, 0) = 1$ and $C(n, 1) = n$ to evaluate $C(5, 3)$.
18. Show that $\sum_{r=0}^{n} C(n, r) = 2^n$ for any positive integer n. [*Hint*: Use the binomial theorem with $(1+1)^n$.]
19. Show that $\sum_{r=0}^{n} (-1)^r C(n, r) = 0$ for any positive integer n. [*Hint*: Use the binomial theorem with $(1-1)^n$.]
20. A basketball coach has seven players available.
    a) How many different teams of five can he form?

b) How many teams are there including the tallest of the seven players? (Assume that one person is taller than the rest.)

21. Given 12 points in the plane no three of which are on the same straight line, how many lines are determined by these points?

22. How many five-card poker hands (see Example 5) can be formed that have "four of a kind," that is, four cards all having the same face value plus one additional card?

23. How many five-card poker hands can be formed having a "flush," that is, five cards all of the same suit?

24. There are 48 students in each of the freshman and sophomore classes, each class having 24 male and 24 female students. In how many ways can six representatives be selected so that there are three females and two freshmen?

25. Suppose that 25 automobiles are to be selected from 100 to determine whether they meet air pollution standards. An additional 25 will be selected to see if they meet safety requirements. If no two cars will have both inspections, in how many ways can the selections be done?

26. How many ways can the cars be selected in Exercise 25 if exactly 10 of the 50 selected cars will have both inspections?

3.3 REPETITIONS AND PARTITIONS

In the previous two sections we have been concerned with making selections from among distinct objects. But suppose we wanted to staff a project, and we needed to select two engineers, three programmers, and one statistician. Would our needs be met by the following set?

$$L = \{\text{engineer, engineer, programmer, programmer, programmer, statistician}\}.$$

No, they would not, because we know that repetitions within a set do not matter. The foregoing describes a set consisting of one engineer, one programmer, and one statistician. However, as a "multiset" L gives us the correct description, namely, two engineers, three programmers, and one statistician, since a **multiset** is defined to be a collection of objects that are not necessarily distinct.

To take another example, the multiset

$$\{M, I, S, S, I, S, S, I, P, P, I\}$$

gives us the letters in the word Mississippi, where there are four repetitions of the letter S, four repetitions of I, two of P, and only one occurrence of M. Our multiset contains 11 elements. The corresponding set

$$\{M, I, S, P\},$$

however, has only four elements.

It is not unusual to have a counting problem in which selections are made from a multiset, rather than from a set, as in the following examples.

Example 1 How many ways can we obtain distinct arrangements of the letters of the word MISSISSIPPI?

There are a total of 11 letters in this word, so our first guess might be 11! However, this is incorrect. First of all, anywhere that the four (indistinguishable) S's appear in a permutation, they can be interchanged for a total of 4! ways. In the same permutation, the four I's can also be interchanged 4! ways, while the two P's can be interchanged 2! ways. This means that a single permutation really accounts for (4!)(4!)(2!) duplications among our 11! alleged permutations. Hence the number of *distinct* permutations is actually

$$\frac{11!}{(4!)(4!)(2!)} = 34{,}650. \qquad \square$$

The rule given in Theorem 3.5 is just a simple generalization of the previous example.

Theorem 3.5 ***Permutations With Repetitions*** If a multiset of n objects has r_1 repeated (or indistinguishable) objects of the first kind, r_2 of the second kind, . . . , and r_k of the kth kind, where $r_1 + r_2 + \cdots + r_k = n$, then the number of distinct permutations of the n objects is equal to

$$n!/[(r_1!)(r_2!) \ldots (r_k!)].$$

Example 2 Suppose we have a suite of 10 offices and we want to paint one of them beige, two of them brown, three of them gray, and the remaining four white. The number of ways to select rooms for each color is

$$10!/[(1!)(2!)(3!)(4!)] = 12{,}600.$$

It does not seem like a permutations problem at first, but think instead of arranging four white marbles, three gray marbles, two brown marbles, and one beige marble in a row. Letting the kth position correspond to the kth room, the number of ways to arrange the marbles equals the number of ways to paint the rooms. $\square$

The next type of problem that we consider shares characteristics of both the permutations and combinations problems. It is also related to the kind of problem we have just considered.

Recall that combinations problems are concerned with the number of subsets of a given size from a given set. When we divide a set into several subsets that do not overlap, moreover, we said that we "partitioned" the given set.

Theorem 3.6 ***Ordered Partitions I*** The number of ways to partition a set of n distinct objects into r distinct categories so that n_1 are in the first category, n_2 in the second, . . . , and n_r in the rth, where $n_1 + n_2 + \cdots + n_r = n$, is

$$\frac{n!}{n_1!n_2!\ldots n_r!}.$$

Proof: Let us suppose that the objects are numbered positions and the ith category is represented by n_i identical symbols. Then our problem amounts to finding the number of permutations of all n symbols, where in general the ith symbol is repeated exactly n_i times, $i = 1, 2, \ldots, r$. As we saw in Theorem 3.5, the correct number is $n!/((n_1!)(n_2!)\ldots(n_r!))$. ■

Another way to prove Theorem 3.6 is suggested in the following example.

Example 3 A package of 12 computer programs is assigned to a team of three programmers. The most experienced programmer writes five of them, the next most experienced writes four, and the least experienced programmer writes only three. Then the number of ways to partition the 12 programming jobs, according to Theorem 3.6, is

$$\frac{12!}{5!\cdot 4!\cdot 3!} = 27{,}720.$$

Here is another way to do this problem: The number of ways to assign five of the 12 jobs to the first programmer is $C(12, 5)$. We then assign four of the remaining seven jobs to the next programmer in $C(7, 4)$ ways and the remaining three jobs in $C(3, 3)$ ways. Multiplying these binomial coefficients gives us

$$\frac{12!}{\cancel{7!}\cdot 5!}\cdot\frac{\cancel{7!}}{4!\cdot\cancel{3!}}\cdot\frac{\cancel{3!}}{3!\cdot 0!} = \frac{12!}{5!\cdot 4!\cdot 3!} = 27{,}720$$

as before. □

Let us emphasize that the objects we were selecting in the previous problem were distinguishable. In the following problem, however, we consider partitioning a multiset in which *none* of the elements are distinguishable from one another. The categories, on the other hand, will be identified by the order in which they appear.

Example 4 The computer science department has just received permission to buy six new microcomputers from any of three different manufacturers, Commodore, Apple, and IBM. If the units do not all have to be purchased from the same company, in how many ways can we make our choices?

The purchasing department considers the units to be indistinguishable. Hence one possible order is

"two from Commodore, three from Apple, and one from IBM"

Commodore	Apple	IBM
xx	xxx	x

This scheme suggests the following way of looking at the problem in terms of combinations. Write a list of six x's and two "slashes" for separating the different companies. The number of ways to choose six out of the eight positions for the x's equals the number of ways to divide up the sales among the companies, for a total of $C(8, 6) = 28$ ways. □

Theorem 3.7 ***Ordered Partitions II*** The number of ways to divide up n indistinguishable objects into r distinct categories is

$$C(n + r - 1, n).$$

The proof of Theorem 3.7 follows the same lines as Example 4 except that the number of x's is n and the number of slashes is now $r - 1$. (See Exercise 3.3-18.)

Example 5 In the game of "Yatzee" five indistinguishable dice are rolled. The number of different outcomes (such as a "1," a "2," and three "6's") is

$$C(5 + 6 - 1, 5) = C(10, 5) = 252$$

because we can think of the indistinguishable objects as the five dice and the categories as the numbers 1, 2, 3, 4, 5, and 6. □

After doing several problems involving partitions, it is natural to try to conceive of the next problem in the same way. Although this may be a possible way to attack the problem, it may not be the best way. But doing a problem in two different ways does have this advantage: It often reveals an unexpected relationship.

Example 6 In how many ways can one form an ordered partition of a set of n distinct objects into two distinct categories so that r objects are in the first and $n-r$ are in the second for all $r=0, 1, 2, \ldots, n$?

The answer for each r is, of course, $C(n, r)$, since taking r members to form one set automatically leaves $n-r$ members for the second set. If we do this for each r and add the sum

$$C(n, 0)+C(n, 1)+C(n, 2)+\cdots+C(n, n),$$

we will have our answer. But this is a tedious calculation, even for, say, $n=20$.

A different way to think of partitioning our set S having n members into two subsets in all possible ways is this: Just think of forming all different subsets of size $r=0, 1, 2, \ldots, n$ of the set S. Recall that the set of all subsets of S, $P(S)$, is called the "power set" of S, and that

$$|P(S)|=2^n.$$

This then solves our problem again. For $n=20$, there are $2^{20}=1{,}048{,}576$ ways to partition the set S into two subsets, for example. But we have also proved that, in general,

$$C(n, 0)+C(n, 1)+C(n, 2)+\cdots+C(n, n)=2^n.$$ □

Completion Review 3.3

Complete each of the following.

1. A collection of objects that are not necessarily all distinct is called a________________.
2. If a multiset of n objects has r_1 repeated of the first kind, r_2 repeated of the second kind, $\ldots$, and r_k repeated of the kth kind, where $r_1+r_2+\cdots+r_k=n$, then the number of distinct permutations of the n objects is equal to________________.
3. The number of ways to partition a set of n distinct objects into r distinct categories so that n_1 are in the first category, n_2 are in the second, $\ldots$, and n_r are in the rth, where $n_1+n_2+\cdots+n_r=n$ is________________. Each of these ways is called an ________________ partition.
4. The number of ways to divide n indistinguishable objects into r distinct categories is ________________. These partitions are also considered to be________________.

Answers: **1.** multiset. **2.** $n!/[r_1!r_2!\ldots r_k!]$. **3.** $n!/[n_1!n_2!\ldots n_r]$; ordered. **4.** $C(n+r-1, n)$; ordered.

Exercises 3.3

1. How many elements are in the multiset {M, I, S, S, O, U, R, I}?
2. How many distinct permutations are there of the letters in the following words? (a) MISSOURI; (b) FOOTBALL; (c) ZOOLOGY.
3. How many distinct permutations are there of the letters in the following words? (a) KNITTING; (b) CANADA; (c) BAMBOOZLE.
4. How many signals can be made by hanging 10 flags in a vertical line if two are red, four are white, three are green, and one is blue?
5. In how many ways can a job order of 14 computer programs be divided among four programmers A, B, C, and D so that A writes five, B writes four, C writes three, and D writes two?
6. In how many ways can we paint 12 houses on a block so that two are white, two are yellow, three are blue, and the rest are brown?
7. How many ways can we divide 12 indistinguishable pieces of candy among three children?
8. Do Exercise 7 assuming the pieces of candy are all different and each child gets four pieces.
9. How many different outcomes are possible if one rolls four indistinguishable six-sided dice?
10. Do Exercise 9 assuming we have four indistinguishable "Dungeons and Dragons" dice with 12 sides each.
11. A domino consists of two squares, each marked 1 through 6 or left blank. How many different dominoes are there?
12. Show that $C(n+r-1, n) = C(n+r-1, r-1)$.
13. Show that $\sum_{r=0}^{k} C(m, r)C(n, k-r) = C(m+n, k)$ for all positive integers n, m, and r by a combinational argument. (*Hint*: Show that the left-hand side is the number of ways one can form a subset of size k from a set of size $m+n$, where k is less than or equal to m and n.)
14. Suppose we have a multiset containing infinitely many copies of the letters $a, b, \ldots, z$. How many four-letter words can we make by selecting four elements of this multiset in some order?
15. Give an algorithm for calculating $n!/(n_1!n_2! \ldots n_r!)$ by alternately multiplying and dividing, where $n_1 + n_2 + \cdots + n_r = n$.
16. Show that $(a+b+c)^n =$ the sum of all the terms of the form

$$\frac{n!}{r!s!t!} a^r b^s c^t$$

where $r+s+t=n$ for nonnegative integers r, s, t, and n. (*Note*: The $n!/r!s!t!$ are called *multinomial coefficients*.)

17. Use the "multinomial theorem" of Exercise 16 to expand $(a+b+c)^4$.
18. Write out a formal proof of Theorem 3.7.
19. How many five-card poker hands can one form having two cards of one face value, two of a second face value, and the fifth card of a third face value?

3.4 INCLUSION–EXCLUSION

Partitioning a finite set A into mutually disjoint subsets $A_1, A_2, \ldots, A_n$ so that

$$A = \bigcup_{i=1}^{n} A_i$$

can be used to help us find the number of elements in A, since Theorem 2.7 tells that we can write

$$|A| = \sum_{i=1}^{n} |A_i|$$

in this case. For example, to determine how many students attend your college, you could count the populations of the male and female students, respectively, and then add your two sums. We obtain the correct total population because our classification of the students by gender partitions the total population into two disjoint sets.

It is often the case that a "natural" classification of the elements of a set does not, unfortunately, partition the set, even when all of the elements are accounted for. If we want to know how many elements are in the set when some elements fall into more than one category, or subset, we must use formulas of the following type. (See Exercise 2.2-15.)

Theorem 3.8 ***Addition Rule*** If A and B are finite sets, then

$$|A \cup B| = |A| + |B| - |A \cap B|.$$

Example 1 In Example 4 of Section 2.2 we were given, in part, the following information:

$$|A| = 40; \quad |B| = 24; \quad |A \cap B| = 8; \quad \text{and} \quad |U| = 80,$$

where

U = the set of computer science majors,

A = the subset of those who completed "Data Structures I,"

B = the subset of those who completed "Assembler I."

If we wanted to know how many students had completed *neither* "Data Structures I" *nor* "Assembler I," we could go back to Fig. 2.7 and consult our "natural partition" of U. But instead we will proceed in another, more direct, way: Just subtract the students in $A \cup B$ from the total of U. However, *to find $|A \cup B|$, we cannot just add the numbers $|A|$ and $|B|$, because this would count the elements in the set $A \cap B$ twice.* Thus

$$|A \cup B| = |A| + |B| - |A \cap B| = 40 + 24 - 8 = 56$$

is the correct way, since it eliminates the "double count" of the elements in $A \cap B$. (See Theorem 3.8.)
Hence

$$|\overline{A \cup B}| = |U| - |A \cup B| = 80 - 56 = 24,$$

and there are 24 students who took neither "Data Structures I" nor "Assembler I." □

The following is just a restatement of the kind of result we found in the previous example (where $\overline{A \cup B} = \bar{A} \cap \bar{B}$ by De Morgan's laws).

Theorem 3.9 ***Inclusion-Exclusion (Version 1)*** Let A and B be subsets of U. Then

$$|\bar{A} \cap \bar{B}| = |U| - |A| - |B| + |A \cap B|$$

Example 2 We say that two integers are **relatively prime** if they have no divisors ($\neq \pm 1$) in common. In particular any two different prime numbers are relatively prime.

Let us ask, "How many positive integers less than 35 are relatively prime to 35?" To solve this problem:

Let U = the set of integers x such that $1 \leqslant x \leqslant 35$,

A = the set of integers in U divisible by 5,

B = the set of integers in U divisible by 7.

Since $35 = (5)(7)$, the problem asks for $|\bar{A} \cap \bar{B}|$. However, it is easy to determine that

$$A = \{5, 10, 15, 20, 25, 30, 35\}, \quad \text{and so } |A| = 7;$$

$$B = \{7, 14, 21, 28, 35\}, \quad \text{and so } |B| = 5;$$

$A \cap B = \{35\}$, and so $|A \cap B| = 1$. Since $|U| = 35$, Theorem 3.9 implies

$$|\bar{A} \cap \bar{B}| = 35 - 7 - 5 + 1 = 24$$

positive integers less than 35 are relatively prime to 35. □

Both Theorems 3.8 and 3.9 can be generalized for more than two subsets of U. Let us see how to generalize them for three subsets, A, B, and C, and then state the result more generally for n subsets of U.

First of all, recall that $A \cup B \cup C = (A \cup B) \cup C$ by Definition 2.2. This means that we can work with just two sets at a time. Hence

$$\begin{aligned} |A \cup B \cup C| &= |(A \cup B) \cup C| \\ &= |A \cup B| + |C| - |(A \cup B) \cap C| \end{aligned} \tag{1}$$

by Theorem 3.8. Moreover,

$$|A \cup B| = |A| + |B| - |A \cap B| \tag{2}$$

and

$$\begin{aligned} |(A \cup B) \cap C| &= |(A \cap C) \cup (B \cap C)| \qquad \text{(by the distributive law)} \\ &= |A \cap C| + |B \cap C| - |A \cap B \cap C|. \end{aligned} \tag{3}$$

(See Exercise 3.4-18.) Substituting Eqs. (2) and (3) into the right-hand side of (1), we get part (a) of the following rule.

Theorem 3.10 Let A, B, and C be subsets of a finite set U. Then

a) $|A \cup B \cup C| = (|A| + |B| + |C|) - (|A \cap B| + |A \cap C| + |B \cap C|) + |(A \cap B \cap C)|$.

b) $|\bar{A} \cap \bar{B} \cap \bar{C}| = |U| - (|A| + |B| + |C|) + (|A \cap B| + |A \cap C| + |B \cap C|)$
$- (|A \cap B \cap C|)$.

Proof: We obtain part (b) by subtracting part (a) from $|U|$, observing that De Morgan's laws (see Theorem 2.10) give us

$$(\overline{A \cup B \cup C}) = \bar{A} \cap \bar{B} \cap \bar{C}.$$

Part (a) was, of course, proved above. ■

Part (b) of Theorem 3.10 is yet another "inclusion–exclusion" formula. It may be rewritten as a procedure as follows:

Inclusion-Exclusion Procedure To find the number of elements in a finite universe that have none of three given properties:

1. Count all the elements in the universe.
2. Subtract the number of elements having (a) the first property; (b) the second property; (c) the third property.
3. Add the number of elements having each pair of properties, that is, (a) the first and second; (b) the first and third; (c) the second and third.
4. Subtract the number of elements having all three properties.
5. Stop: Your answer is the final result.

Example 3 The ancient Babylonians used a number system that is called "duodecimal" because it uses a base of 12 rather than base 10 as in the decimal system. Such a system would require 12 digits, such as 0, 1, 2, 3, 4, 5, 6, 7, 8, 9, *T*, and *E*, for example, where *T* and *E* stand for 10 and 11 respectively. Let us calculate the number of duodecimal digit sequences of length k that contain at least the digits 1, 2, and 3.

So that we can apply Theorem 3.10, let $A=$ the set of sequences without a 1, $B=$ the set of sequences without a 2, and $C=$ the set of sequences without a 3. Then $\bar{A}\cap\bar{B}\cap\bar{C}=$ the set of sequences of length k that contain at least one 1, one 2, and one 3. Moreover,

$$|A|=|B|=|C|=11^k,$$

since for each of A, B, and C only one of the 12 digits is excluded as a possibility in each of the k places of the duodecimal sequence. Similarly,

$$|A\cap B|=|A\cap C|=|B\cap C|=10^k,$$

since two digits are excluded in these cases; and

$$|A\cap B\cap C|=9^k,$$

because all three digits are excluded from the sequences in the set $A\cap B\cap C$. By Theorem 3.10 we have

$$|\bar{A}\cap\bar{B}\cap\bar{C}|=12^k-(3)(11^k)+(3)(10^k)-9^k$$

because there are 12^k duodecimal sequences of length k in which we can use any of the digits 0, 1, 2, 3, 4, 5, 6, 7, 8, 9, *T*, and *E*. □

Example 4 Let us calculate how many five-card hands can be dealt from an ordinary deck of playing cards in which there is at least one picture card of each type: jack, queen, and king.

Let K be the set of hands with no king, let Q be the set of hands with no queen, and let J be the set of hands with no jack.

Then

$$|K| = |Q| = |J| = C(48, 5),$$

and

$$|K \cap Q| = |K \cap J| = |Q \cap J| = C(44, 5),$$

and

$$|K \cap Q \cap J| = C(40, 5).$$

Hence

$$|\bar{K} \cap \bar{Q} \cap \bar{J}| = C(52, 5) - (3)(C(48, 5)) + (3)(C(44, 5)) - C(40, 5) = 62{,}064$$

is the number of five-card hands with a picture card of every kind. □

The pattern of subtracting and adding that we established in Theorems 3.9 and 3.10 carries over to the general case in which we have subsets of a set U. In fact a proof by induction of the following principle is very much like the one we gave for Theorem 3.10. (See Exercise 3.4-20.)

Theorem 3.11 ***Inclusion-Exclusion (General)*** Let $A_1, A_2, \ldots, A_m$ be subsets of a finite set U. Then $|\bar{A}_1 \cap \bar{A}_2 \cap \ldots \cap \bar{A}_m| =$

$$|U| - \sum_i |A_i| + \sum_{i<j} |A_i \cap A_j| - \sum_{\substack{i<j\\ j<k}} |A_i \cap A_j \cap A_k| + \cdots$$
$$+ (-1)^m \cdot |A_1 \cap A_2 \cap \ldots \cap A_m|.$$

Example 5 Suppose that five people went into a dark cloakroom to get their hats, and when they emerged each person was wearing one of the other four persons' hats. Let us calculate how many different ways this could happen. (These ways are sometimes called the "derangements" of the hats.)

First we shall label the different people 1, 2, 3, 4, and 5. Then the order in which they should get their hats is $h_1h_2h_3h_4h_5$. *We are looking for the permutations of the h_i's so that no h_i is in position i.*

Let A_i = the set of permutations in which h_i is in position i. Then $|A_i| = 4!$, since we are permuting the other four h's. Similarly,

$$|A_i \cap A_j| = 3! \quad \text{if } i<j;$$
$$|A_i \cap A_j \cap A_k| = 2! \quad \text{if } i<j<k;$$
$$|A_i \cap A_j \cap A_k \cap A_p| = |A_1 \cap A_2 \cap A_3 \cap A_4 \cap A_5| = 1, \quad \text{if } i<j<k<p.$$

Thus

$$|\bar{A}_1 \cap \bar{A}_2 \cap \bar{A}_3 \cap \bar{A}_4 \cap \bar{A}_5| =$$
$$= 5! - C(5, 1)\cdot 4! + C(5, 2)\cdot 3! - C(5, 3)\cdot 2! + C(5, 4)\cdot 1! - C(5, 5) = 44,$$

since each k-fold intersection can be found $C(5, k)$ ways. □

Completion Review 3.4

Complete each of the following.

1. The formula for adding the sum of the elements in a union of two finite sets is $|A \cup B| =$________________.
2. The inclusion–exclusion formula for two subsets of a finite universe U is $|\bar{A} \cap \bar{B}| =$________________.
3. The inclusion–exclusion formula for three subsets of a finite universe U is $|\bar{A} \cap \bar{B} \cap \bar{C}| =$________________.
4. The inclusion–exclusion formula for m subsets of a finite universe U is $|\bar{A}_1 \cap \bar{A}_2 \cap \dots \cap \bar{A}_m| =$________________.

Answers: **1.** $|A| + |B| - |A \cap B|$. **2.** $|U| - |A| - |B| + |A \cap B|$. **3.** $|U| - (|A| + |B| + |C|) + (|A \cap B| + |A \cap C| + |B \cap C|) - |A \cap B \cap C|$. **4.** $|U| - \sum_{\text{all } i} |A_i| + \sum_{i \neq j} |A_i \cap A_j| + \cdots + (-1)^m |A_1 \cap A_2 \cap \dots \cap A_m|$.

Exercises 3.4

1. Suppose that $|A| = 5$, $|B| = 7$, $|A \cap B| = 3$, and $|U| = 10$.
Find **(a)** $|A \cup B|$; **(b)** $|\bar{A} \cap \bar{B}|$.

2. Suppose that $|S| = 40$, $|T| = 50$, $|S \cap T| = 25$, and $|U| = 100$.
Find **(a)** $|S \cup T|$; **(b)** $|\bar{S} \cap \bar{T}|$.

3. Which of the following pairs of integers are relatively prime?
   (a) 15, 108; **(b)** 25, 12; **(c)** 18, 200; **(d)** 19, 2001.
4. How many positive integers less than 55 are relatively prime to 55?
5. How many positive integers less than 39 are relatively prime to 39?
6. If $|A| = 15$, $|B| = 12$, $|C| = 10$, $|A \cap B| = |A \cap C| = |B \cap C| = 5$, $|A \cap B \cap C| = 2$, and $|U| = 25$, find **(a)** $|A \cup B \cup C|$; **(b)** $|\bar{A} \cap \bar{B} \cap \bar{C}|$.
7. How many positive integers less than 105 are relatively prime to 105?
8. How many r-digit quintary sequences (that is, using the digits 0, 1, 2, 3, 4) are there that contain at least a 0, a 1, and a 2?
9. How many r-digit hexadecimal sequences (that is, using 16 digits, or base 16) are there that contain at least 0, 1, 2, and 3?
10. How many seven-card hands can be dealt from a deck of 52 playing cards in which there are at least one heart, one spade, one diamond, and one club?
11. In how many ways can a waiter serve six people at a table so that each person gets someone else's order?
12. Do Exercise 11 for n people. Show that one obtains

$$n!\left(\sum_{k=0}^{n} \frac{(-1)^k}{k!}\right) = n!\left(\frac{1}{0!} - \frac{1}{1!} + \frac{1}{2!} - \frac{1}{3!} \pm \cdots + \frac{(-1)^n}{n!}\right).$$

13. How many ways are there to distribute 20 distinct objects into four distinct boxes with at least one empty box?
14. How many ways are there to distribute 20 identical objects into four distinct boxes with at most three objects in any of the first two boxes?
15. How many ways are there to assign each of four computer science professors to three courses in the fall semester and then three courses in the spring semester so that no professor teaches the same three courses both semesters? (*Note*: Assume a total of 12 different courses.)
16. How many ways are there to arrange (permute) the 26 letters of the alphabet that do not contain any of the words (sequences) "WORD," "MATH," "PUNCH," or "CARD"?
17. How many ways can one paint the rooms of the house whose floor plan is given in the following with n colors if rooms with a common doorway must have different colors?

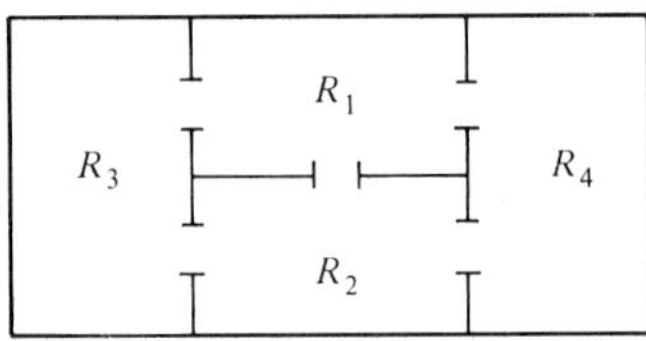

18. In the course of proving Theorem 3.10 we made use of the fact that $(A \cap C) \cap (B \cap C) = A \cap B \cap C$. Prove this.

19. To obtain part (b) of Theorem 3.10, one needs the fact that $\overline{(A \cup B \cup C)} = \bar{A} \cap \bar{B} \cap \bar{C}$. Prove this.

20. Prove Theorem 3.11, the general inclusion–exclusion rule, by mathematical induction. (*Hint*: Follow our proof of Theorem 3.10.)

3.5 APPLICATIONS OF COUNTING FROM PROBABILITY TO PIGEONHOLES

Probability theory has been developed to model events that are more predictable in the long run than in the short run. For example, if you flip a coin once, it is uncertain whether it will show a head, but if you flip it 1000 times, it will probably show heads "about" half the time. That is, the relative frequency of heads would be "about" $500/1000 = 1/2$. If a fair (cubical) die is tossed, the probability that a "2" will show is 1/6. Why?

Figure 3.7 Tossing a die is a typical probability experiment. Here we obtain outcome of "4."

Because we can think of a "mental experiment" in which the outcomes form a set of numbers $\{1, 2, 3, 4, 5, 6\}$, each with an equal chance of occurring. Since the outcome "2" is among six possibilities, it is natural to divide 1 by 6 and say that 1/6 is the probability of getting a "2." Thus *our idea of probability depends very heavily upon counting*, and that is why we include this section here. Before we make any definitions let us consider the following "probability experiment."

Example 1 Suppose we toss a fair coin three times and ask for the probability that a certain number of heads is obtained. What does this mean exactly? It is clear that we can obtain either zero, one, two, or three heads. Is the probability of each of these the same, that is, 1/4? No. The correct way to think about this experiment is in terms of the head–tail sequences one could get: These sequences form the following set.

$$S = \{\text{HHH, HHT, HTH, HTT, THH, TTH, THT, TTT}\}$$

These outcomes describe exactly what could happen in each case. It is reasonable to assume that each of these cases is equally likely. But the "event" of getting no heads really consists of the single case TTT, while that of getting one head consists of the three distinct cases, or "basic" outcomes HTT, THT, and TTH. Thus getting one head ought to be three times as "probable" as getting no heads. In fact since there are exactly eight "basic" outcomes in this experiment, most people would accept 1/8 as the probability of getting no heads and 3/8 as the probability of getting one head. □

Using Example 1 as our guide, we are now ready to formulate the following definitions.

Definition 3.3 A set of possible outcomes of a probability experiment is called the (a) **sample space** of the probability experiment if the outcomes are (b) **exhaustive**—one of the outcomes must occur each time the experiment is performed; and (c) **mutually exclusive**—no two outcomes can occur at the same time. The members of a sample space of a probability experiment are called the (d) **basic outcomes** of the experiment. An (e) **event** is a subset of the sample space.

Definition 3.4 A (a) **probability distribution** is a rule that assigns a number called a (b) **probability** to each basic outcome of a sample space so that (i) the probability of each basic outcome is between 0 and 1, and (ii) the sum of the probabilities of all the basic outcomes is 1.

Theorem 3.12 ***Probability Rule*** If a sample space consists of n basic outcomes, all of which are equally likely, and an event E from the sample space consists of m basic outcomes, then the probability of E occurring is m/n. [We will write $\Pr(E) = m/n$.]

In other words, Theorem 3.12 describes the probability distribution that we will use to determine probabilities in cases where we know (a) how many basic outcomes are in our sample space, (b) how many basic outcomes make up the event in which we are interested, and that (c) all the basic outcomes in the sample space are equally likely.

Example 2 In five-card poker the hand called "two pairs" consists of two cards of one face value, two cards of a second face value, and a fifth card with a third distinct face value. If a person is dealt five cards at random from an ordinary deck of playing cards, what is the probability of getting "two pairs"?

The appropriate sample space in this example is the set of all poker hands, namely, five-card subsets of the deck of 52 cards. (The order in which one is dealt the cards is not important.) There are $C(52, 5) = 2{,}598{,}960$ members in this sample space. To determine the number of members of $E = \{x : x \text{ is a two-pair poker hand}\}$, we reason as follows: There are 13 ways to pick the face value of the first pair and $C(4, 2) = 6$ ways to pick the two suits in that pair. Next there are now 12 ways to pick the face

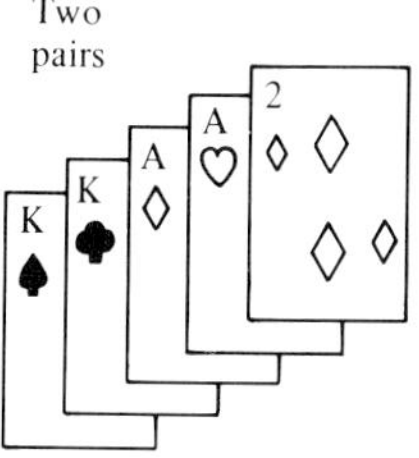

Figure 3.8 A two-pair poker hand consists of two cards of one face value, two cards of a second face value, and a third card with still another face value.

value of the second pair, and $C(4, 2) = 6$ ways to pick the suits for that pair. Finally, there are $52 - 8 = 44$ ways to pick the fifth card. By the multiplication rule,

$$|E| = [(13)(6)(12)(6)(44)]/2 = 123{,}552;$$

where we divide by 2 because the order in which we choose the face values of the pairs is irrelevant. ($\{K_S, K_C, A_D, A_H, 2_S\} = \{A_D, A_H, K_S, K_C, 2_S\}$, for example.) Hence

$$\Pr(E) = \frac{123{,}552}{2{,}598{,}960} \doteq 0.047539 \text{ (to the nearest millionth).}$$ □

Sometimes it is easier to find the probability of the complement $\bar{E}$ of E than of E itself. In such cases we rely on the simple rules given in Theorem 3.13.

Theorem 3.13 ***Addition Rules*** Let A and B be two mutually exclusive events from the finite sample space U ($\neq \varnothing$). Then

1. $\Pr(A \cup B) = \Pr(A) + \Pr(B)$.
2. In particular, $\Pr(\bar{A}) = 1 - \Pr(A)$.

Proof: Since A and B are disjoint sets, we have

$$|A \cup B| = |A| + |B|.$$

We now obtain the first part of Theorem 3.13 by dividing both sides of the foregoing equation by $|U|$. The second part of Theorem 3.13 is implied by the fact that $\Pr(U)=1$, and $U=A\cup\bar{A}$, where the sets A and $\bar{A}$ are mutually exclusive. ■

Here is a surprising application of Theorem 3.13.

Example 3 Suppose that 25 people are chosen at random. What is the possibility that at least two of these people have the same birthday, that is, that they were born on the same day of the year? It seems unlikely, but it is not.

Let us assume that all birthdays are equally likely and just ignore "leap years." Since there are 365 days from which to choose, we can let our sample space be all possible sequences of length 25, using the integers 1 through 365. (Each term in a sequence corresponds to one of the 25 people.) There are 365^{25} basic outcomes in this sample space. If

$E=$ the set of sequences of length 25 having at least two terms equal,

we want to calculate $\Pr(E)$. We can do this if we can find $|E|$. But $|E|$ is relatively hard to calculate directly. It is easier to calculate

$$|\bar{E}|=(365)(364)\ldots(341),$$

since

$\bar{E}=$ the set of sequences of length 25 having no two terms equal.

Hence

$$\Pr(\bar{E})=\frac{(365)(364)\ldots(341)}{(365)(365)\ldots(365)}\doteq 0.43 \text{ (to the nearest 100th)},$$

and so

$$\Pr(E)=1-0.43\doteq 0.57 \text{ (to the nearest 100th)}.$$ □

One can give an algorithm, based upon our "multiply–divide" method for calculating combinations, to determine the probability that at least two people in randomly chosen sets of from one to 365 people have the same birthday. (See Exercise 3.5-11.) Clearly, as the number of people in the sets increase, the probability that at least two people have the same birthday increases as well.

Now let us suppose, however, that we have chosen 367 people at random. We claim *it is certain* (*probability* $=1$) *that at least two of these* 367 *people have the same birthday*, even if we count leap years. This claim is not based upon any of our previous counting formulas, but rather on the following simple-minded, but far-reaching, idea.

Theorem 3.14 ***Pigeonhole Principle*** If there are more than k times as many pigeons as pigeonholes, then some pigeonhole must contain at least $k+1$ pigeons. In particular, if there are more pigeons than pigeonholes, then some pigeonhole must contain at least two pigeons.

One can prove the pigeonhole principle by induction on the number of pigeonholes. (See Exercise 3.5-12.) To apply it to our 367 people, we proceed by analogy. Let the people correspond to 367 "pigeons" and let the 366 possible birthdays correspond to "pigeonholes." Then, by Theorem 3.14, at least two "pigeons" are in some "pigeonhole" that is, at least two people have the same birthday. (See Exercise 3.5-13.)

Here is another surprising application of the pigeonhole principle.

Example 4 Suppose that we have six randomly chosen dinner guests seated at the same table. Then it is certain that *either three of these guests are mutual strangers, or three of these guests are mutual acquaintances.* To see this let p_6 be any one of the six guests. (See Fig. 3.9.) This leaves a set A of the remaining five guests,

$$A = \{p_1, p_2, p_3, p_4, p_5\}.$$

Figure 3.9 Person p_6 is either acquainted with or a stranger to at least three of five neighbours.

Let us partition the set A into the mutually exclusive sets B and C, which are the acquaintances of and the strangers to p_6 respectively. Since this is like putting *$5(=2\cdot 2+1)$ pigeons into two pigeonholes*, the pigeonhole principle tells us that *there are either three people in set B or three people in set C.*

If there are three people in set B, then some p_i, p_j, and p_k are acquaintances of p_6. If two people in the set $\{p_i, p_j, p_k\}$ know each other, then with p_6 we have a set of three mutual acquaintances. Otherwise the set $\{p_i, p_j, p_k\}$ consists of three mutual strangers. The other cases are similar (see Exercise 3.5-19), and so this completes the proof. □

Completion Review 3.5

Complete each of the following.

1. A set of possible outcomes of a probability experiment is called the sample space of the experiment if the set is ________________ and ________________.
2. The members of the sample space are called ________________.
3. An event is a ________________ of the sample space.
4. A probability distribution assigns a number x called a probability to each basic outcome such that ________________ and ________________ over all basic outcomes.
5. If a sample space consists of n outcomes that are all equally likely, and an event E from the sample space consists of m basic outcomes, then $\Pr(E) =$ ________________.
6. If A and B are mutually exclusive events from the finite sample space U, then $\Pr(A \cup B) =$ ________________.
7. In particular, $\Pr(\bar{A}) =$ ________________.
8. If there are more than k times as many pigeons as pigeonholes, then some pigeonhole must contain at least ________________ pigeons. This is known as the ________________ principle.

Answers: **1.** exhaustive; mutually exclusive. **2.** basic outcomes. **3.** subset. **4.** $0 \leqslant x \leqslant 1$; $\sum x = 1$. **5.** m/n. **6.** $\Pr(A) + \Pr(B)$. **7.** $1 - \Pr(A)$. **8.** $k+1$; pigeonhole.

Exercises 3.5

1. Suppose that we toss a fair coin four times and ask for the number of heads and tails.
- **a)** Make a tree diagram illustrating the outcomes.
- **b)** What is a suitable sample space for this probability experiment?
- **c)** What is the event of getting three heads?
- **d)** What is the probability of getting three heads?

2. In the previous exercise what is the probability of getting **(a)** exactly two heads; **(b)** exactly one head; **(c)** no heads; **(d)** four heads; **(e)** at least three heads; **(f)** at most three heads?

3. Two cubical dice are tossed, each with from one to six dots on its six faces. What is the probability of getting a sum of **(a)** 0; **(b)** 1; **(c)** 2; **(d)** 5; **(e)** 12?

4. In Exercise 3, what is the probability of getting **(a)** a sum that is even; **(b)** a sum that is at least 7?

5. In a genetics experiment involving the characteristics of wrinkled versus smooth, green versus black, and tall versus short in pea plants, it was found that 60 were black; 78 were wrinkled; 48 were both short and black; 64 were short and wrinkled; 42 were black and wrinkled; 40 were black, wrinkled, and short; and eight were green, smooth, and tall. If one of these plants were selected at random, what is the probability that it would be short, smooth, and green, given that a total of 74 were short?

6. A basketball coach has seven players out of which to form a team of five. If all players are of different heights, what is the probability that **(a)** the tallest player gets picked; **(b)** the tallest and the shortest players are picked?

7. The letters in the word COMPUTER are rearranged at random. What is the probability that the resulting sequence **(a)** again spells COMPUTER; **(b)** contains the word COMPUTE?

8. A contest will determine first, second, and third prizes from among 10 contestants including Erny, Ike, and Mike. If it turns out that the prizes will be awarded at random, what is the probability that our three heroes **(a)** win all three prizes; **(b)** Erny wins first, Ike wins second, and Mike wins third prize?

9. In a certain game of chance one must choose six different numbers out of the integers 1, 2, 3, . . . , 36 correctly. What is the probability of winning if **(a)** the order in which the numbers are chosen is essential; **(b)** the order is irrelevant?

10. What is the probability of getting each of the following hands of five-card poker hands?

- **a)** A pair (two of one face value, and three having three different face values).
- **b)** Three of a kind (three of one face value and two with two additional face values).
- **c)** A flush (all the same suit not in sequence).
- **d)** A full house (a pair and three of a kind).
- **e)** Four of a kind.
- **f)** A royal flush (ace, king, queen, jack, and 10 of the same suit).
- **g)** A straight flush (five in sequence and the same suit, but not a royal flush).
- **h)** A straight (all in sequence, not all the same suit).

11. **a)** Give an algorithm, based upon our "multiply–divide" method for calculating combinations, to determine the probability that at least two people in a set having from one to 365 people have the same birthday.

- **b)** Make a table of n persons versus P_n, the probability that no two of n randomly chosen persons have the same birthday for $n = 10, 15, 20, 25, 30, 35, 50$.

12. Prove the pigeonhole principle by induction on the number of pigeonholes.

13. How many people are needed so that the probability that at least three of them have birthdays that fall on the same day of the year is 1? (*Hint*: Assume 366 possible birthdays, and use the pigeonhole principle.)

14. **a)** Why must there be at least one pair of matching shoes among 10 shoes chosen at random from nine matched pairs? Explain carefully using the pigeonhole principle.

- **b)** How many shoes would one need to guarantee at least two matching pairs? Why?

15. Given a group of 100 husbands and their wives, how many people from this group must be chosen to guarantee at least one married couple? Why?

16. Show that any function mapping the set $\{1, 2, 3, 4, 5, 6, 7\}$ into the set $\{0, 1\}$ must map at least four numbers into 0 or at least four numbers into 1.

17. Given a group of seven people. Must there be either four mutual strangers or four mutual acquaintances? Explain.

18. Given n students who are assigned to 20 computer terminals. How large must n be to ensure that some three students are assigned to the same terminal?

19. **a)** What are the "other cases" referred to in Example 4?

b) Finish the proof by discussing these "other cases" in detail.

c) Give a counterexample to the following claim: "Given a set of five randomly chosen people, then it is certain that either three are mutual acquaintances or three are mutual strangers."

3.6 RECURRENCE MODELS

Suppose that we are studying a sequence of numbers $a_1, a_2, \ldots, a_n, \ldots$ that arises in the course of trying to compute, or just count, something, and where the calculation of each term a_n depends upon the previous terms of the sequence. We call such a sequence a **recurrence model** of whatever it is we are trying to compute. You have already seen such models in previous sections of this book, where we discussed, for example, the Fibonacci numbers, permutations, and sums. In the following two sections, we expand upon this idea and its role in counting and algorithms.

Example 1

a) To show how to sum n numbers $f(1), f(2), \ldots, f(n)$ and thereby calculate $s_n = \sum_{k=1}^{n} f(k)$, we give the **initial condition** $s_1 = f(1)$ and then write

$$s_n = s_{n-1} + f(n)$$

for each $n = 2, 3, \ldots$. This amounts to our definition of the "sum notation" in Section 1.3.

b) To show how to form an arithmetic progression, $a_1, a_2, a_3, \ldots$ with first term a and common difference d, we give the initial condition $a_1 = a$ and then write

$$a_n = a_{n-1} + d$$

for all $n = 2, 3, \ldots$.

c) To show how to form a geometric progression $a_1, a_2, a_3, \ldots$ with first term a and common ratio r, we again give the first term $a_1 = a$ and then write

$$a_n = r a_{n-1},$$

$n = 2, 3, \ldots$.

d) The initial condition $a_0 = 1$ together with

$$a_n = n \cdot a_{n-1}, \; n = 1, 2, \ldots$$

is another way to define $a_n = n!$ ("n factorial," as in Section 2.1), since $a_1 = 1$, $a_2 = 2 \cdot 1$, $a_3 = 3 \cdot 2 \cdot 1, \ldots,$ and, in general, $a_n = n(n-1)(n-2) \ldots 2 \cdot 1$. It also defines $P(n, n) =$ the number of permutations of n things taken n at a time. □

The recursion formulas (a)–(d) are sometimes called **difference equations**, in part because they can be written (a) $s_n - s_{n-1} = f(n)$; (b) $a_n - a_{n-1} = d$; (c)$a_n - r \cdot a_{n-1} = 0$; and (d) $a_n - n \cdot a_{n-1} = 0$ respectively.

Each of the recursion formulas, or **recurrence relations**, in Example 1 is relatively simple, in that each a_n only depends upon the preceding a_{n-1} and, perhaps, some function of n. Here are some more complicated examples.

Example 2

a) $a_n = a_{n-1} - a_{n-2}$. Here a_n depends on two previous terms.

b) More generally $a_n = k_1 a_{n-1} + k_2 a_{n-2} + \cdots + k_r a_{n-r} + f(n)$, where the k_i are constants and f is a function of n, is called a "linear recurrence relation."

c) $a_{n,r} = a_{n-1,r} + a_{n-1,r-1}$ has double subscripts. If we specified $a_{n,n} = a_{n,0} = 1$, you might recognize the recurrence relation in (c) as giving the number of combinations of n things taken r at a time, $r = 0, 1, \ldots, n$. □

To calculate a_n, where a_n is defined in terms of r former terms, we generally need r initial conditions, for example, the values of $a_1, a_2, \ldots,$ and a_r.

To illustrate with Example 2(a), we can calculate a_5, given $a_1 = 3$ and $a_2 = 1$, by writing

$$a_3 = 1 - 3 = -2; \qquad \text{hence } a_4 = -2 - 1 = -3; \qquad \text{and } a_5 = -3 - (-2) = -1.$$

More generally, here is an algorithm for calculating the terms in the linear recurrence relation of Example 2(b).

Algorithm 3.3 To calculate the mth term, a_m, given by the recurrence relation $a_n = k_1 a_{n-1} + k_2 a_{n-2} + \cdots + k_r a_{n-r} + f(n)$, the r coefficients $k_1, \ldots, k_r$, and the r initial conditions $t_0, t_1, \ldots, t_{r-1}$,

1. [a_m is given.] If $m < r$, output t_m and stop. (Otherwise)
2. [Initialize a_i's, n.] Set $a_0 \leftarrow t_0$, $a_1 \leftarrow t_1, \ldots, a_{r-1} \leftarrow t_{r-1}$, and $n \leftarrow r$.
3. [Recurrence loop.] Repeat step 4 for $n = r, r+1, \ldots, m$.
    4. [Use relation.] Set $a_n \leftarrow k_1 a_{n-1} + k_2 a_{n-2} + \cdots + k_r a_{n-r} + f(n)$. (If $n < m$, then we repeat this step after raising n by 1. Otherwise continue with step 5.)
5. [Done.] Output a_m and stop.
6. End of Algorithm 3.3.

If we wanted to determine a_{500} in Example 2(a), it would be a tedious exercise indeed. But $n=500$ presents no problem at all for today's high-speed computers, most of which can be programmed to perform steps analogous to our steps 3–6 of Algorithm 3.3.

Recurrence relations are sometimes very useful in *modeling* combinatorial problems, that is, counting problems of the kind we discussed in Chapter 2. The counting problem may be phrased in such a way that it can be written directly in terms of a recurrence relation just by translating the words of the problem into our language of "a_n's." One such classic problem was given by Leonardo of Pisa (known as "Fibonacci") in the year 1202 A.D.

Example 3 Fibonacci's problem was to model the growth of an unchecked population of rabbits. More precisely, we want to find the number of (mated) pairs of rabbits after n months (let this be a_n) if each month every pair of rabbits over one month old has a (mated) pair of offspring, and we begin with one newborn pair. (Hence $a_0=1$.)

The reproductive pattern and population growth of our hypothetical rabbit pairs can be seen in the decision tree of Fig. 3.10. Notice that each mated pair (represented by a single rabbit in the figure) takes one month to mature and an additional month to reproduce, giving us a_n pairs after n months. It would seem that a_n satisfies the relationship defining the Fibonacci numbers,

$$a_n = a_{n-1} + a_{n-2},\ n \geqslant 2,$$

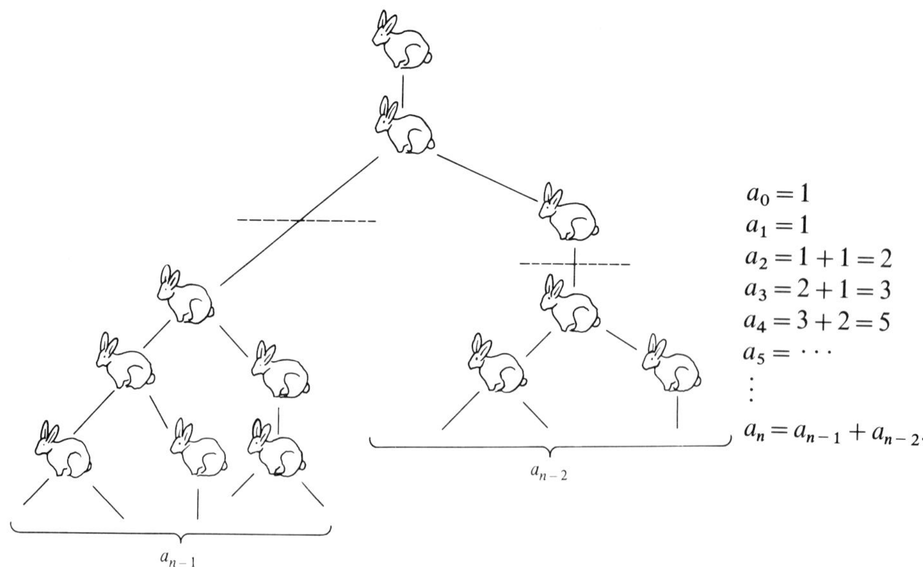

Figure 3.10 In this diagram of Fibonacci's rabbit population model, each rabbit represents a pair, and new pairs branch off to the right.

when we have the initial conditions $a_0 = a_1 = 1$. To calculate the number of pairs after 12 months, say, we can use Algorithm 3.3, given $m = 12$, $r = 2$, and $t_0 = t_1 = 1$. We obtain

$$1, 1, 2, 3, 5, 8, 13, 21, 34, 55, 89, 144, \underline{233}.$$

Aside from modeling the remarkable fecundity of rabbits, the Fibonacci numbers exhibit many interesting mathematical properties. (See Exercise 3.6-20.) So much active research is going on concerning the Fibonacci numbers, in fact, that a journal, the *Fibonacci Quarterly*, is devoted entirely to this subject.

Looking back at Fig. 3.10, you might have noticed that after the first reproduction, the right-hand branch of the tree repeats the pattern of the whole tree. The descendants of the first offspring will amount to a_{n-2} pairs in the nth month. The branch on the left, however, begins this pattern one month before its offspring. Hence the left-hand pairs' descendants will amount to a_{n-1} individuals in the nth month. (In each case the original pair is included in the total.) Adding the descendants from each branch, we obtain

$$a_n = a_{n-1} + a_{n-2},\ n \geqslant 2,$$

as before. □

The following examples develop recurrence relation models using the very idea we have just illustrated: *Divide the counting problem into a first and second step, where the second is just a smaller version of the first.*

Example 4 In how many ways, a_n, can we give away a \$1 bill or a \$5 bill on successive days, until we have given away a total of n dollars?

As the decision tree of Fig. 3.11 shows, if we give a \$1 bill on the first day, we will have $n - 1$ dollars left, which we can give away in a_{n-1} ways. But if we give a \$5 bill on the first day, this will leave us with $n - 5$ dollars to dispose of on subsequent days in a_{n-5} ways. Hence

$$a_n = a_{n-1} + a_{n-5},\ n \geqslant 5.$$

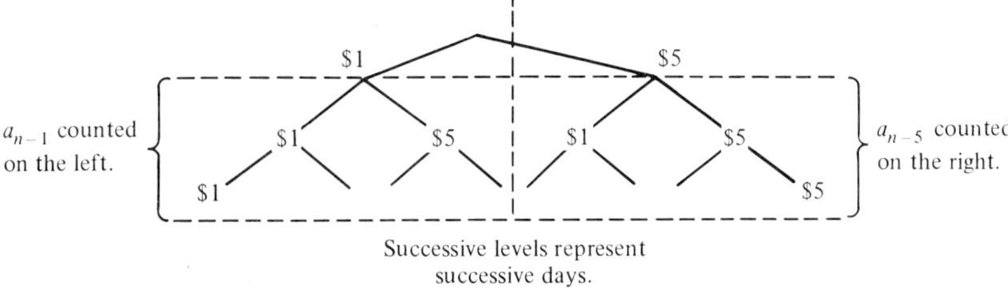

Figure 3.11 Giving a total of $\$a_n$ with \$1 or \$5 bills on successive days.

Notice that $a_1 = a_2 = a_3 = a_4 = 1$, since we cannot give away a \$5 bill unless $n \geqslant 5$. Moreover, $a_0 = 1$, since there is exactly one way to give away nothing. These five initial conditions permit us to calculate $a_5 = a_4 + a_0 = 1 + 1 = 2$, $a_6 = a_5 + a_1 = 2 + 1 = 3$, and so on. □

To find the remaining recurrence models in this section, we again reduce our problem to a smaller version of the original, but we have to take into account how the smaller version is *different* from the original.

Example 5 Suppose that we decide to start putting money away for our retirement 10 years hence. We open a savings account that pays a guaranteed 10 percent interest per year on one's savings deposits, with the intention of depositing \$1000 at the beginning of each year. Let us find the amount a_n that we will have after we have made our nth deposit, where $a_0 = 1000$. (The total amount is called the *amount of an annuity*.)

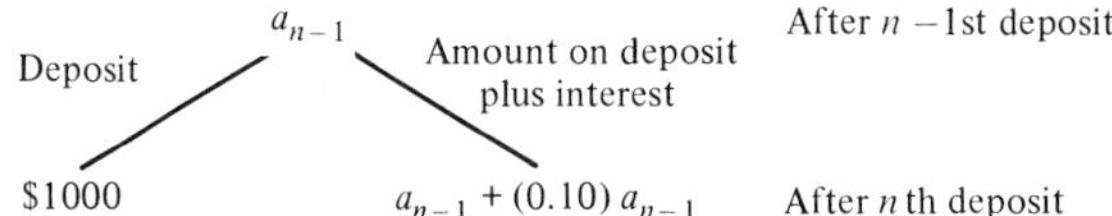

Figure 3.12 The amount a_n after the nth deposit of \$1000 into an annuity yielding 10 percent interest per year is the sum $1000 + (a_{n-1} + (0.10)a_{n-1})$.

In the previous year we will have had a_{n-1} dollars on deposit, and this will be carried over into the present year. But there are two additions: the 10 percent interest, giving us $0.10a_{n-1}$, and the \$1000 deposited at the beginning of the year. Hence

$$a_n = a_{n-1} + 0.10a_{n-1} + 1000 = (1.10)a_{n-1} + 1000$$

for $n \geqslant 1$.

Notice that we can apply Algorithm 3.3 to the result we obtained in Example 5. We will leave as an exercise (Exercise 3.6-18) the question as to what inputs are suitable here in order to find, for example, that

$$a_{10} = \$18{,}531.17 \quad \text{to the nearest cent.}$$ □

The final model of this section is of a geometric nature. However, the method of reduction to a simpler problem is similar to the previous examples.

Example 6 Let us count the number of regions a_n into which the plane is divided by n lines, such that no two lines are parallel and no three intersect at a point.

If we have no lines, clearly $a_0 = 1$, while one line gives us $a_1 = 2$ regions. Now suppose we have a_{n-1} regions from $n-1$ lines, and we draw the nth line. (See Fig. 3.13.)

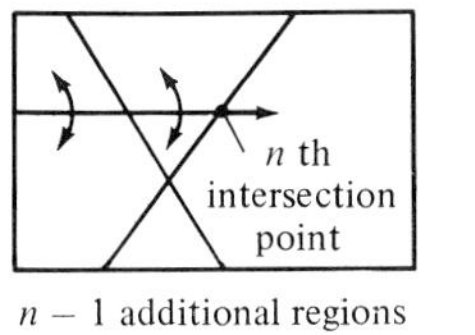

$n-1$ additional regions

(a)

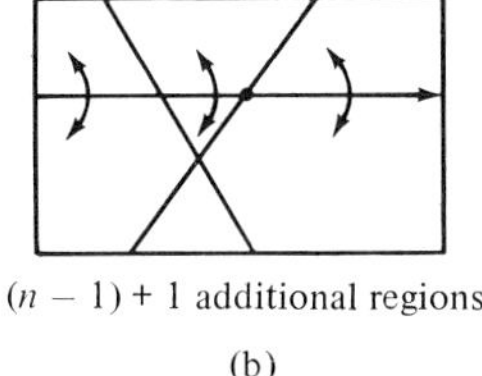

$(n-1)+1$ additional regions

(b)

Figure 3.13 The nth line adds $(n-1)+1$ additional regions to the previous a_{n-1} regions.

This line intersects each of the previously drawn lines in exactly one point, and when it does so, the nth line divides a region in two. This gives $n-1$ additional regions (Fig. 3.13a). But when the nth line continues onward from the $(n-1)$th intersection point, it divides another region into two parts, increasing the number of regions again by 1 (Fig. 3.13b). So we get a total of n more regions, and

$$a_n = a_{n-1} + n, \; n \geqslant 1. \qquad \square$$

Completion Review 3.6

Complete each of the following.

1. Another name for a recursion formula is a ______________ equation.
2. A sequence of numbers that arises in the course of a computation, where each term in the sequence depends upon previous terms, is called a ______________ model.
3. Specification of the first few terms in a recurrence model is called giving ______________ conditions.
4. The Fibonacci numbers are given by the recurrence formula $a_n =$ ______________ with the initial conditions ______________.
5. A geometric progression with common ratio r may be given by the recurrence relation $a_n =$ ______________ and an initial condition specifying the ______________ ______________.
6. An arithmetic progression with common difference d may be given by the recurrence relation ______________ and an initial condition specifying the ______________.
7. A sum s_k of terms $f(k)$ may be given by the recurrence relation ______________.
8. The sum of an annuity a_n after n years of depositing \$1000 at the beginning of each

year at 10 percent interest compounded annually may be given by the recurrence relation____________________.

9. The number of ways a_n to give away a \$1 bill or a \$5 bill on successive days, until we have given away a total of n dollars is determined by the recurrence relation $a_n =$____________________.

10. The number of regions into which the plane is divided by n lines, no two of which are parallel and no three of which intersect in a point, is $a_n =$____________________.

Answers: **1.** difference. **2.** recurrence. **3.** initial. **4.** $a_{n-1} + a_{n-2}$; $a_0 = a_1 = 1$. **5.** ra_{n-1}; term a_0. **6.** $a_{n-1} + d$; term a_0. **7.** $s_k = s_{k-1} + f(k)$. **8.** $(1.10)a_{n-1} + 1000$. **9.** $a_{n-1} + a_{n-5}$. **10.** $a_{n-1} + n$.

Exercises 3.6

In Exercises 1 to 7 use the given recurrence relation and initial condition to calculate the term a_5.

1. $a_n = a_{n-1} + n^2$, $a_0 = 1$.
2. $a_n = a_{n-1} - 2$, $a_0 = -1$.
3. $a_n = 10a_{n-1}$, $a_3 = 1$.
4. $a_n = a_{n-1} + a_{n-2}$, $a_0 = 1$, $a_1 = 3$.
5. $a_n = (n-1)(a_{n-1} + a_{n-2})$, $a_1 = 0$, $a_2 = 1$.
6. $a_n = (1.06)a_{n-1} + 100$, $a_0 = 100$.
7. $a_n = a_{n-1} + n$, $a_1 = 1$.
8. **(a)** How many pairs of rabbits would we have after two years according to the Fibonacci model? **(b)** Is this a reasonable model?
9. Suppose that each mated pair of rabbits produces a litter each month consisting of two additional mated pairs, and that a pair must be over one month old to mate. Find a recurrence model for the number of pairs a_n after n months. How many pairs are there after one year? Two years?
10. If a species of bacteria reproduces by dividing into two organisms once every hour, give a recurrence model for the number a_n of bacteria after n hours. If we start with 10 bacteria, how many will there be after 24 hours?
11. A person has to climb a staircase of n stairs. If the person can climb either one or two stairs with each step, find a recurrence relation giving $a_n =$ the number of ways this person can climb the staircase. (*Hint*: What happens after the first step?)
12. Find a recurrence relation that tells the number of ways a_n to give away n cents in pennies, nickels, dimes, and quarters. How many initial conditions will you need, in general, to determine a_n?
13. Find a recurrence relation for the number of sequences of 0's and 1's of length n with no pair of consecutive 0's.
14. Jane Doe deposits \$P into a savings account at the beginning of each year, and

the account pays $100r$ percent in interest per year. Find a recurrence relation for the amount a_n that Jane will have in her account after she makes her payment at the beginning of the nth year.

15. Find a recurrence relation for the amount a_n outstanding on a \$50,000 mortgage after n months of payments of \$600 made at the beginning of each month if the bank charges 1 percent per month interest on the outstanding debt. How many months remain until the mortgage is fully paid?

16. How many regions a_n are formed in the plane by n circles each pair of which intersect in exactly two points, and such that no three circles intersect in a point. [*Hint*: Each of the $2(n-1)$ arcs on the nth circle divide each of the a_{n-1} arcs on the previous circles in two parts.]

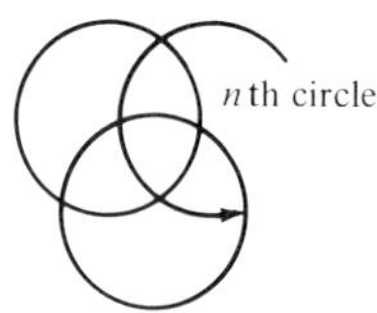

17. The Tower of Hanoi consists of r circular disks placed on an upright peg in decreasing order of sizes. There is another peg onto which the disks are to be transferred so that they are in the same order. The disks must be transferred one at a time, and there is a third peg to enable you to do the transferring so that a larger disk is never placed on a smaller one. Give a recurrence relation for the minimum number of moves a_r required to do the transferring of r disks. (*Hint*: Begin by transferring the $r-1$ smallest disks onto the third peg.)

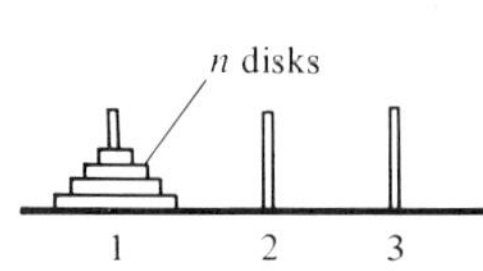

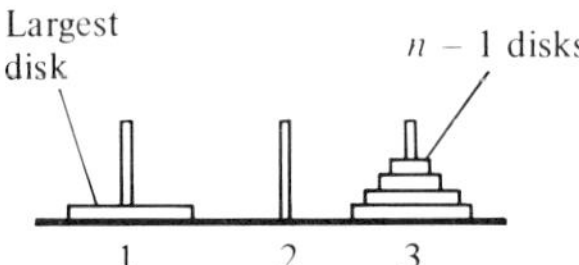

18. Write out an algorithm to find a_{10} if

$$a_n = (1.10)a_{n-1} + 1000, \; n \geqslant 1; \; a_0 = 1000.$$

19. The number of **derangements** of $i = 1, 2, \ldots, n$, that is, the number of ways to permute the i's so that no i is in its natural position, p_i, is given by

$$d_n = (n-1)d_{n-1} + (n-1)d_{n-2} \text{ if } n > 2,$$

$d_1 = 0$ and $d_2 = 1$. Show this. (*Hint*: Use a decision tree to consider the case when there is an $i = 1, 2, \ldots, n-1$ so that i goes into position p_n and n goes into position p_i, and the case when there is no such i. Also see Section 3.4, Exercise 10.)

20. Let F_k denote the kth Fibonacci number. Verify the following.

a) $\sum_{k=0}^{n} F_{2k} = F_{2n+1}$.

b) $\sum_{k=0}^{n} F_{2k+1} = F_{2n+2} - 1$.

c) $\sum_{k=0}^{n} F_k = F_{n+2} - 1$.

d) $\sum_{k=1}^{2n} F_k(-1)^{k+1} = -F_{2n-1}$.

e) $\sum_{k=0}^{n} F_k^2 = F_n F_{n+1}$.

f) $F_k = \frac{1}{\sqrt{5}}\left(\frac{1+\sqrt{5}}{2}\right)^{k+1} - \frac{1}{\sqrt{5}}\left(\frac{1-\sqrt{5}}{2}\right)^{k+1}$.

$\left[\textit{Hint}\text{: Do this one by induction, expanding } \left(\frac{1+\sqrt{5}}{2}\right)^2 \text{ and } \left(\frac{1-\sqrt{5}}{2}\right)^2.\right]$

3.7 CLOSED FORMS AND ANALYSIS OF ALGORITHMS

The idea of recursion is inherent in many algorithms, especially those that reduce a problem to one or more "subproblems" that resemble the original but are reduced in size. We have already applied this idea to certain counting problems in the previous section. In the present section we indicate how recurrence can be used to search through and reorder records. We also introduce an elementary way to obtain more explicit formulas for terms a_n in some recurrence relations.

You will recall that we first used recurrence relations in Example 1 to define or compute things for which we already had an explicit formula. For example, we showed how to find the nth term $a_n = r^{n-1} \cdot a$ of a geometric progression recursively, that is, by means of the formula $a_n = ra_{n-1}$ and the initial condition $a_1 = a$. The original formula $a_n = r^{n-1} \cdot a$ is called the **closed form**, or the **solution**, of the corresponding recurrence relation and initial condition.

We generally left our solutions to counting problems in Section 3.6 in the form of a recurrence relation. But, as in our "annuity problem," Example 5, where we obtained

$$a_n = (1.10)a_{n-1} + 1000 \qquad \text{if } n \geq 1,$$

$$a_0 = 1000,$$

it is sometimes desirable to have the closed-form solution, especially if one is going to perform the calculation of a_n frequently.

Example 1 Let us solve the recurrence relation

$$s_n = r \cdot s_{n-1} + a_0,\ n \geqslant 1;\ s_0 = a_0$$

by *(1) working backward to establish a pattern, (2) guessing at a general formula, and (3) proving our formula by mathematical induction.* What we mean by "working backward" is to write s_{n-1} in terms of s_{n-2}, and then substitute in the formula for s_n; then write s_{n-2} in terms of s_{n-3}, and so on back to s_0. Using this method of *backward substitution,* we get

$$s_n = rs_{n-1} + a_0 = r(\overbrace{rs_{n-2} + a_0}^{s_{n-1}}) + a_0 = r^2 s_{n-2} + ra_0 + a_0.$$

Continuing a little further we get

$$s_n = r^2(\overbrace{rs_{n-3} + a_0}^{s_{n-2}}) + ra_0 + a_0 = r^3 s_{n-3} + r^2 a_0 + ra_0 + a_0.$$

The pattern seems to lead us to *the sum of a geometric progression* with a first term of a_0 and common ratio r. In other words, we seem to get

$$s_n = r^n a_0 + r^{n-1} a_0 + r^{n-2} a_0 + \cdots + a_0,$$

which has the closed form

$$s_n = \frac{a_0 - a_0 r^{n+1}}{1 - r} \qquad \text{[See Example 2(d), Section 1.3.]}$$

Let us prove this formula by induction: for $n = 0$, clearly $s_0 = a_0$. Assume the formula for s_{n-1}. We can then write

$$s_n = rs_{n-1} + a_0 = r \cdot \frac{a_0 - a_0 r^n}{1 - r} + a_0 = \frac{ra_0 - a_0 r^{n+1} + (1 - r)a_0}{1 - r}$$

$$= \frac{\cancel{ra_0} - a_0 r^{n+1} + a_0 - \cancel{ra_0}}{1 - r} = \frac{a_0 - a_0 r^{n+1}}{1 - r}$$

as claimed for s_n, completing the induction. □

Example 2 Applying the formula we derived in the last example to our annuity problem with $a_n = s_n$, $n = 10$, and $r = 1.10$, we get

$$a_n = \frac{1000 - 1000(1.10)^{n+1}}{1 - (1.10)}$$

and

$$a_{10} = \frac{1000 - 1000(1.10)^{11}}{1 - (1.10)} \doteq \$18{,}531.17,$$

just as we found in Example 5, Section 3.6. □

It is not always possible to obtain the closed form of a recurrence relation, but we present more advanced methods for obtaining these solutions in Section 3.8 and Chapter 8. For now we will use the method of backward substitution to analyze certain algorithms having to do with the arranging of and searching through files.

Sorting and Searching

Let us assume that we have a file consisting of n records in fixed-storage locations $r_1, r_2, \ldots, r_n$. Suppose that the record in r_i contains a single identifying item, k_i, such as a name or a number, and that each record can be moved about from location to location. The k_i's, called **keys**, can be ordered in a natural way, either alphabetically or numerically, so that we may refer to a "natural ordering" $k_1 \leqslant k_2 \leqslant \ldots \leqslant k_n$ of the keys in the records without ambiguity.

In Fig. 3.14 we can see part of a file in which the records containing keys k_1 = "Allan," k_2 = "Bob," k_3 = "Carol", . . . , and k_n = "Zee" have apparently been arranged in "ascending" order. Observe that if we interchanged the records in locations r_1 and r_2, for example, we would then have k_1 = Bob and k_2 = Allan.

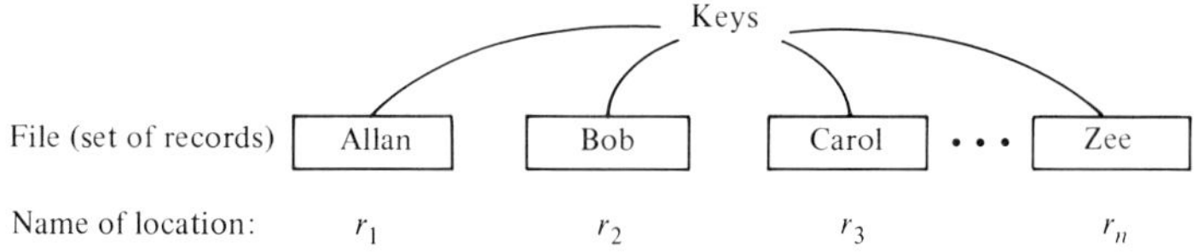

Figure 3.14 Key k_i is found in record (or register) r_i.

Definition 3.5 A (a) **sorting algorithm** is one that tells how to (re)arrange the records in fixed locations $r_1, r_2, \ldots, r_n$ according to the numerical or alphabetical order of their keys. A (b) **search algorithm** tells how to locate a piece of data or key k, subject to certain conditions, from among the records in fixed locations $r_1, r_2, \ldots, r_n$.

One of the easiest sorting algorithms to describe (though not the most efficient) is often called the "bubble-sort" method, because the record with the "largest" key "bubbles up" to the top location, r_n, through a process of comparisons and exchanges.

For example, if we begin with the sequence of numbers 6, 4, 2, then comparing and exchanging the first two terms gives us 4, 6, 2; comparing and exchanging the second and third terms in the last sequence yields 4, 2, 6; finally, a second round of comparing and exchanging the first two terms in the last sequence puts all the terms in ascending order and gives us 2, 4, 6.

Here is a formal version of the bubble-sort algorithm.

Algorithm 3.4 ***Bubble-Sort Algorithm*** Suppose that we are given records in locations $r_1, r_2, \ldots, r_n$. To sort these records in ascending order of their keys $k_1, k_2, \ldots, k_n$;

1. [Boundary loop.] Repeat steps 2 and 3 for $b = n, n-1, n-2, \ldots, 2$. (Each iteration puts a record into the appropriate location r_b. Records in $r_{b+1}, r_{b+2}, \ldots$ are unaffected.)
2. [Comparison loop.] Repeat step 3 for $i = 1, 2, \ldots, b-1$. (Notice that the i's increase by 1 on each pass. Moreover, i goes from 1 to $b-1$ for each value of b.)
    3. [Comparison–exchange.] If $k_i > k_{i+1}$, then interchange the records in locations r_i and r_{i+1}.
4. [Done.] Output the sorted file.
5. End of Algorithm 3.4.

Step 3, the "comparison–exchange," will be denoted by the symbol $c(r_i, r_{i+1})$. More generally, $c(r_a, r_b)$ will denote instructions to put the record with the smaller of keys k_a and k_b into location r_a and the record with the larger key into location r_b.

Notice how the problem is reduced in size by step 2. After the first pass, instead of sorting n records, we only have to sort $n-1$ records. We then sort the $n-1$ records in essentially the same way we sorted the previous n records, namely, by means of the comparison steps $c(r_i, r_{i+1})$. Here is a small-scale example to help us clarify these ideas.

Example 3 Let us sort (in ascending order) the integers 1 through 4 when they are arranged 4, 3, 2, 1:

$(b=4; j=1)$	$(b=4; j=2)$	$(b=4; j=3)$	*Result*
4, 3, 2, 1	3, 4, 2, 1	3, 2, 4, 1	3, 2, 1, 4
$(b=3; j=1)$	$(b=3; j=2)$	*Result*	
3, 2, 1, 4	2, 3, 1, 4	2, 1, 3, 4	
$(b=2; j=1)$	*Final result*		
2, 1, 3, 4	1, 2, 3, 4		

□

We can describe the sequence of steps in the first row by "concatenating" the comparison–exchange steps (or just "comparisons"), $c(r_j, r_{j+1})$, that is, applying one after the other. Thus the first line in Example 3 is

$$c(r_1, r_2)c(r_2, r_3)c(r_3, r_4).$$

Using the idea of concatenation, we can give a very succinct description of the bubble sort $b(r_1, r_2, \ldots, r_n)$ of n records by means of Theorem 3.15.

Theorem 3.15 If "$b(r_1, r_2, \ldots, r_n)$" denotes the bubble sort of n records $r_1, r_2, \ldots, r_n$, then $b(r_1, r_2) = c(r_1, r_2)$, and, for $n \geqslant 3$:

$$b(r_1, r_2, \ldots, r_n) = c(r_1, r_2)c(r_2, r_3) \ldots c(r_{n-1}, r_n)b(r_1, \ldots, r_{n-1}).$$

While Theorem 3.15 lacks the detail of Algorithm 3.4, it displays the recursive nature of that algorithm more explicitly. In fact if we *let a_n denote the number of comparison steps needed to sort n records with the bubble sort*, then Theorem 3.15 immediately gives us the recursive relation

$$a_n = (n-1) + a_{n-1}, \text{ if } n > 2 \qquad (*)$$

$$a_2 = 1.$$

In fact we can go further and state Theorem 3.16.

Theorem 3.16 The bubble-sort algorithm requires $n(n-1)/2$ comparison steps to sort n records.

Proof: If we set $n \leftarrow n-1$ in the recursive relation (*), we obtain

$$a_{n-1} = a_{n-2} + (n-2).$$

Substitute this in (*), and we obtain

$$a_n = a_{n-1} + (n-1) = (a_{n-2} + (n-2)) + (n-1) = a_{n-2} + (n-1) + (n-2).$$

Similarly,

$$a_n = a_{n-3} + (n-1) + (n-2) + (n-3), \ldots, \text{ and finally}$$

$$a_n = (n-1) + (n-2) + (n-3) + \cdots + 1.$$

By Example 2(c), Section 1.3, the sum of the first $n-1$ positive integers can be written $(n-1)n/2$. Hence the bubble sort requires

$$a_n = n(n-1)/2$$

comparison steps to sort n records, $n \geqslant 1$.

We will leave the induction proof that our formula is correct as an exercise. (See Exercise 3.7-10.) ■

We can use Theorem 3.15 to compare the efficiency of our bubble-sort algorithm with other sorting algorithms, provided we know the number of comparison steps in the other algorithms.

Example 4 The Bose-Nelson sorting algorithm is one of a class of "divide-and-conquer" methods that sorts a file by (a) dividing the file into two halves, (b) sorting each half, and (c) merging the two sorted files into one sorted file. (See Exercises 3.7-14 and 15 for a more explicit description.) It can be shown that this procedure requires no more than $3^k - 2^k$ comparison steps to sort $n = 2^k$ records. How does this compare with the bubble-sort method?

If we have $n = 2^7 = 128$ records to sort, then the bubble-sort method requires

$$\frac{2^7(2^7-1)}{2} = 8128 \text{ comparisons,}$$

while the Bose-Nelson procedure needs at most

$$3^7 - 2^7 = 2059 \text{ comparisons,}$$

or a little more than one-fourth the number required by the bubble sort. The *ratio* in the number of comparisons that the two methods require is

$$\frac{3^k - 2^k}{\frac{1}{2}(2^k(2^k-1))} = \frac{\text{Comparisons needed by Bose-Nelson}}{\text{Comparisons needed by bubble sort}},$$

which, for large k, is approximately equal to

$$\frac{2}{1}\left(\frac{3^k - 2^k}{4^k}\right) = 2((3/4)^k - (2/4)^k).$$

Since the fractions $(3/4)^k$ and $(2/4)^k$ get closer and closer to zero as k increases, it follows that the bubble sort suffers more and more in comparison with the Bose-Nelson procedure as the number of records gets larger. □

Let us now consider the problem of *searching n records for the largest and smallest keys* k_{max} and k_{min}. The concatenation of $n-1$ comparison steps

$$c(r_1, r_2)c(r_2, r_3) \ldots c(r_{n-1}, r_n)$$

that we used in the bubble sort will find k_{max} and put its record into location r_n. The concatenation of $n-2$ comparison steps

$$c(r_{n-2}, r_{n-1})c(r_{n-3}, r_{n-2}) \ldots c(r_1, r_2)$$

will find k_{min} among the records in locations $r_1, r_2, \ldots, r_{n-1}$ and put its record into location r_1. (See Exercise 3.7-9.) Hence we can state Theorem 3.17.

Theorem 3.17 The algorithm described by the concatenation

$$c(r_1, r_2)c(r_2, r_3) \ldots c(r_{n-1}, r_n)c(r_{n-2}, r_{n-1})c(r_{n-3}, r_{n-2}) \ldots c(r_1, r_2)$$

requires $(n-1)+(n-2)=2n-3$ comparison steps to put the record of k_{max} into location r_n and the record of k_{min} into location r_1.

We can give a divide-and-conquer procedure that does somewhat better than this. The procedure, which we define by the letter M, divides $n=2^k$ records into two half-files, each of size $n/2=2^{k-1}$. It then finds the largest and smallest keys in each half. Finally it chooses the largest and smallest from among these four keys.

For example, given the sequence 4, 3, 2, 1, we first determine the largest and smallest numbers in *each* of the half sequences 4, 3 and 2, 1. We then find that 4 is the larger of the maximums (4 and 2) and that 1 is the smaller of the minimums (3 and 1).

Theorem 3.18 The recursive procedure M defined by $M(r_1, r_2)=c(r_1, r_2)$, and

$$M(r_1, r_2, \ldots, r_n)=M(r_1, \ldots, r_{n/2})M(r_{(n/2)+1}, \ldots, r_n)c(r_1, r_{(n/2)+1})c(r_{n/2}, r_n)$$

requires $\frac{3}{2}n-2$ comparison steps to put k_{min} in r_1 and k_{max} in r_n, where $n=2^k$.

Proof: The required number of steps a_k is given by the recurrence relation

$$a_k = 2a_{k-1} + 2 \text{ if } k > 1;$$

$$a_1 = 1.$$

(We could have written $a_n = a_{n/2} + 2$, but it is easier to work with k.) By backward substitution we obtain

$$\begin{aligned} a_k &= 2a_{k-1} + 2 = 2(2a_{k-2} + 2) + 2 = 2^2 a_{k-2} + 2^2 + 2 \\ &= 2^2(2a_{k-3} + 2) + 2^2 + 2 = 2^3 a_{k-3} + 2^3 + 2^2 + 2 = \dots \\ &\dots\dots\dots\dots\dots\dots \\ &= 2^{k-1}a_1 + 2^{k-1} + 2^{k-2} + \cdots + 2^2 + 2 \\ &= 2^{k-1} + (2^{k-1} + 2^{k-2} + \cdots + 2^2 + 2) \\ &= 2^{k-1} + (2^k - 2), \end{aligned}$$

where we summed the geometric progression in parentheses to obtain the last expression. Since $n = 2^k$, we get

$$a_k = \tfrac{1}{2}n + n - 2 = \tfrac{3}{2}n - 2,$$

as required. ■

We can now see that the procedure described in Theorem 3.18 requires about 3/4 as many steps, in general, as the one we gave in Theorem 3.17, since $(\frac{3}{2}n - 2)/(2n - 3)$ gets closer and closer to $\frac{3}{2}/2 = \frac{3}{4}$ as n increases.

Completion Review 3.7

Complete each of the following.

1. A function of n that gives the nth term in a recurrence relation is called a ________________ of the relation.
2. The method of backward substitution gives the closed-form solution of $s_n = rs_{n-1} + a_0$, $s_0 = a_0$, as $s_n =$ ________________.
3. The bubble sort of n records may be given recursively as $b(r_1, r_2, \ldots, r_n) =$ ________________, where each $c(r_i, r_{i+1})$ is called a ________________.
4. The number a_n of comparison–exchange steps needed by the bubble sort of n records is given by the recursion formula ________________ and in closed form by ________________.
5. The Bose-Nelson sorting procedure of $n = 2^k$ records requires ________________ comparison–exchanges.

6. The modified bubble-sort method for finding the maximum and minimum records in a file of n records may be described by________________. This requires ________________ comparison–exchanges.

7. The divide-and-conquer procedure for finding the maximum and minimum records in a file of $n=2^k$ records requires________________ comparison–exchanges.

Answers: **1.** closed-form solution. **2.** $[a_0-a_0r^{n+1}]/[1-r]$. **3.** $c(r_1, r_2)c(r_2, r_3)\ldots c(r_{n-1}, r_n)$; $b(r_1, r_2, \ldots, r_{n-1})$; comparison–exchange. **4.** $a_n=a_{n-1}+n-1$, $a_2=1$; $n(n-1)/2$. **5.** 3^k-2^k. **6.** **6.** $c(r_1, r_2)c(r_{n-1}, r_n)c(r_{n-2}, r_{n-1})\ldots c(r_1, r_2)$; $2n-3$. **7.** $(3/2)n-2$.

Exercises 3.7

1. Solve each of the following recurrence relations in closed form.
  a) $a_n=n\cdot a_{n-1}$ if $n\geqslant 1$, and $a_0=1$.
  b) $a_n=3a_{n-1}$ if $n\geqslant 1$, and $a_1=\frac{1}{2}$.
  c) $a_n=a_{n-1}+2$, $n\geqslant 1$, and $a_0=1$.
  d) $a_{n,r}=a_{n-1,r}+a_{n-1,r-1}$ for $r, n\geqslant 0$, $n\geqslant r$; $a_{n,n}=a_{n,0}=1$.

2. Use Example 1 to give a closed-form solution to $s_n=\frac{1}{2}s_{n-1}+10$; if $s_0=10$.

3. Give a closed-form solution for $a_n=a_{n-1}+n^2$, $n>1$, and $a_1=1$. (*Hint*: Exercises 2.1-5 and 6.)

4. Give a closed-form solution to $a_n=a_{n-1}+n^3$, $n>1$, and $a_1=1$.

5. Find a closed-form solution to Example 6, Section 3.6, which asked for the number of regions into which the plane is divided by n lines, no two of which are parallel and no three of which intersect at a point.

6. Use backward substitution to find the closed-form solution to $a_n=(1.01)a_{n-1}-600$ if $n>0$, and $a_0=50{,}000$. (This comes from Exercise 3.6-15.)

7. Find a closed-form solution to the Tower of Hanoi problem, Exercise 3.6-17.

8. Show how the bubble-sort algorithm would sort the sequence of numbers 2, 1, 4, 3, 6, 5, 8 into ascending order.

9. How many comparisons would Algorithm 3.4 require to sort 256 records? How does this compare with an algorithm that needs 3^k-2^k comparisons to sort 2^k records?

10. Prove by induction on n that $a_n=n(n-1)/2$ satisfies $a_n=a_{n-1}+(n-1)$, $a_1=0$.

11. Write the search algorithm given in Theorem 3.17 in a form like that of Algorithm 3.4.

12. Test your algorithm (Exercise 11) on the sequence of numbers 3, 6, 7, 5.

13. **a)** Write out what $M(r_1, r_2, r_3, r_4)$ would be in Theorem 3.18, in terms of comparison steps only.
  b) Do the same for $M(r_1, \ldots, r_8)$.
  c) Use $M(r_1, r_2, r_3, r_4)$ to search the sequence 3, 6, 7, 5 for $k_{\max}$ and $k_{\min}$.

14. (Merge) We can define an algorithm $Mg(r_1, \ldots, r_n; r_{n+1}, \ldots, r_{2n})$, $n=2^p$, that sorts the keys in the file $\{r_1, r_2, \ldots, r_{2n}\}$ given that the files $\{r_1, \ldots, r_n\}$ and

$\{r_{n+1}, \ldots, r_{2n}\}$ have already been sorted, by

$$Mg(r_1, \ldots, r_{\frac{n}{2}}; r_{n+1}, \ldots, r_{\frac{3}{2}n})Mg(r_{\frac{n}{2}+1}, \ldots, r_n; r_{\frac{3}{2}n+1}, \ldots, r_{2n}) \ldots$$
$$Mg(r_{\frac{n}{2}+1}, \ldots, r_n; r_{n+1}, \ldots, r_{\frac{3}{2}n}),$$

where $Mg(r_1; r_2) = c(r_1, r_2)$.

a) What is $Mg(r_1, r_2; r_3, r_4)$ in terms of comparison steps?
b) What is $Mg(r_1, .\,;.\,, r_8)$?
c) Find a recurrence relation for the number a_k of comparison steps that Mg requires to merge $n = 2^k$ records.
d) What is its closed form?

15. (Bose-Nelson) The Bose-Nelson procedure for sorting $2n = 2^{k+1}$ records is defined by

$$B(r_1, r_2, \ldots, r_{2n})$$
$$= B(r_1, r_2, \ldots, r_n)B(r_{n+1}, \ldots, r_{2n})Mg(r_1, \ldots, r_n; r_{n+1}, \ldots, r_{2n})$$

and

$$B(r_1, r_2) = c(r_1, r_2),$$

where Mg is defined in Exercise 14.

a) Write out $B(r_1, \ldots, r_4)$ in terms of comparison steps.
b) Write out $B(r_1, \ldots, r_8)$ in terms of comparison steps.
c) Write a recurrence relation for b_k the number of comparison steps B needs to sort $n = 2^k$ records. (Use Exercise 14.)
d) Prove by induction that $b_n = 3^k - 2^k$ if $k \geqslant 1$.

3.8 DIVIDE-AND-CONQUER RELATIONS

Many, if not most, of the methods that can be used to find closed-form solutions to recurrence relations are beyond the scope of this book. (Some additional methods are included in Chapter 8, however.) But there is one class of recurrence relation that frequently arises in the context of computer science, especially when one uses a divide-and-conquer strategy to design an algorithm. We have seen at least one application of this strategy applied in the previous section to the problem of finding the maximum and minimum keys in a file of $n = 2^k$ records. Recall that our procedure required a_k comparison steps to search $n = 2^k$ records in a file, where

$$a_k = 2a_{k-1} + 2 \qquad \text{if } k > 1;$$
$$a_k = 1 \qquad \text{if } k = 1.$$

We were able to find a closed-form solution by adroitly avoiding direct reference to the n in our recurrence relation, which was originally given as

$$a_n = 2a_{n/2} + 2 \qquad \text{if } n > 2,\ n = 2^k;$$

$$a_n = 1 \qquad \text{if } n = 2.$$

This is an example of what are sometimes called "divide-and-conquer" recurrence relations, which are defined in Definition 3.6.

Definition 3.6 A recurrence relation of the form

$$a_n = ka_{n/d} + f(n),$$

where k and d are constants and f is a function of n, will be called a **divide-and-conquer relation**.

Thus $a_n = 2a_{n/2} + 2$ is a divide-and-conquer relation in which $k = 2$, $d = 2$, and $f(n) = 2$ for every positive integer $n > 2$.

The general appearance of the closed-form solutions to divide-and-conquer relations (or at least their general behavior) are well known. Some of these are given in Table 3.1. Indeed one can find the closed-form solution to some divide-and-conquer relations with the aid of this table.

Table 3.1

	k and d	$f(n)$	$a_n = ka_{n/d} + f(n)$
1.	$k = d$	c	$A \cdot n + B$
2.	$k = 1$	c	$c(\log_d n + 1)$, if $a_1 = c$.
3.	$k = d$	$c \cdot n + B$	$O(n \cdot \log_d n)$

Notes: (1) A, B, c, k, and d are constants. (2) $a_n = O(F(n))$, read "big oh of $F(n)$," means that a_n does not grow faster than $F(n)$ for very large values of n. (More precisely, there exist constants K and $N > 0$ such that $a_n < K|F(n)|$ if $n > N$.)

Example 1 Find a closed-form solution to the recurrence relation $a_n = 2a_{n/2} + 2$, where n is a power of 2 for $n > 2$ and $a_2 = 1$, using Table 3.1.

To obtain a solution we observe that form 1 of Table 3.1 is applicable, since $k = d = 2$ and $f(n) = 2$, a constant. Hence there are constants A and B such that

$a_n = A \cdot n + B$. Therefore, $2a_{n/2} + 2 = 2(A \cdot n/2 + B) + 2$, and we may write

$$An + B =$$
$$2a_{n/2} + 2 =$$
$$2(A \cdot n/2 + B) + 2 =$$
$$An + 2B + 2,$$

or

$$An + B = An + 2B + 2.$$

Solving for B, we obtain $B = -2$. The initial condition $a_2 = 1$ gives us

$$1 = A \cdot 2 + (-2).$$

Hence $A = 3/2$, and so $a_n = 3n/2 - 2$, just as we found in the last example of Section 3.7. □

Example 2 Let us illustrate form 2 of Table 3.1 with the recurrence relation

$$a_n = a_{n/2} + 1, \quad \text{where } a_1 = 1.$$

This recurrence relation satisfies form 2 of our table with $k = 1$, $d = 2$, and $f(n) = 1$ $(= c)$. Hence our solution should be of the form

$$a_n = 1(\log_2 n + 1).$$ □

Before we consider the third form in Table 3.1, let us briefly consider the meaning of the "big-oh" notation, $G(n) = O(F(n))$. This notation is used when we want to estimate the size of one function, $G(n)$, in terms of another function, $F(n)$, whose behavior is better understood. Typical choices for the function $F(n)$ are $\log n$, n, $n \log n$, n^2, $n^3, \ldots, n^k, \ldots, e^n$, $n!$, and n^n, each of which is, in turn, a function whose rate of increase is significantly greater than the preceding function. Thus $f(n) = n \log n$ increases much less rapidly than $g(n) = n^2$ when n is large, for example. Consequently algorithms that require approximately $n \log n$ comparison steps are more desirable than those that require, say, n^2 steps for large values of n, if we ignore other considerations.

The following example will illustrate how one might determine that one function is "big-oh" of another.

Example 3 Recall that our modified bubble-sort procedure

$$c(r_1, r_2)c(r_2, r_3) \ldots c(r_{n-1}, r_n)c(r_{n-2}, r_{n-1})c(r_{n-3}, r_{n-2}) \ldots c(r_1, r_2)$$

of Section 3.7 required $2n-3$ comparison steps to find the maximum and minimum keys among n records. Letting $F(n)=2n-3$ and $G(n)=3n/2-2$, show that $G(n)=0(F(n))$.

To solve this problem, it is sufficient to observe that $\frac{3}{2}n-2<2n-3$ for all $n>2$. We can obtain a sharper result, however, if we notice that G/F is approximately equal to $(3n/2)/(2n)=3/4$. Thus for large values of n, $G(n)$ is approximately equal to 3/4 times $F(n)$. □

Notice that in the previous example $G(n)/F(n)$ is never exactly equal to 3/4. But even for $n=100$, we obtain $G(100)/F(100)=0.751269$ to the nearest millionth, and the ratio continues to approach 0.75 as n increases.

We now give an example of a divide-and-conquer relation whose behavior can be described with the big-oh notation. Our relation is obtained by counting the number of comparison steps required by a certain divide-and-conquer strategy for sorting a sequence of numbers, for example, the sequence

1, 3, 5, 7, 2, 4, 6, 8.

Our plan is (1) to divide the sequence into two subsequences of equal length, sort each of the subsequences, and then (2) to merge the two sorted subsequences into one big sorted sequence.

To simplify matters we have arranged things so that the first four numbers 1, 3, 5, 7 and the last four numbers 2, 4, 6, 8 are already sorted. To merge these sorted subsequences, one can repeatedly compare the first two elements of each and then extract and list the minimum of these two elements. Thus comparing 1 and 2, we extract and list 1, leaving us with the sequences 3, 5, 7 and 2, 4, 6, 8. In fact repeating this process leaves us with the sequence pairs (3, 5, 7; 2, 4, 6, 8), (3, 5, 7; 4, 6, 8), (5, 7; 4, 6, 8), (5, 7; 6, 8), (7; 6, 8), (7; 8), (–; 8), and (–; –) when the numbers 1, 2, 3, 4, 5, 6, 7, and 8 are extracted and listed in turn, giving us a completely sorted sequence.

Had we started with the sequence 3, 1, 7, 5, 4, 2, 8, 6, we would see that the first four and the last four numbers are not sorted. We can, however, apply the ideas of the last paragraph to each of the subsequences 3, 1, 7, 5 and 4, 2, 8, 6 separately. Taking the subsequence 3, 1, 7, 5, for example, we can sort each of its subsequences 3, 1 and 7, 5 with a single comparison–exchange step, as in the previous section. Then we can merge the sorted sequences 1, 3 and 5, 7 as in the last paragraph to obtain 1, 3, 5, 7. Similarly the two subsequences 4, 2 and 8, 6 can be sorted individually and then merged to produce 2, 4, 6, 8. This brings us back to where we began in the previous paragraph, that is, to the step in which we had to merge the sorted sequences 1, 3, 5, 7 and 2, 4, 6, 8.

The procedure we have illustrated can be generalized for $n=2^k$ numbers. We will define our procedure recursively, as Definition 3.7.

Definition 3.7 Given a sequence of n numbers $r_1, \ldots, r_n$, where n is an integral power of 2, we define **merge–sort (mgs)** by

$$\begin{aligned}&\text{mgs}(r_1, r_2, \ldots, r_n) =\\&\text{mgs}(r_1, r_2, \ldots, r_{n/2}) \ldots\\&\text{mgs}(r_{(n/2)+1}, \ldots, r_n) \ldots\\&m(r_1, r_2, \ldots, r_{n/2}; r_{(n/2)+1}, r_{(n/2)+2}, \ldots, r_n),\end{aligned}$$

where the procedure $m =$ **merge** repeatedly extracts and lists the minimum of the first two numbers in each of the two subsequences sorted by mgs, thereby merging these subsequences into one sorted sequence [$\text{mgs}(r_1, r_2) = c(r_1, r_2)$].

By our previous examples we see that our merge–sort procedure works by first getting sorted pairs, merging these into sorted quadruples, and so on. We can analyze this sorting procedure as we did in the previous section in terms of the number of comparison–exchange or comparison–extraction steps required to sort a sequence of size n. In fact we may state the following result (Theorem 3.19).

Theorem 3.19 A merge–sort of a sequence of size $n = 2^k$ requires $0(n \log_2 n)$ comparison steps in the worst case.

Proof: If a_n denotes the required number of comparison steps, then the definition of merge–sort immediately gives us the recurrence relation

$$a_n = 2a_{n/2} + (n-1),$$

for $n \geqslant 4$, $a_2 = 1$, since the merging procedure, m, requires at most $n-1$ comparison steps, at which point or sooner one of the two subsequences of size $n/2$ must be entirely exhausted. Our recurrence relation is of the form found on line 3 of Table 3.1, with $k = d = 2$, and $f(n) = n - 1$. (Hence $c = 1$ and $B = -1$.) According to our table the closed-form solution is $0(n \log_2 n)$, as asserted in the theorem. ■

Notice that we did not say that $a_n = n \log_2 n$ in the last example. But the two quantities are of the same order of magnitude for large values of n. The relationship between their sizes is illustrated in the next example.

Example 4 Compare the values obtained from $a_n = 2a_{n/2} + (n-1)$ with $n \log_2 n$ for $n = 2, 4, 8, 16, 32, 64, 128$, and 256.

Table 3.2

n	$F = 2a_{n/2} + (n-1)$	$G = n \log_2 n$	G/F
2	1	2	2.00
4	5	8	1.60
8	17	24	1.41
16	49	64	1.31
32	129	160	1.24
64	321	384	1.20
128	769	896	1.16
256	1793	2048	1.14

□

We remark that the ratio G/F in our last example very slowly gets closer and closer to 1 as n gets larger and larger. (See *The Art of Computer Programming*, volume 3, by D. E. Knuth, Addison-Wesley, 1973.) Moreover, it can be shown that any sorting procedure that relies on comparisons must take at least $0(n \log_2 n)$ comparisons to sort n records or numbers in the worst case. (See *Discrete Mathematics in Computer Science*, by D. F. Stanat and D. F. McAllister, Prentice-Hall, 1977, for example, for a proof of this and for the proofs of the facts in Table 3.1.) Consequently one may say that, in this sense, merge–sort is an optimal sorting procedure.

Completion Review 3.8

Complete each of the following.

1. A divide-and-conquer relation is one of the form $a_n =$ ________________.
2. Merge–sort is a divide-and-conquer sorting procedure of $n = 2^k$ records that may be defined by $\text{mgs}(r_1, \ldots, r_n) =$ ________________.
3. A recursive relation for a_n, the number of comparisons needed by merge–sort to sort n records, is ________________. The closed-form solution has behavior described by ________________, where $0(f(x))$ means ________________ when x gets large.

Answers: **1.** $ka_{n/k} + f(k)$. **2.** $\text{mgs}(r_1, \ldots, r_{n/2})\text{mgs}(r_{(n/2)+1}, \ldots, r_n)m(r_1, \ldots, r_{n/2}; r_{(n/2)+1}, \ldots, r_n)$. **3.** $a_n = 2a_{n/2} + n - 1$, $a_2 = 1$; $a_n = 0(n \log_2 n)$; grows no faster than $f(x)$.

Exercises 3.8

Solve Exercises 1 and 2 by use of Table 3.1, line 1.

1. $a_n = 2a_{n/2} + 3$, n a power of 2, $a_2 = 1$.
2. $a_n = 3a_{n/3} + 2$, n a power of 3, $a_3 = 1$.

For Exercises 3 to 7 find the form of the recurrence relation in Table 3.1 and use this form to describe the behavior of the recurrence relation.

3. $a_n = a_{n/4} + 1$.
4. $a_n = 4a_{n/4} + 1$.
5. $a_n = 4a_{n/4} + n$.
6. $a_n = 2a_{n/2} + n$.
7. $a_n = 4a_{n/4} + n + 1$.
8. Calculate a_{256} for Exercises 3 to 5 if $a_1 = 1$ and compare with the result obtained from Table 3.1.
9. Use merge–sort to arrange the following sequence in ascending order: 16, 15, 14, 13, 12, . . . , 3, 2, 1. How many comparisons were actually required?
10. Show that the "merge" part of merge–sort can merge two sorted lists of length m and n into a single sorted list using no more than $m + n - 1$ comparisons.
11. What is meant by the statement "the bubble sort is a $0(n^2)$ algorithm" on n records. Justify this statement.
12. In a tennis tournament involving $n = 2^k$ players, all the winners from one round move up to the next round, until a single winner is left. Find a recurrence relation for the number of rounds required to determine a winner, and show that this recurrence relation yields $a_n = \log_2 n$.

COMPUTER PROGRAMMING EXERCISES

(Also see the appendix and Programs A10 to A13.)

3.1. Write a program that calculates $n!$ by implementing Algorithm 3.1.
3.2. Write a program that calculates $P(n, r)$, that is, the number of permutations of n things taken r at a time.
3.3. Write a program that finds all the permutations of the sequence of letters ABCDEF taken r at a time, for any $r = 1, 2, 3, 4, 5$, and 6.
3.4. Write a program that finds $C(n, r)$, that is, the number of combinations of n things taken r at a time, by means of Algorithm 3.2.
3.5. Write a program that generates Pascal's triangle for $n = 20$.
3.6. Write a program that expands $(x + y)^n$ for an arbitrary positive integer n.
3.7. Write a program that will calculate the probability of a randomly selected set of n people having at least two people with the same birthday, as in Example 3, Section 3.5.
3.8. Write a program that will calculate the nth Fibonacci number, where $a_n = a_{n-1} + a_{n-2}$, and $a_0 = a_1 = 1$.

3.9. Write a program that calculates the number of ways, a_n, to give away \$100 by giving away a \$1 bill or a \$5 bill on successive days, if $a_n = a_{n-1} + a_{n-5}$, and $a_0 = a_1 = a_2 = a_3 = a_4 = 1$.

3.10. Write a recursive program to calculate the number of moves, a_n, needed to solve the Tower of Hanoi problem (Exercise 3.6-17), if $a_n = 2a_{n-1} + 1$, $a_0 = 1$, and $n = 100$. Compare the time it takes the computer to compute this recursively with a computation using the closed-form solution $a_n = 2^n - 1$.

3.11. Write a program that implements the bubble-sort algorithm, Algorithm 3.4, Section 3.7, and sorts the sequence 256, 255, 254, . . . , 3, 2, 1.

3.12. Write a recursive program that implements the sorting procedure "merge–sort," Definition 3.7, Section 3.7, and sorts the sequence 256, 255, 254, . . . , 3, 2, 1.

3.13. Write a program that finds the minimum and maximum number in a sequence using the modified bubble-sort algorithm defined in Theorem 3.17.

3.14. Write a recursive program that finds the minimum and maximum number in a sequence using the "divide-and-conquer" algorithm defined in Theorem 3.18.

3.15. Write a program that will extend Table 3.2 through $n = 2^{20}$.

3.16. Write a program that will print a table comparing the values of the functions $\log_2 n$, n, $n \log_2 n$, n^2, 2^n, $n!$, and n^n for $n = 1, 2, 5, 10, 15, 20, 25$.

CHAPTER 4

INTRODUCTION TO GRAPH THEORY

The word "graph" often brings to mind coordinate systems and plotting points that satisfy equations. In the next three chapters, however, we will be more concerned with objects that are closely related to the tree diagrams we introduced when we studied counting and recursion. Indeed the widespread applications of tree diagrams and other graphs have made graph theory an area of great interest in recent years.

Pictures, and geometric descriptions in general, play a major role in graph theory. Visual intuition often enables us to discover principles that seem otherwise to elude us. But pictures only supplement language, and one must be able to communicate the meaning of a mathematical idea in words. Therefore, a large part of any introduction to graph theory necessarily includes many definitions. The sooner the reader learns these definitions, the sooner he or she will be able to apply graph theoretic concepts and methods in a meaningful way.

4.1 GRAPHS AND WALKS

The origin of the subject of graph theory can be traced to the Swiss mathematician Leonard Euler when, in 1736, he published an elegant solution to the following problem.

Example 1 In the city of Königsberg (now Kaliningrad) in the year 1736 there were seven bridges. These connected the four regions separated by the river Pregel. (See Fig. 4.1.) The people of Königsberg liked to stroll about their city. Being of a methodical nature, however, many were disturbed by the fact that no one had figured out a way to take a walk that would cross every bridge exactly once. And would it not be nice, they said, if such a walk would bring one right back home again after a circuit of the city? Well, Leonard Euler solved their problem by reducing the map of Königsberg to its bare essentials. Can you see the solution?

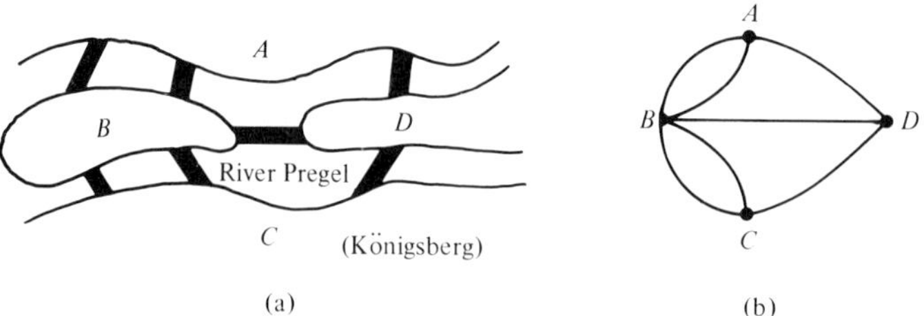

Figure 4.1 Leonard Euler reduced the map of Königsberg to a multigraph to assist in his analysis of the Königsberg bridges problem. □

It is easier to explain the rather simple solution to the Königsberg bridge problem once one has learned the words that properly describe Fig. 4.1. Therefore, we now give some formal definitions.

Definition 4.1 A (a) **multigraph**, denoted by "G" or "$G(V, E)$," consists of a set V of objects called (b) **vertices** and another set E of objects called (c) **edges**, such that each edge is (d) **incident** with (or "on") exactly two (possibly identical) vertices called its (e) **endpoints**. A (f) **graph** is a multigraph in which every edge has two distinct endpoints and no two vertices are incident with (or "connected by") more than one edge.

We now see that Euler's diagram in Fig. 4.1 is a multigraph having four vertices and seven edges. It is not a graph, however, because it has **multiple edges** connecting the pair of vertices $\{A, B\}$, for example.

Both graphs and multigraphs are commonly represented by diagrams such as Euler's, with a dot representing a vertex and a straight-line segment or an arc standing for an edge.

Figure 4.2 gives us further examples of multigraphs. Figure 4.2(a) is also a graph $G(V, E)$, in which we have $V = \{X, Y, Z, W, U\}$ and $E = \{e_1, e_2, e_3\}$.

In Fig. 4.2(b) we not only have multiple edges f_3 and f_2, but one of the edges, f_1, does not have distinct endpoints. We will call edges such as f_1 **loops**. By Definition 4.1(f), graphs cannot have loops. Thus Fig. 4.2(b) is a multigraph, but it is not a graph.

You may have noticed that the vertex U in Fig. 4.2(a) is not on any edge. We say that U is an **isolated vertex**. If we think in terms of our Königsberg bridges problem, we might say that there is no way to walk from vertex U to any of the other vertices by means of a sequence of edges, and this "disconnects" the graph. The adjacent multigraph, by comparison, is "all of one piece" or "connected." The need for more precise language to describe these ideas adequately compels us to form definitions such as Definition 4.2.

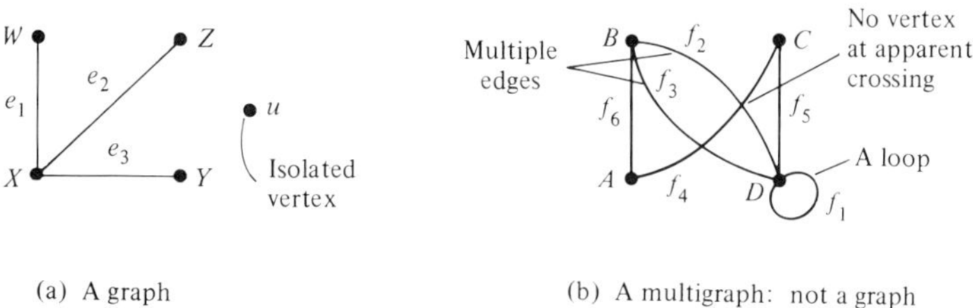

(a) A graph (b) A multigraph: not a graph

Figure 4.2 Two multigraphs.

Definition 4.2 A (a) **walk** in a multigraph is an alternating sequence of its vertices and edges of the form

$$v_0, e_1, v_1, e_2, v_2, \ldots, e_n, v_n,$$

where vertices v_{i-1} and v_i are endpoints of edge e_i for each i. We say that (b) **vertices v_0 and v_n are connected** if there is such a walk beginning at v_0 and ending at v_n. (Every vertex is connected to itself.) A (c) **connected multigraph** is one in which every pair of vertices is connected. A (d) **disconnected multigraph** is one that is not connected.

Combining all the vertices and edges in Fig. 4.2 gives us another example of a disconnected multigraph. Not only is U isolated, but there is no walk between vertices A and Y, for example.

The Königsberg bridges multigraph is, of course, connected. If we label its edges, as in Fig. 4.3(a), then Fig. 4.3(b) shows one of many walks from A to D, namely, $Ae_1Be_2Ce_3D$. Since there is also a walk Ae_4D from A to D having only *one* edge, we say that A and D are **adjacent vertices**. In other words, two vertices are adjacent when they are joined by an edge.

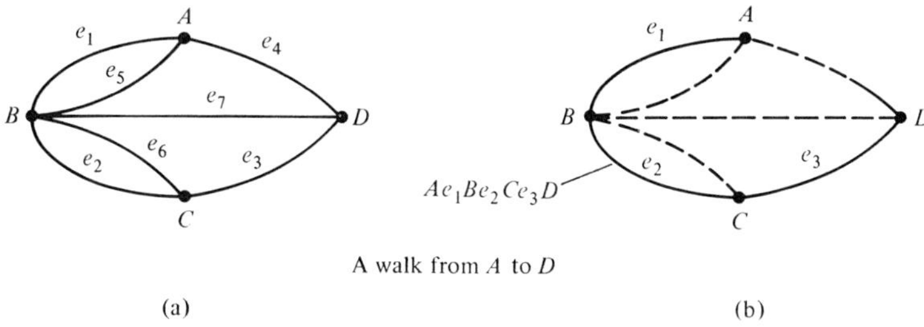

Figure 4.3 $Ae_1Be_2Ce_3D$ is a walk in the multigraph of Fig 4.3(a) from A to D.

Definition 4.3 A walk is a (a) **path** if all its vertices are distinct. A walk is a (b) **circuit** if $v_0 = v_n$. A walk is a (c) **trail** if all of its edges are distinct. An (c) **Euler trail** is one that includes every edge of the multigraph. And, finally, a walk will be called an (d) **Euler circuit** if it is both an Euler trail and a circuit.

Thus $Ae_1Be_2Ce_3D$ is both a path and a trail through the Königsberg bridges, since all the vertices and edges in it are distinct. Moreover, $Ae_1Be_2Ce_3De_4A$ is an example of a trail (but not a path—why?) that is also a circuit; but it is neither an Euler trail nor an Euler circuit, since it does not include every edge in the multigraph.

To say why the people of Königsberg could not find an Euler trail or Euler circuit, we need just one more definition.

Definition 4.4 The **degree of a vertex v**, denoted by **deg(v)**, is the number of edges on v, where each loop on v counts twice.

In Fig. 4.3(a), for example, we have

$$\deg(A)=\deg(C)=\deg(D)=3, \quad \text{and} \quad \deg(B)=5,$$

which is why there is no Euler circuit or Euler trail on that multigraph. In an Euler circuit one must enter a vertex v as many times as one leaves. Hence $\deg(v)$ = an even number, for every vertex in the multigraph having an Euler circuit. If we begin and end our Euler trail at different vertices, v_1 and v_2, respectively, then these, and only these, vertices can have odd degree, since, for example, we leave v_1 one more time than we come back to it. Thus *Fig. 4.3(a) does not have an Euler trail because it has four vertices of odd degree*, two too many!

Euler found that these vertex degree conditions not only were necessary, but also sufficient, to ensure the existence of Euler trails or Euler circuits in a connected multigraph. (Connectedness is, of course, a necessary condition!)

Theorem 4.1 A connected multigraph G has an Euler circuit if and only if every one of its vertices has an even degree. The connected multigraph G has an Euler trail that is not a circuit if and only if precisely two vertices of G have odd degree.

Proof: We have already proved the necessity conditions. Therefore, suppose that every vertex of G has even degree, and consider the *construction of an Euler circuit* in Algorithm 4.1.

Algorithm 4.1 ***Euler Circuits***

1. [Initial vertex.] Choose any vertex v_0 in G.
2. [Initial circuit.] Construct a trail T: $v_0 e_1 \ldots v_0$ as follows: Find an edge e_1 adjacent to v_0 and adjacent to a second vertex v_1; find a new edge e_2 adjacent to v_1, and so on, until no new edges can be added to the trail. (This trail ends at v_0 since the even degrees permit us to exit each vertex we enter—except, at last, for v_0.)
3. [Subcircuit loop.] Repeat steps 4a to 4c until T contains every edge of G.
   - 4a. [New initial vertex.] Find the first vertex, v_f, in T adjacent to an unused edge of G. (If there are none, then T is an Euler circuit or G is disconnected.)
   - 4b. [Construct S.] Construct a circuit S: $v_f \ldots v_f$ as in step 2, using only edges that have not yet been chosen. (Follow step 2 with v_f in place of v_0.)
   - 4c. [Enlarge T.] Set $T \leftarrow T + S$. ($T + S$ is formed by taking the edges of T up to v_f, then taking the edges of S, and finishing with the remaining edges of T.)
5. [Done.] Output T.
6. End of Algorithm 4.1.

With minor changes in wording, our algorithm will find an Euler trail in a connected multigraph having exactly two vertices of odd degree. (Let v_0 be one of these vertices, and see Exercise 4.1-9. For another method see Exercise 4.1-10.) Since we have shown how to produce Euler circuits and trails (Exercise 4.1-9) given the stated conditions, *we have proved Theorem 4.1.* ■

Figure 4.4 shows how a circuit T might be enlarged to an Euler circuit by adding two subcircuits and relabeling the edges.

Our proof of Euler's theorem, Theorem 4.1, tells us how to find Euler circuits when they exist. But do not underestimate the value of the statement of the theorem itself. With the statement alone we can see at a glance whether or not a rather complicated multigraph has an Euler circuit or trail. For example, each vertex in graph G_1

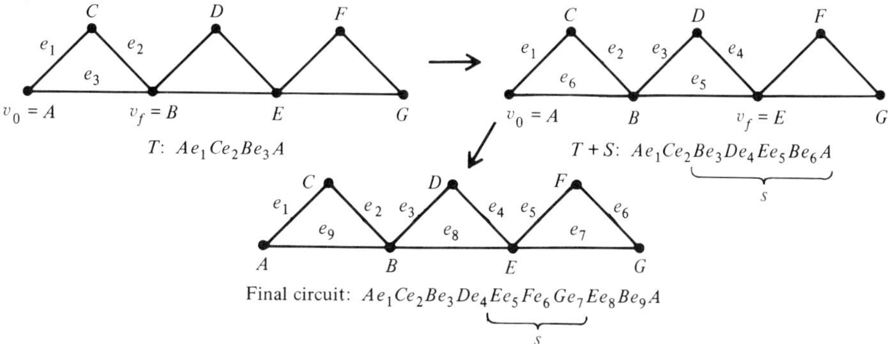

Figure 4.4 Finding a Euler circuit by means of adding subcircuits.

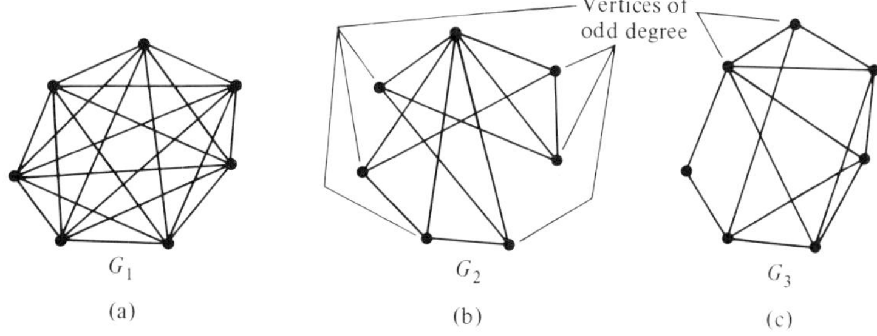

Figure 4.5 In this figure (a), graph G_1 has an Euler circuit, since each vertex has even degree. (b) Graph G_2 has no Euler trail since there are more than two vertices of odd degree, but (c) graph G_3 has an Euler trail.

of Fig. 4.5 has degree 6. *Thus we do not have to find an Euler circuit in graph* G_1 *in order to know that there is such a circuit.*

Likewise one need not bother looking for an Euler trail in graph G_2; its six odd vertices tells us that there are none. The graph G_3, on the other hand, has exactly two vertices of odd degree. Therefore, we could find an Euler trail, but not an Euler circuit, if we choose to do so.

While we are on the subject of "existence," the following theorem implies that multigraphs having exactly one vertex of odd degree do not exist.

Theorem 4.2 (a) The number of *vertices of odd degree* in a multigraph G is *even.* (b) The *sum of the degrees of the vertices* of G is also an even number, namely, *twice the number of edges.*

Proof: We will prove part (b) first, since it can be used to prove part (a). First we notice that every edge contributes "1" to the degree of each of its distinct endpoints (or "2" to its one endpoint if the edge is a loop). Therefore, adding two for each edge sums the degrees of all the vertices once, proving part (b).

For part (a), recall that a sum of even numbers is always even. Hence if we add the degrees of only those vertices with even degree, we must obtain an even number. Therefore, the following equations show that the sum of degrees of the rest of the vertices is also an even number, since the sum of all the degrees, by part (b), is even.

$$\begin{aligned} &\textit{Sum of even degrees} + \textit{Sum of odd degrees} = \textit{Sum of degrees},\\ &\textit{Even number} \qquad\quad + (?)\ \textit{number} \qquad\quad = \textit{Even number}. \end{aligned}$$

We see that if there were an odd number of vertices of odd degree, then the second number in the sum would be odd. This is true since the sum of an odd number of odd numbers is odd. (See Exercise 4.1-5.) Hence we have proved Theorem 4.2. ■

Hamiltonian Paths and Circuits

Until this point we have largely been concerned with trails. Recall, however, that we defined a "path" as a walk in which the vertices are all distinct. But if the vertices in a walk are distinct, then, clearly, so are the edges. In fact, we can state the following.

Theorem 4.3 If there is a walk in multigraph G from vertex v_0 to vertex v_n, then there is also a path from v_0 to v_n. ($v_0 \neq v_n$.)

We will leave the proof of Theorem 4.3 for you to do in Exercise 4.1-10, as it is both easy and instructive. But many problems involving paths, though easy to state, are hard to solve. One very famous problem seems, at first, to be much like Euler's problem. We need the following definitions in order to state it.

Definition 4.5 A path is called a (a) **Hamiltonian path** if it includes every vertex in the multigraph. A circuit that includes every vertex except for the initial vertex exactly once is called a (b) **Hamiltonian circuit**.

The name "Hamiltonian circuit" is attributable to a game invented by the mathematician William Rowan Hamilton. It was called "Around the World," and it consisted of a solid whose vertices represented 20 well-known cities [see Fig. 4.6(a)]. The object of the game was to make what we have called a "Hamiltonian circuit" of the cities along the edges of the solid, as in the simplified Fig. 4.6(b), which, of course, is a graph.

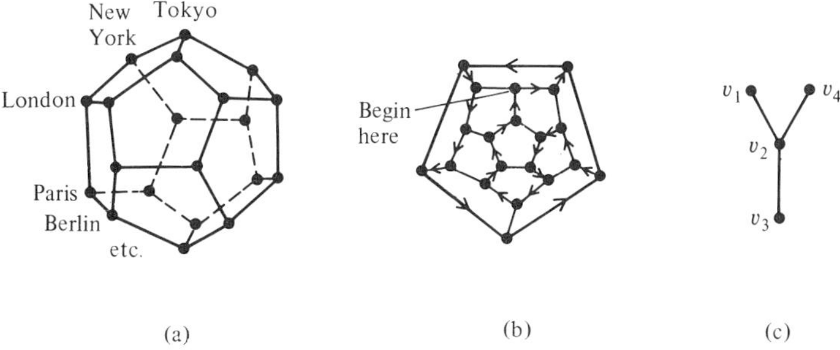

Figure 4.6 The Hamiltonian path problem on the "dodecahedron of Fig 4.6(a) "around the word" is reduced to a problem on a graph by reducing the edges on the solid to a network, as in part (b). A hamiltonian circuit is indicated by the arrows. The graph in part (c) has no Hamiltonian circuit or path.

There is no simple solution to the problem of whether a connected multigraph has a Hamiltonian circuit or path. We can sometimes find one or the other when they exist, or prove that a graph, such as the one in Fig. 4.6(c), has neither by discussing the particular arrangement of vertices and edges. It is clear from Fig. 4.6(c), for example, that *a graph having more than two vertices of degree 1 cannot have a Hamiltonian path or circuit*. (See Exercise 4.1-20.)

One can also give sufficient conditions for a graph to have a Hamiltonian circuit or path. We state the following without proof.

Theorem 4.4 Let G be a connected graph with $n(\geqslant 3)$ vertices.

a) G has a Hamiltonian path if the sum of the degrees of each pair of its vertices is at least $n-1$.

b) G has a Hamiltonian circuit if the sum of the degrees of each pair of its nonadjacent vertices is at least n.

Example 2 The graph in Fig. 4.7 satisfies Theorem 4.4(a), since the sum of the degrees of each pair of its vertices is at least $4=5-1$. Hence this graph must have a Hamiltonian path ($ACBED$, for example). It does not, however, have a Hamiltonian circuit, since edges AB, BC, and AC would have to be included. In fact Theorem 4.4(b) is not satisfied by our graph. It is easy to show, however, that neither part (a) nor part (b) of Theorem 4.4 is a necessary condition. (Consider any "polygon"—a connected graph with every vertex having degree 2.) □

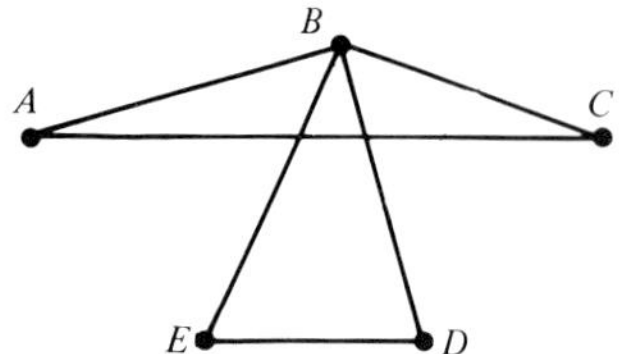

Figure 4.7 *ABCDE* is a Hamiltonian path on the given graph, satisfying the conclusion of Theorem 4.4(c). Notice that the hypothesis is satisfied because the sum of the vertex degrees of any pair of vertices in this graph is at least four and there are five vertices.

Completion Review 4.1

Complete each of the following.

1. A multigraph $G(V, E)$ consists of a set V of__________ and a set E of __________ such that each element in E is__________ with exactly two elements in V. A graph is a multigraph having no__________ edges or__________.
2. A vertex that is not on any edge is called__________.
3. A walk is an alternating sequence of__________ such that __________.
4. Two vertices v_0 and v_n are connected if__________. They are adjacent if __________.
5. A multigraph is connected if__________. Otherwise we say that the multigraph is__________.
6. A walk that does not repeat any edges is called a__________. A walk that does not repeat any vertices is called a__________. A walk in which the first and last vertices are the same is called a__________.
7. A walk that includes every edge of the multigraph once and only once is called an __________. If such a walk is also a circuit, then it is called an__________ __________.

8. The number of edges on a vertex v is denoted by ________ and called the ________ of v. Each loop on v counts ________.

9. The number of vertices of odd degree in a multigraph is ________. The sum of the degrees in a multigraph equals twice the ________.

10. A path that includes every edge of a multigraph is called a ________ path. A circuit that includes every vertex of the multigraph once and only once, except for the initial vertex, is called a ________.

11. If G is a connected multigraph, then G has an Euler circuit iff every vertex in G has ________. Furthermore, G has an Euler trail that is not a circuit iff there are exactly two ________.

12. A connected graph G on n (≥ 3) vertices has a Hamiltonian path if the sum of the ________ is at least ________. Furthermore, G has a Hamiltonian circuit if the sum of the ________ is at least ________.

Answers: **1.** vertices; edges; incident; multiple; loops. **2.** isolated. **3.** vertices and edges $v_0, e_1, v_1, \ldots, \ldots, e_n, v_n$; v_{i-1} and v_i are endpoints of e_i for each i. **4.** there is a walk from v_0 to v_n; there is an edge connecting v_0 and v_n. **5.** every pair of its vertices are connected; disconnected. **6.** trail; path; circuit. **7.** Euler trail; Euler circuit. **8.** deg(v); degree; twice. **9.** even; sum of the edges. **10.** Hamiltonian; Hamiltonian circuit. **11.** even degree; vertices of odd degree. **12.** degrees of each pair of vertices; $n-1$; degrees of each pair of nonadjacent vertices in G; n.

Exercises 4.1

1. Which of the following multigraphs is a graph?

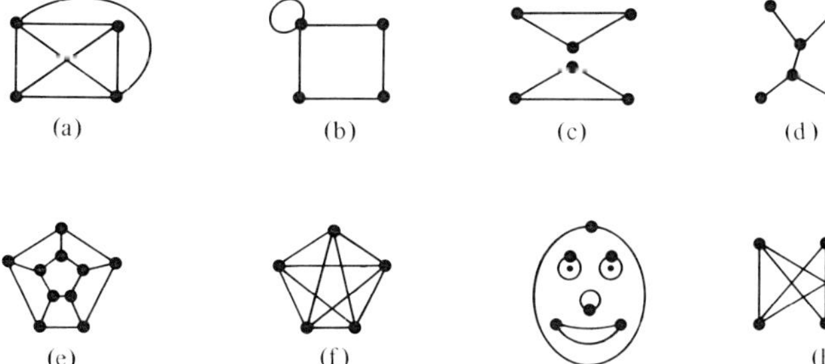

2. Which of the multigraphs in Exercise 1 are connected?
3. Which of the multigraphs in Exercise 1 have an Euler trail?

4. In the following graph **(a)** list all paths from vertex A to vertex E; **(b)** list all trails from A to E; **(c)** tell why you can represent each walk in a graph in terms of vertices alone.

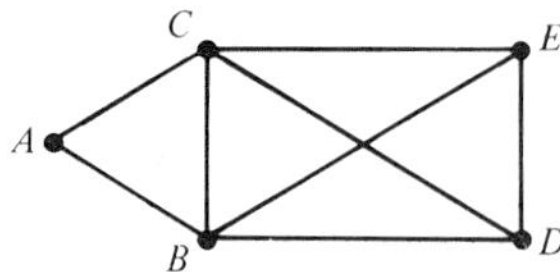

5. Prove that the sum of an odd number of odd numbers is odd.
6. Find an Euler circuit and an Euler trail, respectively, for the graphs G_1 and G_3 in Fig. 4.5.
7. Show that a graph having every pair of its n vertices connected by an edge has a total of $n(n-1)/2$ edges.
8. Find the minimum number of edges in **(a)** An Euler trail on n vertices; **(b)** an Euler circuit on n vertices.
9. **a)** Suppose that we have a connected graph in which exactly two vertices, v_1 and v_2, have odd degree. Explain how a minor rewording of step 2 in the algorithm in the proof of Theorem 4.1 gives us an Euler trail from v_1 to v_2 if we let v_0 be, say, v_1.
   b) Apply this method to graph G_3 in Fig. 4.5.
10. If G has exactly two odd vertices v_1 and v_2, can one find an Euler trail on G by **(a)** adding a fictitious edge between v_1 and v_2, obtaining a new graph G^*; then **(b)** finding an Euler circuit on G^*; **(c)** finally, removing the fictitious edge? Why? Apply this method to G_3 in Fig. 4.5.
11. Prove that a graph with a Hamiltonian path has no more than two vertices of degree 1.
12. Prove that if there is a walk in multigraph G from vertex v_0 to vertex v_n, then there is a path from v_0 to v_n. ($v_0 \neq v_n$.)
13. Show that the number of Hamiltonian circuits on a graph having every pair of its n vertices connected is $(n-1)!/2$, if we neglect the direction of the walk.
14. Prove that the number of graphs on n distinct vertices is $2^{n(n-1)/2}$.
15. A certain neighborhood consists entirely of two-way streets that are connected to each other so that one can drive between any two points of the neighborhood. Show that a street sweeper can clean each side of every street without going over any side twice, and arrive at the starting point.
16. Consider an eight-by-eight chessboard on which we place a knight, and make a graph with the squares as vertices as follows: Two vertices are connected by an edge if the knight can move from one square to the other in a single move. Show that this graph does not have an Euler circuit or Euler trail.
17. There is a Hamiltonian path (called a "knight's tour") on the graph of Exercise 16. Try to find such a path.
18. Suppose we have seven examinations to give on seven consecutive days. If no instructor has more than four exams to give, and we do not want to schedule

two exams by the same instructor on two consecutive days, explain whether the scheduling can be done. (*Hint*: Make a graph in which each exam is a vertex, and two vertices are connected by an edge only if the corresponding exams are given by a different instructor.)

19. The following illustration shows the floor plan of a house with doorways connecting rooms to each other or to the outside. Can one start outside the house and walk through each doorway exactly once? (*Hint*: Consider a multigraph in which the rooms are the vertices and vertices are connected by a distinct edge for every doorway connecting the rooms. The outside is considered a room.)

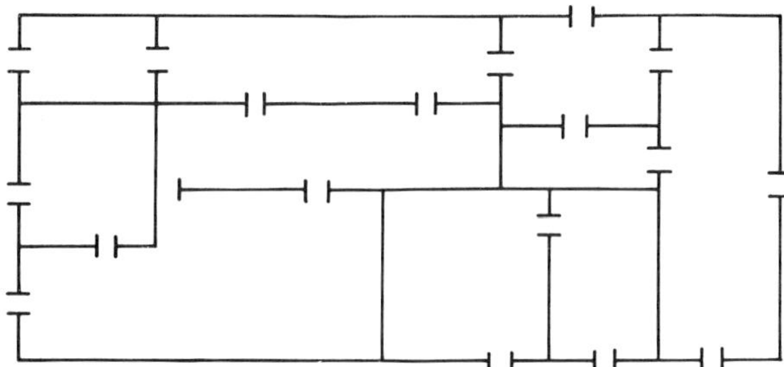

20. Prove by induction that a connected graph on n vertices has at least $n-1$ edges. (*Hint*: Find a maximal path beginning at, say, v_1. If the path has less than $n-1$ edges, then removing v_1 and all of its incident edges leaves a connected graph on $n-1$ vertices. Why?)

4.2 CLASSIFICATION OF GRAPHS

In the preceding section we discussed graphs that had Eulerian or Hamiltonian circuits. Such graphs are called **Eulerian graphs** and **Hamiltonian graphs** respectively. These are but two illustrations of how one might classify graphs in terms of some outstanding property. Sometimes it is useful to give a different formulation of the same property, perhaps in some form that it is more easily verified. Euler's theorem does just this for Eulerian graphs, telling us that these are precisely the graphs that are connected and for which every vertex has an even degree. As we have already pointed out, no one at this time knows a simple alternate characterization of Hamiltonian graphs.

While it is natural to think of graphs and multigraphs in terms of modeling transportation-related problems, modern applications of graph theory are very diversified. Moreover, certain types of problems give rise to closely related graphs, and this in turn creates a need to pinpoint their similarities and differences. For example, in Chapter 3 we occasionally made use of what we called "tree diagrams" to help keep track of things we were counting. The graph theoretic name for such an object is a "tree." We usually drew the "tree" with its "branches" hanging down, as in

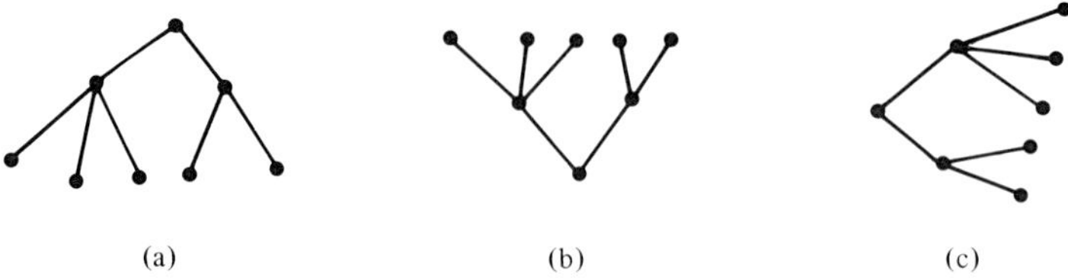

Figure 4.8 The "same" tree is drawn three different ways.

Fig. 4.8(a). But it would still be the same graph if drawn upside down as in Fig. 4.8(b), or even sideways, as in Figure 4.8(c).

How can one characterize these graphs? They are clearly connected graphs, but to describe properly the "branching-out" property that seems so essential, we need to introduce a key definition.

Definition 4.6 A (a) **cycle** is a circuit of the form

$$v_0, e_1, v_1, e_2, v_2, \ldots, v_{n-1}, e_n, v_0$$

in which all the vertices $v_i \neq v_0$ and edges e_i are distinct. A (b) **tree** is a connected graph that has no cycles. A (c) **forest** is a graph (connected or not connected) that has no cycles.

In Fig. 4.9 we show *all* the different possible trees having exactly four or fewer vertices. (We justify this remarkable statement later in the section.)

If we think of Fig. 4.9 as being all one graph, having 14 vertices and nine edges, we have our first example of a forest. Taken as a whole there is no cycle anywhere in Fig. 4.9, and, of course, there is not requirement that the graph be connected. Let us call this forest F. We would say that each tree in the forest F is a **subgraph** of F, meaning that *each consists of vertices from the vertex set of F, and two vertices in each may be connected by an edge only when they are connected by an edge in F.* Moreover, each of the trees in Fig. 4.9 is a **connected component** of F. This means that each is a *connected subgraph of F which is not contained in any larger connected subgraph of F.* We discuss trees and forests at greater length in Chapter 5, but some of their properties can be found in the exercises. (See Exercises 4.2-15, -22, and -23.)

Figure 4.9 Tress on four or fewer vertices.

Graphs have been used to study social relationships. For example, one can let different individuals be represented by distinct vertices, where two vertices are joined by an edge if the corresponding individuals are mutually friendly.

Example 1 The following graph gives the "friendship graph," G, of the set of six people {Andy, Bob, Chuck, Dina, Ed, Flo}, where we have assigned vertices to each person in the obvious way.

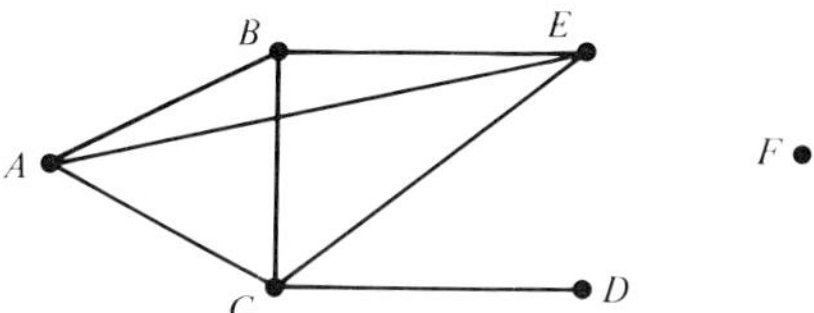

Figure 4.10 A "friendship" graph; adjacent vertices signify friends.

We can tell at a glance from this graph that Andy, Bob, and Ed each have three friends in the group, since $\deg(A) = \deg(B) = \deg(E) = 3$. We can also see that Chuck is the person with the greatest number of friends here, since $\deg(C) = 4$. Indeed Chuck is Dina's only friend, and Flo has no friends at all in the group. Sociologists would describe Flo as an "isolate," corresponding to the isolated vertex F. Moreover, the four mutual friends, Andy, Bob, Chuck, and Ed, would be called a "clique." In graph theory we say that a **clique** is a *maximal subgraph of a graph having an edge joining each pair of its vertices*. By "maximal" we mean that it is not contained in a larger subgraph having each pair of vertices connected by an edge. For example, the subgraph consisting of A, B, and C and the edges connecting them are not a clique, even through vertex pairs A and B, B and C, and C and A are joined by edges. The set of vertices A, B, C, D, E and all the edges in the graph are not a clique, since there is no edge in the original graph joining D and E. □

Definition 4.7 A **complete graph on n vertices**, denoted K_n, is a graph in which each pair of distinct vertices are joined by an edge.

In Fig. 4.11, we have complete graphs on three, four, five, and six vertices respectively.

This might be a good time to recall that when dealing with a graph, as opposed to a multigraph, every edge connects exactly two distinct vertices. Hence one can denote edges by the pair of vertices joined. For example, in Fig. 4.11 we can refer to edge AB, say, without any ambiguity. We will continue this practice hereafter when we discuss graphs.

Observe that when two edges in a diagram of a graph cross, the apparent point

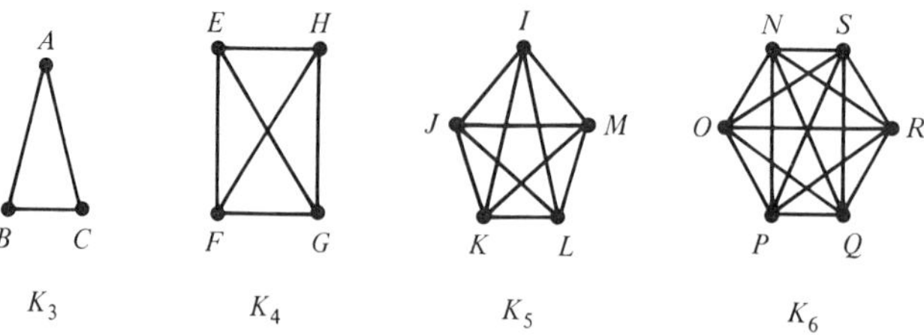

Figure 4.11 The complete graphs on three, four, five, and six vertices.

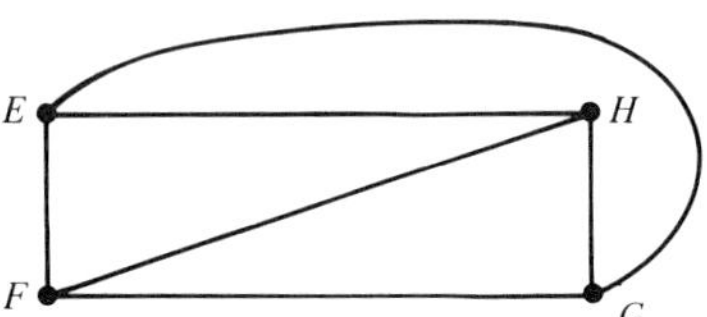

Figure 4.12 The complete graph K_4 is drawn without intersecting edges.

of intersection might not be a vertex of the graph. Thus in our picture of K_4, for example, edges EG and FH cross each other, but these edges do not have any vertex in common. One can draw a complete graph on four vertices in which the edges intersect only at the vertices of the graph, as in Fig. 4.12.

Another interesting class of graphs arises in the context of committee or job assignments.

Example 2 In the graph of Fig. 4.13 we have partitioned the vertex set into the disjoint sets $\{P_1, P_2, P_3, P_4, P_5\}$ and $\{J_1, J_2, J_3\}$.

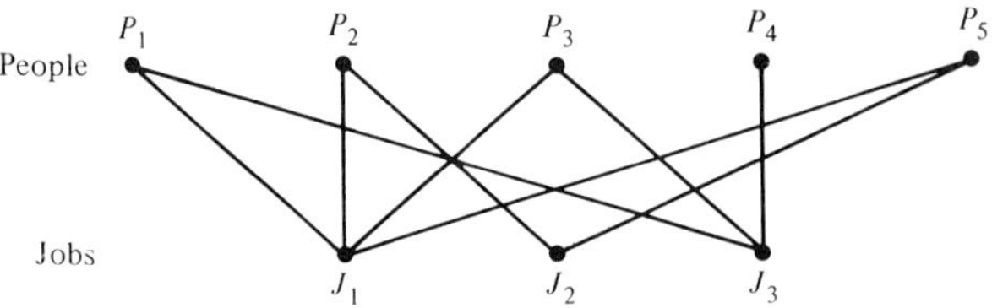

Figure 4.13 A bipartite graph and job assignments.

Figure 4.13 could be used to model a situation in which five people are competing for three jobs. Both the people and the jobs are assigned vertices, but we must distinguish between the "people" vertices and the "job" vertices. A person's vertex is joined by an edge to a job's vertex if and only if the person is qualified for that job. There are no edges of the form P_iP_j or J_rJ_s. □

Another situation that we could model with our graph is one in which *five people serve on three committees*. Once again the vertices of two people are never joined, nor are the vertices of committees.

We call this kind of graph "bipartite."

Definition 4.8 A (a) **bipartite graph is** one in which the vertex set can be partitioned into two disjoints sets $A=\{A_1, A_2, \ldots, A_m\}$ and $B=\{B_1, B_2, \ldots, B_n\}$ such that no two vertices in A and no two vertices in B are joined by an edge. A (b) **complete bipartite graph**, denoted by $K_{m,n}$, is a bipartite graph in which each vertex in A is joined to every vertex in B.

It is sometimes possible to partition the vertices of a bipartite graph in more than one way, so that vertices within the same set are not joined by an edge. For the bipartite graph in Fig. 4.14, we can either partition the vertices into sets $\{A, B, E\}$ and $\{C, D, F, G\}$ or into sets $\{A, B, G, F\}$ and $\{C, D, E\}$, so that there are no edges between different vertices in the same set.

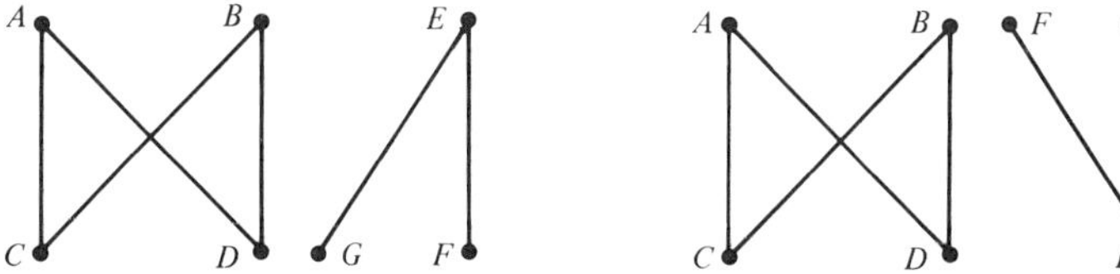

Figure 4.14 A bipartite graph partitioned in two different ways.

Isomorphism of Graphs

The fact that we can draw the same graph in many different ways raises the question: *When are two graphs essentially the same and when are they essentially different?* Evidently if two graphs have different numbers of vertices or different numbers of edges then they must be different graphs. But the names or labels we assign to the edges and vertices are not an essential feature of the graph, since the same graph can be used to model different situations. The two representations of the graph in Fig. 4.14 can be regarded as interchanging the labels F and G and keeping the vertices in places, rather than interchanging the vertices. One could say that the essential feature of a graph is *how*, rather than which, vertices are joined by edges. This idea is made more precise in the following.

Definition 4.9 Given two graphs G_1 and G_2 with vertex sets $\{A_1, A_2, \ldots, A_n\}$ and $\{B_1, B_2, \ldots, B_n\}$, respectively, we will say that an (a) **isomorphism** between G_1 and G_2 is a one-to-one correspondence between their vertex sets such that whenever A_i and A_j correspond to B_i and B_j, respectively, then *A_i and A_j are connected by an edge if and only if B_i and B_j are connected by an edge*. In that case we will say that the graphs G_1 and G_2 are (b) **isomorphic.**

Hence two graphs cannot be isomorphic unless they have *the same number of vertices*, as the one-to-one correspondence criterion clearly implies. It is not hard to see that they must also have *the same number of edges*. But there is more to Definition 4.9, as the following examples illustrate.

Example 3 Consider the following four graphs, G_1, G_2, G_3, and G_4. Graphs G_1 and G_2 have five vertices and five edges each, while graphs G_3 and G_4 each have four vertices and four edges. Hence *neither G_1 nor G_2 is isomorphic to either graph G_3 or G_4*; no one-to-one correspondence between the vertex sets can be found in these cases. (See Fig. 4.15.)

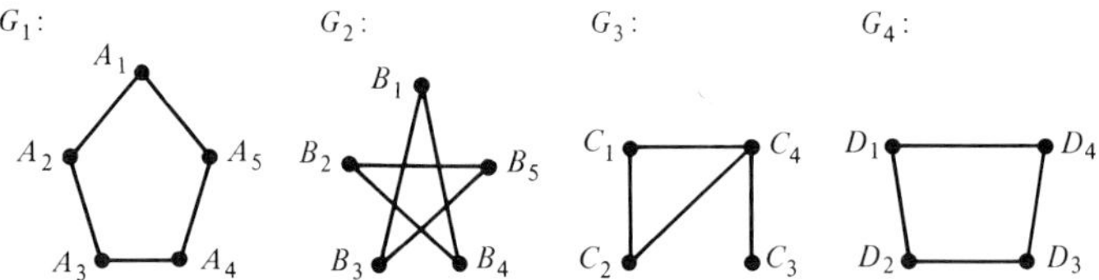

Figure 4.15 G_1 is isomorphic to G_2, but G_3 is not isomorphic to G_4.

The graphs *G_1 and G_2, however, are isomorphic*, as the correspondence $\{(A_1, B_1), (A_2, B_3), (A_3, B_5), (A_4, B_2), \text{and } (A_5, B_4)\}$ shows. Notice how the "edge connections are preserved" by this correspondence. Where vertices A_1 and A_2 are joined by an edge, so are the corresponding vertices B_1 and B_3; where vertices A_1 and A_3 are not joined by an edge in the first graph, the corresponding vertices B_1 and B_5 are not joined by an edge in the second.

Even though *graphs G_3 and G_4* have the same number of vertices and edges, they *are not isomorphic*. This means that one will not be able to find the right kind of correspondence between their vertices. One way to see this is to notice that $\deg(C_4) = 3$. Hence if there were an isomorphism making C_4 correspond to some vertex D_i in the fourth graph, then D_i would have to be joined to all the other vertices in G_4, just as C_4 is joined to all the other vertices in G_3. But $\deg(D_i) = 2$ for every vertex in G_4. Therefore, there cannot be an isomorphism between G_3 and G_4. These graphs are essentially different. □

It is in the sense of isomorphism that we can say that "all the trees in Fig. 4.8 are the same," for example. We can also make statements such as, "Fig. 4.9 gives all the trees having one, two, three, or four vertices," or, "For each positive integer n there is essentially only one complete graph K_n." We mean that all the others in the given category are just isomorphic replicas of each other. For example, the essential feature of being K_n is that every vertex has degree $n-1$. This property must be preserved under an isomorphism correspondence. Any two K_n's must share this distinguishing feature: The isomorphism between them, we say, "preserves" it.

Let us list a number of features that isomorphic graphs must share.

Theorem 4.5 ***Isomorphism Preservation*** If two graphs G_1 and G_2 are isomorphic, then G_1 and G_2 have

1. the same number of vertices and edges;
2. the same degrees for corresponding vertices;
3. the same number of connected components.

Moreover,

4. G_1 has an Euler trail (circuit) iff G_2 has an Euler trail (circuit);
5. G_1 has a Hamiltonian path (circuit) if and only if G_2 has a Hamiltonian path (circuit);
6. if vertices A_1 and B_1 in G_1 correspond to vertices A_2 and B_2 in G_2 and there is a trail (path, cycle, or circuit) of length n from A_1 to B_1, then there is a trail (path, cycle, or circuit respectively) of length n from A_2 to B_2.

The list given in Theorem 4.5 is not complete by any means. But it gives us several ways to discover quickly, in many cases, when two graphs are *not* isomorphic. However, *none of the six conditions in Theorem 4.5, by themselves, are sufficient to ensure that two graphs are isomorphic.*

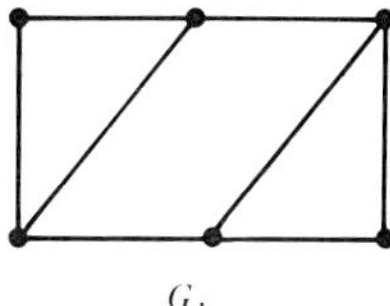

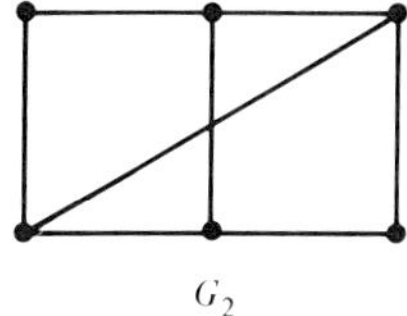

Figure 4.16 G_1 and G_2 are not isomorphic even though they have many properties in common.

Example 4 By now you should be cautious enough not to jump to the conclusion that the two graphs in Fig. 4.16 are not isomorphic simply because they are drawn differently. In fact both graphs have the same number of vertices and edges, and both have two vertices of degree 2 and four vertices of degree 3. Hence neither graph has an Euler path, but both graphs have Hamiltonian circuits. And yet the graphs are not isomorphic. There is something more subtle that makes these graphs essentially different.

One way to see why the graphs are not isomorphic is to look at the cycles and at condition (6) of Theorem 4.5. Vertices of degree 3 must correspond under any isomorphism. But each vertex of degree 3 in G_1 is on a cycle having only three vertices; no vertex of degree 3 in graph G_2 is on a cycle having only three vertices. Hence the graphs G_1 and G_2 cannot be isomorphic. □

Completion Review 4.2

Complete each of the following.

1. A graph is Eulerian if it has an ________________. A graph is Hamiltonian if it has a ________________.
2. A cycle is a circuit in which ________________. A tree is a ________________ graph having no ________________. A forest is a ________________.
3. A graph H is a subgraph of a graph G if all the vertices in H are ________________ and if two vertices in H are adjacent only if they are ________________.
4. In a complete graph, every pair of vertices is ________________. A clique in a graph G is a maximal ________________ of G. The graph K_n is a ________________.
5. In a bipartite graph G the set of ________________ of G is partitioned into two sets A and B such that ________________. If G is a complete bipartite graph, then each vertex in A or B is adjacent to every ________________.
6. Two graphs G_1 and G_2 are isomorphic if there is a ________________ between their vertex sets that preserves ________________.
7. If two graphs have the same number of vertices and edges, then they must be isomorphic. (*True or False?*)
8. If G_1 is a connected graph and G_2 is not connected, then G_1 and G_2 cannot be isomorphic. (*True or False?*)

Answers: **1.** Euler circuit; Hamiltonian circuit. **2.** no two vertices are equal, except the first and last; connected; cycles; connected graph. **3.** vertices of G; adjacent in G. **4.** connected by an edge; complete subgraph; complete graph on n vertices. **5.** vertices; no vertex in A or B, respectively, is adjacent to another vertex in A or B; vertex in B or A respectively; vertex in B or A. **6.** one-to-one correspondence; vertex adjacencies and nonadjacencies. **7.** False. **8.** True.

Exercises 4.2

For Exercises 1 to 9 refer to the following graphs.

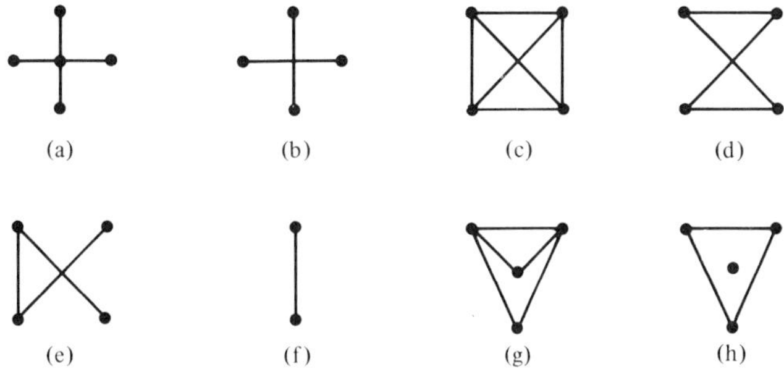

1. Which of the graphs is a tree?
2. Which of the graphs is a forest?
3. Which of the graphs is complete?
4. Which of the graphs is bipartite? (*Hint*: A bipartite graph cannot have a cycle of length 3. Why not?)
5. Which of the graphs is complete bipartite?
6. Which of the graphs is Eulerian?
7. Which of the graphs is Hamiltonian?
8. Which of the graphs have more than one connected component?
9. Neglecting the order of vertices, how many cycles does graph (a) have? graph (c)? graph (d)? graph (g)?
10. When will the graph K_n, $n \geqslant 3$, be **(a)** Hamiltonian? **(b)** Eulerian? Why?
11. When will the graph $K_{m,n}$, for m and $n \geqslant 2$, be **(a)** Hamiltonian? **(b)** Eulerian? Why?
12. How many cliques does K_4 have? K_4 minus an edge? K_n? $K_{n,m}$?
13. Show that $K_{m,n}$ has exactly mn edges.
14. Show that K_n has exactly $n(n-1)/2$ edges.
15. Every forest having n vertices and k connected components has $n-k$ edges. (We will prove this later.)
  a) How many vertices does a forest have if there are three components and 10 edges?
  b) How many edges does a tree on n vertices have?

For Exercises 16 to 19 refer again to graphs (a) to (h). Give a reason why the graphs named are *not* isomorphic to one another.

16. Graph (a) and graphs (b) through (h). (One reason for all.)
17. Graph (c) and all the rest. (One reason for all.)
18. Graphs (d) and (e).
19. Graphs (e) and (h).
20. Show that the following two graphs are not isomorphic.

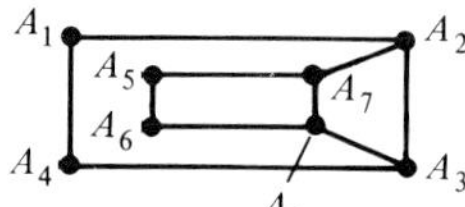

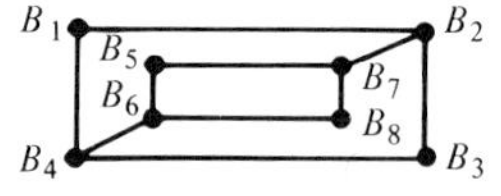

21. Show that the following two graphs are isomorphic.

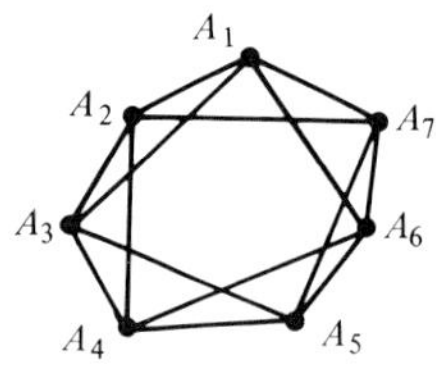

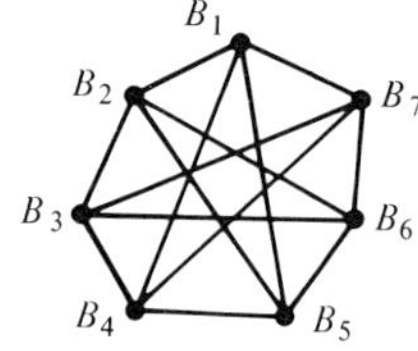

22. a) Prove that if there are two different paths between distinct vertices in a graph, then the graph must contain a cycle.
b) Prove that a tree has a *unique* path between any pair of vertices.
23. Show that every tree is a bipartite graph. (*Hint*: Color alternate vertices in the tree red and green.)
24. Show that if some vertices of K_6 are red and the rest are green, then there is either a subgraph K_3 with three red vertices or a subgraph K_3 with three green vertices.
25. Consider a bipartite graph that matches men to jobs, as in the following. Is it possible to find a different job for each man?

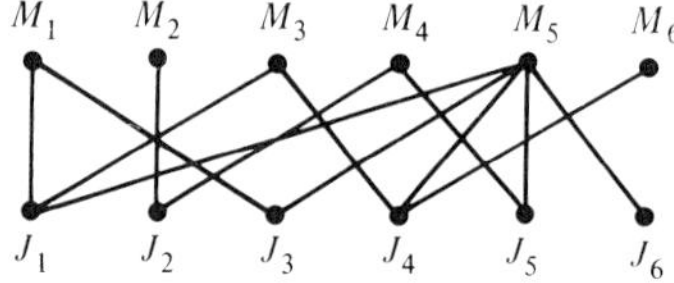

26. The following graph can be thought of as a map of a rice farm in which certain rice paddies are surrounded by earthen dams, and the entire farm is surrounded by a lake. The edges represent the walls of each dam. What is the least number of walls that must be broken in order to flood every rice paddy? Show that, in general, if there are n edges and m vertices, one must break at least $n - m + 1$ edges to remove every cycle.

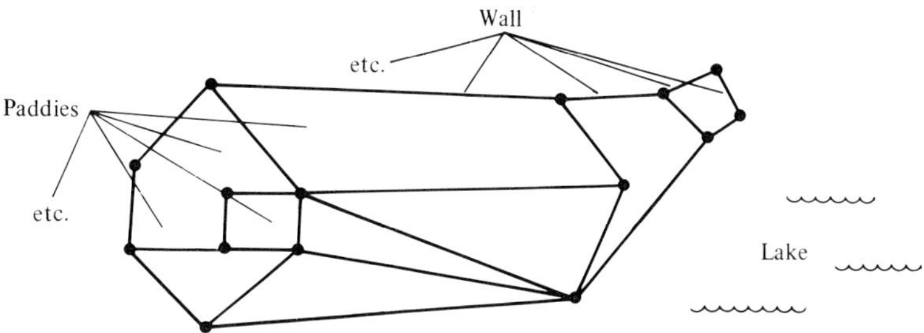

4.3 PLANAR GRAPHS AND EULER'S FORMULA

In the Königsberg bridges problem of Section 4.1 we used a map of the city of Königsberg to obtain a multigraph. Multigraphs can be obtained from ordinary maps and their boundary lines between geographical regions in a more natural way. Where three or more regions meet, natural vertex points are formed. These divide the boundary lines into segments that we can identify with edges. [See Fig. 4.17(a).] Since the map is often drawn on a flat surface, what we obtain can be thought of as a "planar graph."

We could, on the other hand, proceed more in the spirit of Euler's construction as follows: Identify each region with a distinct vertex [the *'s in Figure 4.17(b)] and connect a pair of these vertices with an arc wherever the corresponding regions have a common boundary "edge." This multigraph also forms a "map" in which the edges separate distinct regions [see Fig. 4.17(c)], since the edges only meet at *-vertices. Each of the two maps can be obtained from the other. We say that the corresponding multigraphs are (geometric) **duals** of each other. Moreover, we will formulate the following definition.

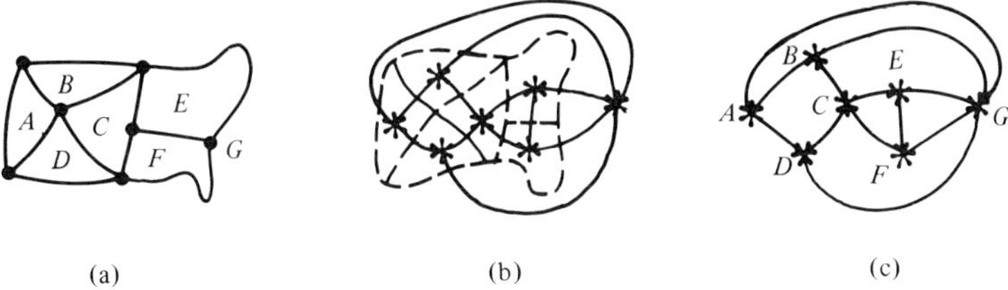

Figure 4.17 The dual of a planar map is formed, in part, by letting each region correspond to a vertex. Two vertices are adjacent if the corresponding regions have an edge in common.

Definition 4.10 A multigraph is called (a) **planar** if it can be drawn in the plane so that its edges do not intersect (except at their common vertices). Such a representation will be called a (b) **planar map**.

Example 1 The complete graph K_4 on four vertices is a planar graph. Figure 4.18(a) gives the usual representation of K_4 with intersecting "diagonals." In Fig. 4.18(b) we show how to modify this representation so that no edges intersect. In Fig. 4.18(c) we have added a loop and a multiple edge to K_4, which clearly do not affect the planarity of the multigraph. Indeed *if when you delete loops and multiple edges from a multigraph you get a planar graph, then clearly the original multigraph was also planar,* since parallel edges, for example, can be taken as close together as one likes. □

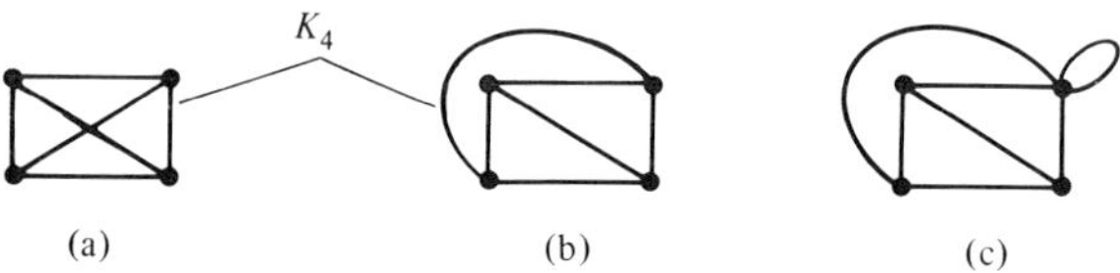

Figure 4.18 Adding a loop and making an edge into a multiple edge does not destroy a graph's planarity.

The importance of planarity is most easily seen in connection with printed electrical circuits, where the intersection of one conducting trail with another could easily mean a short-circuited component. Another application of planar graphs is to the following *utility problem.*

Example 2 Suppose that each of three houses must be connected to a water main, sewer, and electrical line, all underground and all at exactly the same depth. This situation can be modeled by the bipartite graph $K_{3,3}$, but not as pictured in Fig. 4.19. In fact *it is impossible to connect the three houses as required so that edges joining houses to utilities never intersect* (except at vertices). In other words, $K_{3,3}$ is not a planar graph (see Example 4). Of course, if we are not restricted to a plane, it is easy to make the necessary connections. □

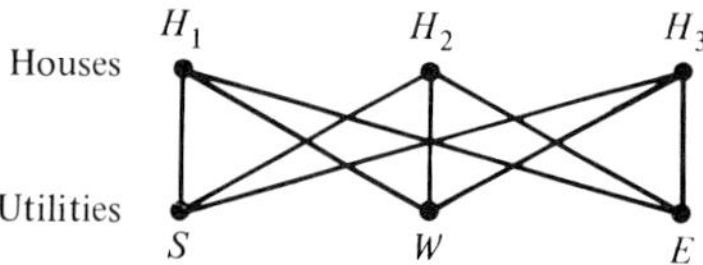

Figure 4.19 The "utility problem" is modeled with the graph $K_{3,3}$, a nonplanar graph.

A planar map can be described as having "regions," as well as the usual vertices and edges. Intuitively speaking, these are like the territories that are all of one piece on an ordinary map. Even a tree, a connected graph with no cycles, gives us one region when it is represented by a planar map. Planar maps with at least one cycle divide the plane into at least two regions.

Definition 4.11 The **degree of a region** r in a planar map, denoted by **deg(r)**, is the number of edges, counting repeats if necessary, in the smallest circuit including each edge in the border of r.

Example 3 We have labeled the regions of the planar map in Fig. 4.20(a) r_1, r_2, and r_3. It is clear that $\deg(r_1) = 3$. But we also have $\deg(r_2) = 7$, since $ABCDAEFA$ is a smallest circuit around the entire border of r_2 that includes the triangle AEF. There are seven edges in this circuit, with none repeated. Moreover, $\deg(r_3) = 6$, since the circuit $ABCDAGA$ includes every edge in the border of r_3. Edge AG has to be counted twice to get a circuit that includes every edge of the border. □

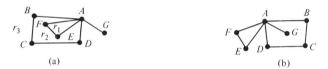

Figure 4.20 Different planar maps may arise from the same planar graph. Here the infinite regions have different degrees in (a) and (b).

Different planar maps can be obtained from the same graph. In Fig. 4.20(b) we have a different planar map obtained from the graph in Example 3. It is natural to ask what else the two planar maps obtained from the same multigraph have in common. We might begin with the following interesting analog of Theorem 4.2(a).

Theorem 4.6 The sum of the degrees of the regions of a planar map is twice the number of edges.

Proof: To prove Theorem 4.6, one notices that each edge is on the border of exactly two regions or, as with edge AG in Fig. 4.20(a), it will be crossed twice in a circuit around the border of one region. Hence if we take the sum of the degrees of the regions, each edge is counted twice, proving Theorem 4.6 ■

For example, in Fig. 4.20(a), the sum of the degrees of the regions is $3 + 7 + 6 = 16$, and the number of edges is eight. In Fig. 4.20(b) the reader should check that the sum of the degrees of the regions is $3 + 6 + 7 = 16$ once again.

We will say that a **planar map is connected** if the underlying multigraph is connected. The following formula was discovered by Leonard Euler while he was studying three-dimensional solids called "convex polyhedra" (see Exercises 4.3-10 to 4.3-13), but it is stated here for planar maps.

Theorem 4.7 ***Euler's Formula*** If a connected planar map has V vertices, E edges, and R regions, then

$$V - E + R = 2.$$

(In Example 3 we have $V = 7$, $E = 8$, and $R = 3$. By putting these in the left side of Euler's formula, we can verify that $7 - 8 + 3 = 2$. *Euler's formula implies that the number of regions one will obtain from any planar map of a graph is always the same!*)

Proof: To prove Euler's formula we may proceed by induction on the number E of edges of the planar map.

If $E = 0$, we have a single isolated vertex, no edges, and one region: $V - E + R = 1 - 0 + 1 = 2$, as claimed. (For $E = 1$ we have two cases that do not, according to our principles of mathematical induction, have to be checked. However, see Exercise 4.3-19.)

We will assume that Euler's formula is true for planar maps having $E - 1$ edges and show that it holds for one having E edges. First we notice that one of the regions will always be infinite. If the boundary of this region has an edge that also borders a finite region, then removing that edge decreases the number of edges E and regions R by one each. (See Fig. 4.21.) The resulting planar map, by our induction assumption, satisfies Euler's formula. Hence $V - (E - 1) + (R - 1) = 2$; thus $V - E + R = 2$.

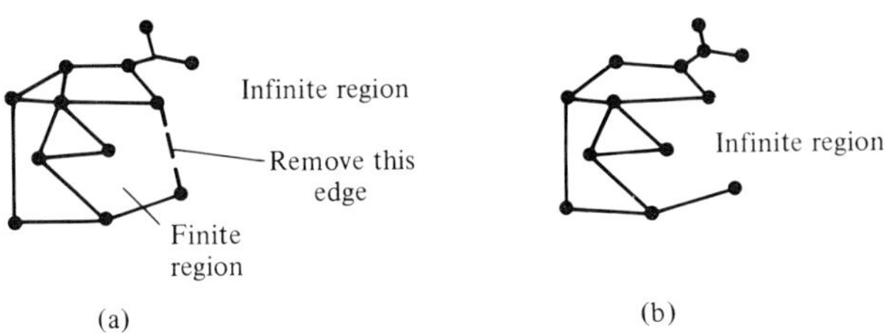

Figure 4.21 Proving Euler's formula by induction on the number of edges.

If none of the edges in the boundary of the infinite region are on two distinct regions, our planar map must come from a tree, where, as we know, $E = V - 1$ and $R = 1$. But $V - (V - 1) + 1 = 2$, and this verifies Euler's formula once again, and completes the proof. ■

Example 4 *We can use Euler's formula to show that $K_{3,3}$ is not planar.* In looking back at Fig. 4.19 we see that there are $V = 6$ vertices and $E = 9$ edges. Euler's formula says that any planar representation of $K_{3,3}$ must have $R = 5$ regions, since

$$V - E + R = 6 - 9 + 5 = 2.$$

But no three of the vertices in $K_{3,3}$ form a triangle, and so each region would have to have four or more edges in its boundary. That is, for each of the five regions r we must have $\deg(r) \geqslant 4$. Hence the sum of the degrees of the regions is at least $(4)(5)=20$. But Theorem 4.6 tells us that this is twice the number of edges. So we would have to have at least $20/2=10$ edges, giving us a contradiction! We have exactly nine edges. We conclude that $K_{3,3}$ is not planar. □

Here is another useful condition that a planar map must necessarily satisfy. We will sketch out its proof in the exercises. (See Exercises 4.3-14 and 4.3-15.)

Theorem 4.8 A connected planar map with V vertices, E edges ($E \geqslant 2$), and R regions without loops or multiple edges satisfies

$$\tfrac{3}{2}R \leqslant E \leqslant 3V-6.$$

Example 5 The complete graph K_5 on five vertices is not planar. To see this we first observe that for K_5 we have $E=10$ and $V=5$. Proceeding as in Example 4, Euler's formula tells us that if K_5 were planar, we must have $R=7$ regions, since

$$5-10+7=2.$$

But $\tfrac{3}{2}(7)=21/2>10$, contradicting $\tfrac{3}{2}R \leqslant E$ of Theorem 4.8. Hence we conclude that K_5 is not planar. □

It is clear that any graph containing either $K_{3,3}$ or K_5 as a subgraph is not planar either. (See Exercises 4.3-6 and 4.3-7.) As for the opposite case, when a graph *is* planar, we can state the following.

Theorem 4.9 ***Kuratowski's Theorem*** A graph is planar if and only if it does not contain K_5, or $K_{3,3}$, or any graph obtained from these by "inserting vertices of degree 2 into their edges" as a subgraph.

Kuratowski's theorem gives a simply stated and complete characterization of planar graphs, but its proof is fairly long and intricate. We will content ourselves with an illustration of what is meant by "inserting vertices of degree 2 . . . ."

Example 6 In Fig. 4.22(a) we have an example of a nonplanar graph that has neither K_5 nor $K_{3,3}$ as a subgraph, although it does resemble K_5. In fact it can be obtained from K_5

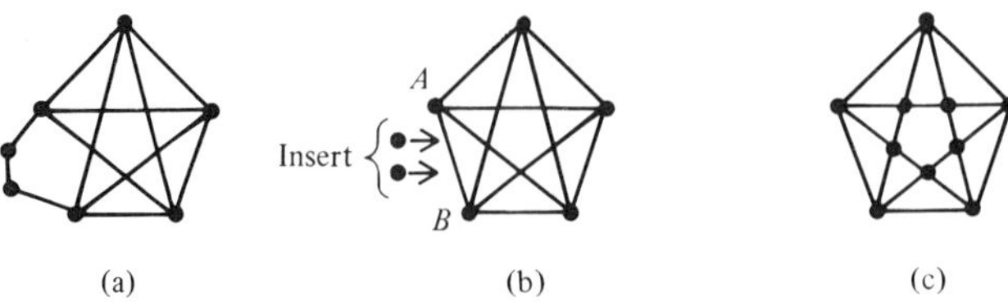

Figure 4.22 The graphs in (a) and (b) are nonplanar, but the graph in (c) is planar.

by inserting two vertices of degree 2 in edge AB, as indicated in Fig. 4.22(b). This is the kind of procedure that is covered in Kuratowski's theorem, and the given graph is not planar.

However, the graph represented by Fig. 4.22(c) is planar, in spite of its resemblance to K_5. Indeed this picture is a planar map of the graph. We can think of obtaining our map by inserting vertices of degree 4 into the edges of K_5 precisely where we expected the edges of K_5 to intersect. Hence we no longer have intersecting edges, and the map is planar. □

Completion Review 4.3

Complete each of the following.

1. A planar multigraph is one that can be drawn in the plane such that ________________ ________________.
2. A planar map is ________________. The same planar ________________ can be represented by different ________________.
3. If we draw a graph in the plane so that two of its edges appear to intersect at points that are not vertices, then the graph must be nonplanar. (*True or False*?)
4. The graphs $K_{3,3}$ and K_5 are examples of ________________ graphs. A theorem that uses these graphs to characterize planar graphs is called ________________ theorem.
5. Euler's formula says that if a connected planar map has V vertices, E edges, and R regions, then ________________.
6. An inequality concerning the number of vertices V, edges E ($E \geqslant 2$), and regions R in a connected planar map without loops or multiple edges states that ________________.
7. The degree of a region in a planar map is the length of a minimal ________________.
8. The sum of the degrees of the regions in a planar map is equal to ________________.

Answers: **1.** edges intersect only at their endpoints. **2.** a particular way to draw a planar graph without edge intersections; graph; planar maps. **3.** False. **4.** nonplanar graphs; Kuratowski's. **5.** $V-E+R=2$. **6.** $(3/2)R \leqslant E \leqslant 3V-6$. **7.** circuit around its boundary. **8.** twice the number of edges in the map.

Exercises 4.3

For Exercises 1 to 4 refer to the following planar maps.

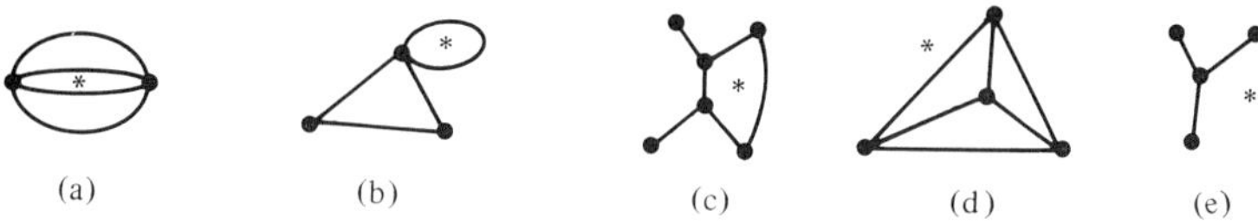

1. Find the degree of each of the regions indicated by an asterisk in each map.
2. What is the sum of the degrees of the regions in each map?
3. Give the dual of each of the planar maps.
4. Verify Euler's formula in the maps.
5. Which of the following combinations of vertices, edges, and regions are not possible for connected planar graphs? **(a)** $V=12$, $E=15$, $R=3$; **(b)** $V=8$, $E=12$, $R=6$; **(c)** $V=20$, $E=12$, $R=10$; **(d)** $V=13$, $E=18$, $R=8$. Use Euler's formula.
6. **a)** Is K_6 a planar graph?
   b) Is K_n for n greater than 5 a planar graph?
   c) Explain.
7. **a)** Is $K_{3,2}$ planar?
   b) Is $K_{n,2}$ for n greater than 3 planar?
   c) How about $K_{n,3}$?
8. Show that each of the following graphs is planar.

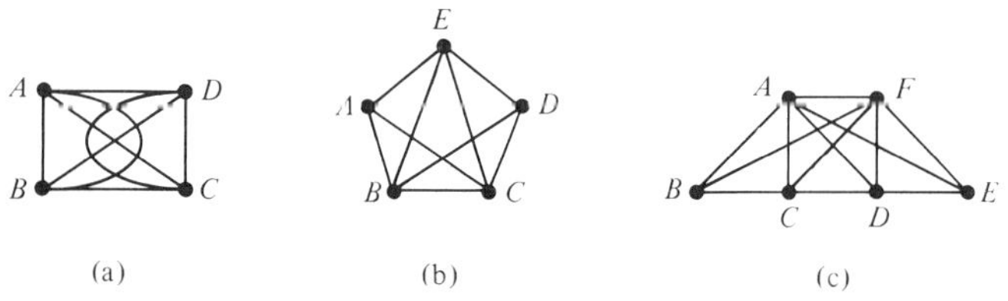

9. If the utility problem of Example 2 involved two houses and four utilities, what would the solution, if any, be?
10. The following solids are some examples of convex polyhedra. If we think of their vertices and edges in the obvious way and identify their faces with "regions," show that Euler's formula applies to each.

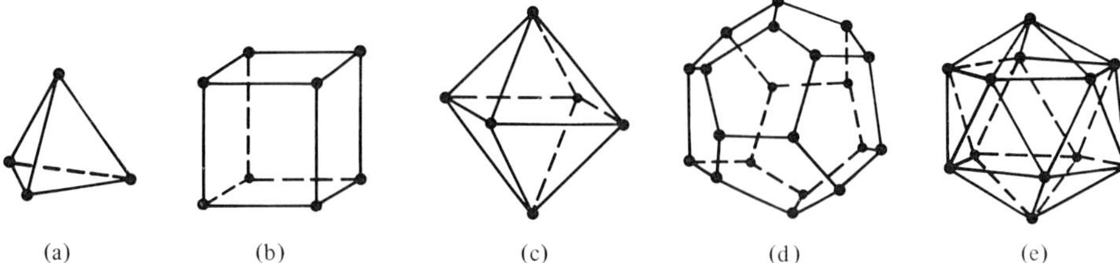

11. The solids shown are sometimes called "Platonic solids" because the philosopher Plato referred to them in his work *Timaeus*. The corresponding graphs are planar. Show this.
12. The solids in Exercise 10 are called a tetrahedron, cube, octahedron, dodecahedron, and icosahedron respectively. Show that the graph of the octahedron is the dual of the graph of the cube. Can you find other duals here?
13. A planar graph is called **completely regular** if both it and its dual are **regular**. (All vertices have the same degree.) Verify that the graphs of the Platonic solids are completely regular. [In fact they are the only completely regular graphs for which $\deg(v) \geqslant 3$.]
14. Prove the $\frac{3}{2}R \leqslant E$ part of Theorem 4.8 using the sum of the degrees of the regions, Theorem 4.6, and our assumption about $\deg(r)$.
15. Prove the $E \leqslant 3V - 6$ part of Theorem 4.8 by writing the first part $3R \leqslant 2E$ and using this together with Euler's formula.
16. Show that K_5 is not planar using only Theorem 4.8.
17. Verify that the following graph is not planar, and yet $E = 3V - 6$.

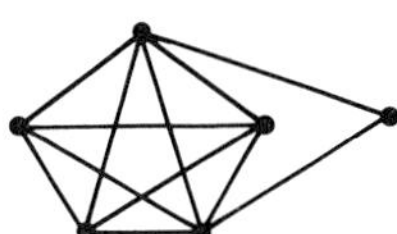

18. Show that the following graph (called Peterson's graph) has a subgraph that can be obtained from $K_{3,3}$ by inserting vertices of degree 2.

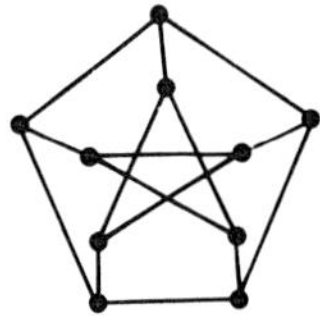

19. Verify the special case "$n = 1$ edge" in the proof of Euler's formula. (*Hint*: There are two possibilities.)
20. Suppose that a planar map has exactly two connected components, V vertices, E edges, and R regions. State and prove an analog of Euler's formula. Do the same assuming it has n connected components.

4.4 GRAPH COLORING

One of the most famous problems associated with planar graphs and, indeed, with all of graph theory, is known as the *four-color problem*. It was first mentioned in a letter from Augustus De Morgan (De Morgan's laws) to William Rowan Hamilton (Hamiltonian circuits) in 1852. In it De Morgan *asks whether any planar map can have its regions colored with four or fewer colors so that regions with a common boundary edge have a different color*. This is called **map coloring**. (The first statement of this problem has been traced back to the brother of a student of De Morgan's, Francis Guthrie.)

Example 1 The planar map in Fig. 4.23(a) requires only two different colors (say, red and green), one for the interior of the triangle and one for its exterior. No fewer than four colors can be used for the planar map in Fig. 4.23(b), however. And while we can use five colors to distinguish among the five regions in Fig. 4.23(c), the reader is invited to show that only three colors are necessary.

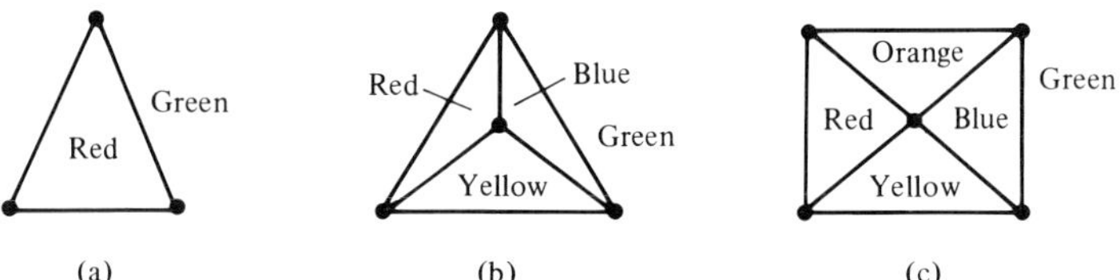

Figure 4.23 Some map colorings. Figure 4.23(c) can be colored with fewer colors. □

The astonishing thing about the four-color problem is not its statement, which is easy to understand, or even its solution (*yes, four colors suffice for every planar map*). It is that a correct proof of the affirmative answer had to wait until 1976, when K. Appel and W. Haken, two mathematicians at the University of Illinois, announced their conclusion after about 1200 hours of computer-assisted, case-by-case analysis. In light of this monumental effort, it is perhaps not surprising that so many earlier attempts had failed.

It may seem a bit silly to have spent so much precious computer time, not to mention over a hundred years of very active research, on such a problem. However, this research produced many reformulations of the problem that had great interest, in

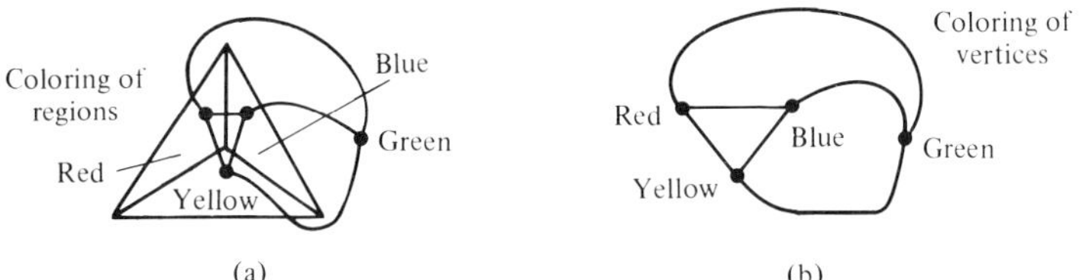

Figure 4.24 (a) Map coloring versus (b) graph coloring.

themselves, and many related ideas, so it was not a waste at all. (Some people are even a little sorry that the problem has been solved!) One such reformulation involves coloring the vertices of a graph rather than the regions. In fact as we can see in Fig. 4.24, *coloring the regions of a planar map is equivalent to coloring the vertices of its dual.*

Definition 4.12 A (a) **coloring of a graph** G is an assignment of colors to the vertices of the graph so that adjacent vertices have different colors. A graph is (b) ***n*-colorable** if it can be colored (or has a coloring) with n distinct colors. The (c) **chromatic number**, denoted $\chi(G)$, of G is the minimum number of colors needed to color G.

We can state Appel and Haken's result as Theorem 4.10.

Theorem 4.10 Every planar graph G has a chromatic number $\chi(G) \leq 4$. That is, every planar graph is four-colorable.

Recall that the regions of a planar map of a graph correspond to the vertices of the dual of the graph. Since the dual of every planar map (of a graph) is a planar graph, Theorem 4.10 implies that the regions of every planar map can be colored with four or fewer colors.

We are tempted to present a proof that every planar graph can be colored with five or fewer colors, since such a proof can be given in about two pages. (For a proof of this fact, just open most books on graph theory.) But we believe that the reader will gain more by attempting such a proof himself or herself, with helpful hints provided by the author. (See Exercise 4.4-13.)

Let us consider the colorability of two nonplanar graphs.

Example 2

a) *The chromatic number* $\chi(K_5) = 5$. Each of the five vertices K_5 is adjacent to each of the other four and, therefore, must receive a different color.

b) *The chromatic number* $\chi(K_{3,3}) = 2$. Even though we have six vertices in $K_{3,3}$, these are partitioned into two sets of three nonadjacent vertices. If we color the vertices in one set red and in the other set green, this two-coloring is obviously the best one can do. □

This example illustrates some rather obvious facts about the coloring of complete and bipartite graphs. We state these in Theorem 4.11, together with some other facts about graph coloring.

Theorem 4.11 ***Graph Coloring***

a) A graph G is two-colorable if and only if it is bipartite.

b) Every tree is two-colorable.

c) The complete graph K_n has chromatic number n.

d) If a graph G contains a clique on n vertices, then $\chi(G) \geqslant n$.

e) If the maximum degree of all the vertices in G is d, then $\chi(G) \leqslant d+1$.

Proof: To show that every bipartite graph is two-colorable, one proceeds as in Example 2(b). To prove the converse, if the vertices of G have been colored, say red and green, this produces a natural partition of the vertices of G, where no two red vertices are adjacent and no two green vertices are adjacent. Hence G is bipartite.

For part (b) begin by coloring any vertex v_0 of the tree red. Next recall that, by Exercise 4.2-22, there is a unique path between any two vertices of a tree. Hence the number of edges that a vertex is away from v_0 is well defined. In particular if u and v are both an even (or odd) number of vertices away from v_0, then u and v cannot be connected by an edge. Hence we can color all those vertices red that are an even number of edges away from v_0 and we can color the rest green without two adjacent vertices having the same color.

Part (c), regarding $\chi(K_n)=n$, is proved just like $\chi(K_5)=5$. As for part (d), we merely note that the chromatic number of a graph is clearly at least as large as any of its subgraphs. Finally a proof of part (e) is sketched out in Exercise 4.4-9. ■

Example 3 a) In the graph K_{10} the maximum vertex degree is $d=9$, which is the same for every vertex. By Theorem 4.11(c), however, we reach the upper bound given in part (e):

$$\chi(K_{10})=10=9+1=d+1.$$

b) The complete bipartite graph $K_{10,10}$ also has each of its vertices achieving the maximum vertex degree $d=10$. But Theorem 4.11(a) tells us that

$$\chi(K_{10,10})=2<10+1=d+1.$$

□

Example 3 shows us that, while the upper bound of $d+1$ is sometimes reached by the chromatic number of a graph, $d+1$ may greatly overestimate $\chi(G)$. Sometimes a "best" coloring can be obtained by inspection.

Example 4 In Fig. 4.25 we have a clique on the five vertices A, B, C, D, and E. Hence the chromatic number is at least 5, and we need at least the five "colors" a, b, c, d, and e. We cannot

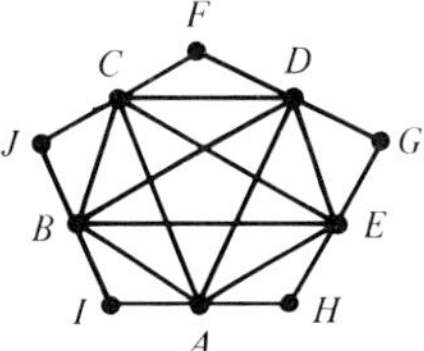

Figure 4.25 The chromatic number of this graph is 5.

use any of these colors to color *all* of the five remaining vertices, since one of the remaining vertices is always adjacent to exactly two of the original five. We do not need an additional color, however. Just assign colors a, b, c, d, and e to vertices F, G, H, I, and J, respectively, and the graph is colored with five colors. Hence its chromatic number is 5. □

The following algorithm, attributable to D. J. A. Welsh and M. B. Powell, can be used to color the vertices of any graph efficiently without exceeding the upper bound of Theorem 4.11(e), that is, the maximum vertex degree plus one.

Algorithm 4.2 ***Coloring Algorithm*** To color the vertices $v_1, v_2, \ldots, v_n$ of a graph $G = G(V, E)$ with a subset of the colors $c_1, c_2, \ldots, c_n$,

1. [Initialize S, *i*.] Set $S \leftarrow V$, $i \leftarrow 1$. (S is the set of uncolored vertices.)
2. [Coloring loop.] Repeat steps 3 through 6 until $S = \varnothing$.
    3. [List S.] List the vertices in S in order of decreasing degree. (This order may not be unique.)
    4. [Color subset of S.] Assign color c_i to the first vertex on the list and then, in sequence, to each vertex not adjacent to one that has been colored with c_i.
    5. [Update *i*.] Set $i \leftarrow i + 1$.
    6. [Update S.] Set $S \leftarrow S -$ the set of vertices colored in step 3.
7. [$S = \varnothing$.] Output the assignment of colors to vertices.
8. End of Algorithm 4.2.

If the Welsh-Powell algorithm is applied to the graph in Example 4, with its vertices listed $A, B, C, D, E, F, G, H, I, J$, one obtains $\chi(G) \leqslant 5$, which is less than the upper bound of $7 = 6 + 1$ indicated by the vertices of degree 6. (See Exercise 4.4-12.) This algorithm does not, however, always determine the chromatic number of a graph. (See Exercise 4.4-13.) Indeed an efficient algorithm has not yet been found that will determine $\chi(G)$ for an arbitrary finite graph.

We close this section by showing how coloring numbers can be used to model an important scheduling problem in computer science.

Example 5 One can sometimes store the values of variables that are used over and over again in a computer program in sections of a computer called registers. If this is possible, then the execution of a program may be speeded up. These variables are used at possibly overlapping periods of time during the program's execution, as in Fig. 4.26(a), where variable X needs three storage periods, Y needs two periods, and so on. The variable may be stored in different registers at different periods, but we can only store one variable at a time in each register.

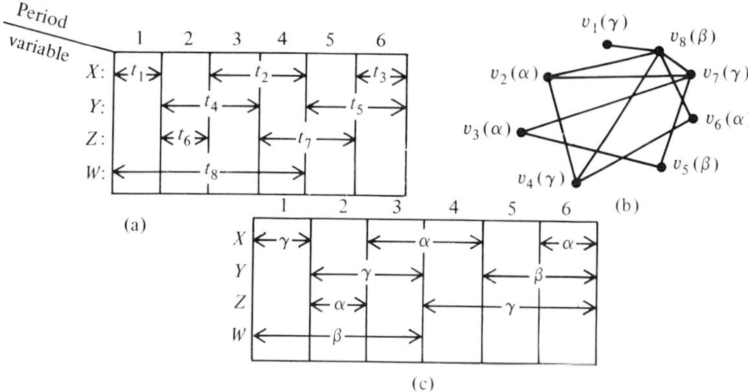

Figure 4.26 Each time interval t_i in (a) that a variable X, Y, Z, or W is used corresponds to a vertex v_i in (b). Vertices are adjacent when the corresponding time intervals overlap. Figure 4.26(c) gives a minimum allocation of registers corresponding to the minimum coloring of the graph in part (b).

To determine how many registers we will need during the course of the program, we may form a graph in which the vertices correspond to storage periods of each variable. These are joined by an edge when the corresponding periods overlap. Then we let each available register correspond to a color. The chromatic number of our graph will then correspond to the minimum number of registers that will be needed.

The largest clique in the graph we have constructed has three vertices. Hence $\chi(G) \geq 3$. After we use colors α, β, and γ to color vertices v_2, v_7, and v_8, respectively, in Fig. 4.26(b), it is easy to use these same colors to color the graph completely. Hence $\chi(G) = 3$, and a minimum of three registers is needed. [See Fig. 4.26(c).] □

Completion Review 4.4

Complete each of the following.

1. When coloring a planar map, one assigns different colors to ________________.
2. When coloring a graph, one assigns different colors to ________________.
3. The coloring number of graph G is the ________________. The coloring number of G is denoted ________________.
4. The coloring number of every planar graph is ________________, a fact that is called the ________________ theorem, which has been proved by ________________. This is also the greatest number of colors needed to color any ________________.
5. The coloring number of a nontrivial bipartite graph (with at least one edge) is ________________.
6. The coloring number of every tree having at least one edge is ________________.
7. A graph that can be colored with n colors is called ________________.
8. A graph having a clique on n vertices has coloring number ________________.
9. If the maximum degree of all the vertices of G is n, then the coloring number of G is ________________.
10. An algorithm that can be used to color G, in Problem 9, with $n+1$ colors or fewer, was found by ________________.

Answers: **1.** regions having a common boundary edge. **2.** adjacent vertices. **3.** least number of colors required to color G; $\chi(G)$. **4.** less than or equal to 4; four-color; Appel and Haken; planar map. **5.** 2. **6.** 2. **7.** n-colorable. **8.** at least n. **9.** less than or equal to $n+1$. **10.** Welsh and Powell.

Exercises 4.4

1. Color the regions of each of the following planar maps with a minimum number of colors.

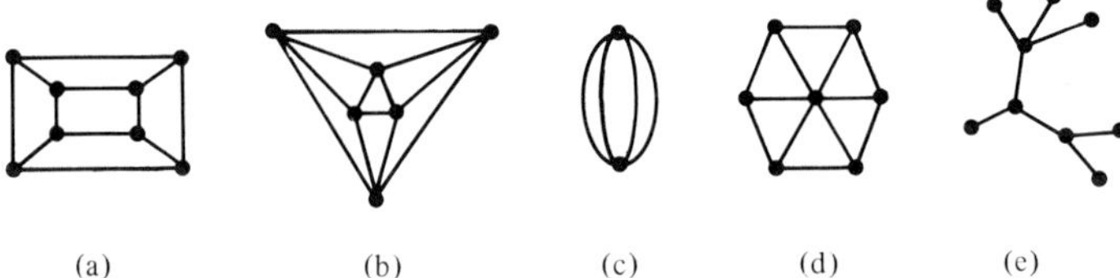

2. Color the vertices of each of the planar *graphs* in Exercise 1 with a minimum number of colors. What is $\chi(G)$ in each case?
3. What is $\chi(G)$ when $G=$ **(a)** K_{100}; **(b)** $K_{100,100}$; **(c)** a tree on 100 vertices?
4. "In light of the four-color theorem, if a graph has a chromatic number of 4 or less, it must be planar." Explain why this is false.
5. Supply an argument showing that $\chi(K_n)=n$.
6. Use Theorem 4.11 to show that every tree is bipartite.
7. Suppose that a graph G has two connected components, H and J. If $\chi(H)=m$ and $\chi(J)=n$, what is $\chi(G)$? Explain.
8. Suppose that a graph consists of a single polygon having n sides. When is the corresponding graph two-colorable? three-colorable?
9. Show that if the maximum degree of the vertices in a graph G is d, then $\chi(G)\leqslant d+1$. (*Hint*: Use induction on the number of vertices in the graph. Remove a vertex v_0 of maximum degree and its d edges. Why can one color the remaining graph with $d+1$ or fewer colors? Why can one of these $d+1$ colors be used for v_0, thereby coloring the entire graph? How does this complete the proof?)
10. Use the Welsh-Powell algorithm, Algorithm 4.2, to color the graph of Example 4.
11. Use the Welsh-Powell algorithm to find a coloring of each of the following graphs. Use the ordering of vertices in (c) to show that this algorithm may not give $\chi(G)$.

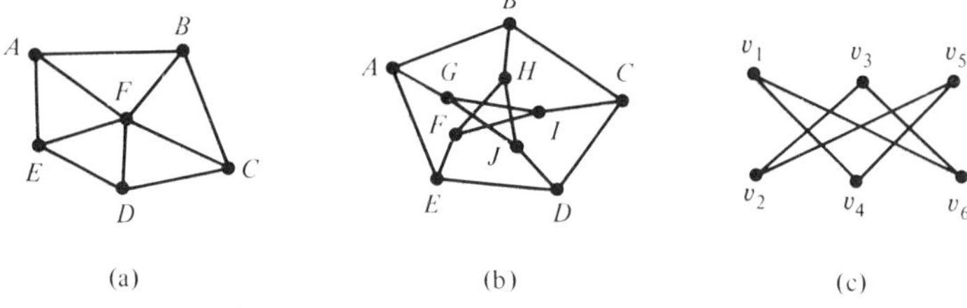

12. During a computer program's execution, we would like to store the values of five variables in a minimum number of registers (see Example 5) when the variables are used as in the periods of time in the following illustration. Model this as a graph-coloring problem; find the chromatic number of your model and the minimum number of registers.

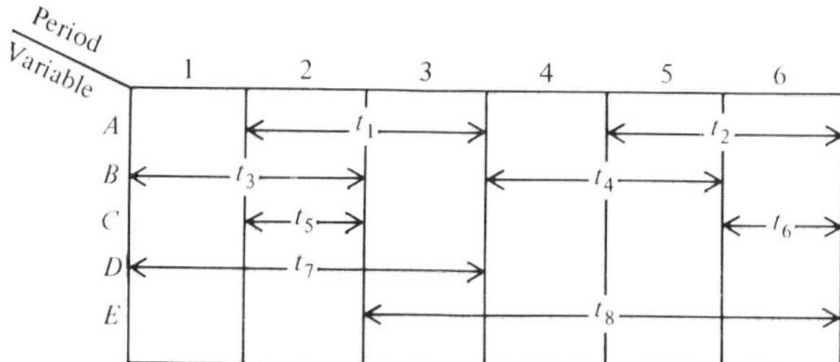

13. In this exercise we will sketch out a proof that every planar graph is five-colorable.

a) First show that a finite connected planar graph with at least three vertices has a vertex of degree 5 or less. (*Hint*: Use Theorem 4.9, the right-hand inequality.)

b) Begin by induction on the number of vertices n of G, with $n \leqslant 5$.

c) Suppose $n > 5$ and that the theorem holds for graphs with less than n vertices. What can you say about the graph G-v, where v is a vertex of degree 5 or less?

d) Do the case where the vertices adjacent to v have been colored with fewer than five colors.

e) In the remaining case, assume the vertices v_1, v_2, v_3, v_4, v_5 adjacent to v are colored with five different colors a, b, c, d, e and are arranged counterclockwise about G. Let H be the subgraph of $G - v$ generated by v_1 and v_3 (colored a and c). Do the case where v_1 and v_3 are in different components of H.

f) If v_1 and v_3 are in the same component of H, why is there a cycle C including the edges vv_1 and vv_3 that encloses v_2 or v_4?

g) Consider the subgraph generated by vertices colored b or d. Why are v_2 and v_4 in different components of K?

h) Complete the proof by interchanging colors b and d in the component containing v_2 and using b to color vertex v.

4.5 MULTIGRAPHS AND MATRICES

Diagrams are very useful for the purpose of defining graphs and multigraphs. They enable us to absorb a great deal of geometric information all at once, including which edges are incident with which vertices, which pairs of vertices are adjacent, and so on. But diagrams are not the only way that one can give the vertex–edge relationships that define a multigraph. In fact one can store the information necessary for defining a multigraph in a computer by means of an array of numbers called a "matrix."

Definition 4.13 A (a) **matrix** is a rectangualr array of numbers. If the matrix has m rows and n columns of entries, we say that it has (b) **dimensions *m* by *n***. Two matrices are (c) **equal** if and only if they have the same dimensions and corresponding entries are equal.

Example 1 Let

$$A = \begin{bmatrix} a_{11} & a_{12} & a_{13} \\ a_{21} & a_{22} & a_{23} \end{bmatrix}, \quad B = \begin{bmatrix} 2 & 1 & 0 \\ -1 & 0 & 3 \end{bmatrix}, \quad \text{and} \quad C = \begin{bmatrix} 2 & 1 \\ -1 & 0 \\ 0 & 4 \end{bmatrix}.$$

Then the dimensions of A, B, and C are 2 by 3, 2 by 3, and 3 by 2 respectively. $B \neq C$, since these matrices do not have the same dimensions. (It does not matter that some of the corresponding entries are equal.) $A = B$ if and only if $a_{11} = 2$, $a_{12} = 1$, $a_{13} = 0$, $a_{21} = -1$, $a_{22} = 0$, and $a_{23} = 3$. □

We say that the general entry of matrix A is a_{ij}, where the two subscripts tell us that a_{ij} is in the ith row and jth column of the matrix A. Where there is no danger of confusion, we sometimes write $A = (a_{ij})$, $B = (b_{ij})$, and so forth.

Definition 4.14 The n-by-m **incidence matrix** B of a multigraph G having vertices $v_1, v_2, \ldots, v_n$ and edges $e_1, e_2, \ldots, e_m$ is defined by

$$b_{ij} = \begin{cases} 1 \text{ if vertex } v_i \text{ is incident with edge } e_j. \\ 0 \text{ otherwise.} \end{cases}$$

Notice that the rows of the incidence matrix correspond to the vertices of G and the columns correspond to the edges.

Example 2 Given the multigraph G in Fig. 4.27 with vertices v_1, v_2, v_3 and edges e_1, e_2, e_3, e_4, we construct the incidence matrix B, shown to the right of G, having three rows and four columns. We have labeled the rows and columns of the matrix to make it easier to see how they correspond to vertices and edges of G.

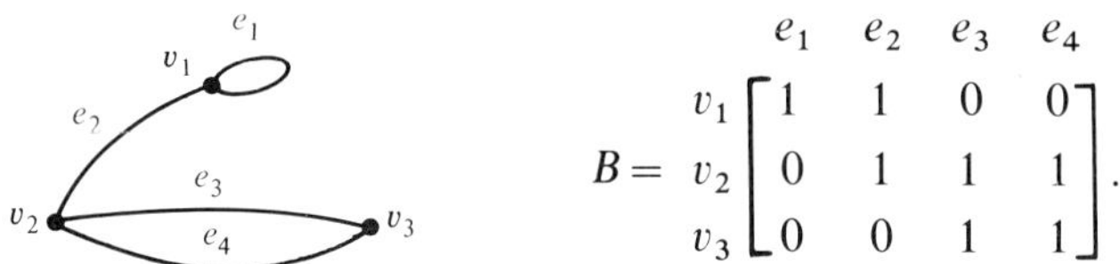

$$B = \begin{array}{c} \\ v_1 \\ v_2 \\ v_3 \end{array} \begin{array}{c} \begin{array}{cccc} e_1 & e_2 & e_3 & e_4 \end{array} \\ \begin{bmatrix} 1 & 1 & 0 & 0 \\ 0 & 1 & 1 & 1 \\ 0 & 0 & 1 & 1 \end{bmatrix} \end{array}.$$

Figure 4.27 A multigraph and its incidence matrix B.

We can obtain a great deal of information about the multigraph G just by looking at the matrix B. For example, it is clear that edge *e_1 is a loop, since there is only a single 1 in the first column.* This means that e_1 is incident with only one vertex. Moreover, since the 1's in the third and fourth columns are in the same rows, we can see that *e_3 and e_4 are multiple edges connecting the same pair of vertices*, and we can see this without looking at G. Matrix B also tells us that $\deg(v_2) = 3$, for example, since there are exactly three 1's in the second row and none of edges e_2, e_3, or e_4 is a loop. □

The incidence matrix of a multigraph is not unique, since a different labeling of the edges and vertices of the multigraph will, in general, produce a different incidence matrix. (See Exercise 4.5-1c.) Nevertheless, given any n-by-m matrix of 0's and 1's, having no more than two 1's in any one column, there corresponds one and only one multigraph. (Any two multigraphs having the given matrix as their incidence matrix are isomorphic, as defined in Section 4.3. See Exercise 4.5-17.)

One can give a more compact matrix description of vertex–edge incidence relationships in a multigraph using a two-by-m or an m-by-two "edge matrix," which may be defined as follows.

Definition 4.15 An **edge matrix** of a multigraph G having vertices $v_1, \ldots, v_n$ and edges $e_1, \ldots, e_m$ is an m-by-two (or two-by-m) matrix whose ith row (or ith column respectively) is the ordered pair (k, j), $k \leq j$, corresponding to the edge e_i joining vertices v_k and v_j.

Definition 4.15 requires that the edges be listed in lexicographic (or "dictionary") order, according to the pairs of vertices that they join. For example, an edge joining v_1 to v_2 is listed before an edge joining v_1 to v_3 or v_2 to v_3. Multiple edges and loops are easily represented, as the following example illustrates.

Example 3 The multigraph of Example 2 is given once again in Fig. 4.28, together with its two edge matrices, denoted by B_1 and B_2. The loop on vertex v_1 can be seen from the ordered pair (1, 1), for example. The multiple edges correspond to the identical third and fourth rows in B_1 (or third and fourth columns in B_2). □

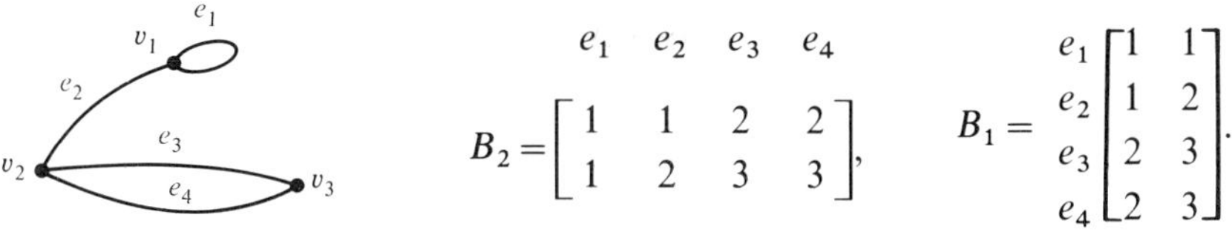

Figure 4.28 A multigraph and its edge matrices B_1 and B_2.

A multigraph is not, in general, defined by its edge matrices, since isolated vertices are not accounted for. (However, see Exercises 4.5-3 and 4.5-5.) Moreover, we say that the two edge matrices of a multigraph are *transposes* of one another, because *one may be obtained from the other by writing the ith row as the ith column, for every row*. We then write $B_1 = B_2^t$ (read "B_2 transpose": The "t" is not an exponent).

One can define a multigraph simply by telling how many edges are incident with each pair of vertices. For a graph this amounts to telling which pairs of distinct vertices are adjacent. The following matrix describes these adjacencies.

Definition 4.16 Let G be a graph having n vertices, labeled $v_1, v_2, \ldots, v_n$. Then the **adjacency matrix** A of G is the n-by-n matrix whose ij entry $a_{ij} = 1$ if vertex v_i is connected by an edge to vertex v_j; $a_{ij} = 0$ otherwise. (For the adjacency matrix of a multigraph, see Exercise 4.5-15.)

Example 4 Each of the graphs on four vertices in Fig. 4.29 has its adjacency matrix given directly beneath it. Notice that the *main diagonal entries, a_{ii}, in each matrix are all zero,* since a graph has no loops. Also observe that *each row has the same entries as the corresponding column,* since v_i is connected to v_j if and only if v_j is connected to v_i. Hence $A_1 = A_1^t$ and $A_2 = A_2^t$. In general we say that a matrix is **symmetric** if it equals its own transpose. Thus *every adjacency matrix is symmetric.* (See Exercise 4.5-8.) □

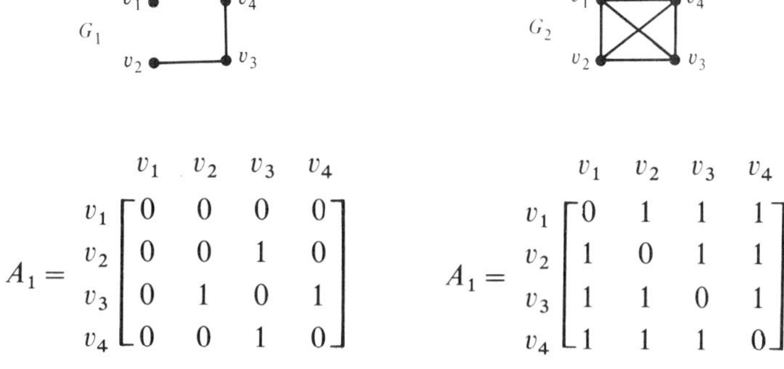

$$A_1 = \begin{array}{c} \\ v_1 \\ v_2 \\ v_3 \\ v_4 \end{array} \begin{array}{c} \begin{array}{cccc} v_1 & v_2 & v_3 & v_4 \end{array} \\ \begin{bmatrix} 0 & 0 & 0 & 0 \\ 0 & 0 & 1 & 0 \\ 0 & 1 & 0 & 1 \\ 0 & 0 & 1 & 0 \end{bmatrix} \end{array} \qquad A_1 = \begin{array}{c} \\ v_1 \\ v_2 \\ v_3 \\ v_4 \end{array} \begin{array}{c} \begin{array}{cccc} v_1 & v_2 & v_3 & v_4 \end{array} \\ \begin{bmatrix} 0 & 1 & 1 & 1 \\ 1 & 0 & 1 & 1 \\ 1 & 1 & 0 & 1 \\ 1 & 1 & 1 & 0 \end{bmatrix} \end{array}$$

Figure 4.29 Two graphs and their adjacency matrices.

It is particularly easy to draw a graph when we are given its adjacency matrix. We simply connect v_i to v_j with an edge whenever there is a 1 in the i–jth entry of the matrix. We can determine the vertex degrees, however, without drawing the graph. *To find the degree of vertex v_i, we count the number of 1's in row i (or column i) of the adjacency matrix.* For example, in matrix A_1 of Example 4, row 1 has no 1's; hence $\deg(v_1) = 0$. But row 3 has two 1's; hence $\deg(v_3) = 2$.

Matrix Addition and Multiplication by a Scalar

The utility of the adjacency matrix in studying a graph can only be appreciated if one knows how to perform the operations of matrix addition, multiplication by a scalar, and matrix multiplication. We briefly discuss the first two operations in this section, leaving matrix multiplication for Section 4.6.

Definition 4.17 Let $A = (a_{ij})$ and $B = (b_{ij})$ denote two m-by-n matrices, and let c be any real number. Then we define

$$A + B = (a_{ij} + b_{ij}) \qquad \text{and} \qquad cA = (ca_{ij})$$

as (a) **matrix addition** and (b) **matrix multiplication by a scalar** respectively.

Definition 4.17 says that each operation is carried out term by term. This is why it is important that the two matrices in a matrix sum have the same dimensions. Moreover, we may combine the two operations given in Definition 4.17, as in the following example.

Example 5 Let

$$A=\begin{bmatrix}1 & 3 & -1\\ 0 & 1 & 2\end{bmatrix},\quad B=\begin{bmatrix}2 & 1 & 0\\ 4 & -1 & 6\end{bmatrix},$$

and the constant $c=5$. Then $A+5B=$

$$\begin{bmatrix}1+5(2) & 3+5(1) & -1+5(0)\\ 0+5(4) & 1+5(-1) & 2+5(6)\end{bmatrix}=\begin{bmatrix}11 & 8 & -1\\ 20 & -4 & 32\end{bmatrix}.$$

□

Matrix addition and scalar multiplication obey many (but not all) of the familiar laws of algebra of real numbers. Here are a few examples.

Theorem 4.12 Let A, B, and C be any m-by-n matrices, and let r and s be any real numbers. Then:

a) $A+B=B+A$ (commutative law).

b) $A+(B+C)=(A+B)+C$ (associative law).

c) $r(A+B)=rA+rB$ (distributive law).

d) $(r+s)A=rA+sA$ (distributive law).

e) $(A+B)^t=A^t+B^t$ (law of transposes).

f) $(1)\,A=A$ (identity law).

Moreover, if Z is the m-by-n matrix that has only zero entries, then

g) $A+(-1)A=Z$.

h) $A+Z=A$.

You are asked to illustrate these laws in Exercises 4.5-10 and 4.5-11.

We end this section by giving an algorithm for the addition of two matrices having dimensions 2 by 3. This will give you an idea of how a computer might perform this operation.

Algorithm 4.3 Given matrices $A = (a_{ik})$ and $B = (b_{ik})$ having dimensions 2 by 3, we find $C = A + B = (c_{ik})$ by doing the following.

1. [Row changing loop.] Do steps 2 and 3 for $i = 1, 2$.
   2. [Column changing loop.] Repeat step 3 for $k = 1, 2, 3$.
      3. [Add corresponding entries.] Set $c_{ik} \leftarrow a_{ik} + b_{ik}$.
4. [Done.] Output the c_{ik}'s.
5. End of Algorithm 4.3.

Algorithm 4.3 calculates the entries of $A + B$ one row at a time, beginning with the first. One could give an algorithm that calculates the entries of $A + B$ one column at a time by simply modifying the "loop within a loop" pattern of this algorithm. (See Exercise 4.5-13.) Moreover, a similar algorithm could be developed for scalar multiplication of a matrix. (See Exercise 4.5-14.)

Completion Review 4.5

Complete each of the following.

1. A matrix is a ________________ of numbers.

2. A matrix has dimensions n by m if it has ________________ rows and ________________ columns.

3. Two matrices $A = (a_{ij})$ and $B = (b_{ij})$ are equal if A and B have the same ________________, and for all i and j, we have ________________.

4. Two matrices $A = (a_{ij})$ and $B = (b_{ij})$ can be added if they have the same ________________. In that case, if $A + B = C = (c_{ij})$, then $c_{ij} =$ ________________ for all i and j, and the dimensions of C are ________________.

5. If $A = (a_{ij})$ and c is a number, then $cA =$ ________________.

6. If $A = (a_{ij})$ has dimensions n by m, then A transpose, denoted ________________ and defined to be ________________, has dimensions ________________.

7. A matrix B is symmetric if $B =$ ________________.

8. If a multigraph G has n vertices and m edges, then the incidence matrix $B=(b_{ij})$ of G will have dimensions________________. Moreover, $b_{ij}=1$ if________________, and $b_{ij}=$________________ otherwise.

9. The edge matrix of G in Exercise 8 has dimensions________________.

10. The adjacency matrix A of a graph G as in Exercise 8 has dimensions ________________. Moreover, $a_{ij}=1$ if________________ and $a_{ij}=$________________ otherwise. The adjacency matrix of a graph is always a ________________ matrix.

Answers: **1.** rectangular array. **2.** n; m. **3.** dimensions; $a_{ij}=b_{ij}$. **4.** dimensions; $a_{ij}+b_{ij}$; the same as those of A and B. **5.** $c \cdot a_{ij}$. **6.** A^t; (a_{ji}); m by n. **7.** B^t. **8.** n by m; vertex i is incident with edge j; 0. **9.** 2 by m or m by two. **10.** n by n; vertex i is adjacent to vertex j; 0; symmetric.

Exercises 4.5

1. Find the **(a)** incidence matrix and **(b)** the edge matrix for the following multigraph. **(c)** Give the incidence matrix if the labels e_1 and e_2 are switched.

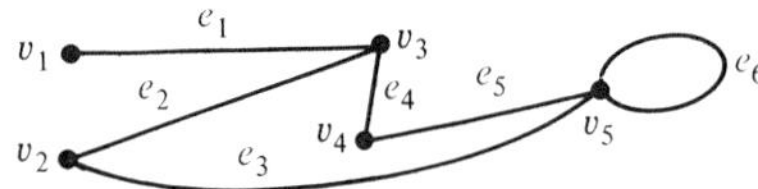

2. Draw a multigraph having the following incidence matrix.

	e_1	e_2	e_3	e_4	e_5
v_1	1	0	1	0	1
v_2	0	1	0	1	1
v_3	1	1	0	1	0

3. Draw a multigraph with the following edge matrix in addition to an isolated vertex.

$$\begin{bmatrix} 1 & 1 & 2 & 2 & 2 & 3 & 3 & 5 \\ 1 & 1 & 3 & 5 & 6 & 4 & 6 & 6 \end{bmatrix}.$$

4. Find the incidence matrix for **(a)** $K_{3,3}$; **(b)** K_5.

5. Find a rule for determining the degree of a vertex from the edge matrix of a multigraph. Prove it.

6. Find the adjacency matrix for each of the following graphs.

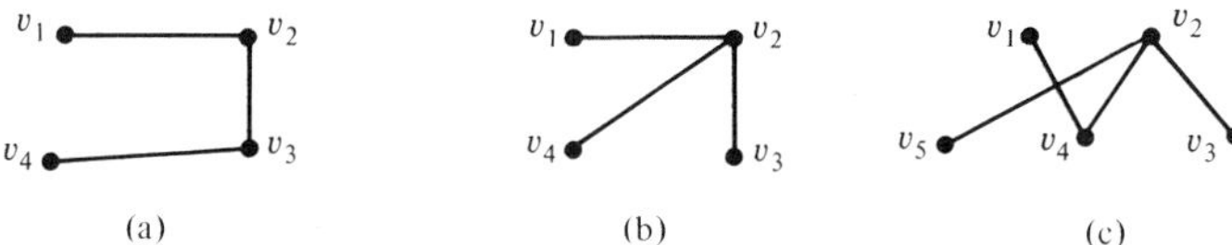

7. Draw the graph whose adjacency matrix is

$$A = \begin{array}{c} \\ v_1 \\ v_2 \\ v_3 \\ v_4 \\ v_5 \end{array} \begin{array}{c} \begin{array}{ccccc} v_1 & v_2 & v_3 & v_4 & v_5 \end{array} \\ \begin{bmatrix} 0 & 1 & 1 & 0 & 1 \\ 1 & 0 & 1 & 0 & 0 \\ 1 & 1 & 0 & 1 & 0 \\ 0 & 0 & 1 & 0 & 1 \\ 1 & 0 & 0 & 1 & 0 \end{bmatrix} \end{array}.$$

8. Prove that $A = A^t$ for the adjacency matrix of any graph.

In Exercises 9 to 11 let

$$A = \begin{bmatrix} 1 & 0 \\ 2 & -1 \end{bmatrix}, \quad B = \begin{bmatrix} 3 & 1 \\ 0 & 4 \end{bmatrix}, \quad \text{and} \quad C = \begin{bmatrix} -1 & 2 \\ 1 & 3 \end{bmatrix}.$$

9. What are the dimensions of these matrices?

10. Find **(a)** $A+B$; **(b)** $B+C$; **(c)** $(A+B)+C$; **(d)** $A+(B+C)$.

11. Letting $r=2$ and $s=-3$, illustrate each part of Theorem 4.12.

12. Modify Algorithm 4.3 so that it will calculate the sum of two matrices that have dimensions 4 by 5.

13. Modify Algorithm 4.3 so that it will calculate the sum of two matrices by columns instead of by rows.

14. Invent an algorithm, patterned after Algorithm 4.3, that will find the scalar product cA, where A is a 4-by-5 matrix.

15. The **adjacency matrix of a multigraph** is defined by $A=(a_{ij})$, where $a_{ij}=k$ iff vertices v_i and v_j of the multigraph are connected by k edges. Find the adjacency matrix of the Königsberg bridges multigraph.

16. Draw the multigraph whose adjacency matrix is the following.

$$A = \begin{array}{c} \\ v_1 \\ v_2 \\ v_3 \\ v_4 \\ v_5 \end{array} \begin{array}{c} \begin{array}{ccccc} v_1 & v_2 & v_3 & v_4 & v_5 \end{array} \\ \begin{bmatrix} 0 & 1 & 2 & 0 & 1 \\ 1 & 0 & 3 & 1 & 0 \\ 2 & 3 & 0 & 1 & 1 \\ 0 & 1 & 1 & 0 & 0 \\ 1 & 0 & 1 & 0 & 0 \end{bmatrix} \end{array}.$$

17. a) Show that any two graphs having the same incidence matrices are isomorphic.
b) Show that the converse is false.

18. a) Show that any two graphs having the same adjacency matrices are isomorphic.
b) Show that the converse is false.

4.6 MATRIX MULTIPLICATION AND CONNECTEDNESS

Connectedness properties of a multigraph may be studied by means of an operation called matrix multiplication, in which one matrix is multiplied by another. Those who are familiar with this operation can skip ahead to the part of this section entitled "Connectedness." For those who are not familiar with this operation, or would like a review, we present the following definition.

Definition 4.18 Let A denote an m-by-n matrix and let B denote an n-by-r matrix. Then the **matrix product**

$$AB = C = (c_{ij})$$

is the m-by-r matrix whose entries are defined by

$$c_{ij} = \sum_{k=1}^{n} a_{ik}b_{kj} = a_{i1}b_{1j} + a_{i2}b_{2j} + \cdots + a_{in}b_{nj}$$

for all $i = 1, 2, \ldots, m$ and all $j = 1, 2, \ldots, r$.

The double subscripts in our definition can be a little confusing, and so it might help to think of the sum in Definition 4.18 in terms of "the ith row of A times the jth column of B," where corresponding terms in the row and column are multiplied and the products are then added. (See Fig. 4.30.)

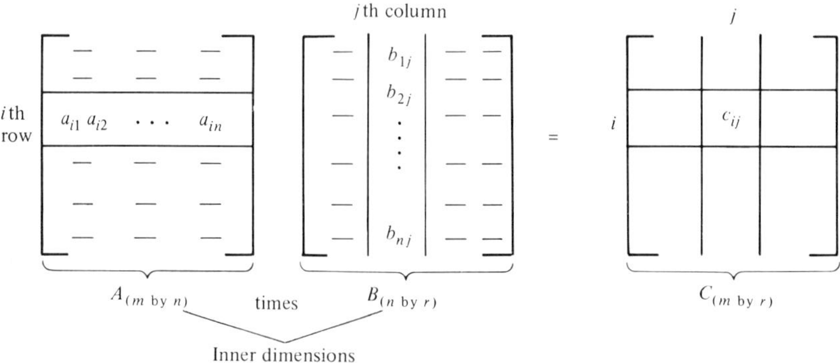

Figure 4.30 Determining the entry $c_{ij} = \sum_{k=1}^{n} a_{ik}b_{kj}$ in the matrix product AB.

Notice both in Definition 4.18 and Fig. 4.30 how the inner dimensions of the matrices A and B in the product AB must agree. Thus we cannot even form the matrix product BA unless the dimensions r and m are the same.

Example 1 Let $A = \begin{bmatrix} 2 & 1 & 0 \\ -1 & 0 & 3 \end{bmatrix}$ and $B = \begin{bmatrix} 2 & 1 \\ 4 & 5 \\ 1 & 6 \end{bmatrix}$.

The dimensions of A and B are 2 by 3 and 3 by 2 respectively.

First observe that we cannot form the product $A \cdot A$, since a 2-by-3 matrix can only be multiplied by a 3-by-r matrix. However, both matrix products AB and BA are defined. Before we even calculate these products, we can see that they are not the same.

If $AB = C$ and $BA = D$, let us compare c_{11} with d_{11}.

$c_{11} = (2)(2) + (1)(4) + (0)(1) = 8.$ (First row of A times first column of B.)

$d_{11} = (2)(2) + (1)(-1) = 3.$ (First row of B times first column of A.)

Hence the corresponding matrices are not equal. In fact

a) $$AB = \begin{bmatrix} (2)(2) + (1)(4) + (0)(1) & (2)(1) + (1)(5) + (0)(6) \\ (-1)(2) + (0)(4) + (3)(1) & (-1)(1) + (0)(5) + (3)(6) \end{bmatrix} = \begin{bmatrix} 8 & 7 \\ 1 & 17 \end{bmatrix},$$

and

b) $$BA = \begin{bmatrix} (2)(2) + (1)(-1) & (2)(1) + (1)(0) & (2)(0) + (1)(3) \\ (4)(2) + (5)(-1) & (4)(1) + (5)(0) & (4)(0) + (5)(3) \\ (1)(2) + (6)(-1) & (1)(1) + (6)(0) & (1)(0) + (6)(3) \end{bmatrix} = \begin{bmatrix} 3 & 2 & 3 \\ 3 & 4 & 15 \\ -4 & 1 & 18 \end{bmatrix}.$$ □

One can give an algorithm for calculating matrix products, as in Algorithm 4.4.

Algorithm 4.4 To calculate the product $AB = (c_{ij})$ of matrices $A = (a_{ik})$ and $B = (b_{kj})$, where A and B have dimensions 2 by 3 and 3 by 3, respectively,

1. [Rows loop.] Repeat steps 2 through 5 for $i = 1, 2$.
    2. [Column loop.] Repeat steps 3 through 5 for $j = 1, 2, 3$.
        3. [Initialize c_{ij}.] Set $c_{ij} \leftarrow 0$.
        4. [Sum products loop.] Repeat step 5 for $k = 1, 2, 3$.
            5. [Sum products.] Set $c_j \leftarrow c_{ij} + a_{ik}b_{kj}$. (We continue with step 6 when $i = 2$, $j = 3$, and $k = 3$.)
6. [Done.] Output the c_{ij}'s.
7. End of Algorithm 4.4.

The reader should refer to the calculation of AB in Example 1 and see what i, j, and k are at each step of the calculation of, for example, c_{21}. (Also see Exercises 4.6-5 and 4.6-6.)

If A is an n-by-m matrix, one can multiply it on the left by an n-by-n square matrix and on the right by an m-by-m square matrix. If we choose matrices, in each case, whose **main diagonal** entries (entries $\mathbf{a}_{ii}$) are 1 and whose other entries are 0, we just obtain A again. Hence these special matrices are called n-by-n and m-by-m **identity matrices**, I_n and I_m respectively. For example,

$$I_3 = \begin{bmatrix} 1 & 0 & 0 \\ 0 & 1 & 0 \\ 0 & 0 & 1 \end{bmatrix} \quad \text{and} \quad I_2 = \begin{bmatrix} 1 & 0 \\ 0 & 1 \end{bmatrix}.$$

If we then take matrix $A = \begin{bmatrix} 2 & 1 & 0 \\ -1 & 0 & 3 \end{bmatrix}$ of Example 1, we get

$$I_2 \cdot A = A \quad \text{and} \quad A \cdot I_3 = A.$$

Matrix multiplication is not a commutative operation, as we have already illustrated in Examples 1(a) and (b). However, this operation does possess certain useful algebraic properties that are much like ordinary multiplication of real numbers. We list the following for future reference.

Theorem 4.13 Let A, B, and C be matrices, and let k be a real number. When the dimensions of these matrices permit the indicated operations, the following identities hold.

a) $A(BC) = (AB)C$ (associative law).

b) $A(B + C) = AB + AC$ and $(B + C)A = BA + CA$ (distributive laws).

c) $AI = IA = A$ (identity matrix law).

d) $(AB)^t = B^t A^t$ (law of transposes).

e) $k(AB) = (kA)B = A(kB)$ (scalar factorization law).

The reader is asked to verify these properties in Exercise 4.6-9 for 2-by-2 matrices, that is, those of the form

$$A = \begin{bmatrix} a_{11} & a_{12} \\ a_{21} & a_{22} \end{bmatrix}.$$

Let us point out once again that "A^t" denotes, not a power of the matrix A, but rather its transpose, as defined in the preceding section. One can define powers of a matrix, however, for square (n-by-n) matrices, by the following rule.

Definition 4.19 If A is an n-by-n matrix, then $A^0 = I_n$, and $A^{k+1} = A^k \cdot A$, for $k \geqslant 0$.

As an illustration, if $A = \begin{bmatrix} 1 & 0 \\ 2 & 3 \end{bmatrix}$, then $A^0 = \begin{bmatrix} 1 & 0 \\ 0 & 1 \end{bmatrix}$, and

$$A^1 = A^0 \cdot A = IA = A.$$

Hence

$$A^2 = A \cdot A = \begin{bmatrix} 1 & 0 \\ 8 & 9 \end{bmatrix},$$

$$A^3 = A^2 \cdot A = \begin{bmatrix} 1 & 0 \\ 26 & 27 \end{bmatrix},$$

and so on. Observe that $A \cdot A^k = A^k \cdot A$ for all integers $k \geqslant 0$. (See Exercise 4.6-22.)

Connectedness

Recall that a walk $v_0 e_1 v_1 e_2 \ldots e_n v_n$ in a multigraph is an alternating sequence of incident vertices and edges. The **length of this walk** is defined to be n, the number of edges, where some edges may be repeated. Thus there is a walk of length 1 connecting two distinct vertices v_i and v_j in a graph if and only if these vertices are connected by an edge. As we discussed in the previous section, this edge is indicated in the adjacency matrix of the graph by an entry of "1" in the ijth position.

The adjacency matrix of a graph can also be used to study walks of length greater than 1 between the graphs vertices. More precisely, we need the *powers of the adjacency matrix*, as the following theorem indicates.

Theorem 4.14 If A is the adjacency matrix of a graph G, then the number of walks in G from vertex v_i to vertex v_j of length k ($k \geqslant 1$) is given by the ijth entry of the matrix A^k.

Proof: We prove this theorem by induction on k. In fact we have already verified the case $k = 1$ in the preceding paragraphs.

Hence let us assume that A^r gives us the number of walks of length r from vertex v_i to vertex v_j and, therefore, that this number is the ijth entry of A^r, which we denote by $a_{ij}^{(r)}$. Now if A has dimensions n by n, so do its powers, A^k. (See Exercise 4.6-11.) Moreover,

$$a_{ij}^{(r+1)} = \sum_{s=1}^{n} a_{is}^{(r)} a_{sj}^{(1)}.$$

Each of the products in the foregoing summation will be nonzero iff there is a walk of length r from vertex v_i to vertex v_s and an edge connecting v_s to vertex v_j. This is precisely how we get walks of length $r+1$ from vertex v_i to vertex v_j. Since, by assumption, $a_{is}^{(r)}$ = the number of walks of length r from vertex i to some vertex s, and $a_{sj}^{(1)} = 0$ or 1, we see that the summation adds up the number of walks from vertex i to vertex j of length $r+1$. This shows that A^{r+1} satisfies the statement of Theorem 4.19, completing the proof by mathematical induction. ■

In Exercise 4.6-12 you are asked to prove the corresponding theorem for multigraphs.

Example 2 Consider the graph whose adjacency matrix A and its square A^2 are given by

$$A = \begin{bmatrix} 0 & 1 & 0 & 1 & 0 \\ 1 & 0 & 1 & 0 & 0 \\ 0 & 1 & 0 & 1 & 1 \\ 1 & 0 & 1 & 0 & 0 \\ 0 & 0 & 1 & 0 & 0 \end{bmatrix} \quad \text{and} \quad A^2 = \begin{bmatrix} 2 & 0 & \boxed{2} & 0 & 0 \\ 0 & 2 & 0 & 2 & 1 \\ 2 & 0 & \boxed{3} & 0 & 0 \\ 0 & 2 & 0 & 2 & 1 \\ 0 & 1 & 0 & 1 & 1 \end{bmatrix}.$$

Then the entry $a_{13}^{(2)} = 2$, for example, tells us that there are two walks of length 2 from v_1 to v_3. Likewise, $a_{33}^{(2)} = 3$ indicates that there are three walks of length 2 from v_3 to v_3. If we draw the graph G corresponding to the adjacency matrix A, as in Fig. 4.31, we can verify that the walks from v_1 to v_3 are $v_1v_2v_3$ and $v_1v_4v_3$, and the walks from v_3 to v_3 are v_3,v_4v_3, $v_3v_5v_3$, and $v_3v_2v_3$. (The vertex sequences are sufficient to define a walk in a graph. Why?)

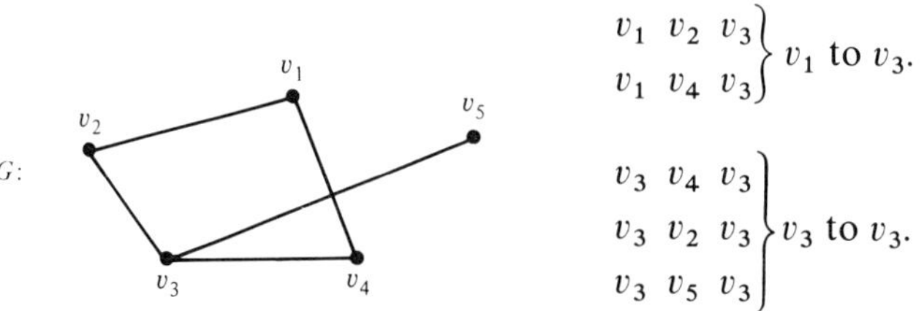

Figure 4.31 Some walks of length 2.

Any two vertices in a connected graph having n vertices are connected by a walk of length $n-1$ or less. (See Exercise 4.6-19.) Therefore, A is the adjacency matrix of a connected graph if and only if at least one of the matrices $I, A, A^2, A^3, \ldots, A^{n-1}$ has a positive integer in the ijth position for some i and j from 1 to n. Since each of these matrices has only nonnegative entries, this amounts to the following.

Theorem 4.15 Let A be the adjacency matrix of a graph G having n vertices. Then

a) G is connected iff $I + A + A^2 + \cdots + A^{n-1}$ has only positive entries.

And

b) G is connected iff $(I + A)^{n-1}$ has only positive entries.

Proof: Part (a) is obvious from what we have already discussed, since a sum of nonnegative numbers (corresponding to the entries of the A^k) is nonnegative. The sum is positive iff at least one addend is positive.

Part (b) is seen by writing

$$(I + A)^{n-1} = \sum_{k=0}^{n-1} C(n-1, k)A^k I^{n-1-k} = \sum_{k=0}^{n-1} C(n-1, k)A^k$$

and observing that each of the binomial coefficients is positive. This completes the proof. ∎

Example 3 The adjacency matrix A of the graph in Fig. 4.31 and its powers A^2, A^3, and A^4 are given in the following.

$$A = \begin{bmatrix} 0 & 1 & 0 & 1 & 0 \\ 1 & 0 & 1 & 0 & 0 \\ 0 & 1 & 0 & 1 & 1 \\ 1 & 0 & 1 & 0 & 0 \\ 0 & 0 & 1 & 0 & 0 \end{bmatrix}, \quad A^2 = \begin{bmatrix} 2 & 0 & 2 & 0 & 0 \\ 0 & 2 & 0 & 2 & 1 \\ 2 & 0 & 3 & 0 & 0 \\ 0 & 2 & 0 & 2 & 1 \\ 0 & 1 & 0 & 1 & 1 \end{bmatrix}, \quad A^3 = \begin{bmatrix} 0 & 4 & 0 & 4 & 2 \\ 4 & 0 & 5 & 0 & 0 \\ 0 & 5 & 0 & 5 & 3 \\ 4 & 0 & 5 & 0 & 0 \\ 2 & 0 & 3 & 0 & 0 \end{bmatrix},$$

$$A^4 = \begin{bmatrix} 8 & 0 & 10 & 0 & 0 \\ 0 & 9 & 0 & 9 & 5 \\ 10 & 0 & 13 & 0 & 0 \\ 0 & 9 & 0 & 9 & 5 \\ 0 & 5 & 0 & 5 & 3 \end{bmatrix}.$$

Hence

$$I + A + A^2 + A^3 + A^4 = \begin{bmatrix} 11 & 5 & 12 & 5 & 2 \\ 5 & 12 & 6 & 11 & 6 \\ 12 & 6 & 17 & 6 & 4 \\ 5 & 11 & 6 & 12 & 6 \\ 2 & 6 & 4 & 6 & 5 \end{bmatrix},$$

which has only positive entries, as guaranteed by Theorem 4.15(a). The calculation of $(I + A)^4$ requires fewer additions. (Why?) We obtain

$$(I + A)^4 = \begin{bmatrix} 1 & 1 & 0 & 1 & 0 \\ 1 & 1 & 1 & 0 & 0 \\ 0 & 1 & 1 & 1 & 1 \\ 1 & 0 & 1 & 1 & 0 \\ 0 & 0 & 1 & 0 & 1 \end{bmatrix}^4 = \begin{bmatrix} 21 & 20 & 22 & 20 & 8 \\ 20 & 22 & 24 & 21 & 11 \\ 22 & 24 & 32 & 24 & 16 \\ 20 & 21 & 24 & 22 & 11 \\ 8 & 11 & 16 & 11 & 10 \end{bmatrix},$$

whose positive entries were guaranteed by Theorem 4.15(b). □

Storing the entries of even relatively small matrices such as those in Example 3 can take up a rather large amount of computer memory. This problem is compounded when the entries themselves get large. For example, if we write the entries of $I + A + A^2 + A^3 + A^4$, found in Example 3, in binary notation, we obtain the matrix

$$\begin{bmatrix} 1011 & 101 & 1100 & 101 & 10 \\ 101 & 1100 & 110 & 1011 & 110 \\ 1100 & 110 & 10001 & 110 & 100 \\ 101 & 1011 & 110 & 1100 & 110 \\ 10 & 110 & 100 & 110 & 101 \end{bmatrix},$$

which has a total of 82 bits of information. (We count each digit as 1 bit.) A savings can be gained in terms of storage in memory if we define the following.

Definition 4.20 The **path matrix** of a graph having adjacency matrix A is the matrix obtained from $(I + A)^{n-1}$ by making every positive entry of $(I + A)^{n-1}$ a 1 and leaving every zero entry 0.

Thus the path matrix of the graph in Fig. 4.31 is

$$\begin{bmatrix} 1 & 1 & 1 & 1 & 1 \\ 1 & 1 & 1 & 1 & 1 \\ 1 & 1 & 1 & 1 & 1 \\ 1 & 1 & 1 & 1 & 1 \\ 1 & 1 & 1 & 1 & 1 \end{bmatrix},$$

which contains only 25 bits of information, compared with the 82 bits in the matrix $I + A + A^2 + A^3 + A^4$.

It is clear that *the path matrix of a graph has only positive entries if and only if the graph is connected.* (See Exercise 4.6-20.)

Completion Review 4.6

Complete each of the following.

1. If A has dimensions n by m and B has dimensions r by s, then we can form the product AB if and only if________________.
2. If A and B are as in Exercise 1 and $AB = C = (c_{ij})$, then C has dimensions ________________ and $c_{ij} =$________________.
3. In general $AB = BA$ for matrices A and B (*True or False*?)
4. The identity matrix I_n has all 1's on its________________ and________________ elsewhere.
5. If A is an n-by-n matrix, then we define $A^0 =$________________, and $A^n =$________________ for positive integers n.
6. If A is the adjacency matrix of a graph G and $A^k = (a_{ij})$, then a_{ij} equals the number of ________________.
7. If A is the n-by-n adjacency matrix of G, then $(I + A)^{n-1}$ and $I + A + A^2 + \cdots + A^{n-1}$ have no positive entries iff________________.
8. If $P = (p_{ij})$ is the path matrix of a graph G, then $p_{ij} = 1$ iff________________ and $p_{ij} =$________________ otherwise.

Answers: **1.** $m = r$. **2.** n by s; $\sum_{k=1}^{m} a_{ik}b_{kj}$. **3.** False. **4.** major diagonal; 0's. **5.** I_n; $A^{n-1}A$. **6.** walks of length k from vertex i to vertex j. **7.** G is connected. **8.** vertex i is connected (by a path) to vertex j; 0.

Exercises 4.6

For each pair of matrices A and B in Exercises 1 to 4, find AB and BA whenever possible.

1. $A=\begin{bmatrix}1 & -1 & 0\\ 0 & 1 & 2\end{bmatrix}$ and $B=\begin{bmatrix}0 & 3\\ 1 & 1\\ 2 & 4\end{bmatrix}$.

2. $A=\begin{bmatrix}1 & 2 & 3\\ 4 & 5 & 6\end{bmatrix}$ and $B=\begin{bmatrix}0 & 1 & 0\\ 4 & 1 & 5\end{bmatrix}$.

3. $A=\begin{bmatrix}1 & 2 & 3\\ 4 & 5 & 6\\ 7 & 8 & 9\end{bmatrix}$ and $B=\begin{bmatrix}1 & 0 & 0\\ 0 & 1 & 0\\ 0 & 0 & 1\end{bmatrix}$.

4. $A=[1\ \ 1\ \ 1\ \ 1\ \ 1]$ and $B=\begin{bmatrix}1\\ 1\\ 1\\ 1\\ 1\end{bmatrix}$.

5. Use Algorithm 4.4 to calculate the product AB in Example 1. In particular, find the values of i, j, and k in the calculation of c_{21} at each step.

6. Modify Algorithm 4.4 so that it will calculate the product of a 3-by-2 times a 2-by-3 matrix.

7. Write an algorithm for calculating the kth power A^k of an n-by-n matrix in terms of Definition 4.19.

8. Verify that all parts of Theorem 4.13 hold if

$$A=\begin{bmatrix}2 & -1\\ 0 & 1\end{bmatrix},\quad B=\begin{bmatrix}1 & 2\\ -2 & 0\end{bmatrix},\quad C=\begin{bmatrix}0 & 1\\ 3 & 4\end{bmatrix},\quad \text{and } k=3.$$

9. Verify that all parts of Theorem 4.13 hold for general 2-by-2 matrices A, B, and C.

10. Find A^3, if A is the matrix in Exercise 8.

11. Prove by mathematical induction: If A has dimensions n by n, then so do all its powers A^k.

12. Modify the proof of Theorem 4.14 so that it proves a version of Theorem 4.14 for multigraphs.

For Exercises 13 to 16 refer to the following adjacency matrices. Vertices correspond to row numbers. Graph G_i corresponds to matrix A_i.

$$A_1=\begin{bmatrix}0 & 1 & 1 & 0\\ 1 & 0 & 1 & 1\\ 1 & 1 & 0 & 0\\ 0 & 1 & 0 & 0\end{bmatrix},\quad A_2=\begin{bmatrix}0 & 0 & 1 & 1\\ 0 & 0 & 0 & 1\\ 1 & 0 & 0 & 1\\ 1 & 1 & 1 & 0\end{bmatrix},\quad A_3=\begin{bmatrix}0 & 1 & 1 & 0 & 1\\ 1 & 0 & 0 & 1 & 1\\ 1 & 0 & 0 & 1 & 0\\ 0 & 1 & 1 & 0 & 1\\ 1 & 1 & 0 & 1 & 0\end{bmatrix},\quad A_4=\begin{bmatrix}0 & 1 & 0 & 0 & 0 & 1\\ 1 & 0 & 1 & 0 & 0 & 0\\ 0 & 1 & 0 & 1 & 0 & 0\\ 0 & 0 & 1 & 0 & 1 & 0\\ 0 & 0 & 0 & 1 & 0 & 1\\ 1 & 0 & 0 & 0 & 1 & 0\end{bmatrix}.$$

13. Calculate A_i^2 for each matrix and use it to calculate the number of walks from v_1 to v_2 in each graph of length 2.
14. For what n must you calculate $(I + A_i)^{n-1}$ in order to find out if G_i is connected for each matrix?
15. Use Theorem 4.15(b) to calculate the path matrix corresponding to each A_i.
16. How many bits of information are stored in the entries of **(a)** $(I + A_i)^3$ for A_1; **(b)** the path matrix of G_1?
17. Show that $a_{ii}^{(2)} = 0$ only if v_i is an isolated vertex. What does this tell you about $I + A + \cdots + A^{n-1}$? (*Hint*: Is I really needed? Explain.)
18. How can one determine the connected component containing a vertex v_i by looking at the path matrix of the graph?
19. Prove that any two vertices in a connected graph having n vertices are connected by a walk of length $n - 1$ or less.
20. Prove that the path matrix of a graph has all positive entries if and only if the graph is connected.
21. Prove that the connected component of a graph containing vertex v_i corresponds to those vertices v_j for which there is a 1 in row i column j of the path matrix of G. (Now look at Exercise 18 again!)
22. Prove by induction that $A^k \cdot A = A \cdot A^k$ for all integers $k \geqslant 0$. (*Hint*: Use the associative law, Theorem 4.13.)
23. The path matrix of a graph G is the adjacency matrix of the **transitive closure** of G. Draw the transitive closure of each of the graphs G_i referred to in Exercises 13 to 16.

COMPUTER PROGRAMMING EXERCISES

(Also see the appendix and Programs A14 to A19.)

4.1. Write a program that will read in a symmetric matrix (a two-dimensional array) of zeros and ones and print out, for each i and j, "vertex i is (or is not)" adjacent to vertex j according to whether the ijth entry is or is not a 1 respectively.

4.2. Write a program that will print the sum of an n-by-m matrix A with an n-by-m matrix B. (See Algorithm 4.3.)

4.3. Write a program that will print the product of an n-by-m matrix A times an m-by-r matrix B. (See Algorithm 4.4.)

4.4. Write a program that will find $I + A + A^2 + \cdots + A^n$ by nesting, for an n-by-n matrix A.

4.5. Write a program that will determine the vertex degrees of the vertices and the number of edges of a multigraph that is inputted as an adjacency matrix.

4.6. Write a program that calculates the path matrix of a graph, given its adjacency matrix. Have your program determine if the graph is connected.

4.7. Write a program that uses Euler's theorem, Theorem 4.3 to determine whether a graph (or multigraph) has an Euler trail or circuit.

4.8. Write a program that finds an Euler trail or circuit in a graph or multigraph, given its adjacency matrix.

4.9. Write a program that finds a Hamiltonian path or circuit in a graph, or tells if there are not any, given the adjacency matrix of the graph.

4.10. Write a program that decides if a graph is bipartite and, if it is, lists the disjoint sets of edges.

4.11. Write a program that implements Welsh and Powell's graph-coloring algorithm, 4.2, and colors a graph inputted as an adjacency matrix.

4.12. Write the programs of Exercises 8 to 11 with the multigraph inputted as an edge matrix.

CHAPTER 5

TREES AND ALGORITHMS

The class of graphs called trees (connected graphs having no cycles) occupies a position of considerable importance in the theory and applications of graphs. They are of special interest to computer scientists, especially those who are engaged in the day-to-day application and analysis of data structures and algorithms. The first two sections in this chapter deal with what are called "minimization algorithms," in which we try to find the "shortest way." Section 5.3 is concerned with organization of data by means of trees, and in Section 5.4 we return to the problem of searching a file for a particular "key," or content.

5.1 WEIGHTED GRAPHS AND THE CONNECTOR PROBLEM

Many problems having to do with graphs, and with trees in particular, have numbers called **weights** assigned to the edges of the graph. If e belongs to $E(G)$, the set of edges of a graph G, we will write $w(e)$ for the weight assigned to e. In most cases it will be natural to assume that the weights of any edges are positive, especially when, as in Fig. 5.1, these weights represent the *distance between locations*. Other typical interpretations of these weights are *cost*, *time*, or *capacity* (as in, for example, the capacity of a pipeline to carry natural gas).

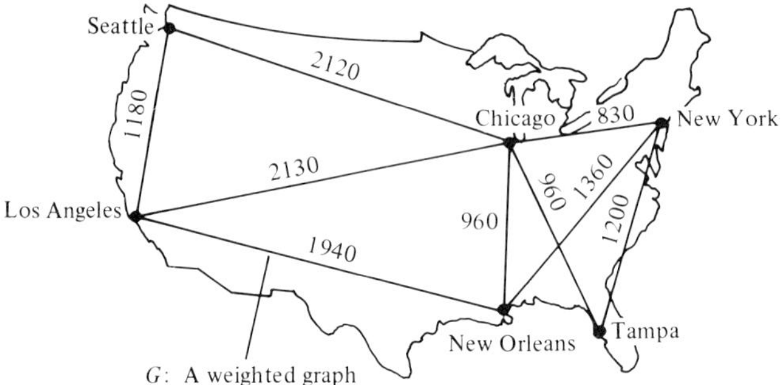

Figure 5.1 A road map of the United States can be made into a weighted graph by letting distances between locations (vertices) correspond to weights.

It is sometimes appropriate to sum the weights on the edges of a subgraph of a given weighted graph. For example, we might interpret the weights on the edges of the subgraph as the length of a path, or the cost of building a communications network. In Fig. 5.1 the sum of the weights of the edges linking New York with Chicago and Chicago with Los Angeles is

$$830 + 2130 = 2960,$$

giving the length of this particular path from New York to Los Angeles. So that we may sum the weights on the edges of our graphs, we will assume that, as usual, our graphs have only finitely many vertices and edges.

Definition 5.1 If each edge e in the edge set $E(G)$ of a graph G is assigned a number $w(e)$, we will say that G is a (a) **weighted graph**. If H is a subgraph of G, we will say that the (b) **weight of H, $w(H)$** is the sum of all the weights on the edges of H, and we will write

$$w(H) = \sum_{e \in E(H)} w(e).$$

Let us take the weighted graph in Fig. 5.1 for G. The graph in Fig. 5.2, then, is a subgraph of G linking Chicago with all the other cities (vertices). Calling this subgraph H, we have

$$w(H) = 2120 + 2130 + 960 + 830 + 960 = 7000.$$

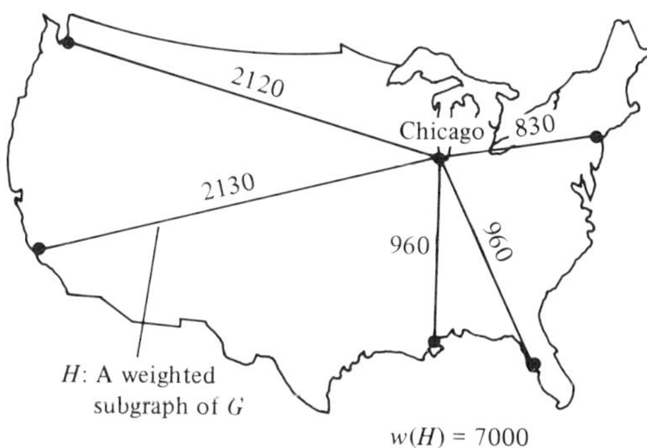

Figure 5.2 H is a weighted subgraph of G, the graph in Fig. 5.1.

The Connector Problem

Suppose that we want to build a communications network that links together several cities. Since modern communications are nearly instantaneous, via telephone lines for example, there is no need to look for shortest routes between the cities. In fact let us suppose that *our problem is to construct this network for the minimum possible cost*. Then what we have to know is not the distance between pairs of cities, but rather the cost of building a direct line of communications between each pair. More precisely we have Definition 5.2.

Definition 5.2 If G is a weighted graph and $w(e)$ is taken to be the cost of each edge, then the problem of finding a connected subgraph H of G that includes each vertex of G and such that the cost of H,

$$w(H) = \sum_{e \in E(H)} w(e),$$

is a minimum for all such subgraphs of G is known as the **connector problem**.

Looking back at Figs. 5.1 and 5.2, we see that in Fig. 5.2 we have certainly found a subgraph connecting all the cities. If we interpret the weights as costs, rather than distances, then the cost of this network is 7000. We cannot be sure that it is a minimum cost, however, because there are many other ways to connect all the cities with the edges given in Fig. 5.1. (Take the path visiting the cities of Tampa, New York, Chicago, Seattle, Los Angeles, and New Orleans in that order, for example. What is its cost?)

Example 1 Let us consider a weighted graph in which the weights are somewhat smaller numbers as in Figs 5.3(a) and (b). We have indicated two subgraphs, H_1 and H_2, of this graph that link together all the vertices of the graph. Now

$$\begin{aligned} w(H_1) &= w(EF) + w(AC) + w(CD) + w(CJ) + w(DJ) + w(EJ) + w(FB) \\ &= 1 + 1 + 3 + 1 + 2 + 1 + 1 \\ &= 10. \end{aligned}$$

whereas

$$\begin{aligned} w(H_2) &= w(AC) + w(CJ) + w(DJ) + w(EJ) + w(EF) + w(FB) \\ &= 1 + 1 + 2 + 1 + 1 + 1 \\ &= 7. \end{aligned}$$

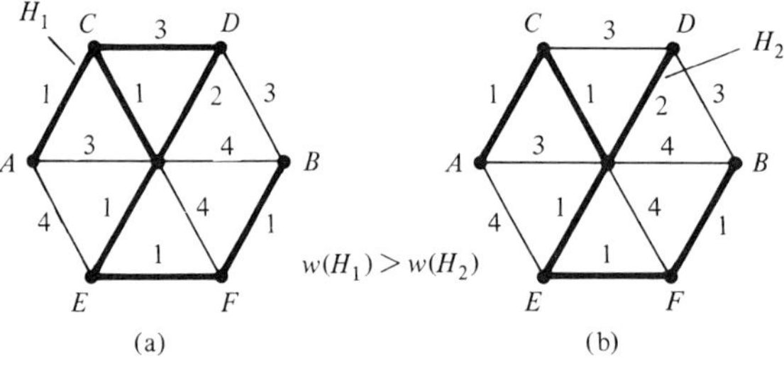

Figure 5.3 Both subgraphs H_1 and H_2 span the graph, but only H_2 is a tree.

We have reduced the cost of H_1 simply by removing edge CD. This broke the cycle CDJ and left the remaining subgraph H_2 connected. It also subtracted the cost $w(CD)$ from the cost $w(H_1)$, giving us the reduced cost $w(H_2)$. Whether this yields a minimum possible cost is still open to question, however, because, once again, we have only compared two ways of connecting all vertices. □

The example discussed should bring out at least one useful piece of information: *Any subgraph of G that solves the connector problem will not have any cycles* because, just as in the foregoing, we can remove an edge of any existing cycles and obtain a subgraph connecting all the vertices of the graph and having a lower cost. But this means that *our solution must be a tree that includes all the vertices of the graph.*

Definition 5.3 Let G be a graph. Then a subgraph T of G is called a **spanning tree of G** if T is a tree containing all the vertices of G.

In other words a solution of the connector problem is a **minimum spanning tree of the graph**, that is, a spanning tree with a *minimum weighted edge sum*. Of course, if our original graph is not connected, it will not have a spanning tree and there will be no solution for that graph. In what follows, however, we will assume that our graphs are complete. (For graphs that are not complete, we can add missing edges, weighting these with prohibitively high costs.)

Now all we have to do is to find the right spanning tree. We could conceivably find a minimum spanning tree by comparing the weighted edge sums of all the spanning trees of our graph and taking those with the lowest sum. Unfortunately the following theorem makes this procedure very unattractive. (See Exercise 5.1-16 for the proof.)

Theorem 5.1 A complete graph on n vertices has n^{n-2} spanning trees.

Example 2 By Theorem 5.1 K_5 has $5^{5-2} = 5^3 = 125$ spanning trees. K_{10} has $10^8 = 100{,}000{,}000$ spanning trees. Even a very fast digital computer would be hard pressed to examine all of these for the purpose of solving the connector problem. And this is for a graph having only 10 vertices! □

Fortunately one does not have to examine all the spanning trees of a graph to find the shortest one. The following common-sense procedure turns out to be rather effective.

Algorithm 5.1 ***Minimum Spanning Tree Algorithm*** Suppose that the edges e of the complete graph K_n have each been given a positive weight $w(e)$. To find a minimum spanning tree on the graph,

1. [Initial edge.] Choose an edge e_1 so that $w(e_1)$ is as small as possible.
2. [Next edges.] Repeat step 3 as long as possible.
   3. [Tree loop.] Having chosen edges $e_1, e_2, \ldots, e_k$, next choose edge e_{k+1} so that (a) the subgraph on edges $e_1, e_2, \ldots, e_{k+1}$ has no cycles, and (b) $w(e_{k+1})$ is as small as possible subject to (a).
4. [Done.] Output the subgraph on the selected edges.
5. End Algorithm 5.1.

This procedure, sometimes called **Kruskal's algorithm** (after J. B. Kruskal), or the "greedy algorithm," will stop after $n-1$ edges have been chosen to link together all n vertices of the complete graph K_n. This is a vast improvement over the examination of all n^{n-2} spanning trees of K_n.

A good first step in implementing this algorithm might be to *list the edges in order of increasing weight*, perhaps by using one of the sorting algorithms we discussed earlier (Section 3.7). If the number of vertices is not large, say, 10 or less, one can skip this step and implement the method by hand, using only a sketch of K_n to check conditions (a) and (b) of step 3. One might have to obtain this sketch from a matrix, called the "distance matrix of the graph," which gives the weights of each of the edges.

Definition 5.4 A **distance matrix** D of a weighted graph K_n is the n-by-n matrix whose entries are defined by $d_{ij} = w(v_iv_j)$ for every edge v_iv_j in K_n, and where $d_{ii} = 0$ for all $i = 1, 2, \ldots, n$.

As indicated in Definition 5.4, for a graph G that is not complete we can add all missing edges and give these outstandingly high weights that indicate their special "fictitious" status. In this way we can define a distance matrix for any graph.

Example 3 Let us find a minimum spanning tree for a complete graph whose vertices are A, B, C, D, E, and F and whose distance matrix is

$$\begin{array}{c} \\ A \\ B \\ C \\ D \\ E \\ F \end{array}\begin{array}{c} \begin{array}{cccccc} A & B & C & D & E & F \end{array} \\ \begin{bmatrix} 0 & 4 & 6 & 3 & 8 & 5 \\ 4 & 0 & 2 & 7 & 5 & 7 \\ 6 & 2 & 0 & 4 & 6 & 9 \\ 3 & 7 & 4 & 0 & 7 & 8 \\ 8 & 5 & 6 & 7 & 0 & 6 \\ 5 & 7 & 9 & 8 & 6 & 0 \end{bmatrix} \end{array} \qquad \text{Examples} \begin{cases} w(AB)=4 \\ w(AC)=6 \\ w(BD)=7 \end{cases}$$

We can sketch this graph, showing the weights on each edge, as in Fig. 5.4. Here we have six vertices and $6^4 = 1296$ spanning trees—but we will only need five steps to find one having the minimum weighted edge sum, which in this case is

$$2+3+4+5+5=19.$$

[See Fig. 5.5(a) to (e).]

Notice that in Fig. 5.5(c), for example, $w(CD) < w(AF)$, but we cannot choose edge CD without forming the cycle $ABCDA$. □

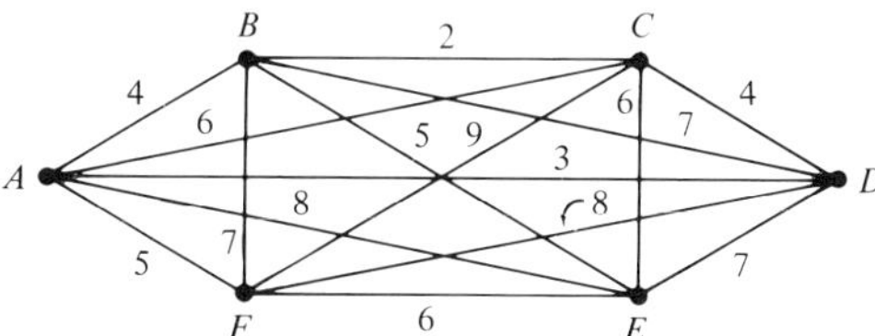

Figure 5.4 A weighted complete graph.

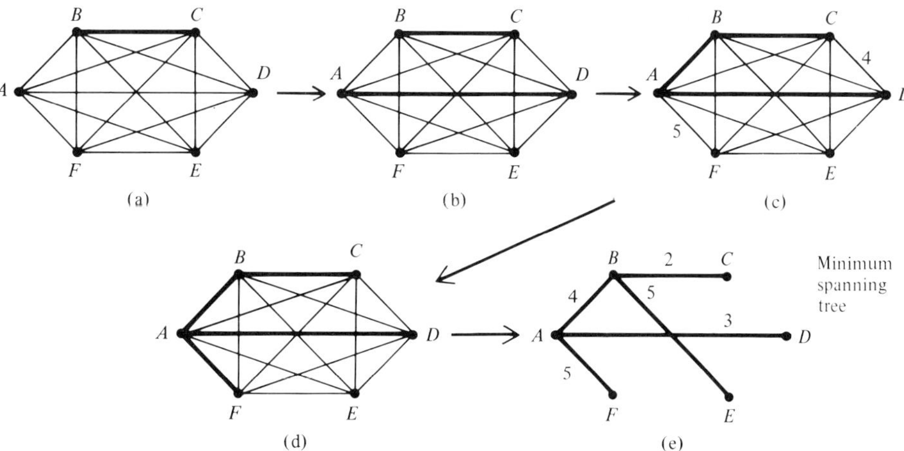

Figure 5.5 Determining a minimum spanning tree in the graph of Fig. 5.4 using Kruskal's greedy algorithm.

Proof of Kruskal's Algorithm

A proof of Algorithm 5.1's effectiveness can be based upon the following facts about trees. We will leave some of the details of the proofs of these facts to the reader.

Theorem 5.2 Every tree on n vertices ($n \geqslant 2$) has at least one vertex of degree 1.

Proof: Convince yourself that if this were not the case, then some vertex of the tree would be on a cycle, which is impossible. (See Exercise 5.1-14.) ■

Theorem 5.3 Every tree on n vertices has exactly $n-1$ edges.

Proof: We can prove this by induction on n. If $n=1$, there are $1-1=0$ edges. Suppose Theorem 5.3 is true for some $n=k-1 \geqslant 1$, and consider a tree with k vertices. Find a vertex v having degree 1. Remove this vertex and its one edge from the tree. What remains is a tree on $k-1$ vertices. By our induction assumption this tree has $(k-1)-1=k-2$ edges. Hence our original tree had $(k-2)+1=k-1$ edges, and we are done. ■

Theorem 5.4 A connected graph on n vertices having $n-1$ edges must be a tree.

Proof: If our connected graph were not a tree, we could remove edges one at a time and break all cycles, leaving a connected subgraph at each stage. When we have broken the last cycle, we have a tree on n vertices that has fewer than $n-1$ edges. But this contradicts Theorem 5.4, and so we must have had a tree to begin with. This completes the proof. ■

To prove that Kruskal's algorithm works, let T^* be a spanning tree we obtain with its aid, where we have chosen the edges $e_1, e_2, \ldots, e_{n-1}$ in that order. Also let $w(T)$ denote the weighted-edge sum of any spanning tree T of our graph G. If S is any minimum spanning tree of G and $S \neq T^*$, let e_i be the first edge in the list $e_1, e_2, \ldots, e_{n-1}$ of edges of T^* that is not in S. If $e_i = AB$ and $P(AB)$ is a path in S connecting A and B, then $P(AB)$ together with AB form a cycle. This cycle, in turn, must have an edge e_s that does not belong to T^*. Otherwise T^* would have a cycle. If we add edge e_i and take away edge e_s from tree S, we will have a connected graph S^* with $n-1$ edges. By Theorem 5.4, S^* is a tree.

What is the weighted-edge sum of S^*? It is given by

$$w(S^*) = w(S) + w(e_i) - w(e_s).$$

Remember that we are assuming S is a minimum spanning tree, and therefore that it has the smallest possible weighted-edge sum. So $w(S^*) \geqq w(S)$ and $w(e_i) - w(e_s) \geqq 0$. In other words,

$$w(e_i) \geqq w(e_s).$$

But e_i was added to the list $e_1, e_2, \ldots, e_{i-1}$ because it had the smallest weight among those that did not produce a cycle. So

$$w(e_i) \leqq w(e_s)$$

and we conclude that

$$w(e_i) = w(e_s).$$

Hence

$$w(S^*) = w(S).$$

But S^* has one more edge in common with T^* than S. We can obviously continue substituting edges in this way until we finally obtain tree T^* and $w(T^*) = w(S)$; that is, T^* has a minimum weighted-edge sum too. ■

Completion Review 5.1

Complete each of the following.

1. If each edge of a graph G is assigned a number $w(e)$, then G is called a ________________. The number $w(e)$ is called the ________________ of the edge e.
2. The weight $w(H)$ of a subgraph H of G is defined to be the ________________.
3. The connector problem is that of finding a ________________ subgraph of G having the ________________ weight that includes ________________.
4. The subgraphs of a connected graph G that are solutions of the connector problem are the ________________ of G. They may be found by means of an algorithm by ________________.

5. The number of different labeled spanning trees of the graph K_n is______________.

6. Every tree must have at least one vertex of degree______________.

7. Every tree on n vertices has exactly______________ edges.

8. A connected graph on n vertices having exactly $n-1$ edges must be a

______________.

Answers: **1.** weighted graph; weight. **2.** sum of the weighted edges in H. **3.** connected; minimum; all vertices of G. **4.** minimum spanning trees; Kruskal. **5.** n^{n-2}. **6.** 1. **7.** $n-1$. **8.** tree.

Exercises 5.1

1. Find all the spanning trees of the following graphs.

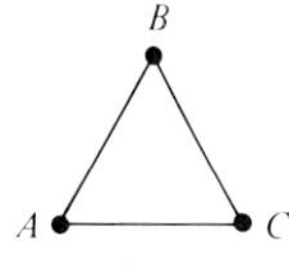

(a)

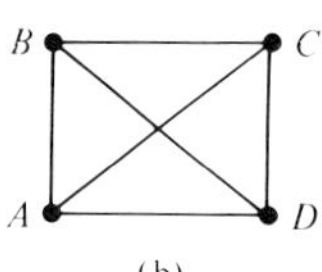

(b)

2. If a connected graph has n vertices, how many edges will each spanning tree of the graph have?

3. Translate your answer to Exercise 2 into an improved version of step 3 of Kruskal's greedy algorithm.

4. How many spanning trees does the graph K_3 have? K_6? K_{16}?

5. A **spanning forest** consists of the union of the spanning trees of each of the connected components of a graph. If a graph G consists of two connected components K_n and K_m, respectively, how many spanning forests does G have?

6. Find a minimum spanning tree for each of the following distance matrices. (You may assume that the vertices are labeled $v_1, v_2, \ldots$. Begin by drawing the graph, and give the minimum "cost" at the end.)

(a) $\begin{bmatrix} 0 & 3 & 5 & 4 \\ 3 & 0 & 6 & 2 \\ 5 & 6 & 0 & 2 \\ 4 & 2 & 2 & 0 \end{bmatrix}$; **(b)** $\begin{bmatrix} 0 & 1 & 3 & 4 & 6 & 5 \\ 1 & 0 & 5 & 7 & 3 & 4 \\ 3 & 5 & 0 & 6 & 2 & 3 \\ 4 & 7 & 6 & 0 & 1 & 5 \\ 6 & 3 & 2 & 1 & 0 & 4 \\ 5 & 4 & 3 & 5 & 4 & 0 \end{bmatrix}$; **(c)** $\begin{bmatrix} 0 & 3 & 2 & 7 & 9 & 8 & 6 \\ 3 & 0 & 6 & 5 & 4 & 5 & 6 \\ 2 & 6 & 0 & 1 & 2 & 3 & 9 \\ 7 & 5 & 1 & 0 & 6 & 7 & 8 \\ 9 & 4 & 2 & 6 & 0 & 2 & 5 \\ 8 & 5 & 3 & 7 & 2 & 0 & 1 \\ 6 & 6 & 9 & 8 & 5 & 1 & 0 \end{bmatrix}$.

7. The following table gives the airline distances between the six cities Tokyo, New York, Paris, London, Mexico City, and Peking to the nearest mile. Find a minimum spanning tree on the corresponding weighted graph and the weight of this tree.

	New York	London	Paris	Mexico City	Tokyo	Peking
New York	—	3469	3636	2090	6757	6844
London	3469	—	214	5558	5959	5074
Paris	3636	214	—	5725	6053	5120
Mexico City	2090	5558	5725	—	7035	7753
Tokyo	6757	5959	6053	7035	—	1307
Peking	6844	5074	5120	7753	1307	—

8. a) Find the distance matrix for the following graph, assigning a weight of 10^6 to any "missing edges."

b) Find a minimum spanning tree for this graph.

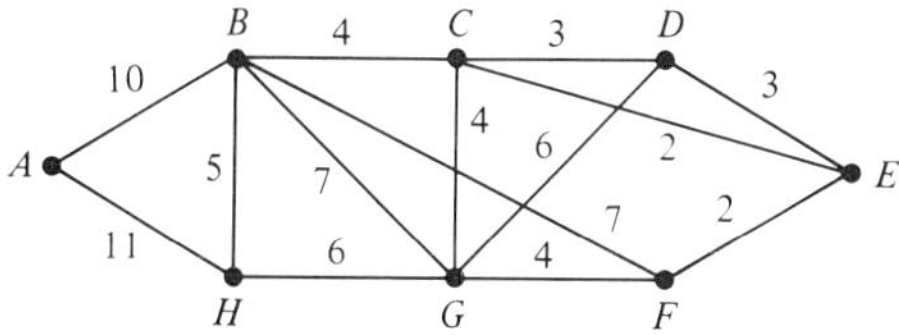

9. Modify Kruskal's algorithm so that it will find a spanning tree in an arbitrary (unweighted) graph. (Also see Exercise 20.)
10. Try to find an example to show that the following modification of Kruskal's algorithm will *not* produce a minimum spanning tree: "Just list the edges of G in increasing order of their weights, and then choose the first $n - 1$ edges."
11. Try to find an example to show that the following modification of Kruskal's algorithm will *not* produce a minimum spanning tree: "Just list the edges of G in increasing order of their weights and then choose every edge in sequence until all the vertices of G are connected.
12. Explain why the following modification of Kruskal's algorithm produces a spanning tree on K_n having a maximum weighted-edge sum: "First change all the weights to their negatives. Then find a minimum spanning tree. This is a maximum spanning tree of the original graph."
13. Find a maximum weight spanning tree for the graph of Exercise 6(b).
14. Show that if a finite tree (with at least two vertices) did not have at least one vertex of degree 1, then some vertex would be on a cycle. (*Hint*: Try to give a proof by contradiction by following a path starting at some vertex.)
15. Which trees are regular graphs? (Recall that a graph is regular if all its vertex degrees are equal.)

16. Show that K_n has n^{n-2} spanning trees as follows:

a) First verify that the number of sequences of the form $(x_1, x_2, \ldots, x_{n-2})$, where the x_i can be any integer such that $1 \leqslant x_i \leqslant n$ is n^{n-2}.

b) Suppose the vertices of the tree are labeled $1, 2, \ldots, n$. Remove the lowest labeled vertex v_1 of degree 1 and call the adjacent vertex x_1. Repeat this process obtaining $x_1, x_2, \ldots, x_{n-2}$, the vertices adjacent to those in the sequence removed: $v_1, v_2, \ldots, v_{n-2}$. Show that a single edge is left. (See Exercise 17.)

c) Show that given any sequence $x_1, x_2, \ldots, x_{n-2}$ as in the foregoing, we determine a unique spanning tree. (However, see Exercise 19.)

d) How do (a) to (c) prove our result?

17. Find the sequences determined in Exercise 16(b) for the following trees on vertices labeled 1, 2, 3, 4, 5.

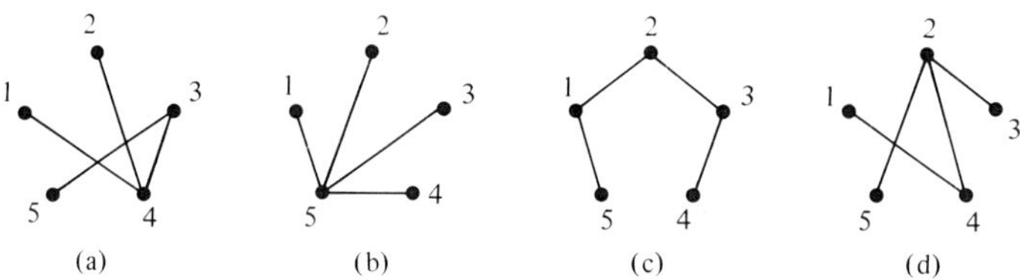

18. Find the trees on five vertices labeled 1, 2, 3, 4, 5 corresponding to the following sequences. **(a)** 5, 1, 4; **(b)** 1, 1, 2; **(c)** 2, 2, 2. (*Hint*: Find the first positive integer v_1 less than 6 and not on list x_1, x_2, x_3. Then (v_1, x_1) is an edge in the tree. Form the list x_2, x_3, v_1 and continue finding the edges of the tree. See Exercise 16.)

19. Show that there are only two nonisomorphic spanning trees on K_4, and only three nonisomorphic spanning trees on K_5.

5.2 THE SHORTEST PATH PROBLEM

It is a commonplace assumption that the shortest path between two points is achieved by traveling along the straight-line segment joining those points. People sometimes forget, however, that a straight-line route may be out of the question in some contexts. If, for example, a person wants to travel from Tampa, Fla., to New Orleans, La., by means of a transportation network such as that given in Fig. 5.6, there is no direct connection. *Distances must be calculated according to the numbers and routes given by the carrier.* One can take one of several indirect routes, such as the one that begins at Tampa, passes through New York, and then goes directly to New Orleans. A somewhat shorter route can be found using Chicago as the intermediate destination.

We could define the distance between Tampa and New Orleans on this network as the length of the shortest route between these two cities, understanding that we would have to add up the lengths of the various legs of our journey. In general this is how we define the distance between any two vertices in a weighted graph.

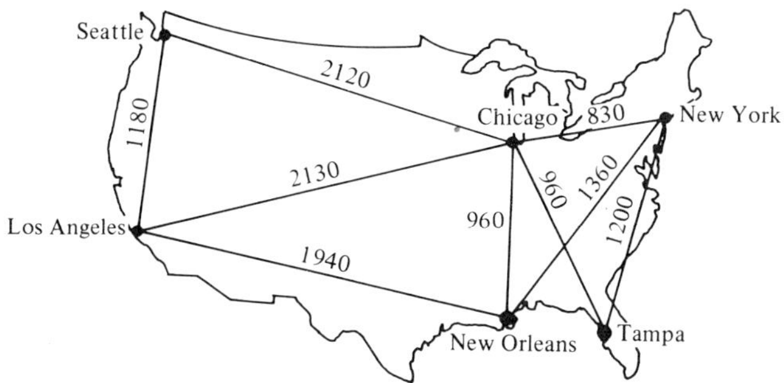

Figure 5.6 A weighted graph.

> **Definition 5.5** Let G be a weighted graph. Then the (a) **distance between two connected vertices u and v** in the vertex set $V(G)$ of G is the minimum of the sums $\sum_{e \in P} w(e)$ over all paths P in G from u to v. We will denote this distance by $d(u, v)$, and we will call each sum (b) **the length, $L(P)$, of the path P.** [*Note:* If u and v are not connected, we sometimes write "$d(u, v) = \infty$."]

Thus in Fig. 5.6 we can say that the distance from Tampa to New Orleans is no more than $960 + 960 = 1920$ on this network, since we have found a path, Tampa–Chicago–New Orleans, that long. In this rather simple situation, one would probably feel confident that we have found the shortest path between these two cities.

We would like a general procedure for finding the *distance between any two vertices in a weighted graph and the paths on which this is achieved.* Most algorithms that solve this **shortest path problem** do so concentrating on one of the two vertices, v_0, and finding the distance from it to many, if not every one, of the vertices in the graph connected to v_0 by a path. Indeed one usually winds up with a spanning tree on the graph, if it is connected, along which one can find the desired paths and distances. However, it is *not* often the case that this is *a minimum spanning tree*, as the following example shows.

Example 1 The spanning tree in Fig. 5.7(a) gives shortest paths from the vertex v_0 to each of the other three vertices in the graph. (One can see this at a glance, we hope, without

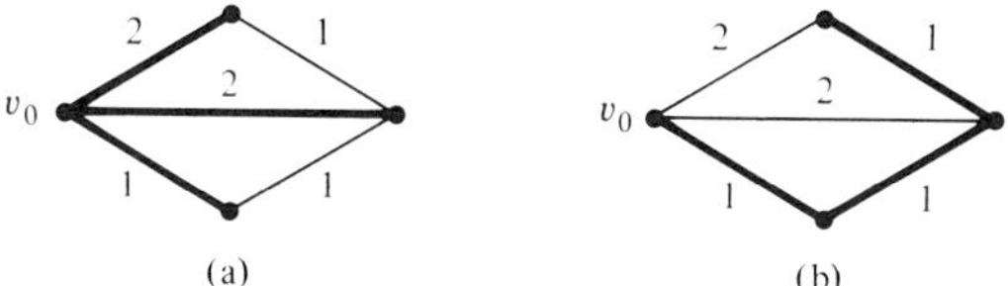

Figure 5.7 A shortest v_0-path tree, such as (a), is not necessarilly a minimum spanning tree (b), or vice versa.

enumerating all the possible paths.) But the spanning tree in Fig. 5.7(b) is clearly a minimum spanning tree of the graph, and the unique paths one must take along its edges to the other vertices are not always shortest paths from v_0. Moreover, since the spanning tree in (b) is minimal, with a total weight of 4, the tree in Fig. 5.7(a) is not minimal. □

The following algorithm finds the distances from a given vertex v_0 to every other vertex in a connected graph G (where G is assumed to have positive weights on its edges). It does so by constructing a spanning tree on the graph G called a **shortest v_0-path spanning tree**. The idea of the algorithm is to begin with the set of vertices $S = \{v_0\}$ and enlarge this set, vertex by vertex, so that at each stage one knows the distance $d(v_0, v)$, as defined by a path of shortest length in the tree, from v_0 to every other vertex v in S.

Algorithm 5.2 ***Shortest Paths*** Let positive weights $w(v, u)$ be assigned to each edge of the connected graph G. *To find the distances $d(v_0, v)$ and a shortest v_0-path spanning tree T from v_0 to every other vertex v in the vertex set V of G,*

1. [Initialize vertex and edge sets, S and T.] Set $S \leftarrow \{v_0\}$, $T \leftarrow \varnothing$, $d(v_0, v_0) \leftarrow 0$, $i \leftarrow 0$.
2. [Tree loop.] Repeat steps 3 through 5 until S includes all the vertices of G.
   3. [Increment i.] Set $i \leftarrow i + 1$.
   4. [Next vertex.] Set $v_i \leftarrow$ any vertex u in $V - S$ for which $d(v_0, v) + w(v, u)$ is a minimum for all vertices v in S and u in $V - S$. Set $d(v_0, v_i) \leftarrow$ this minimum value.
   5. [Increment S and T.] Set $S \leftarrow S \cup \{v_i\}$ and $T \leftarrow T \cup \{vv_i\}$, where v gave us the minimum value in step 4. (If $S = V(G)$, then we continue with step 6. Otherwise we repeat from step 3.)
6. [Done.] Output T and the tree distances $d(v_0, v_i)$.
7. End Algorithm 5.2.

To see the reasoning behind this algorithm, suppose that we have a set of vertices $S \neq V$, where S contains v_0, and, furthermore, that $P = v_0 u_1 u_2 \ldots u_k u$ is a minimum over all paths that begin at v_0 and end in $V - S$. Then one can show that u_k must be in the set S. (See Exercise 5.2-10.) Moreover, the path $v_0 u_1 \ldots u_k$ must be a shortest path to the vertex u_k, and this path must have all of its vertices in the set S. (See Exercise 5.2-11.) This tells us that the distance from v_0 to u is $d(v_0, u_k) + w(u_k, u)$, and this in turn is the minimum of all the sums $d(v_0, v) + w(v, u)$ taken over all vertices v in S and u in $V - S$. But this is exactly what step 4 tells us to calculate at each stage of our algorithm. And so we must end up by finding $d(v_0, u)$ for each vertex u in the vertex set V, adding an edge vu whenever we find $d(v_0, u)$.

Let us take a rather small-scale example to see how the steps in Algorithm 5.2 may be carried out.

Example 2 In Fig. 5.8 we see how the vertices, edges, and consequently paths, are added one at a time until we have a tree spanning the graph.

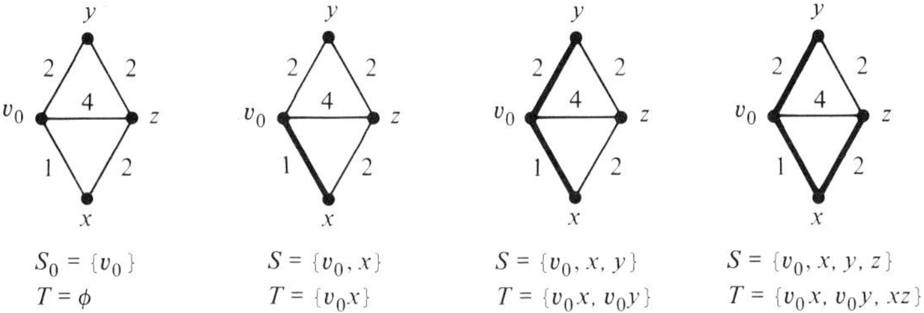

Figure 5.8 Finding a shortest v_0-path spanning tree.

At the last iteration, for example, we have a choice of adding one of three edges to the tree. We calculate

$$d(v_0, y) + w(yz) = 2 + 2 = 4$$

$$d(v_0, v_0) + w(v_0z) = 0 + 4 = 4$$

$$d(v_0, x) + w(xz) = 1 + 2 = 3.$$

Since the last sum is the smallest, we set $d(v_0, z) = 3$, add the edge xz to T, thereby spanning the graph, and we are done. □

The pictorial scheme of Fig. 5.8 works well enough for a small number of vertices and edges. But it is clear that keeping track of the paths, and even determining the distances after the final spanning tree has been found, can become rather cumbersome. The following labeling procedure is one of many that can be used to assist us with the calculations. (See Exercise 5.2-12.)

Labeling Procedure for Shortest v_0 Paths To keep track of the paths and the distances along them from v_0 to each vertex v_j in Algorithm 5.2, we label each vertex v_j that we add to the tree with an ordered pair (v_i, d_j), where

1. v_i is the next to last vertex on the path from v_0 to v_j, and
2. $d_j = d(v_0, v_j)$.

The label (v_i, d_j) *means that the distance along the tree to the labeled vertex* v_j *is* d_j. Moreover, the path along the tree from v_0 to v_j has v_i as its next to last vertex. Using this cue we can find our way back from vertex v_j to vertex v_0 along the path given by the tree. Since we always proceed from labeled to unlabeled vertices, we never form a cycle. This, in fact, will ensure getting a tree in the first place! Let us take a somewhat larger example to illustrate our labeling procedure.

Example 3 We will apply our shortest v_0-path algorithm and labeling procedure to the weighted graph in Fig. 5.9(a). We may begin by giving v_0 the label $(v_0, 0)$. This vertex is incident with three edges, and we choose edge v_0v_1, which has a minimum weight of 1. Therefore, v_1 is labeled $(v_0, 1)$.

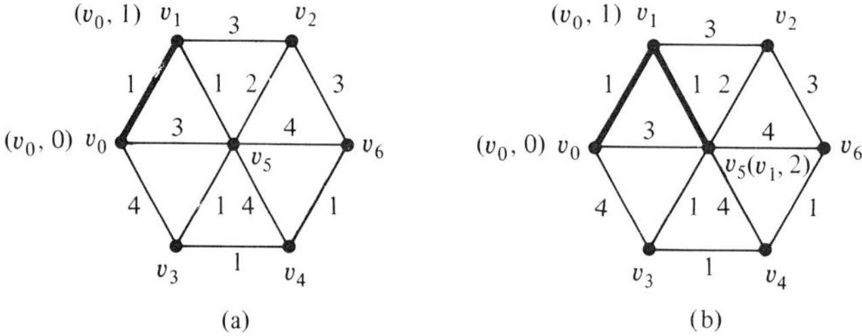

Figure 5.9 Vertex v_5 is labeled $(v_1, 2)$ because it is two units from its predecessor v_1 in the tree.

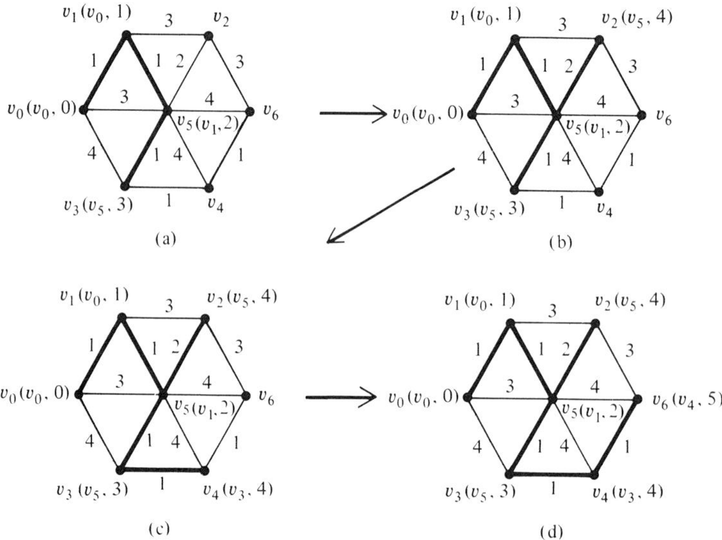

Figure 5.10 Determining a shortest v_0-path spanning tree in the graph of Fig. 5.9 with the aid of our labeling procedure.

We next consider the most economical way to go from the labeled vertices v_0 and v_1 to the unlabeled ones. Our choice is to go from v_1 to v_5. We can check that our choice is minimal by adding the second entry of each of vertices v_0 and v_1's labels to the weights of the edges connecting those vertices with unlabeled vertices. Hence edge v_1v_5 is added, and vertex v_5 is given the label $(v_1, 2)$. This shows that the path from v_0 to v_5 includes v_1v_5 as its last edge and has a length of 2, the distance from v_0 to v_5.

At the next stage of the algorithm, the only unlabeled vertex we can reach with one edge from v_1 is v_2 [see Fig. 5.9(b)], giving us a v_0 to v_2 path length of $1+3=4$. From vertex v_0 we can reach the unlabeled vertex v_3, giving a path length of $0+4=4$. However, from vertex v_5 we can reach vertices v_3, v_4, v_6, and v_2, giving path lengths of

$$2+1=3, \quad 2+4=6, \quad 2+4=6, \quad \text{and} \quad 2+2=4,$$

respectively. Hence we add edge v_5v_3, since it gives a minimum sum. The vertex v_3 is labeled $(v_5, 3)$ in Fig. 5.10(a).

The reader can now follow the progression through three more successive steps in the algorithm, until we obtain Fig. 5.10(d) in which all the vertices are labeled. The distance from v_0 to each vertex is, of course, given by the second entry of the label. □

Notice once again that the distance between two arbitrary vertices is not given by the tree. For example, the distance between v_2 and v_6 is no more than 3, and yet the path given by the tree in Fig. 5.10(d) has a length of $2+1+1+1=5$.

Certain refinements of the algorithm have been shown to be rather efficient in terms of the number of additions and comparisons that one needs to do to implement it. Some further information concerning this is contained in the exercises (see Exercise 5.2-12).

Completion Review 5.2

Complete each of the following.

1. The length of a path $L(P)$ in a weighted graph G is defined to be____________.
2. The distance between two vertices u and v in a weighted graph G is the minimum of the ____________ if the vertices are connected.
3. A minimum spanning tree in general gives a shortest path between any two vertices in a weighted graph. (*True or False*?)
4. A shortest path from a vertex v_0 to all the vertices in G connected to v_0 is given by a ____________.

5. At every stage of the shortest v_0 path algorithm, we find the closest vertex in the set $V - S$ to the set S of________________.

6. In our labeling procedure a label (v_i, d_j) on a vertex v means that dj is the ________________ and v_i is________________.

Answers: **1.** $w(P)$. **2.** lengths of paths between u and v. **3.** False. **4.** shortest v_0 spanning tree. **5.** all vertices already added to the tree, including v_0. **6.** distance of v from v_0; the predecessor of v in the tree.

Exercises 5.2

For Exercises 1 to 3 refer to the following weighted graph G.

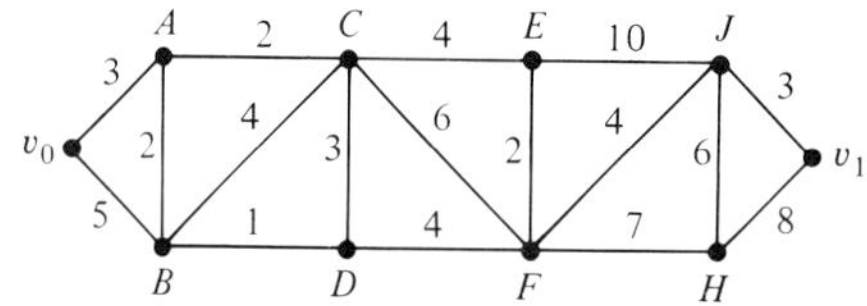

1. Use our shortest v_0-path algorithm, 5.2, to find a minimal v_0-path spanning tree.
   a) What is the distance between v_0 and each of the other vertices in the graph?
   b) What is a shortest path from v_0 to v_1?
2. Use our shortest v_0-path algorithm, 5.2, to find a minimal v_1-path spanning tree.
   a) What is the distance between v_1 and each of the other vertices in the graph?
   b) What is the shortest path given by your tree from v_1 to v_0? Is it the same one as in Exercise 1?
3. **a)** Find a minimum spanning tree of the graph G.
   b) Were either of your trees in Exercise 1 or 2 minimum spanning trees of G? Explain.
   c) What is the length of the path connecting v_0 and v_1?
   d) Is the path in part (c) minimal? Explain.
   e) Will minimum spanning trees contain minimal paths between vertices in general? Explain.
4. **a)** Find the distance between vertices A and F in the weighted graph given by the following distance matrix without drawing a sketch of the graph. (Keep track of the tree by use of our labeling procedure. List all labels at each stage.)
   b) Find a minimum path from A to F, tracing your way back from F by means of your labels.
   c) Was it necessary to find a shortest A-path spanning tree to answer parts (a) and (b)?

$$\begin{array}{c} \\ A \\ B \\ C \\ D \\ E \\ F \end{array} \begin{array}{c} \begin{array}{cccccc} A & B & C & D & E & F \end{array} \\ \begin{bmatrix} 0 & 12 & 20 & 5 & 10 & 25 \\ 12 & 0 & 6 & 2 & 14 & 19 \\ 20 & 6 & 0 & 11 & 22 & 3 \\ 5 & 2 & 11 & 0 & 7 & 18 \\ 10 & 14 & 22 & 7 & 0 & 27 \\ 25 & 19 & 3 & 18 & 27 & 0 \end{bmatrix} \end{array}.$$

5. The following table gives the airline distances between the six cities New York (NY), London (L), Mexico City (MC), Tokyo (T), Paris (Pa), and Peking (Pe).
a) Make a weighted graph corresponding to these data.
b) Find a shortest path tree from New York to the other cities, labeling the tree so that one can find the shortest path and determine the distance in each case.

	NY	L	Pa	MC	T	Pe
NY	—	3469	3636	2090	6757	6844
L	3469	—	214	5558	5959	5074
Pa	3636	214	—	5725	6053	5120
MC	2090	5558	5725	—	7035	7753
T	6757	5959	6053	7035	—	1307
Pe	6844	5074	5120	7753	1307	—

6. Was your answer to Exercise 5 a minimum spanning tree? Explain.

7. Suppose that our shortest path algorithm gives us a Hamiltonian path. (Recall that a Hamiltonian path spans the vertex set.) Give an example to show that this need not be a shortest Hamiltonian path.

8. Suppose that our shortest path algorithm gives us a Hamiltonian path and that it is possible to join the endpoints of this path with a single edge. Give an example to show that the result need not be a shortest Hamiltonian circuit.

9. The problem of finding a Hamiltonian circuit of minimal length on a weighted graph is known as the "traveling salesman problem." Show that if the vertices are n points located on the circumference of a circle, then the salesman's solution is to visit the points in their given order around the circle. The weights are the "ruler" distances between the points. (*Hint*: There cannot be self-intersections. Why?)

10. Suppose that the vertex v_0 of G belongs to a subset $S \subset V(G)$, and suppose that $P = v_0 u_1 u_2 \ldots u_k v$ has minimum length among all paths that begin at v_0 and end in $V - S$. Show that u_k belongs to S.

11. Show that, with regard to Exercise 10, $v_0 u_1 u_2 \ldots u_k$ must be a shortest path to u_k and all the vertices u_i belong to S.

12. The following refinement of our labeling procedure is attributable to E. W. Djikstra: "*Each* vertex v of the graph G is given a label $l(v)$ at *every* stage of the algorithm as follows: (1) $l(v_0) \leftarrow 0$ and $l(v) \leftarrow \infty$ for $v \neq v_0$, $S_0 = \{v_0\}$ and $i \leftarrow 0$. (2)

For every v in $V - S_i$, replace $l(v)$ by the minimum value of $l(v)$ and $l(u_i, v) + w(u_i, v)$ for u_i in S_i. Find the minimum value of $l(v)$ for all v in $V - S_i$ and let u_{i+1} be a vertex for which this minimum is attained. Let $S_{i+1} = S_i \cup \{u_{i+1}\}$. (3) If $i = m - 1$, where $|V| = m$, stop. If $i < m - 1$, set $i \leftarrow i + 1$, and then return to step 2."

Apply Djikstra's algorithm to Exercise 1. In what way, if any, were the computations facilitated?

13. What is the total number of additions and comparisons required by steps 2 and 3 of Djikstra's algorithm for a vertex set of m vertices?

5.3 ROOTED TREES AND POLISH NOTATION

When we first introduced tree diagrams in connection with counting (Section 2.1), we drew these diagrams with a single vertex at the top and followed the edges in a downward direction in order to explore which choices were next available. This procedure motivates the following definitions.

Definition 5.6 A graph is a (a) **rooted tree** if it is a tree having one vertex v_0 singled out to be the (b) **root** of the tree. All vertices of degree 1, except for v_0, are called the (c) **leaves** of the tree; the remaining vertices, including v_0, are called (d) **branch nodes**. Paths from the root to the leaves are called (e) **branches**.

The procedure for locating shortest v_0 paths that we discussed in the previous section gave us rooted trees. The specified vertex v_0 is taken as the root, and the algorithm "grows" branches from v_0 to the branch nodes and leaves. These branches are, of course, shortest paths from v_0 to the other vertices of the graph.

Any tree can be made into a rooted tree simply by designating one of the vertices as the root. This imposes a direction on the edges of the tree *away from the root* and *toward the leaves* because there is a unique path in the tree from the selected vertex to any other vertex. It is, however, unnecessary to indicate this direction once the root has been chosen.

Example 1 In Fig. 5.11 we can see how two different rooted trees can be formed from the same tree [Fig. 5.11(a)]. In Fig. 5.11(b) vertex r was chosen as the root and the leaves are a, b, c, d, and e, with r and s as branch nodes. In Fig. 5.11(c) vertex s is the root; a, b, c, d, and e are the leaves; and both r and s are branch nodes again. Looking at (b) again, we see that there is a branch from the root r to the leaf e passing through the node s. In (c), however, the branches from r only go to leaves a and b. This is implied by having r placed above a and b in the diagram.

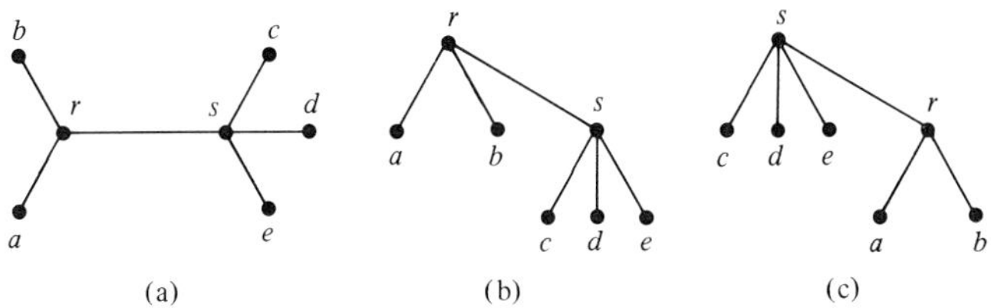

Figure 5.11 Two different rooted trees, (b) and (c), formed from the tree in (a). □

One obvious application of rooted trees is in the representation of "family trees," especially where family names are passed down from generation to generation.

Example 2 Here is the author's own "patrilineal" family tree, having his paternal grandfather PBK as the root, and tracing the generations through the male descendants, who, in a patrilineal society, carry the family name. (Consequently we cannot determine from this tree whether an individual has no descendants or only no male descendants.)

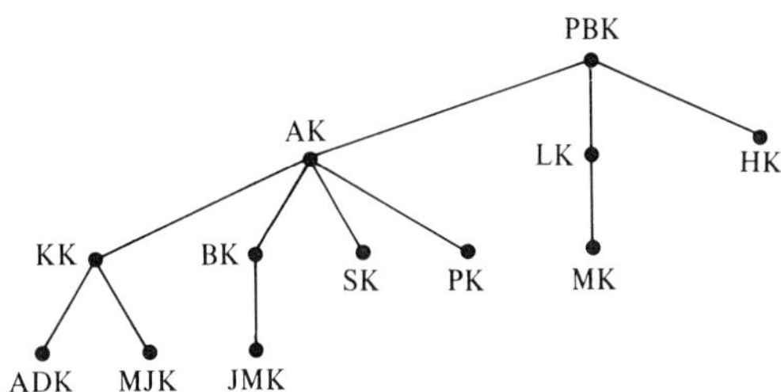

Figure 5.12 A "family name tree" tracing four generations of males. □

Geneological examples such as this have given rise to the following definitions and terminology.

> **Definition 5.7** If f is a branch node in a rooted tree, then we say that f is (a) the **father** of vertex s when there is an edge directed from f to s. We say that s is a (b) **son** of f. Vertex d is (c) a **descendant** of f if there is a path directed from f to d, and f is then called an (d) **ancestor** of d. Two vertices with the same father are called (e) **brothers**.

Hence in Example 2 we would say that KK is the father of ADK and MJK and that KK, BK, SK, and PK are brothers, according to the rooted tree. One could ask whether any order relationship, such as the relative ages of the brothers, is implied by the rooted tree in this example. As it happens, the answer is "yes," in that of the four brothers previously mentioned, KK is the oldest, followed by BK, SK, and PK in order of birth. Not all rooted trees imply such relationships, however. But when they do, we distinguish them with Definition 5.8.

Definition 5.8 A rooted tree that has the sons of each father labeled in a way, as with positive integers 1, 2, . . . , n, that implies order, is called an **ordered rooted tree**.

Example 3 In Figure 5.13(a) we have an ordered rooted tree that we can say is **isomorphic** to the one in Example 2 because the order relationships among the sons is preserved by the obvious correspondence between the vertices. The ordered rooted tree in Fig. 5.13(b) is not isomorphic to these, however, because the third son of the root in this tree does not correspond to the third son of the root in the original tree.

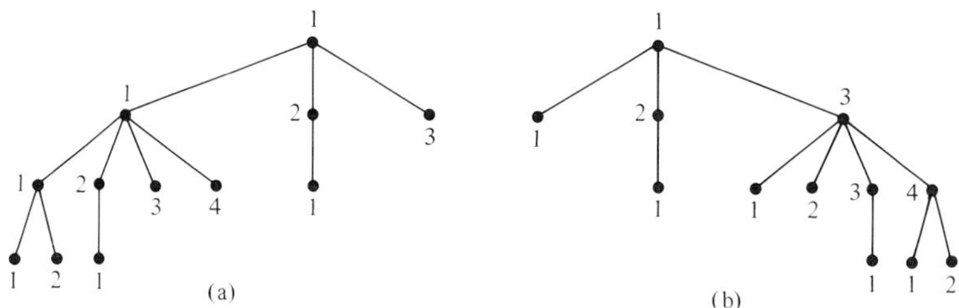

Figure 5.13 Nonisomorphic ordered rooted trees.

If the labels are omitted in an ordered rooted tree, drawn in the plane as in Fig. 5.13, we assume that the vertices would be labeled in increasing order from left to right. No order relation is necessarily implied between sons of different fathers. Thus in Example 2 it happens that MK is older than BK, even though BK occurs to the left of MK. □

One can list the vertices of an ordered rooted tree so that the order in which the vertices appear on the list is "natural" in some sense. But first we will need Definition 5.9.

Definition 5.9 Let T be a rooted tree with root r and vertex v. Then the **level of** $\boldsymbol{v}$ is the length, that is, the number of edges, of the unique path from r to v.

As an illustration, in Example 2 ADK, JMK, and MJK are both on level 3, while KK, BK, SK, PK, and MK are on the second level. Continuing up the tree, AK, LK, and HK are on level 1 and PBK is on level 0. Different levels can be interpreted as different generations in this example.

Definition 5.10 A **universal address system** for an ordered rooted tree is a labeling of the tree as follows: Vertex v is given the label $n_1 . n_2 . \ldots . n_{m-1} . n_m$ if v is a node on the mth level and v is the n_mth son of the vertex labeled $n_1 . n_2 \ldots n_{m-1}$. In particular, a vertex labeled n is the nth son of the root v_0, and v_0 is given the label 0. (We omit the $n_0 = 0$ index.)

This is, of course, a recursive definition, since it depends on the labeling of vertices on the previous level. In comparing labels of different vertices, one can easily determine father–son–brother relationships. For example, 2.1 is a son of 2, the father of 2.1.3, and a brother of 2.3 in Fig. 5.14. (All vertices have the vertex labeled "0" as their ancestor, of course.)

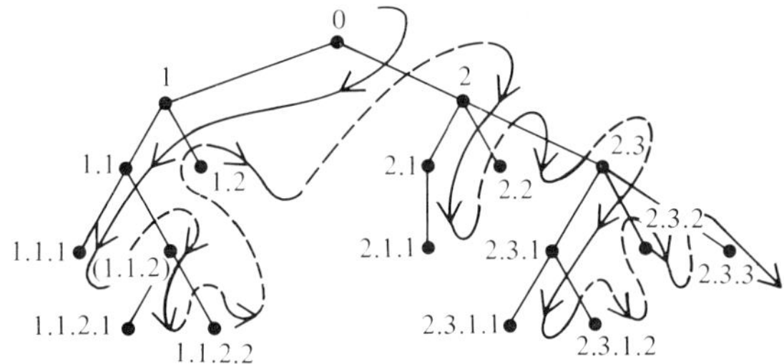

Figure 5.14 Determining a universal address system in an ordered rooted tree.

A universal address system, such as that given in Fig. 5.14, imposes a **lexicographic ordering** upon all the vertices of the tree, that is, a way of writing a list of the vertices that resembles the order in which words occur in a dictionary. To determine whether a vertex v appears before a vertex w we compare their labels level by level from left to right. Vertex v will appear before vertex w if in the first level where the labels disagree, v's index is less than w's. (We can think of the labels as ending in zeros to compare vertices on different levels, if necessary.) Comparing vertices labeled 2.3.1.2 and 2.3.2, for instance, we see that 2.3.1.2 is listed before 2.3.2 because they first differ at the third level, and $1 < 2$.

Example 4 A lexicographic ordering of the vertices in Fig. 5.14 is found by following the pattern indicated by the arrows, thus:

0	1.2	2.3.1.1
1	2	2.3.1.2
1.1	2.1	2.3.2
1.1.1	2.1.1	2.3.3
1.1.2	2.2	
1.1.2.1	2.3	
1.1.2.2	2.3.1	

□

Binary Trees and Polish Notation

Definition 5.11 A **binary tree** is a rooted tree in which the root and every node has at most two sons.

Binary trees have been applied extensively in different areas of mathematics, as an aid to decision-making and searching procedures, and in the study of binary relations and algebraic operations. We consider the first three of these in later sections. For the present we use binary trees to study ways of writing operations, such as the addition of real numbers or the union of sets, which involve two elements at a time. Such operations are called **binary operations**. To see how binary operations can be modeled using binary trees, we define a **subtree rooted at u_0** of a rooted tree T to be the subgraph of T rooted at u_0 that includes only u_0 and the descendants of u_0. (See Fig. 5.15.) The terms **right** or **left** subtree will refer to subtrees of ordered binary trees rooted at a right or left son respectively.

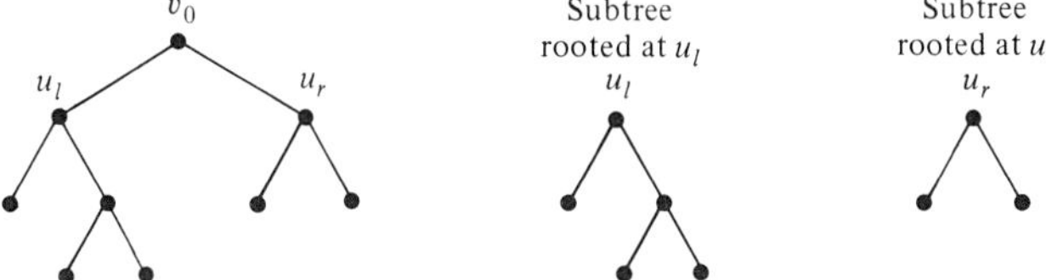

Figure 5.15 Rooted subtrees.

Binary Operation Tree Procedure To construct a binary tree for an algebraic expression involving binary operations:

1. Find the last operation to be performed and make it the root.
2. Let each of the two operands (elements operated upon) correspond to a subtree, and apply step 1 until each node either is an operation or is an operand not involving further operations.

Thus "$a + b$" becomes the tree

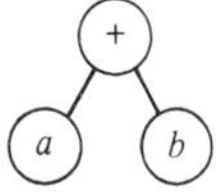

while $a + (b * c)$ becomes

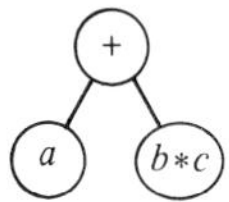

The right node of the last tree can be expanded further by repeating step 1, and we obtain

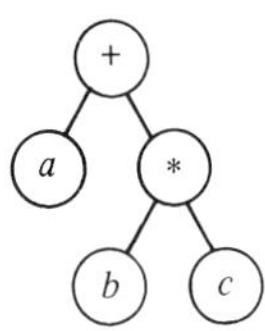

Now we consider a somewhat more complicated example.

Example 5 To construct the binary tree that represents the expression $(A + (C/D)) * (E - (F * G))$, we note that the last operation is an asterisk $(*)$ on the operands $(A + (C/D))$ and $(E - (F * G))$, in that order. These become the left and right sons, respectively, of a tree rooted at a vertex labeled $*$:

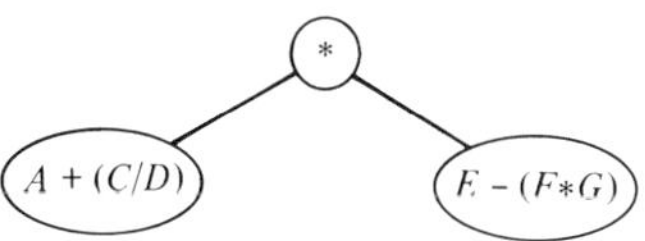

Applying step 1 to $A + (C/D)$ and to $E - (F * G)$, we obtain the subtrees

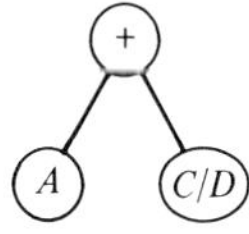

and

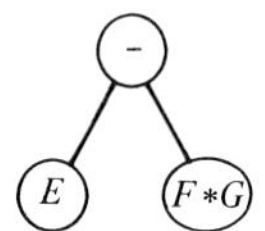

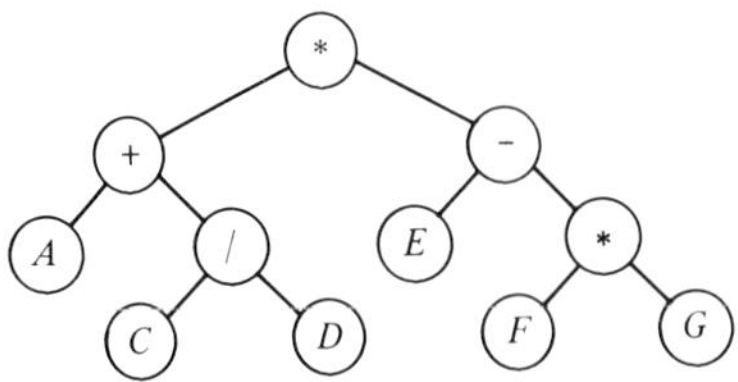

Figure 5.16 Determining a binary tree corresponding to an algebraic expression, $(A+(C/D))*(E-(F*G))$.

respectively. After we have applied step 1 to the expressions C/D and $F*G$, we obtain the tree in Fig. 5.16. □

The Polish mathematician Jan Lucasiewicz has shown, by using binary tree representation, that parentheses are not needed in algebraic expressions involving binary operations. The notation he developed, commonly called *Polish notation*, depends only upon the order in which operations and operands are written down. For example, A/B and $C+D$ are written "$/AB$" and "$+CD$," respectively, in Polish notation.

Definition 5.12 To write an algebraic expression with **Polish notation in prefix form** (or simply **prefix form**—see Exercise 5.3-17 for **postfix form**):

1. Construct the binary tree of the algebraic expression.
2. Write the nodes of this tree in lexicographic order, where the left son is always taken before the right son.

For example, the prefix form of $(A+(C/D))*(E-(F*G))$ is found by considering the tree in Fig. 5.12, and writing the nodes in lexicographic order. This yields $*+A/CD-E*FG$. Further examples, using the operations of set union and intersection, are given in Fig. 5.17.

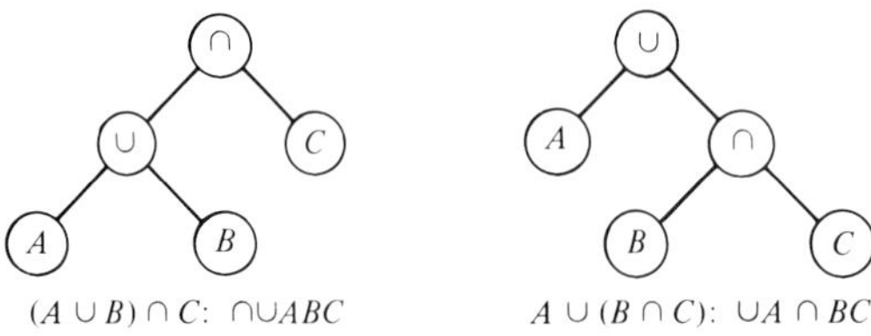

Figure 5.17 Determining prefix forms for set operations.

To change the prefix form of an expression to ordinary algebraic notation or into its binary tree representation, it may be found helpful to proceed as follows:

1. Reading the prefix form *from right to left*, locate the first operations symbol, θ not yet in parentheses.
2. Find the two operands of θ. They are the two symbols to its right. Place parentheses around all three. Treat this as an operand.
3. Repeat steps 1 and 2 until there are no symbols to the left of the last parenthesized expression.

Example 6 The prefix form $+*AB/C-D*EF$ yields in succession, $+*AB/C-D(*EF)$; $+*AB/C(-D(*EF))$; $+*AB(/C(-D(*EF)))$; $+(*AB)(/C(-D(*EF)))$; and $(+(*AB)\times(/C(-D(*EF))))$.

At this point we can easily see how the operations apply to the pairs of operands, beginning with $+$ applied to a product and a quotient. Thus we can either write the binary tree of the prefix form as shown in Fig. 5.18 or we can write out the algebraic notation directly by exchanging the positions of each operation with the operand on its right in the parenthesized expression.

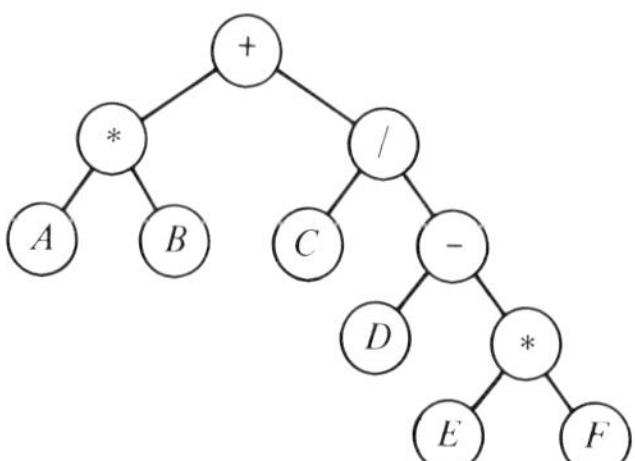

Figure 5.18 A binary tree and its corresponding algebraic and prefix forms. Prefix form: $+*AB/C-D*EF$; parenthesized prefix form: $(+[*AB][/C(-D(*EF))])$, algebraic form: $(A*B)+(C/(D-(E*F)))$. □

We conclude this section by pointing out that every correctly written algebraic expression will have a unique prefix form, which, in turn, can be translated back into the original algebraic form as we have described.

Completion Review 5.3

Complete each of the following.

1. A rooted tree is one in which one vertex v_0 has been singled out to be the ________________ of the tree. All other vertices of degree 1 of a rooted tree are called the ________________ of the tree. The remaining vertices are called the ________________ of the tree. Paths from the root to the leaves are called ________________ of the tree.
2. The direction of edges in a rooted tree is taken from the ________________ and toward the ________________.
3. If there is an edge (path) in a rooted tree directed from vertex x to vertex y, we say that x is the ________________ of y, and that y is the ________________ of x.
4. Two vertices with the same father are called ________________.
5. The number of edges in the path from the root to a vertex v is called the ________________ of v.
6. If the sons of each father in a rooted tree are labeled in a way that implies order, then the tree is called an ________________.
7. In a universal address system the vertices are labeled in ________________ order.
8. A ________________ tree is a rooted tree in which the root and every node have at most two sons.
9. To write an algebraic expression in prefix form, we construct the ________________ of the algebraic expression and then write the nodes of the tree in ________________ order.
10. To change from prefix form to algebraic notation, we can write the ________________ tree of the prefix form, or we can exchange the positions of each operation with the operand on the ________________ in the ________________ expression.

Answers: **1.** root; leaves; branch nodes; branches. **2.** root; leaves. **3.** son (descendant); father (ancestor). **4.** brothers. **5.** level. **6.** ordered rooted tree. **7.** lexicographic. **8.** binary. **9.** binary tree; lexicographic. **10.** binary; right; parenthesized.

Exercises 5.3

Refer to the following diagrams for Exercises 1 to 9.

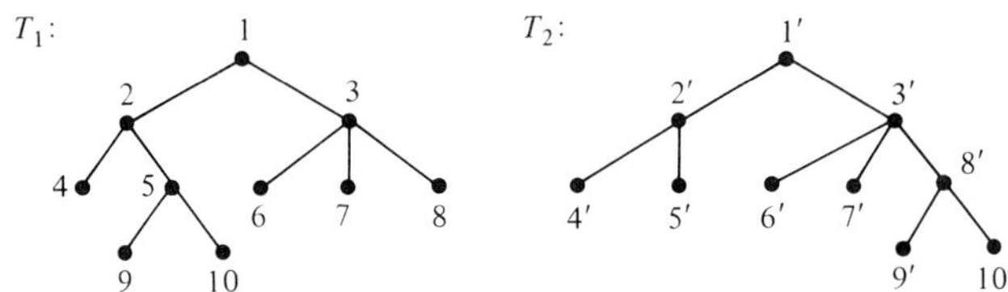

1. Consider the rooted tree T_1 and find **(a)** the root; **(b)** the leaves; **(c)** the sons of vertex 2; **(d)** the level of vertex 5; **(e)** the brothers of vertex 5; **(f)** the descendants of vertex 2.
2. Answer the same questions as in Exercise 1 for T_2.
3. In tree T_1 find the tree rooted in vertex 2.
4. In tree T_2 find the tree rooted in vertex $2'$.
5. Are rooted trees T_1 and T_2 isomorphic to one another? Either find an isomorphism from T_1 to T_2 or explain why there is none.
6. Explain why we can consider trees T_1 and T_2 as ordered rooted trees. Why are they *not* isomorphic ordered rooted trees?
7. Is T_1 a binary rooted tree? Explain.
8. Give a universal address system for T_1 and list the vertices in lexicographic order.
9. Give a universal address system for T_2 and list the vertices in lexicographic order.
10. Suppose that a universal address system contains a vertex labeled 2.3.5.4. What other labels (or "addresses") must be in the system?
11. Write the prefix form and the algebraic expression given by the following ordered rooted tree.

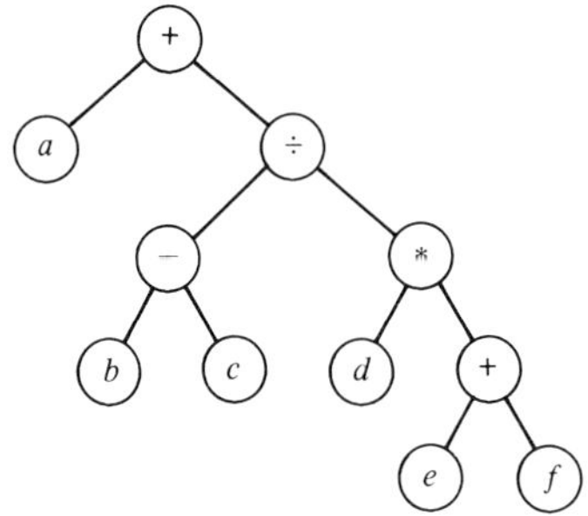

12. Draw a binary tree for each of the following arithmetic, set theoretic, or logical expressions. **(a)** $x+y$; **(b)** $x+(y*z)$; **(c)** $(x+y)*z$; **(d)** $A\cap(B\cup C)$; **(e)** $(A\cap B)\cup C$; **(f)** $p\vee(q\wedge r)$; **(g)** $(p\vee q)\wedge r$.

13. Write each of the expressions in Exercise 12 using Polish notation in prefix form.

14. Draw a binary tree for each of the following expressions and then write the expression in prefix form. **(a)** $(a+(b/(c-d)))*((e+f)/g)$; **(b)** $(p \rightarrow q) \wedge (r \leftrightarrow (s \vee t))$; **(c)** $(A \oplus B) - ((C \cap D) \cup E)$.

15. Modify our binary operation tree procedure so that it also applies to "unary" operations such as taking positive square roots of nonnegative real numbers or complements of sets. Apply your modified algorithm to the construction of binary trees for **(a)** $\bar{A} \cap \bar{B}$; **(b)** $\neg(p \vee q)$;
(c) $\dfrac{-b+\sqrt{b^2-4ac}}{2a}$.

16. Write the following prefix forms in algebraic notation.
(a) $-AB$; **(b)** $*+ABC$; **(c)** $*A+BC$; **(d)** $***XXXX$;
(e) $+*XY*ZW$.

17. Some hand calculators that use Polish notation require that the operands be keyed in before the operation, giving one a **reverse Polish notation**, or **postfix form** ($AB*$ for $A*B$, instead of $*AB$, the prefix form). One can easily change from one form to another, especially after the forms have been parenthesized, as indicated in the following diagram.

$$(+(*AB)(/CD)) \longrightarrow ((AB*)(CD/)+)$$

Prefix form — *Postfix form*

$++AB/CD$ — $AB*CD/+$

Write the prefix forms of Exercise 16 in postfix form.

18. A **regular binary tree** is one in which every branch node has exactly two sons. Show that if a regular binary tree has N branch nodes and L leaves, then $N=L-1$.

19. Regular binary trees can be used to model the schedule of **single-elimination tournaments**, where pairs of players compete in each round of competition, and a player is eliminated after one loss. Ties are not allowed. This is illustrated in the following for a group of six players, where a total of five games must be played to determine a winner. If there is an odd number of players in any round, one player draws a bye until the next round. If 100 players enter a single-elimination tournament, how many matches must be played to determine the champion?

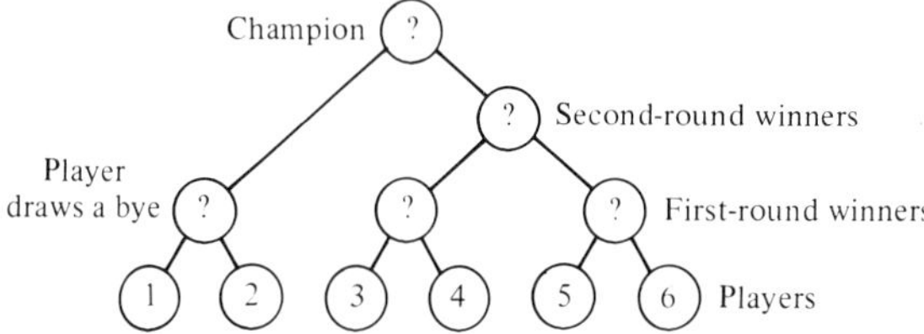

5.4 BINARY SEARCH TREES

In the game of "hi–lo" we are given a number of chances to guess our opponent's number, which is, say, a number between 1 and 7. If we are given just three guesses, can we do it? It could take us as many as seven guesses, unless we are given more information. However, if our opponent agrees to tell us when our guess is too high or too low, the ordered binary tree of Fig. 5.19 shows that three guesses will suffice every time. We make "4" our first guess. If any guess is too low, then the right son is our next choice. And if we guess too high, then we choose the left son next. No path has a length greater than 3, so three choices suffice to guess our opponent's number.

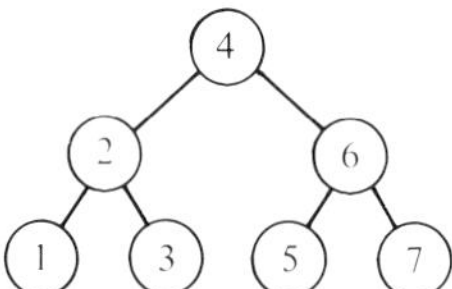

Figure 5.19 Finding n in three guesses ($1 \leqslant n \leqslant 7$).

The problem we just discussed and its method of solution have consequences far beyond childrens' pastimes. The more general problem is that of locating a single piece of data in a veritable sea of information. Binary trees can often assist us by showing how to conduct a systematic and effective search even in the following, more formal, situations.

Suppose we have a file, that is, a set of records in locations $\{r_1, r_2, \ldots, r_n\}$. Each location r_i might be, for example, a box stored in someone's file cabinet or, as is more often the case, locations in the memory bank of a computer. Let us further suppose that the **key** (or content) k_i of each record r_i is either a number, such as someone's bank account number, or an alphabetic string, such as someone's name. In this section we will see how binary trees can be used to illuminate certain algorithms that search a set of records for a particular key.

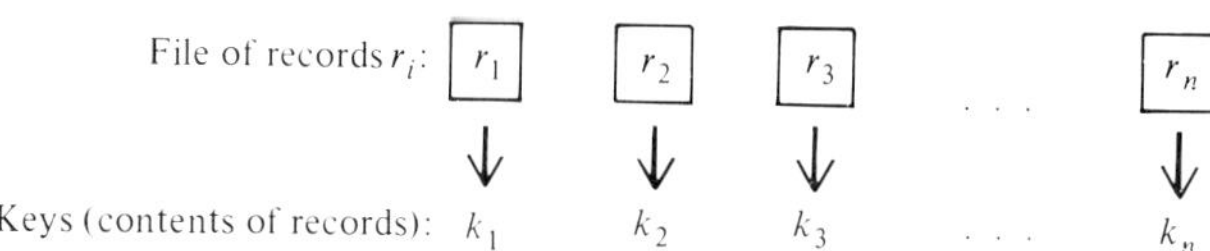

Figure 5.20 Keys in records.

Files, be they in someone's cabinet or in electronic circuitry, often have their keys arranged in alphabetic or ascending order to make the information easier to locate. We assume this to be the case throughout the following discussion. If need be, one could apply one of the sorting algorithms we discussed in Section 3.7 to obtain *the order we assume*:

$$k_1 \leqslant k_2 \leqslant k_3 \leqslant \cdots \leqslant k_{n-1} \leqslant k_n.$$

We will not assume that the name of a record r_i gives us any clues to its key k_i, such as the first letter in a name, for example. These clues are useful secretarial devices, but let us suppose that, like a dictionary without a thumb index, they are simply not available to us.

Our problem, then, is to locate k among the records of our file without having to examine the keys in each and every one. If we were to examine the records $r_1, r_2, \ldots, r_n$ in sequence, it is possible that we would have to compare k with every k_i, a total of n comparisons in all. This is the *worst case* of such a **sequential search**, which, in effect, makes no use of the fact that the records have been sorted. It would be foolish, however, not to use this information. For example, if, after examining record r_1, we find that $k_1 > k$, it should be clear that k is not in any of the remaining records. For each of those remaining records we would have $k_i > k$ as well.

One popular class of search algorithms, of which Fig. 5.19 is a typical example, is known as the "binary search procedures." The details of these procedures can be rather tricky, but we will begin by giving the relatively simple, general idea. We will also make use of a simplifying assumption that we are searching through $n = 2^r - 1$ records where r is a positive integer.

The main idea of binary search procedures is to subdivide the sorted list of keys $k_1, k_2, \ldots, k_n$ into smaller and smaller lists $k_L, k_{L+1}, \ldots, k_{U-1}, k_U$ of about half the previous size, where k_L and k_U denote the keys in the lower and upper records of the sorted file. We will call these lists "search intervals" or merely "intervals," since all the keys originally between k_L and k_U will still appear between them in any sublist. We will denote intervals by "$[k_L, \ldots, k_U]$." One compares k with the key k_M, which is about halfway between k_L and k_U. (Our special assumption will make it exactly halfway between them.) If $k_M \neq k$, we continue by searching either the lower half or the upper half of the interval as before. This is indicated in the binary tree of Fig. 5.21.

We now give a formal statement of our binary search procedure, assuming that we have $n = 2^r - 1$ sorted keys $k_1, k_2, \ldots, k_n$ and that k_i denotes the key in record r_i.

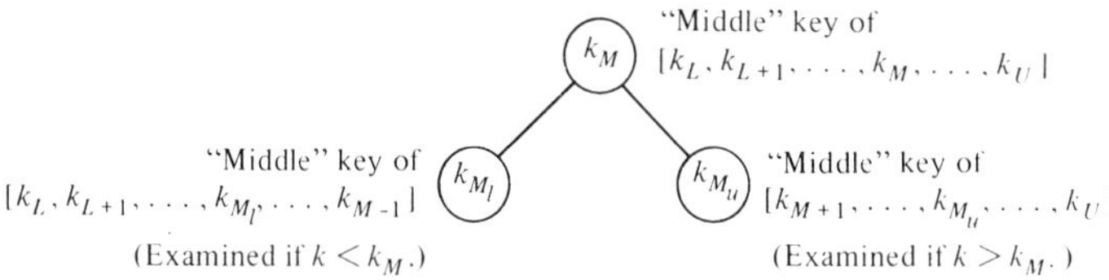

Figure 5.21 Determining the nodes of a binary search tree.

Algorithm 5.3 ***Binary Search Algorithm*** To search $n = 2^r - 1$ sorted records having keys $k_1 < k_2 < \ldots < k_n$ for key k, where r is a positive integer,

1. [Initialize L, U, M.] Set $L \leftarrow 1$, $U \leftarrow 2^r - 1$, and $M \leftarrow 2^{r-1}$.
2. [Done?] If $k_M = k$, then output k_M and stop. (Otherwise we will narrow down the search interval as follows.)
3. [Subinterval loop.] Repeat steps 4(a) or (b) and 5 until $k = k_M$ or $U < L$.
   - 4a. [Lower half.] If $k < k_M$, then set $L \leftarrow L$, $U \leftarrow M - 1$ and continue with step 5. Otherwise:
   - 4b. [Upper half.] If $k > k_M$, then set $L \leftarrow M + 1$, $U \leftarrow U$, and continue with step 5.
   - 5. [Midpoint.] Set $M \leftarrow (L + U)/2$.
6. [Success.] If $k = k_M$, then output k_M and stop.
7. [Failure.] If $U < L$, then output a failure message and stop.
8. End of Algorithm 5.3.

We observe that if the algorithm does not stop at step 1, then the next interval to be examined contains $2^{r-1} - 1$ records, so we have reduced r by 1. Similarly the next nonempty search interval, if there is one, will have $2^{r-2} - 1$ records, and so on. The M in each case will always be an integer due to our assumption that $n = 2^r - 1$. In the more general case one has to make certain adjustments.

Example 1 Let us illustrate the foregoing method with a hypothetical example involving 15 sorted records.

The first interval we examine, according to Algorithm 5.3, runs from k_1 to k_{15} and we compare k with k_8, the key in record r_8.

$L = 1$; $U = 15$; $M = (15 + 1)/2 = 8$: $[k_1, k_2, k_3, \ldots, k_8, \ldots, k_{14}, k_{15}]$

Suppose that $k < k_8$. Then . . .

$L = 1$; $U = 8 - 1 = 7$; $M = (7 + 1)/2 = 4$: $[k_1, k_2, k_3, k_4, k_5, k_6, k_7]$

Suppose that $k > k_4$. Then . . .

$L = 4 + 1 = 5$; $U = 7$; $M = (7 + 5)/2 = 6$: $[k_5, k_6, k_7]$

Suppose that $k > k_6$. Then . . .

$$L = 6 + 1 = 7 = U; \quad M = (7+7)/2 = 7: \qquad [k_7].$$

If $k = k_7$, we are successful. If $k \neq k_7$, then k is not in our file and the algorithm terminates with $L > U$. □

This example does not, of course, cover all the possible eventualities. A binary tree is well suited for illustrating these, as shown in the following example.

Example 2 The tree in Fig. 5.22 shows how the accompanying list of 15 numbers might be searched by our algorithm. The path indicated on the tree and the bracketing of sublists on the right show the pattern of our search in the previous example. But the binary search tree actually shows all possible patterns with $n = 15 = 2^4 - 1$.

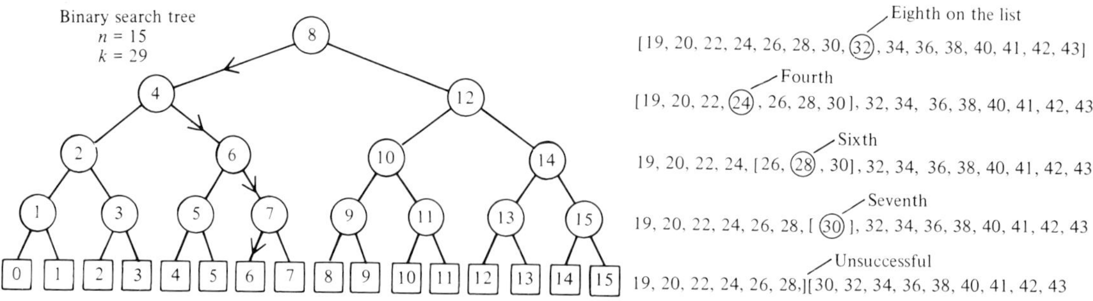

Figure 5.22 A binary search through nodes 8, 4, 6, and 7.

The leaves of our search tree are indicated by the rectangular nodes; leaves to the right are successful and leaves to the left indicate unsuccessful searches, that is, where the final search interval is empty. The worst-case situation in a successful search involving $n = 15$ records requires four comparisons, not 15 comparisons as in a sequential search of the file. □

Theorem 5.5 A successful binary search of $n = 2^r - 1$ sorted records requires r comparisons in the worst case for any positive integer r.

Proof: To see why Theorem 5.5 is true, let a_r = the number of comparisons needed in the worst case successfully to search $2^r - 1$ sorted records. Since each additional power of 2 adds one more level to our tree, we obtain the recurrence relation

$$a_r = a_{r-1} + 1, \quad r \geqslant 2,$$

with the initial condition $a_1 = 1$. Using the methods of Chapter 3, we immediately find that $a_r = r$, proving Theorem 5.5. ∎

The binary tree of Example 2 is formed by choosing the precise midpoint $M = (L + U)/2$ of each subinterval $k_L, \ldots, k_U$ to be a node. This works because the original number of records was assumed to be a power of 2. Hence M will be a whole number. A slight modification of this procedure will enable one to construct a binary search tree for any number of records, not merely powers of 2.

Algorithm 5.4 ***Modification of Binary Search Algorithm*** If n is any positive integer, let $M =$ the greatest integer less than or equal to $(L + U)/2$. Then implement the procedure as before. (M is the "midpoint" of the interval, as before.)

Example 3 Given $n = 18$ records, we choose the approximate midpoint of the interval $k_1, k_2, \ldots, k_{18}$ as k_9, since $(18 + 1)/2 = 9.5$, and the greatest integer less than or equal to 9.5 (written $\lfloor 9.5 \rfloor$) is 9. Similarly $\lfloor (8 + 1)/2 \rfloor = \lfloor 4.5 \rfloor = 4$, and so on. This gives us the binary search tree in Fig. 5.23.

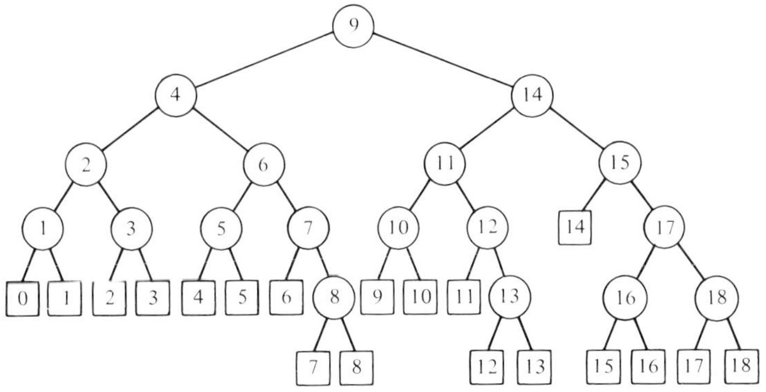

Figure 5.23 A binary search tree with n=18 nodes. □

We can see that no more than five comparisons are necessary to terminate a successful search with $n = 18$ records, the same worst-case number that one obtains when searching $2^5 - 1 = 31$ records. In fact we can state the following generalization of Theorem 5.5.

Theorem 5.6 A successful binary search of n records using Algorithm 5.4, where $2^{r-1} \leqslant n < 2^r$, will require r comparisons in the worst case for any positive integer n.

More subtle analyses of our search algorithms are possible. One can determine the average number of comparisons required to search n files, for example, but such investigations are beyond the scope of this text.

Many variations of binary searching have been invented. Some of these, such as making use of the Fibonacci numbers to avoid the division step, are given in the exercises. (See Exercise 5.4-7.)

Completion Review 5.4

Complete each of the following.

1. Examining the records $r_1, r_2, r_3, \ldots, r_n$ of a file in their given order is known as ______________.
2. In the worst case of the search technique in Question 1, ______________ comparisons are required to find a given key.
3. One can do better, in general, than the worst case in a sequential search if the file has been ______________. In that case one can use a ______________ search, as discussed in the game "hi–lo."
4. A successful binary search of $2^r - 1$ sorted records requires no more than ______________ comparisons.
5. In a binary search of n sorted records, the key selected for comparison at each stage corresponds to $M =$ ______________, where $L =$ the ______________ and $H =$ the ______________ of the search interval.

Answers: **1.** sequential search. **2.** n. **3.** sorted; binary. **4.** r. **5.** the greatest integer not greater than $(L+H)/2$; left endpoint; right endpoint.

Exercises 5.4

1. What is the greatest number of guesses one should need to find someone's number between 1 and n, n an integer, in a game of "hi–lo" if n is **(a)** 15; **(b)** 63; **(c)** 100; **(d)** 1000?
2. Challenge a friend to a game of "hi–lo" and impress your friend with your ability to guess his or her lucky number in a limited number of guesses.
3. What do we mean by the "worst case" in a sequential search of 1000 records? How does this compare with the "worst case" in a binary search of 1000 (sorted) records?

4. Give a binary search tree indicating successful and unsuccessful searches of n sorted records if $n=$ **(a)** 10; **(b)** 20; **(c)** 31.

5. Consider the binary search pattern in the following tree, where we begin by comparing k with the key in the highest order record in our file, rather than the one near the middle.

a) In how many instances does the worst case arise?

b) In how many instances does the worst case arise with $n=8$ using Algorithm 5.4?

c) Try to write a formal algorithm that would describe this version of binary searching.

d) Draw the corresponding binary search tree for $n=16$ sorted records.

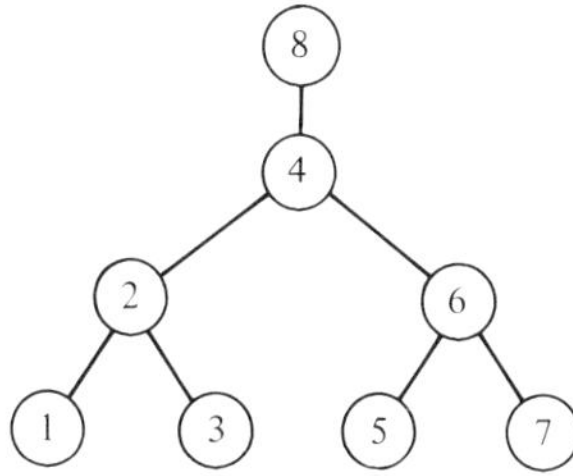

6. Verify the details in the proof of Theorem 5.5, which gives the number of comparisons needed in the worst case for a binary search of $n=2^r-1$ sorted records.

7. **(Fibonaccian search)** Assume that we have $n=F_{r+1}-1$ keys where F_r denotes the rth Fibonacci number and that the keys $k_1, \ldots, k_n$ have been sorted. Begin by setting $M \leftarrow F_r$, $U \leftarrow F_{r-1}$, and $L \leftarrow F_{r-2}$. (1) If $k<M$, go to step 2. If $k>M$, go to step 3. If $k=M$, stop. (2) If $L=0$, then output "unsuccessful" and stop. Otherwise set $M \leftarrow M-L$, and $(U, L) \leftarrow (L, U-L)$. Then go back to step 1. (3) If $U=1$, then output "unsuccessful" and stop. Otherwise set $M \leftarrow M+L$, $U \leftarrow U-L$, and then $L \leftarrow L-U$. Return to step 1.

a) See what might happen in an unsuccessful search with $r=5$.

b) Why are U, L, and M always Fibonacci numbers?

c) How many comparisons are needed in the worst case with Fibonaccian search? (Answer in recursive and closed form.)

COMPUTER PROGRAMMING EXERCISES

(Also see the appendix and Program A21.)

5.1. Write a program that determines whether a graph has any cycles.

5.2. Write a program that tests whether a graph is a tree.

5.3. Write a program that implements Kruskal's "greedy algorithm," Algorithm 5.1, for finding a minimum spanning tree in a complete weighted graph.

5.4. Modify the program that you wrote for the previous exercise so that it will find a minimum spanning tree in an arbitrary graph or print out "not connected."

5.5. Write a program that implements Algorithm 5.2, which finds a shortest v_0-path spanning tree in a graph.

5.6. Write a program that uses a graphics display to draw a rooted tree if it is given the adjacency matrix of a tree and the designated root.

5.7. Write a program that implements our binary search algorithm, Algorithm 5.3. Have it search for a given number in an increasing sequence of 2^n numbers.

5.8. Write a program that uses the Fibonaccian search algorithm described in Exercise 5.4-7. Have it search for a given number in an increasing sequence of F_n numbers, where F_n denotes the nth Fibonacci number ($F_0 = F_1 = 1$).

CHAPTER 6

DIRECTED GRAPHS AND NETWORKS

In the present chapter we take further steps to enhance the modeling capabilities of graphs and multigraphs. Two ideas that we have already introduced are (1) adding weights to the edges of a multigraph and (2) designating a particular vertex of a tree as the root. The second of these techniques had the interesting side effect of imposing an orientation on each edge of the tree away from the root. It is the idea of "orientation" that is the unifying theme of this chapter. Later in the chapter we return to the idea of "special" vertices in connection with objects called "networks."

6.1 DIGRAPHS

We have seen several examples of how multigraphs can be used to model systems of roads, social groups, and other interesting phenomena. One feature that is often missing in multigraph models, however, is the one-way nature of the relationship indicated when two vertices are connected by an edge. For example, person A may be attracted to person B but not the other way around. Here are two further examples in which a one-way relationship should be reflected in one's model.

Example 1 We can sometimes alleviate the heavy traffic on a system of roads by making carefully chosen streets one-way streets. If this is not done with care, however, the city planners might make it impossible to travel between various intersections. For example, in Fig. 6.1(b) one cannot follow the street directions and go from intersection A to intersection B even though it was possible to do so in the original undirected plan of the city's streets. [See Fig. 6.1(a).] □

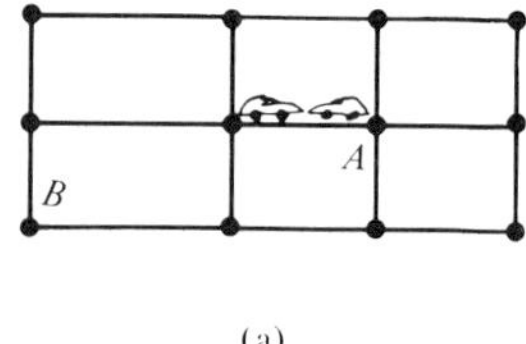

(a)

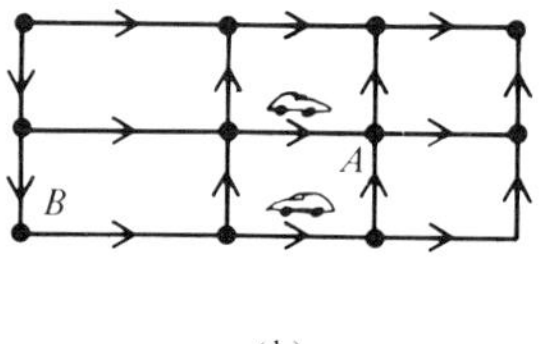

(b)

Figure 6.1 Assigning directions to city streets.
(a) Undirected streets. (b) One-way-street assignment.

Example 2 Graphs can be used to illustrate ecological relationships by connecting organisms that directly interact with one another. [See Fig. 6.2(a).] Orienting the edges from the predator to the prey then makes it clearer as to who eats whom in this "food web." [See Fig. 6.2(b).]

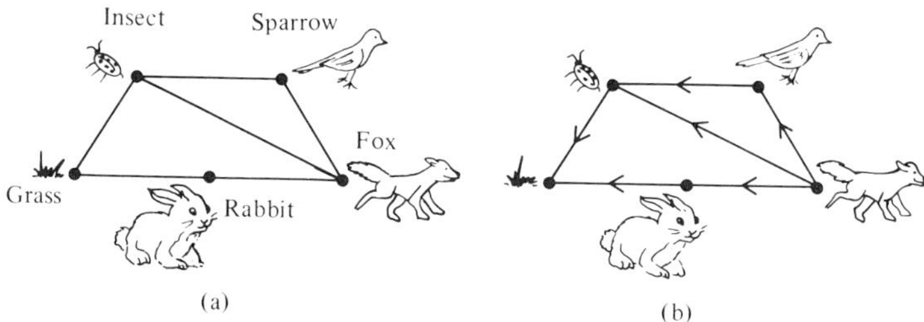

Figure 6.2 In this food web an arrow from the rabbit to the grass indicates, for example, that the rabbit eats the grass. □

The second diagram in each of the two preceding examples is known as a "digraph" or "directed graph." These and related terms are defined in the following.

Definition 6.1 A (a) **digraph** (or **directed graph**) D consists of a set V of (b) **vertices** and a set E of (c) **directed edges** such that to each edge there corresponds an ordered pair of vertices (u, v). The vertex u is called the (d) **initial point** and the vertex v is called the (e) **terminal point** of the edge. Two edges are (f) **parallel** if they have the same initial point and the same terminal point. An edge is a (g) **loop** if its initial and terminal points are identical.

Example 3 Figure 6.3 shows three digraphs. The digraph labeled D_1 has parallel edges e_1 and f_1, both of which have initial point u_1 and terminal point v_1.

The edges e_2 and f_2 in digraph D_2 are not parallel because their initial and terminal points are reversed. Indeed we can write

$$e_2 = (u_2, v_2) \quad \text{and} \quad f_2 = (v_2, u_2)$$

without ambiguity here, a practice we shall continue whenever it is appropriate and convenient.

The digraph labeled D_3 has three edges ("edges" will mean directed edges from now on, unless noted otherwise), including the loop $g = (w, w)$.

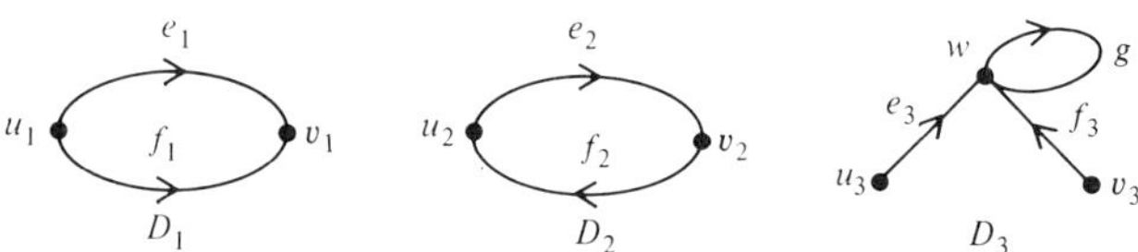

Figure 6.3 Three digraphs. □

We sometimes refer to the "underlying multigraph" of a digraph, meaning the multigraph one obtains by disregarding the orientations of the edges of the digraph. We can use this idea to formulate the following definitions.

Definition 6.2 A (a) **semiwalk** in a digraph is a walk in the underlying multigraph of the digraph. However, a (b) **walk** in a digraph is an alternating sequence of vertices and edges of the form

$$v_0e_1v_1e_2v_2e_3 \ldots e_nv_n,$$

where each edge e_i has initial vertex v_{i-1} and terminal vertex v_i. We will say (c) that **v_n is reachable from v_0** if there is a walk from v_0 to v_n.

In referring to the "food web" of Fig. 6.2(b), for example, we see that vertex "grass" is reachable from the vertex labeled "fox" by three distinct walks. (It is this sort of relationship that makes certain predators so susceptible to soil contaminants.)

Different digraphs can exhibit different degrees of connectivity, even when the underlying multigraphs are all connected, as in Fig. 6.3.

Definition 6.3 A digraph is (a) **strongly connected** if every vertex in the digraph D is reachable from every other vertex in D. We say that D is (b) **unilaterally connected** [or (c) **unilateral**] if, for all vertices u and v in D, either u is reachable from v or v is reachable from u. And D will be called (d) **weakly connected** if its underlying multigraph is connected.

Thus in Fig. 6.3 D_1 is unilaterally connected, D_2 is strongly connected, and D_3 is weakly connected. It should be obvious that if a digraph is strongly connected, then it must be unilateral as well. And a unilaterally connected digraph also must be weakly connected.

We can sometimes determine the connectedness category of a digraph by simply glancing at its diagram. Without the diagram, however, this would seem to be a very difficult job. To check for strong connectedness, for example, it appears that one must find $C(n, 2)\cdot 2 = n(n-1)$ walks, two walks for every pair of distinct vertices.

It is not really necessary to find so many walks, however, in order to check for strong connectedness in a digraph. To explain why this is so, we will have to modify some of our graph theoretic definitions.

Definition 6.4 If D is a digraph, then by a (a) **trail**, a (b) **cycle**, and a (c) **circuit** in D we will mean a walk in D that is a trail, cycle, or circuit, respectively, in the underlying multigraph of D. [The terms (d) **semitrail**, (e) **semicycle**, and (f) **semicircuit** are similarly defined.] A walk (or semiwalk) (g) **spans** $\boldsymbol{D}$ if it includes all the vertices of D.

Thus all the edges in a cycle or a semicycle of a digraph are distinct, the first and last vertices of a circuit are identical, and so on.

Theorem 6.1 Let D be a digraph having finitely many edges and vertices. Then

a) D is strongly connected iff D has a spanning circuit.

b) D is unilateral iff D has a spanning walk.

c) D is weakly connected iff D has a spanning semiwalk.

Proof: We will, in fact, only prove part (a), and we will begin by supposing that D is strongly connected. If we list the vertices of D

$$v_1, v_2, v_3, \ldots, v_n,$$

we observe that for each $i = 1, 2, \ldots n$ there is a walk from v_i to v_{i+1}, where we will identify v_1 with "v_{n+1}." Concatenating these walks in the order given by our list gives us the required spanning circuit of D.

Conversely, if $v_1e_1v_2e_2 \ldots e_{n-1}v_ne_nv_1$ is a spanning circuit of D, we can find a walk from v_i to v_j, where $i < j$, from this circuit by taking the section $v_ie_iv_{i+1}e_{i+1} \ldots e_{j-1}v_j$. If $i > j$, then we just take the part of the circuit from v_i to v_1, and continue by traversing the part of the spanning circuit from v_1 to v_j. Hence any vertices v_i and v_j in D are reachable from one another, since all the vertices of D are included in the spanning circuit.

The remaining parts of Theorem 6.1 are discussed in the exercises. ■

It is easy to translate Theorem 6.1 into a test for strong connectedness in a digraph.

Theorem 6.2 ***Strong Connectedness Test*** To determine whether a digraph with finitely many vertices and edges is strongly connected, it is sufficient to do the following.

a) List the vertices of D in any order: $v_1 v_2 \ldots v_n$.

b) For each $i = 1, 2, \ldots, n$ find a walk in D from v_i to v_{i+1}, where $v_{n+1} = v_1$.

c) If step (b) can be accomplished, then D is strongly connected. Otherwise D is not strongly connected.

The test given in Theorem 6.2 follows directly from the proof of Theorem 6.1(a). To see what an improvement this test makes, notice that for a digraph with 10 vertices we are required to find only 10 walks. Compare this with the $(10)(9) = 90$ walks that it previously seemed we had to find between the pairs of vertices.

There is still the matter of finding those 10 walks, however, and this is the weakest part of our test. But we can use a different approach to test for strong connectedness via matrices, an approach similar to the one we used for graphs. We will need the following definition.

Definition 6.5 **The adjacency matrix $A = (a_{ij})$ of a digraph D** having vertices $v_1, v_2, \ldots, v_n$ and finitely many edges is defined by $a_{ij} = k$, if D has k edges directed from v_i to v_j.

This definition leads to results very much like those we obtained for graphs and their connectivity properties. In particular, we have the following theorem.

Theorem 6.3 If A is the adjacency matrix of a digraph D, then the ijth entry in the matrix A^m, where m is a positive integer, is the number of walks of length m from v_i to v_j. Hence if D has n vertices, then

a) D is strongly connected iff $I + A + A^2 + \cdots + A^{n-1}$ has no zero entries; and

b) D is strongly connected iff $(I + A)^{n-1}$ has no zero entries.

The proof of Theorem 6.3 is just like those we gave for Theorems 4.15 and 4.16 in Section 4.6. Let us give some examples to show how these results are used.

Example 4 We first observe that the adjacency matrix of a digraph, unlike that of a graph, need not be symmetric. This will be seen in both of the following examples.

a) The 3-by-3 matrix

$$A=\begin{bmatrix}0&1&1\\1&0&1\\0&1&0\end{bmatrix}$$

defines a digraph with three vertices, v_1, v_2, and v_3, corresponding to the first, second, and third rows and columns of the matrix. To check this digraph for strong connectedness, we calculate

$$A^2=\begin{bmatrix}1&1&1\\0&1&1\\1&0&1\end{bmatrix}, \quad \text{and so} \quad I+A+A^2=\begin{bmatrix}2&2&2\\1&2&1\\1&1&2\end{bmatrix}.$$

Since the last matrix has no zero entry, Theorem 6.3 implies that the corresponding digraph is strongly connected. We can check this by constructing the digraph from the matrix A. [See Fig. 6.4(a).]

b) The 4-by-4 matrix

$$B=\begin{bmatrix}0&0&0&1\\1&0&0&0\\0&1&0&0\\0&0&1&0\end{bmatrix}$$

defines a digraph having four vertices. To check for strong connectedness, we calculate

$$B^2=\begin{bmatrix}0&0&1&0\\0&0&0&1\\1&0&0&0\\0&1&0&0\end{bmatrix}, \quad B^3=\begin{bmatrix}0&1&0&0\\0&0&1&0\\0&0&0&1\\1&0&0&0\end{bmatrix}, \quad \text{and}$$

$$I+B+B^2+B^3=\begin{bmatrix}1&1&1&1\\1&1&1&1\\1&1&1&1\\1&1&1&1\end{bmatrix}.$$

Once again we find that the last matrix has no nonzero entry, and so the digraph defined by the matrix B is strongly connected. Its diagram is given in Fig. 6.4(b).

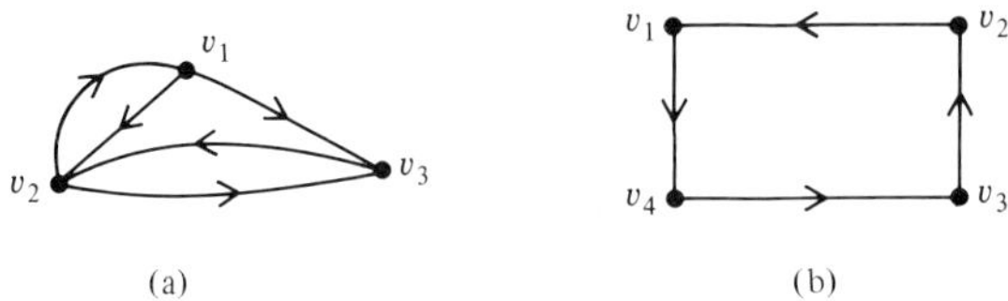

Figure 6.4 Strongly connected digraphs. □

In many of our digraph models of the next few sections we will be studying far more than just properties of connectivity. As indicated at the beginning of this chapter, we can label both the vertices and the edges in order to convey more information. The following example will indicate the dynamic possibilities inherent in labeling a digraph.

Example 5 The digraph in Fig. 6.5 describes a device called a "finite-state machine" that can model the addition of binary numbers. The digraph has three vertices, called the "states" of the machine. They are labeled "*nc*" (no carry), "*c*" (carry), and "*s*" (stop). Each directed edge is labeled near its initial vertex with a pair of binary digits or a "*b*" (blank). The edge is given a second label, for the corresponding output, which appears midway between its two endpoints. Outputs can be either a 1, a 0, or a blank. The initial inputs, a pair of binary digits from the summands or a blank, start at the

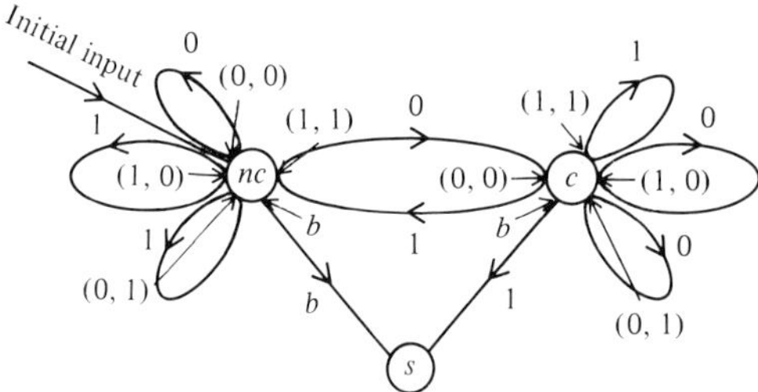

Figure 6.5 A finite-state machine. Inputs are ordered pairs at initial vertices. Outputs are weights on edges. Vertices are labeled with states "no carry," "carry," and "stop."

Table 6.1

State (vertex)	*s*	*c*	*nc*	*c*	*c*	*nc*	
Inputs		*b*	1 1	0 0	1 0	1 1	Initial inputs
Outputs		1	0	1	0	0	

vertex nc. The next input goes wherever the directed edge having the last input requires until, finally, we reach the vertex s, and we stop.

Finite-state machines are theoretical tools used to settle questions about what kinds of algorithms are possible. Although such discussions are beyond the scope of this text, it is fun to contemplate the action of such a device! Table 6.1 shows how our "machine" might calculate the sum $(1011)_2 + (1001)_2$. □

Completion Review 6.1

Complete each of the following.

1. A directed graph, or ________, consists of vertices and ________ edges such that to each edge there corresponds a unique ________ of vertices. The first vertex is the ________ point and the second is the ________ point of the edge.
2. Two directed edges are parallel if they have the same ________ and the same ________. A loop has ________.
3. A semiwalk $v_0e_1v_1e_2 \ldots e_nv_n$ in a digraph corresponds to any ________ in the underlying multigraph. It is a walk if each edge e_i has initial vertex ________ and terminal vertex ________.
4. We say that v_n is reachable from v_0 if there is a ___ ________.
5. If every vertex u in a digraph D is reachable from every other vertex v in D, we say that D is ________. If for every u and v either u is reachable from v or v is reachable from u, we then say that D is ________. D is weakly connected if ________ is connected.
6. A spanning walk in a digraph D is a ________ in D that contains all the ________ of D.
7. A digraph D is strongly connected iff D has a ________. It is unilateral iff D has a ________. It is weakly connected iff D has a ________.
8. To determine whether a digraph having vertices $v_1, v_2, \ldots, v_n$ is strongly connected, it is sufficient to find n walks, one from each ________.
9. The adjacency matrix A of a digraph D has $a_{ij} = k$ iff ________.

10. In a finite-state machine, the vertices are called the ______________ of the machine; outputs are labeled on ______________; inputs are given ______________.

Answers: **1.** digraph; directed; ordered pair (u, v); initial; terminal. **2.** initial point; terminal point; $u = v$. **3.** walk; v_{i-1}; v_i. **4.** walk from v_0 to v_n. **5.** strongly connected; unilateral; the underlying multigraph. **6.** walk; vertices. **7.** spanning circuit; spanning walk; spanning semiwalk. **8.** v_i to v_{i+1}, where $v_{n+1} = v_1$. **9.** there are exactly k walks from v_i to v_j. **10.** states; edges; at the base of each edge.

Exercises 6.1

Refer to the following digraphs for Exercises 1 to 10.

D_1

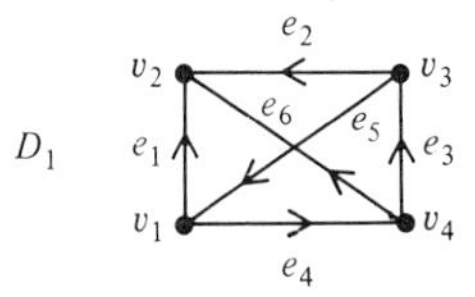

D_2

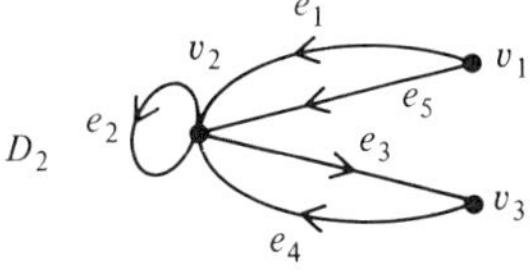

1. Find the initial point u and terminal point v for each edge e in the digraph D_1, and write $e = (u, v)$ in each case.
2. Find the parallel edges, if any, in **(a)** D_1; **(b)** D_2.
3. Find the loops, if any, in **(a)** D_1; **(b)** D_2.
4. Is the sequence of vertices and directed edges $v_1e_1v_2e_2v_3e_3v_4$ a walk, a semiwalk, or neither, in each of **(a)** D_1; **(b)** D_2? Explain.
5. Find a spanning circuit, if any, in each of **(a)** D_1; **(b)** D_2.
6. Is either of the digraphs D_1 or D_2 strongly connected? Explain on the basis of Theorem 6.1.
7. Is either of D_1 or D_2 unilaterally connected? Explain on the basis of Theorem 6.1.
8. Is either of D_1 or D_2 weakly connected? Explain on the basis of Theorem 6.1.
9. Find the adjacency matrix for each of **(a)** D_1; **(b)** D_2.
10. Find the **path matrix** $p = (p, j)$ for each of **(a)** D_1; **(b)** D_2. ($p_{ij} = 1$ if there is a path from v_i to v_j. Otherwise, $p_{ij} = 0$.)
11. **a)** Draw the digraph whose adjacency matrix is

$$A = \begin{bmatrix} 1 & 0 & 0 \\ 0 & 0 & 1 \\ 0 & 1 & 0 \end{bmatrix}.$$

b) Find the path matrix for this graph.
c) Explain on the basis of (b) how many paths of length 2 there are from v_1 to v_3.
d) Explain on the basis of (b) whether the digraph is strongly connected.

12. Repeat Exercise 11 with the adjacency matrix

$$B = \begin{bmatrix} 0 & 1 & 0 & 1 \\ 0 & 0 & 1 & 0 \\ 1 & 0 & 0 & 1 \\ 0 & 1 & 0 & 0 \end{bmatrix}.$$

13. Prove that a digraph D is weakly connected if and only if it has a spanning semi-walk. (*Hint*: Consider the underlying multigraph and our proof of Theorem 6.1a.)

14. Show that if a digraph has a spanning walk, then it is unilaterally connected. (This is half of Theorem 6.1b.)

15. Show that if D is unilaterally connected and S is a nonempty subset of the vertices of D, then there is a particular vertex v in S from which all the other vertices of S can be reached. (*Hint*: Use induction on k, the number of vertices of S.)

16. Use Exercise 15 to prove the converse of Exercise 14, thereby completing the proof of Theorem 6.1(b).

17. Find a way of orienting the edges in Fig. 6.1(a) so that one can travel from any intersection to another in the graph.

18. A **tournament** is a digraph D such that for every distinct pair of vertices u and v in D, either uv or vu is an edge of D, but not both. (This corresponds to every pair of contestants competing in a match and having exactly one winner per match: u if uv is in D, and v if vu is in D.) Given the tournament in the following illustration:

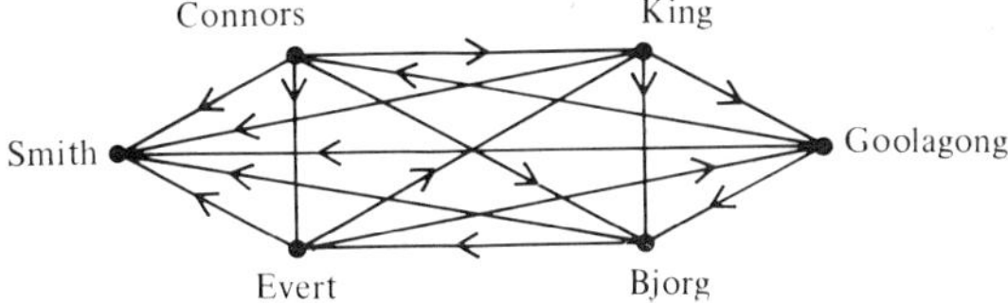

a) Determine who has won the greatest number of games.

b) Show that there is a path in D that includes all the players (vertices) but does not start with the person in (a).

19. Exercise 18 shows that bad rankings are possible within certain tournaments if we rely entirely on the criterion of "who beats whom." Explain this paradox.

20. Prove the following. In a tournament D, if u beats more players than any other player, then for every other player v, either u beats v, or u beats some w who, in turn, beats v.

21. Prove that every tournament D has a path that includes every vertex of D. (*Hint*: Use induction on the number of vertices in D.)

6.2 ORIENTABLE GRAPHS AND TOPOLOGICAL SORTING

The first question we deal with in this section is the one we raised in Example 1 of Section 6.1: When can we direct a network of roads so that they are all one-way roads and so that it is still possible to travel between any two intersections? We can model this network as a graph whose vertices are the intersections of the roads and whose edges are the sections of the roads connecting these intersections. Then what we are trying to do is to *direct the edges of our graph so that the result is a strongly connected digraph.* We say that we are trying to find an **orientation** of the graph. If we can find an orientation, then we say that the graph is **orientable**.

Example 1 The graph in Fig. 6.6(a) is *not* orientable, and the assignment of directions to its edges given in Fig. 6.6(b) is not an orientation of the graph, since the digraph in this figure is not strongly connected.

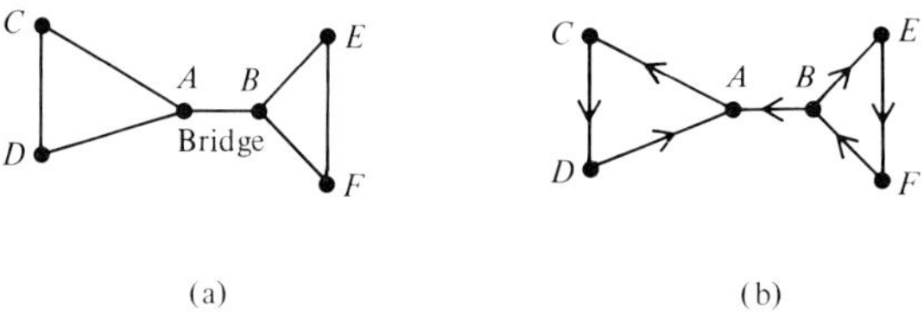

Figure 6.6 The graph in part (a) is nonorientable. Attempts to orient it as in part (b) must confront bridge AB.

Notice that any orientation of the edge connecting A and B either prevents you from going from C to E, as in Fig. 6.6(b), or from E to C if we were to reverse the direction of AB. The reason we are blocked is that any sequence of edges between C and E must include the edge connecting A and B. To use an analogy, we might say that edge AB acts as the only bridge connecting two islands. All traffic between the islands must cross this bridge. □

It is obvious that a graph must be connected in order to be orientable. If we define a **bridge** to be any edge of a connected graph, as in Fig. 6.6(a), whose removal disconnects the graph, then it is equally clear that an orientable graph cannot have any bridges. In fact we can characterize orientable graphs by means of these two properties.

Theorem 6.4 A graph G is orientable if and only if it is both connected and has no bridges.

Proof: Let us suppose that G is a connected graph with no bridges. We will show how to give an orientation of G. We begin by taking an arbitrary edge AB in G and orienting it from A to B. Since G has no bridges, AB must belong to some cycle C_1 of G (see Exercise 6.2-4) whose edges we direct so that we obtain a directed circuit, as in Fig. 6.7(a).

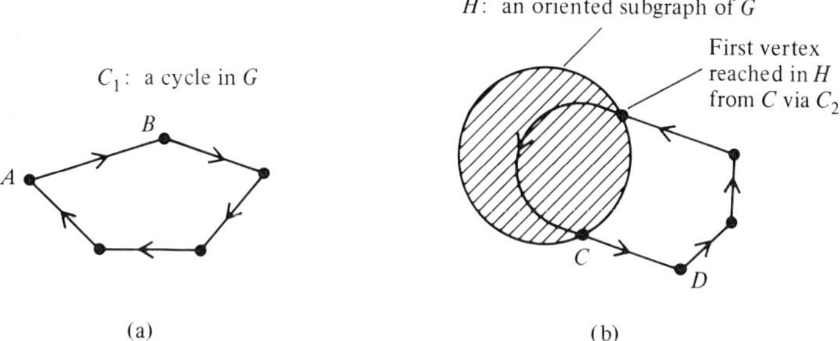

Figure 6.7 Determining an orientation of a connected graph with no bridges.

Now suppose that we have oriented part, but not all, of G. Call the oriented subgraph H. Since G is connected, there must be an edge CD such that C is in H but D is not. This edge must lie on a cycle C_2 of G [see Fig. 6.7(a)], since, as before, G has no bridges. Direct CD from C to D, and continue directing edges along the cycle until the first vertex you reach in H. Since H is oriented, there is a directed path from this first vertex to C, all of whose edges are in H. In this way we eventually direct all the edges of G and obtain a strongly connected digraph.

We have already pointed out the obvious necessity of there being no bridges and the graph being connected. Hence the proof is complete. ■

Our proof contains a method for orienting an orientable graph by finding larger and larger oriented subgraphs. (See Exercise 6.2-5.) Our method does not explain how to locate the cycles such as C_1 and C_2. This is not an insurmountable problem, however, because we can keep track of unused vertices and edges. But there is still the problem of knowing whether our graph is orientable in the first place and, in particular, whether it is connected and has no bridges.

A somewhat different approach is contained in the following algorithm, called a **depth-first search**, which forms the basis for answering our questions concerning orientability, connectivity, and bridges. In fact it will give us a method of orienting a graph when it is possible to do so.

The idea behind our algorithm will be to construct a tree by trying to extend the path that one is currently working on as long as possible. When we can no longer reach new vertices with our current path, we "backtrack" along this path until it is possible to branch out to a new vertex.

Algorithm 6.1 ***Depth-First Search Algorithm*** To construct a **depth-first-search tree** rooted at a vertex x of a connected graph $G = G(V, E)$,

1. [Initialize root, vertex set, and edge set.] Set $v_0 \leftarrow x$, $V_0 \leftarrow \{v_0\}$, $E_0 \leftarrow \emptyset$, and $i \leftarrow 0$.
2. [Tree loop.] Repeat steps 3, 4, and 5 until $V_i = V(G)$.
   3. [Backtrack.] Given V_i and E_i, set $k \leftarrow$ the largest subscript such that v_k is adjacent to a vertex z not in V_i, with v_k in V_i.
   4. [Enlarge tree.] Set $v_{i+1} \leftarrow z$, $V_{i+1} \leftarrow V_i \cup \{v_{i+1}\}$, and $E_{i+1} \leftarrow E_i \cup \{v_k v_{i+1}\}$.
   5. [Update i.] Set $i \leftarrow i + 1$.
6. [Done.] Output E_i and the numbered vertices in V_i. (We discuss this numbering in the following.)
7. End of Algorithm 6.1.

If we begin with a connected graph on n vertices, it is clear that our algorithm terminates when $i + 1 = n$. We will then have $n - 1$ edges. Hence *a depth-first-search tree of a connected graph is a spanning tree of the graph.* Figure 6.8 gives a depth-first spanning tree of a graph. Notice how we must "backtrack" when we reach the vertex labeled v_5 in order to find the first vertex (namely, v_3) adjacent to a vertex not already reached.

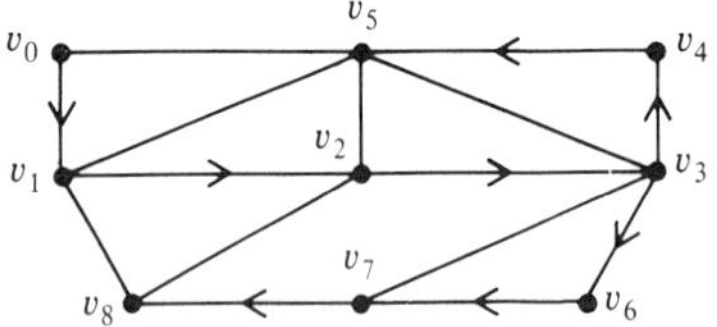

Figure 6.8 The directed edges in the figure are those of a depth-first-search spanning tree, beginning at v_0.

The vertex labeled v_6 is chosen next (although we could have chosen the one labeled v_7), followed by vertices labeled v_7 and v_8. In fact since there are no further vertices in the vertex set, we are done.

In a disconnected graph one must continue the depth-first search by choosing an initial vertex in a different component and labeling this vertex with the next available label.

The numbering of the vertices $v_0, v_1, \ldots, v_n$ given by a depth-first search (DFS) is called a **depth-first numbering (DFN)** of G.

Theorem 6.5 Let G be a connected graph without bridges whose vertices have been given a DFN. An orientation of G is given by directing the DFS tree edges from lower to higher depth-first numbered vertices and directing all other edges of G from higher to lower numbered vertices. (See Exercises 6.2-12–6.2-15.)

Example 2 Notice that when we direct the remainder of the edges of the graph underlying Fig. 6.8 from higher to lower numbers [Fig. 6.9(a)] we obtain a strongly connected digraph. This is easily checked by noting that the two leaves v_5 and v_8 of the DFS tree are adjacent to v_0 and v_1 respectively. This allows us to return to v_0 from any of the ancestors of v_5 and v_8.

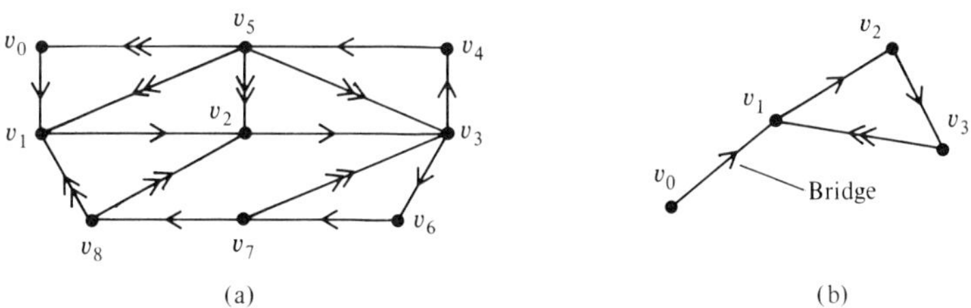

Figure 6.9 The single arrows in parts (a) and (b) indicate *DFS* edges. Double arrows indicate the corresponding orientation of the other edges.

The DFS of the graph underlying Fig. 6.9(b), however, does not lead to an orientation of the graph. Directing the remaining edge from v_3 to v_1, in particular, does not enable one to find a path from v_1, v_2, or v_3 back to v_0. Indeed, *v_0v_1 is identified as a bridge by the fact that no vertex reachable from v_1 along the rooted DFS tree is adjacent to a vertex in the graph having a lower DFN.* (See Exercises 6.2-12 and 6.2-13.) This condition can be used to develop an algorithm that finds the bridges in a graph. (See Exercise 6.2-14.) □

Topological Sorting

Our DFS algorithm can be applied to directed graphs if we are careful to construct our DFS tree or forest so that the edges added in step 2 have the same directions as those given in the given digraph. Therefore, we must begin backtracking the first time we reach a vertex, for example, v_3 in Fig. 6.10(a), that has only incoming edges. (To give a DFN to vertex B we must first backtrack to v_0.) A DFN of the entire digraph is given in Fig. 6.10(b).

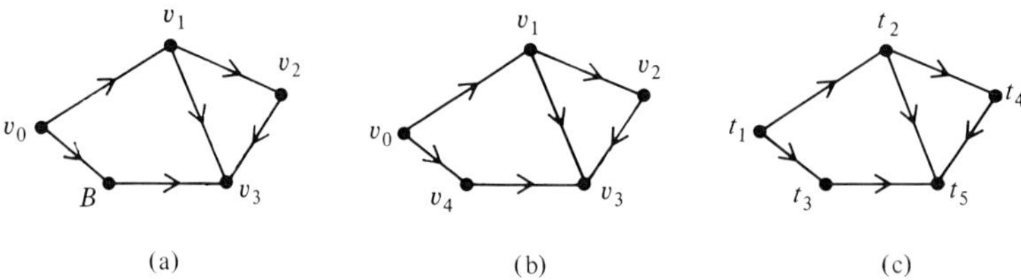

Figure 6.10 The vertices of the digraph in part (a) are topologically sorted in part (c) but not in (b). [$v=\phi$.]

The labeling of vertices of our digraph in Fig. 6.10(c) is, for some purposes, more desirable than the depth-first numbering, since *the numbering in (c) has every edge directed from lower to higher labeled vertices.* This numbering would be desirable if the original digraph had been a model of a complex project in which the vertices represent goals and the edges represent activities leading up to these goals. Such models are useful in, for example, the construction of a building. It is often the case in such projects that certain activities cannot be undertaken until some other activities have been completed. This can be modeled by edges directed out of vertices. The labeling of the vertices v_4 and v_5 in Fig. 6.10(b) would be in conflict with the order in which the corresponding goals could be attained. There are no such conflicts with the labeling given in Fig. 6.10(c), since for every directed edge uw the label on u is less than the label on w.

Definition 6.6 The n vertices of digraph $D=(V, E)$ have been **topologically sorted** if they have been labeled with the integers $1, 2, \ldots, n$ so that for every directed edge uv in E the label on u is less than the label on v.

Figure 6.10(c), therefore, is an example of a digraph whose vertices have been topologically sorted. In fact we know precisely which digraphs can be topologically sorted.

Theorem 6.6 The vertices of a digraph can be topologically sorted if and only if the digraph has no cycles.

Proof: The existence of a topologically sorted digraph with a cycle would give us a sequence of labels in that digraph such as $i<j< \cdots <i$, a contradiction. Hence a topologically sorted digraph must be acyclic.

Conversely suppose that digraph D has no cycles. Then there must be a vertex with no outgoing edges. (See Exercise 6.2-11.) Label this vertex with the integer n, if D has n vertices. Then remove this vertex and all its edges from D. The remaining digraph also has no cycles, and so we can find a second vertex having no outgoing edges. We label this vertex with the integer $n-1$ and repeat this procedure until there are no further vertices. A simple induction argument (see Exercise 6.2-16) shows that this leads to a topologically sorted digraph, and this completes the proof. ■

Algorithm 6.2 ***Topological Sorting*** To topologically sort a digraph $D=(V, E)$ having n vertices, $n \geq 1$, and no cycles,

1. [Initialize i.] Set $i \leftarrow n$.
2. [Sorting loop.] Repeat steps 3 and 4 until $V=\varnothing$. (If $V=\varnothing$, we continue from step 5.)
   3. [Find terminal vertex.] Depth-first search D until a vertex w is found having no outgoing edges.
   4. [Update.] Set $V \leftarrow V-\{w\}$, $E \leftarrow E-\{$all edges directed into $w\}$, $D \leftarrow D(V, E)$, $t_i \leftarrow w$ and $i \leftarrow i-1$. (We now return to step 2. We can use the DFN of any remaining vertices if we apply step 3 again.)
   5. [Done.] Output the sequence of vertices $t_1, t_2, \ldots, t_n$.
6. End Algorithm 6.2.

Example 3 Figure 6.11 shows a sequence of steps whereby Algorithm 6.2 topologically sorts the vertices of a digraph. Notice that no vertex is DFS numbered more than once, an indication of the efficiency of this technique.

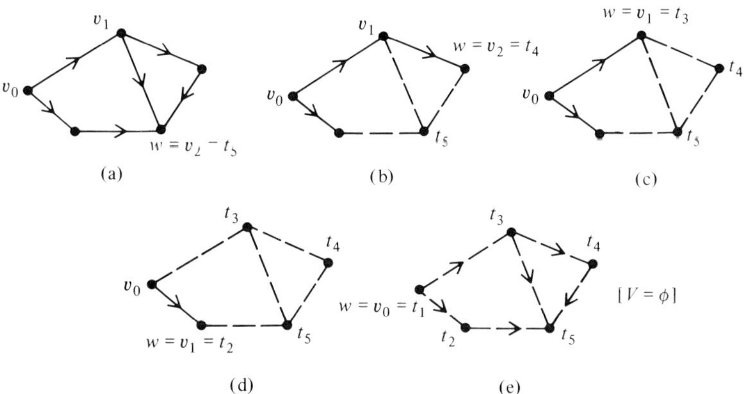

Figure 6.11 Parts (b) to (e) show how to topologically sort the digraph in part (a).

Vertex v_2 is the first vertex found that has no outgoing edges. Hence v_2 is called t_5, since the digraph has five vertices in all. Then vertex v_2 and all the edges going into v_2 are removed from the digraph. The search continues at the vertex previously labeled v_1, with both v_1 and v_0 retaining their depth-first numbers from Fig. 6.11(a). The algorithm ends when v_0 is called t_1 and $V = \varnothing$. □

Completion Review 6.2

Complete each of the following.

1. A graph is orientable if its edges can be directed so that we obtain a ________________. The given directions are called an ________________ of the given graph.
2. A bridge is any edge in a graph whose removal ________________.
3. A graph is orientable iff it is ________________ and has no ________________.
4. One can find an orientation of an orientable graph by using a ________________ search algorithm. This algorithm yields a ________________ tree in which each vertex is given a ________________.
5. One orients a graph having a DFS spanning tree by directing each edge in the tree ________________ and each edge not in the tree ________________.
6. The n vertices of a digraph D have been topologically sorted if they have been labeled $1, 2, \ldots, n$ so that for every directed edge uv in G, ________________.
7. The vertices in a digraph can be topologically sorted iff the digraph has ________________.

Answers: **1.** strongly connected digraph; orientation. **2.** disconnects the graph. **3.** connected; bridges. **4.** depth-first; DFS; DFN. **5.** from lower to higher DFN; from higher to lower DFN. **6.** u has a lower numbered label than v. **7.** no cycles.

Exercises 6.2

1. Find all the bridges in the following graphs.

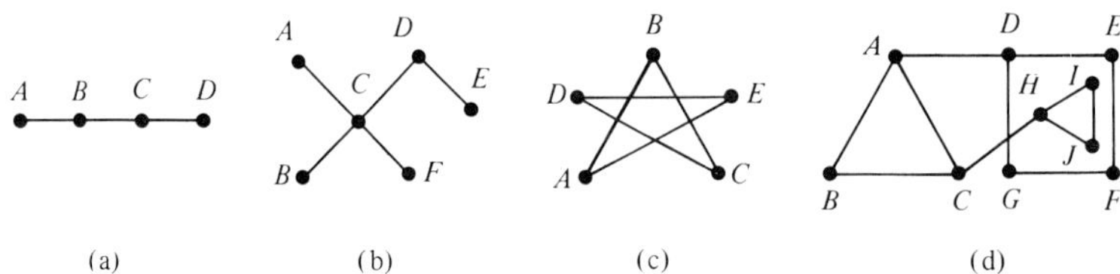

2. How many bridges does a tree on n vertices have? Why?
3. How many bridges does a cycle on n vertices have? Why?
4. Explain why every edge AB of a connected graph G having no bridges must belong to a cycle. (*Hint*: Consider a spanning tree of the graph containing AB.)
5. Write out the orientation algorithm discussed in the proof of Theorem 6.4 as a step-by-step procedure.
6. Use your answer to Exercise 5 to orient each of the following graphs, if possible. If it is not possible, explain why it is not.

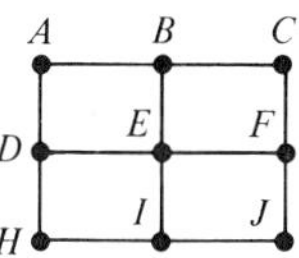

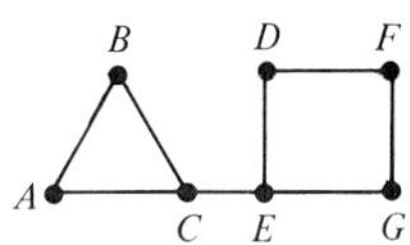

7. The definition of an orientable multigraph is the same as that for a graph. Orient the following model of Yosemite Valley National Park's Yosemite Valley region.

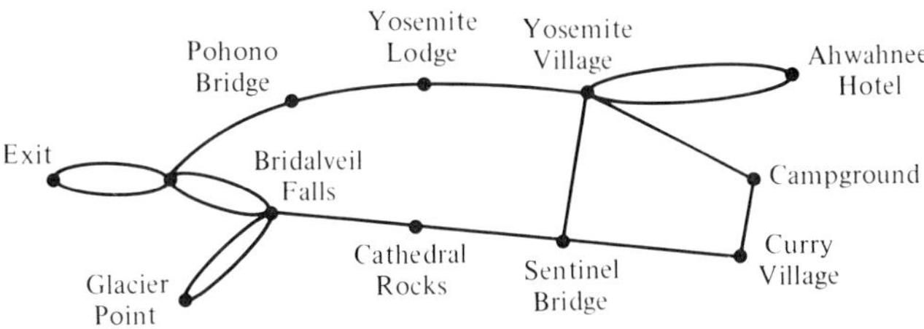

8. Give a depth-first numbering of the vertices of the following graph. Use your DFN to orient the graph.

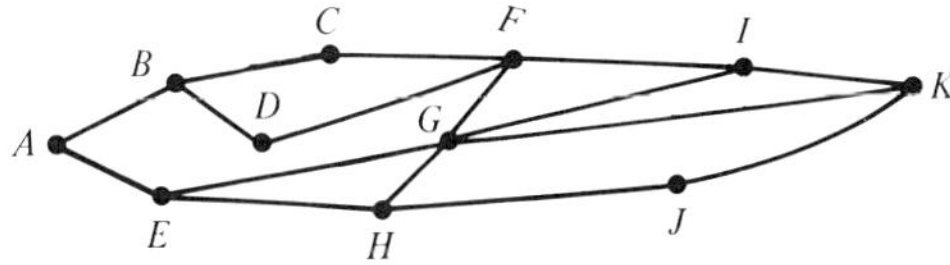

9. Give a DFN for each of the following graphs. Why cannot these be used to orient the graphs?

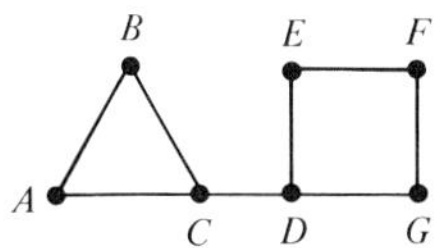

(a)

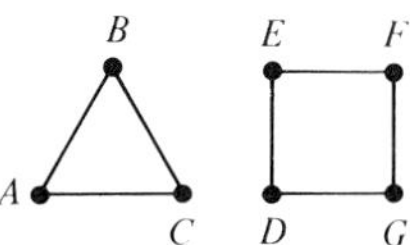

(b)

10. Direct each edge in the graph of Exercise 8 from right to left. Then topologically sort the vertices of the resulting digraph with Algorithm 6.2.

11. Prove that a digraph having no cycles must have a vertex with no outgoing edges. (*Hint*: Prove it by contradiction, assuming that the digraph has n edges.)

12. The **depth-first reach** of a vertex v_i in a DFS tree T is the largest j such that v_j is a descendant of v_i in T. Show that if the depth-first reach of v_i is j, then the vertices reachable from v_i in T are precisely $v_i, v_{i+1}, \ldots, v_j$.

13. Use Exercise 12 to show the following. If T is a DFS tree of the graph G, then an edge of G connecting v_h and v_i is a bridge, where $h < i$, if and only if no vertex reachable from v_i in T is adjacent in G to a vertex v_k with $k < i$.

14. Write an algorithm based on Exercise 13 for finding the bridges in a graph.

15. Use Exercises 12 and 13 to show that our DFS method for orienting a connected graph G having no bridges always works.

16. Complete the proof of Theorem 6.6 with an induction argument showing that Algorithm 6.2 gives a topological sorting of an acyclic digraph.

6.3 ACTIVITY ANALYSIS AND LONGEST PATHS

In the second part of the last section we discussed the method of topologically sorting a set of project-related goals, that is, listing these goals in a way that is consistent with certain precedence relations. Each goal was represented by a vertex and the precedence relations were given by the directed edges of a digraph. Each edge, you will recall, could represent an activity leading up to a goal, but we were not really concerned with the nature of these activities. In any case this type of model might be suitable for determining the sequence in which a student could pursue a course of study, supposing that only one course is taken at a time and the prerequisites for the courses are satisfied.

In many complex projects, such as the construction of a house, it is not unusual for several activities to be going on at the same time. But it is still necessary that certain activities end before others can even begin. For example, the interior walls cannot be erected until the electrical work and the plumbing have been completed, but time can be saved if the electrician and the plumber work simultaneously. This is indicated in Fig. 6.12 where edges AC, BC, and CD represent the activities of electrical work, plumbing, and finishing the interior walls respectively. The vertices merely represent points in time, and the weights on each directed edge are estimates of the

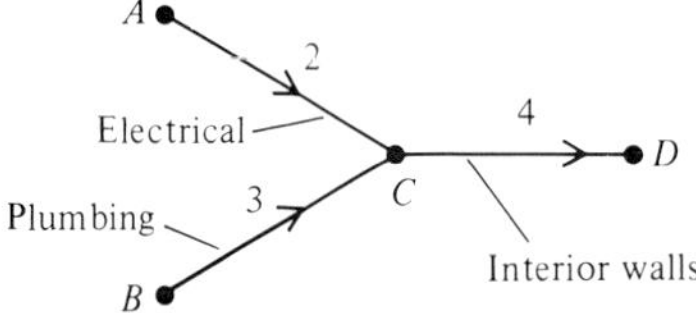

Figure 6.12 Ordered activities and their completion times.

time each activity is expected to take. Thus point C can be reached no sooner than two time units after point A and no sooner than three time units after point B. Similarly, point D can be reached no sooner than four time units after point C.

Digraphs having weighted edges with this kind of interpretation are often used in project activity analysis techniques such as CPM (critical path method) and PERT (program evaluation and review technique). While it is beyond the scope of this text to discuss all the ramifications of these techniques, the following will give the reader a fair idea of what is involved, including some basic definitions.

Definition 6.7 An (a) **activity network** is a nonnegatively weighted digraph having no cycles. Each edge represents an activity. Activity e_1 is an (b) **immediate predecessor** of activity e_2 if the initial vertex of e_2 and the terminal vertex of e_1 coincide. Moreover, there are precisely two vertices, called the (c) **beginning** and (d) **end** [or (e) **source** and (f) **sink**], which have no incoming and no outgoing edges respectively. Finally, no activity may proceed until all of its immediate predecessors have been completed.

Example 1 Let us discuss each of the weighted digraphs in Fig. 6.13.

The digraph in Figure 6.13(a) satisfies our definition of an activity network, apart from the lack of an interpretation of the edges as specific activities. Vertices B and E are the unique beginning and end respectively. There are no cycles, and each of the edges has a positive weight. If we wanted to make both BX and BY immediate predecessors of XE, we could show BX and BY as parallel edges. It is preferable, however, to introduce a fictitious activity having a zero weight, such as YX in (b). While BY is still not an immediate predecessor of XE, the effect is the same, since XE cannot begin until YX (and hence BY) is completed.

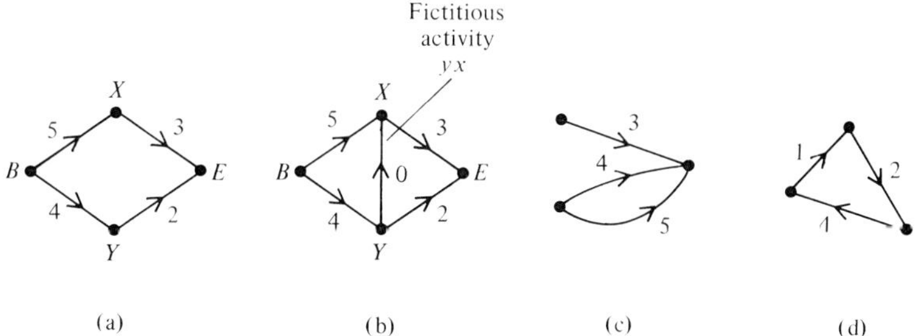

Figure 6.13 Digraphs (a) and (b) are activity graphs, but (c) and (d) are not.

Figure 6.13(c) could not be an activity network because it has two possible beginnings.

Figure 6.13(d) could not be an activity network because it is a cycle, and therefore has neither a beginning nor an end. □

The most important question asked about activity networks concerns *the minimum time required for the project's completion from beginning to end.* In Fig. 6.13(d) we see that this kind of question does not make sense if our digraph has a cycle. (See Exercise 6.3-7.) We also want to identify those activities that might delay the entire project by exceeding their estimated completion times. These are called **critical activities**, and they are found on the **critical paths** of the digraph, that is, the paths in the digraph having the longest possible length. In fact we can state the following theorem about critical paths.

Theorem 6.7 The length of a critical path in an activity network is the minimum time needed to complete every activity. Moreover, every critical path begins at the source and ends at the sink.

Proof: A maximal path (that is, one not contained in a larger path) that did not start at the source must start at some vertex that has an incoming edge. But this means that we could enlarge this path, a contradiction. Similarly, a path of maximum length must end at the sink. Moreover, if the minimum time to complete the project were less than the length of a critical path, then there would not be sufficient time to complete every activity on such a path. Hence the minimum time cannot be less than the length of a critical path.

Now let A be an activity of our activity network and let P denote a path of longest length. We need to show that A can be accomplished in t units of time, where t is the length of P. If A is an edge of P, this is obvious. So we suppose that A is not an edge of P. Then A is an edge of some path Q that begins at the source B and ends at the sink E. (See Exercise 6.3-8.) Since the length of path Q is less than or equal to that of path P, there must be sufficient time to complete activity A. This completes the proof. ∎

Let us now consider a concrete example of an activity network and try to apply Theorem 6.7 to it.

Example 2 Suppose that Table 6.2 describes the major activities associated with the construction of a house.

The first three columns in our table lead to the activity network given in Fig. 6.14. The edge-to-activity correspondences are given in the fourth column of the table. Notice that we use a fictitious activity, WG, so that only one edge is needed for activity 6, namely, GI.

We can see by inspection that there are three paths from B to E, the longest of

Table 6.2

Activity	Time (days)	Immediate Predecessors	Edges
1. Prepare site	3	–	BP
2. Build foundation	2	1	PF
3. Build main structure	12	2	FM
4. Plumbing	6	3	MG
5. Electrical wiring	4	3	MW
6. Complete interior	9	4 and 5	GI
7. Complete exterior	12	2	MX
8. Furnish	3	6	IE
9. Landscape	5	7	XE

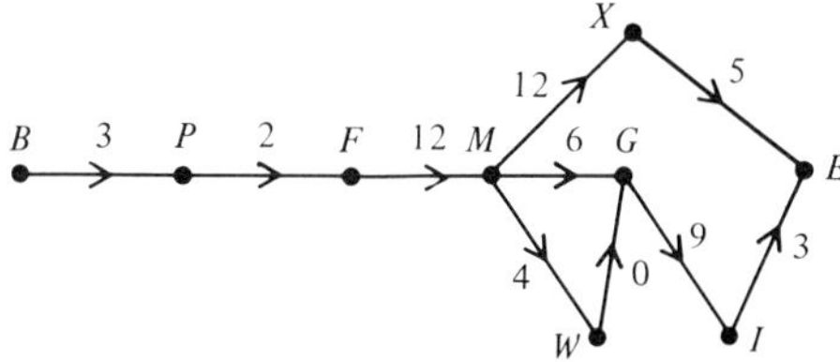

Figure 6.14 The activity graph corresponding to Table 6.2.

which is given by the sequence of vertices $BPFMGIE$ and which has a length of

$$3 + 2 + 12 + 6 + 9 + 3 = 35.$$

This critical path, therefore, tells us that our project will require a minimum of 35 days. Every activity (edge) on this path is critical. This means that every additional day taken by any one of these activities—plumbing, for example—adds an extra day to the project. None of the other activities are critical, however, since there are no other critical paths. Thus we could spend an extra day landscaping (XE) without adding time to the total project. (See Exercise 6.3-3.) □

In actual practice estimates of the times for each activity are often given in terms of probabilities. This is especially true of PERT analysis, which was invented to expedite the development of the Polaris missile system. Moreover, the number of different activities is often enormous, so that the determination of a longest path by simple inspection is out of the question.

Many algorithms have been invented to determine longest paths in acyclic digraphs. The one we present here (Algorithm 6.3) works, in particular, for the activity networks under discussion.

Algorithm 6.3 ***Longest Path Algorithm*** Let D be an activity network with a source B and a sink E in which each edge v_iv_j has a nonnegative weight w_{ij}. To find a longest path in D from B to E,

1. [Find spanning tree.] Let $T \leftarrow$ any spanning tree on D rooted at B.
2. [Initialize slacks.] For every edge v_iv_j in D but not in T set $s_{ij} \leftarrow -1$. (The s_{ij}'s are called "slacks." Negative slacks are a signal that one can increase the length of some path.
3. [Path increasing loop.] Do steps 4 through 6 while at least one $s_{ij} < 0$.
    4. [Find tree distances.] Set $t_i \leftarrow$ the distance from B to each vertex v_i of D along a path in tree T. Label each vertex v_i with the corresponding t_i.
    5. [Find new slacks.] For every edge v_iv_j in D but not in T, set $s_{ij} \leftarrow t_j - t_i - w_{ij}$.
    6. [New tree?] If, for some i and j, s_{ij} is negative, then set $T \leftarrow$ the new tree formed by replacing the edge in (our current tree) T directed toward v_j with the edge v_iv_j.
7. [Done.] Output T and the labels t_i.
8. End of Algorithm 6.3.

Figure 6.15 gives a simplified explanation of why our algorithm increases lengths of paths to each vertex in the digraph. In Fig. 6.15(a) a negative slack $s_{ij} = t_j - t_i - w_{ij}$ implies that w_{ij} is greater than $t_j - t_i$, which in this figure is the sum of the weights of BV and VE. In Fig. 6.15(b), the same negative slack implies that $w_{ij} + t_i$, the sum of the weights of BV and VE, is greater than t_j, which in this figure is just the weight of BE. The edge substitutions indicated by the algorithm increase the tree distances from B to E in both cases.

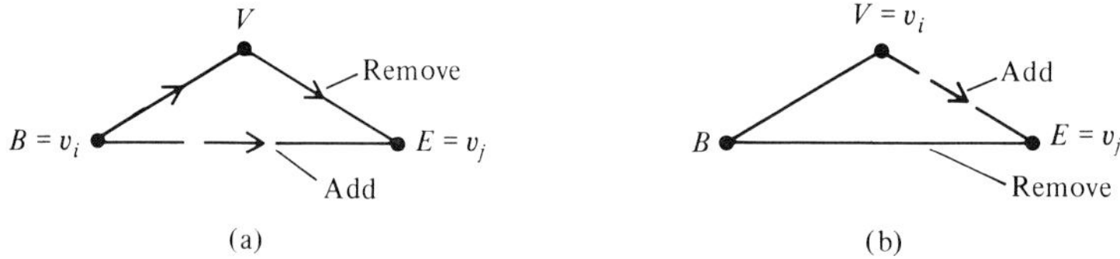

Figure 6.15 The longest path algorithm. (a) $s_{ij} < 0$ implies $w_{ij} > t_j - t_i = BV + VE (t_i = 0)$. (b) $s_{ij} < 0$ implies $BV + VE = w_{ij} + t_i > t_j$.

Let us now apply our algorithm to a specific example.

Example 3 Figure 6.16(a) gives an activity graph on six vertices. In Fig. 6.16(b) we have constructed a depth-first spanning tree and labeled each vertex v_i with its distance t_i from B $(=v_1)$ along the tree. Chord v_3v_2 yields a negative slack, $s_{32}=2-3-4$. Therefore, we substitute edge v_3v_2 for the spanning tree edge v_1v_2 and obtain the tree in Fig. 6.16(c). The vertices in this tree are relabeled with their new distances from B. However, when the slack s_{ij} is calculated for each chord, none are found to be negative. (The slack for each chord is the number in parentheses.) Therefore, the present tree yields a longest path, namely,

$$Bv_3v_2v_4E,$$

having length 16, as given by its label.

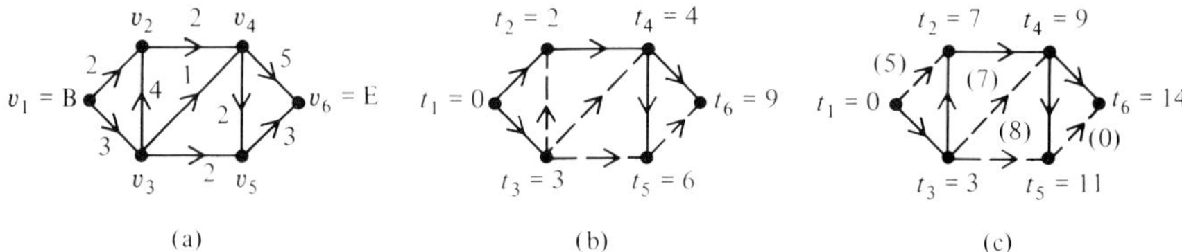

Figure 6.16 The distance of v_i from v_1 along the tree is indicated by t_i. (b) $s_{32}=2-3-4=-5$. (c) $s_{56}=0$.

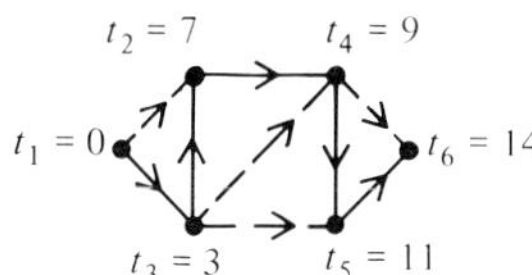

Figure 6.17 A longest t_1-path spanning tree.

The slack $s_{56}=0$ in Fig. 6.16(c) indicates that our longest path is not unique. Indeed if we exchange edge v_4v_6 for edge v_5v_6, we obtain the tree in Fig. 6.17, and this tree contains a different longest path, namely, $Bv_3v_2v_4v_5E$ from B to E. (See Exercises 6.3-14 and 6.3-15.) □

Completion Review 6.3

Complete each of the following.

1. An activity network is a nonnegatively weighted digraph that has__________________.

   Edge e_1 is the immediate predecessor of e_2 if__________________.

2. In an activity network the vertex that has no incoming edges is called the ________________ and the vertex that has no outgoing edges is called the ________________.

3. The path with the longest possible length in an activity network is called a ________________. Edges of these paths are called ________________.

4. The minimum time in an activity network needed to complete every activity is ________________, each of which begins at the ________________ and ends at the ________________.

5. The longest path algorithm increases the length by finding edges having positive ________________.

6. The acronyms CPM and PERT stand for ________________ and ________________ respectively.

Answers: **1.** no cycles; the terminal vertex of e_1 is the initial vertex of e_2. **2.** source or beginning; sink or end. **3.** critical path; critical activities. **4.** the length of a critical path; source; sink. **5.** slacks, s_{ij}. **6.** critical path method; program evaluation and review technique.

Exercises 6.3

1. For each of the following weighted digraphs, explain why it is not an activity network.

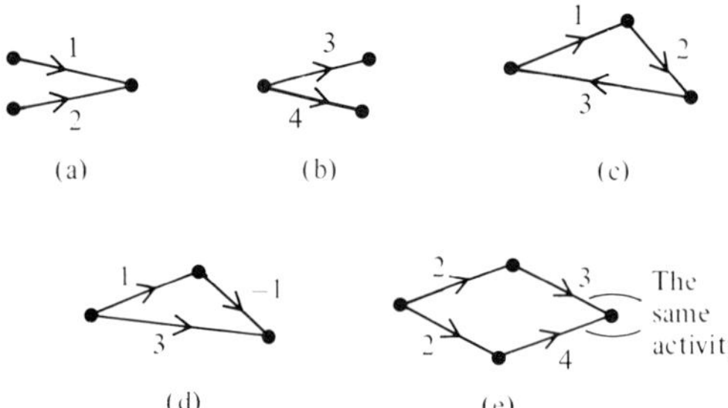

2. With reference to the following activity network:

a) What are the source and the sink?

b) Find a critical path by inspection.

c) Identify the critical activities.

d) Which activity has the most slack?

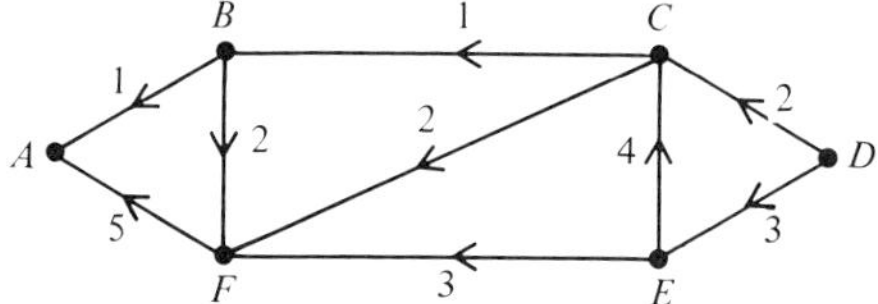

3. In Example 2 we asserted that an extra day could be spent landscaping without adding to the 35-day estimate for completing the house. Explain this.
4. A mathematics professor in a southern university has just been named to a new position up north. She performs an activity analysis to help plan her move to the new location, and she obtains the following table.

Activity	Time (days)	Predecessors
1. Purchase new home	21	None
2. Shop for a moving company	7	1
3. Buy clothing appropriate for colder climate	14	1
4. Obtain boxes for packing	5	2
5. Pack	3	4 and 3
6. Send boxes ahead of movers	1	5
7. Move furniture, etc.	2	2
8. Drive to new home	2	6 and 7

   a) Make an activity network corresponding to the table.
   b) Find a critical path and determine the minimum time it will take her to move.
   c) Identify critical activities.
5. Apply Algorithm 6.3 to the digraph in Exercise 2, beginning with the following tree.

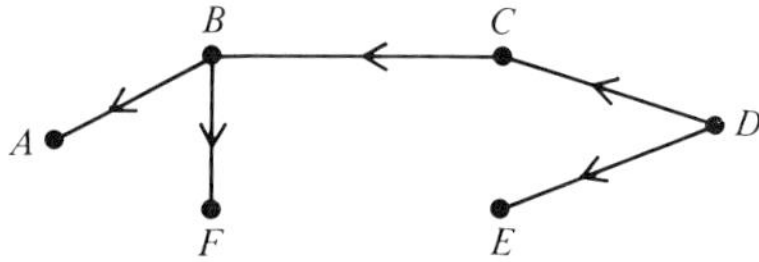

6. Apply Algorithm 6.3 to the following activity network.

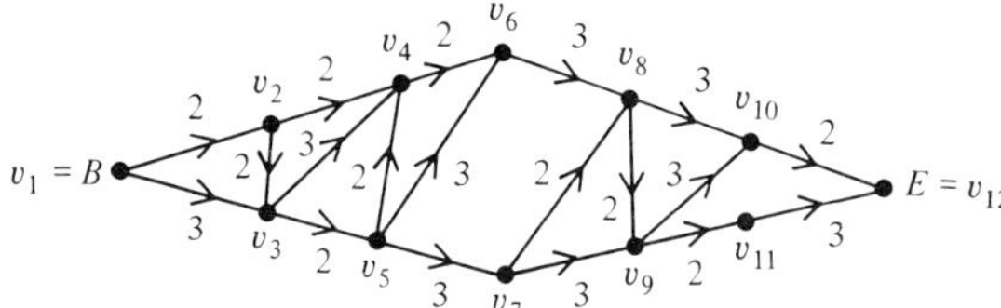

7. What is wrong with trying to find a longest directed walk in a weighted digraph that has a cycle?

8. Complete the proof of Theorem 6.7 by showing that if activity A is not an edge of the critical path P, then A is an edge of some other path that begins at the source B and ends at the sink E.

9. Using the notation of Algorithm 6.2, show that a spanning tree T rooted at B gives a longest path from B to each of the other vertices only if for every chord v_iv_j of the tree, we have $s_{ij}=t_j-t_i-w_{ij}\geqslant 0$. (*Hint*: Show that any spanning tree with a negative s_{ij} can be changed to another spanning tree with a longer path from B to E.)

10. Prove the following. A spanning tree rooted at B is a shortest B-path spanning tree on a weighted digraph or graph only if $s_{ij}=t_j-t_i-w_{ij}\leqslant 0$ for every chord v_iv_j. (*Note*: The converse is also true—but more difficult to prove.)

11. Use Exercise 10 to develop an alternate to Djikstra's algorithm for finding a shortest B-path spanning tree.

12. Apply the algorithm you developed in Exercise 11 to the digraph in Exercise 6.

13. Can Djikstra's shortest path algorithm, 5.2, be used to find longest paths in the following way? "Take a number M that is larger than any edge weight w_{ij}, and use it to give a new edge weight $w_{ij}^*=M-w_{ij}$ to each edge v_iv_j. Find a shortest B-path spanning tree with respect to the new weights. This is a longest B-path spanning tree with respect to the old weights." Prove or disprove this claim.

14. Give a counterexample to show that $s_{ij}=0$ does not necessarily indicate that chord v_iv_j is a critical activity. (*Hint*: See Exercise 15.)

15. Show that if $s_{ij}=0$ and if chord v_iv_j is such that v_j is on a critical path, then v_iv_j is a critical activity.

6.4 TRANSPORT NETWORKS

In this section we consider yet another kind of weighted digraph, one that is suitable for modeling systems of pipelines, for example, that carry natural gas. Let us consider an example to help us decide just which features we would like our model to reflect.

Example 1 In the digraph of Figure 6.18(a) we can think of some material being sent from vertex a along the directed edges and finally exiting at vertex z. The numbers on the edges tell us the most that each edge can carry.

In Fig. 6.18(b) we have added a new feature to our model, namely, a second number on each edge telling us how much material is actually being carried. It is reasonable to assume that the amounts carried in each case do not exceed the first numbers. We can also assume that both the carrying capacities and the actual loads are nonnegative. Moreover, we want the amount flowing into each vertex to be the amount flowing out, except for the special vertices a and z. Notice, for example, that two units are flowing both into and out of vertex c. Finally, we see that the total

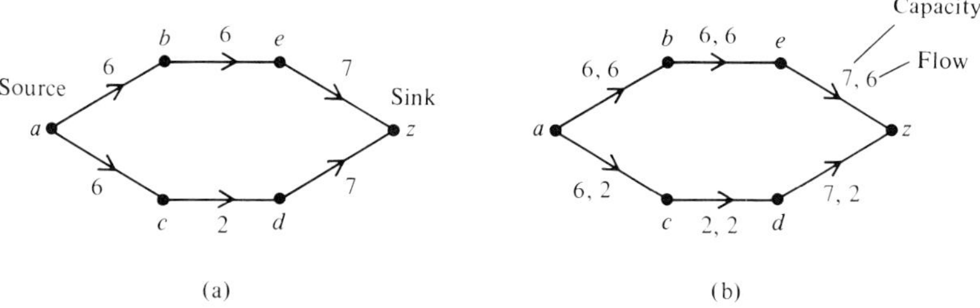

Figure 6.18 Part (a) shows the capacities on the edges of a transportion network. Part (b) shows a network flow.

amount flowing out of vertex a ($6+2=8$ units) equals the total amount flowing into z ($6+2=8$ units).

Let us now give a precise definition of our model, some of whose features have already appeared in earlier sections.

Definition 6.8 A (a) **transport network** is a weighted directed graph without loops and with nonnegative weights on each of its edges. In addition there is exactly one vertex with no incoming edges and another vertex with no outgoing edges called the (b) **source** and the (c) **sink** respectively. The weight on an edge xy from vertex x to vertex y will be denoted (d) $\boldsymbol{c(x, y)}$ and will be called the (e) **capacity** of that edge.

A (f) **flow** in a transport network is a function f that assigns a nonnegative number $f(x, y)$ to each edge xy such that

1. $0 \leq f(x, y) \leq c(x, y)$ for every edge xy, and
2. for every vertex y except the source and the sink, we have

$$\sum_{\text{all } x} f(x, y) = \sum_{\text{all } v} f(y, v).$$

That is, the total flow in equals the total flow out of y.

In looking back at Fig. 6.18(b), a natural question to ask is whether we can increase the total number of units flowing out of vertex a or into vertex z. The answer would seem to be "no," even though the flows assigned to each of edges ac, dz, and ez are less than their capacities. However, any additional flow would be cut off from the sink z by the edges cd and be. The flow in these edges equals their capacities, and removal of these edges from the network puts a and z in different connected components with respect to the underlying graph.

The following definitions will help us to state our ideas more precisely.

Definition 6.9 Let P be a set of vertices in a transport network such that $a \in P$ and $z \in \bar{P}$, the complement of P. Then (a) $(\boldsymbol{P}, \bar{\boldsymbol{P}})$ will denote the set of all edges xy such that $x \in P$ and $y \in \bar{P}$. We will say that $(P, \bar{P})$ is a (b) **cut**. By the (c) **capacity of the cut** we will mean the sum denoted by

$$c(P, \bar{P}) = \sum_{x \in P,\, y \in \bar{P}} c(x, y).$$

Moreover, we will say that the (d) **value of a flow** is the sum

$$f_v = \sum_{\text{all } x} f(a, x)$$

taken over all vertices in the network. A (e) **maximum flow** is one that achieves the highest possible value in the transport network. An edge xy is said to be (e) **saturated** if $f(x, y) = c(x, y)$. Otherwise we say that the edge is (g) **unsaturated**.

Thus in looking back at Fig. 6.18(b), we see that edges ab, be, and cd are saturated because

$$f(a, b) = 6 = c(a, b), \quad f(b, e) = 6 = c(b, e), \quad \text{and} \quad f(c, d) = 2 = c(c, d).$$

The flow given in this network is a maximum because the edges be and cd in the cut $(P, \bar{P})$, where $P = \{a, b, c\}$ and $\bar{P} = \{e, d, z\}$, are saturated. The value of the flow, by definition, is $6 + 2 = 8$, the sum of the flow values assigned to edges ab and ac.

Transport networks can be tailored to fit a variety of situations that are not exactly like the original definition. Here are two examples.

Example 2 Suppose that vertices a_1 and a_2 in Fig. 6.19(a) represent two supply depots having storage capacities of 25 and 35 units respectively. Let vertices z_1 and z_2 stand for

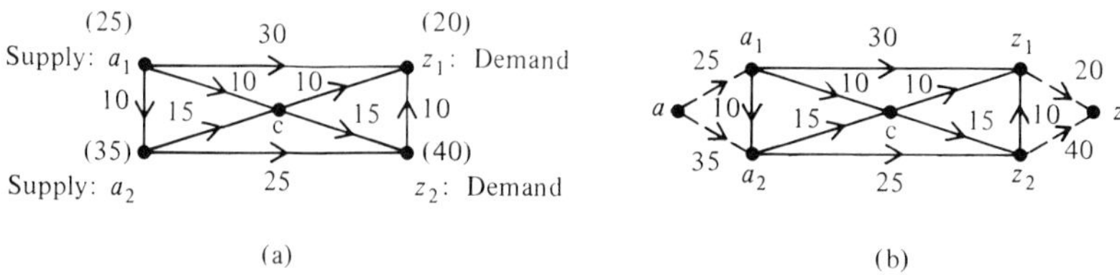

Figure 6.19 Changing (a) a supply—demand model into (b) a network flow model.

factory outlets having demands for 20 and 40 units of the same item. The supply must be sent to the outlets over the given network of capacitated edges. Can the demand be satisfied by the supply?

To change this model into a standard network flow problem, we can connect an artificial source a and a sink z to the supply depots and factory outlets as in Fig. 6.19(b). The directed edges that we added have the capacities (the supplies or demands) of the corresponding vertices a_i and z_i. Clearly we can fill the demands with the given supplies only if we can find a flow whose value is $60 = 25 + 35 = 20 + 40$. □

Example 3 Problems in which objects in one set are matched up with (or mapped one to one into) the objects in another set can be solved using network flows. Figure 6.20(a), for example, shows which boys each of five girls wants to ask her to dance. To determine whether all of the girls can find suitable partners if they all get up to dance, direct one edge to each girl from a source a, and one edge from each boy to a sink z. Give every edge a capacity of 1. Then our problem is reduced to finding a flow in Fig. 6.20(b) that has a value of at least 5.

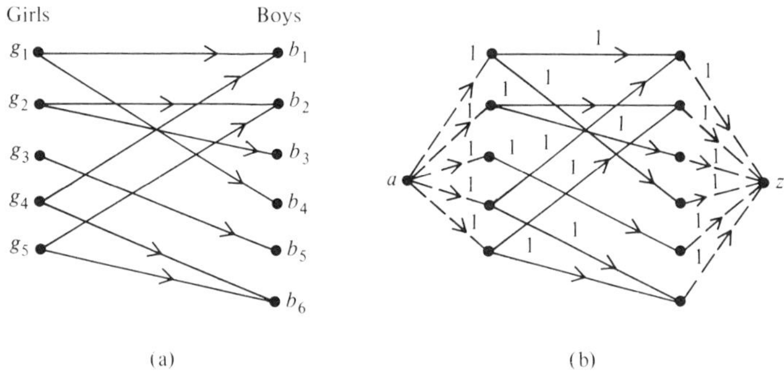

Figure 6.20 Changing (a) a matching model into (b)a network flow model. □

Now let us return to the problem of maximizing a flow in a transport network. This can be done in most realistic problems using an algorithm that we will present a little further in our discussion. But first let us consider another example. The insights we obtain will help us to understand what each step of the algorithm is trying to accomplish.

Example 4 The value of the flow in the network of Fig. 6.21(a) is found by taking the sum of the flows in the edges ab and ac:

$$f_v = 6 + 3 = 9.$$

It would seem difficult, if not impossible, to obtain a larger flow, since all the edges

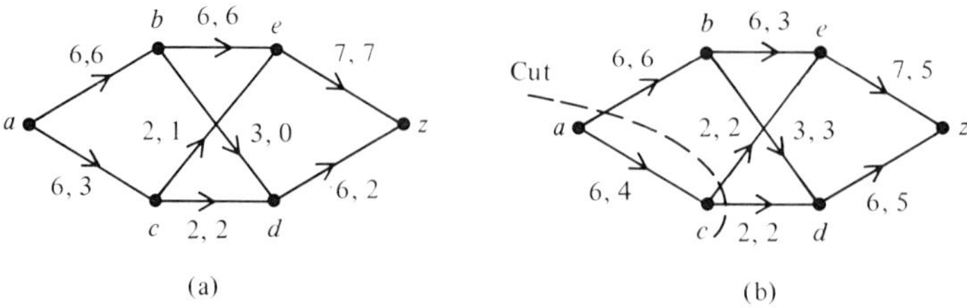

Figure 6.21 The flow in part (a) is not a maximum flow; the flow in part (b) is a maximum since it equals the value of a cut.

in the path *abez* are saturated, as well as the edge *cd* in path *acdz*. This prevents us from increasing the flow across edge *cd* as things now stand.

In spite of the pessimistic observations we have just made, Fig. 6.21(b) gives a greater flow than before, namely,

$$f_v = 5 + 5 = 10.$$

To obtain this improvement we first observe the advantage of decreasing the flow across edge *be* by three units. This allows us to send three of the six units entering vertex b to z by way of path *bdz* and an additional one unit to z by way of path *cez*, for a net increase of $-3 + 3 + 1 = 1$ unit. □

The reader has probably observed that edge *ac*, coming out of the source, is not saturated; nor are edges *ez* and *dz*, the edges going into the sink. Can we increase the flow even more? No, we cannot because the capacities of the edges *ab*, *ce*, and *cd* from the set of vertices $\{a, c\}$ directly to the other vertices is $6 + 2 + 2 = 10$. The significance of this observation is contained in Theorem 6.8.

Theorem 6.8 ***Max Flow—Min Cut Theorem*** In any transport network the value of any flow is less than or equal to the capacity of every cut. Moreover, if the edge capacities are all rational numbers, then the maximum flow is equal to the minimum cut. (See Exercises 6.4-13 through 6.4-18 for the proof.)

As an illustration of the max flow–min cut theorem take the cut $(P, \bar{P})$ where $P = \{a, c\}$ in Fig. 6.21(b). We have already pointed out that the value of this cut is

$$c(a, b) + c(c, d) + c(c, e) = 6 + 2 + 2 = 10.$$

Since the value of the flow in this network is also 10 units, our theorem implies that the given flow is a maximum.

The following algorithm uses the ideas we explored in Example 4 to achieve a maximum flow by augmenting a given flow. It is a systematic labeling procedure in which edges are identified where it might be advantageous to increase the flow directly or where it might be advantageous first to divert some of the present flow to a different edge in order to increase the net flow. The algorithm terminates by identifying a minimum cut, thereby signaling that a maximum flow has been found. The algorithm is given for integral valued capacities, but we will illustrate how it can be applied if there are some fractional capacities as well.

Algorithm 6.4 ***Flow Augmenting Algorithm*** Given a transport network with integral valued capacities $c(x, y)$, a source and a sink denoted by a and z, respectively, and a flow $f \geqslant 0$. Then one can find a maximum flow by means of the following steps.

1. [Label source.] Label a with $(-, \Delta a)$, where Δa is the sum of the capacities of edges adjacent from a. (This is the maximum possible flow out of a.)
2. [Labeling loop.] Repeat steps 3 through 5 until no further vertices can be labeled and, in particular, the sink z cannot be labeled.
   3. [Next labels.] Repeat steps 4(a) and 4(b) until the sink is labeled. If the sink cannot be labeled, then continue from step 6.
      4a. [Edges bv.] If v is an unlabeled vertex adjacent *from* a labeled vertex b and $c(b, v) > f(b, v)$, then label v with $(b^+, \Delta v)$, where Δv is the smaller of Δb and $c(b, v) - f(b, v)$. Otherwise v is not labeled.

      4b. [Edges vb.] If v is an unlabeled vertex adjacent *to* a labeled vertex b and $f(v, b) > 0$, then label v with $(b^-, \Delta v)$, where Δv is the smaller of Δb and $f(v, b)$. Otherwise v is not labeled.
   5. [Sink labeled.] If the sink has been labeled $(x^+, \Delta z)$, then work your way back to the source by following first coordinates of labels, starting with x, and either increasing or decreasing the flows on corresponding edges by Δz units as the first coordinate has a "+" or a "−" sign. Then remove all labels except for the one on the source.
6. [Done.] Output the present flow.
7. End of Algorithm 6.4.

The reason we obtain a maximum flow in step 6 is this: Let P equal the set of labeled vertices. Then $\bar{P}$ is the set of unlabeled vertices, and we have $z \in \bar{P}$ and $a \in P$. Now it can be shown (see Exercise 6.4-19) that for any flow f and cut $(P, \bar{P})$ we have

$$f_v = \sum_{x \in P,\, y \in \bar{P}} f(x, y) - \sum_{x \in \bar{P},\, y \in \bar{P}} f(x, y).$$

(In fact this is a key step in proving Theorem 6.8!) But not being able to label the sink z means that the flows from labeled to unlabeled vertices equal the capacities of the edges $c(x, y)$, while the flows from unlabeled to labeled vertices equal zero. Therefore in the foregoing equation we must have

$$f_v = \sum_{x \in P,\, y \in \bar{P}} c(x, y) - 0 = c(P, \bar{P}).$$

By Theorem 6.8, f_v must be a maximum flow.

The algorithm eventually terminates because step 4 ensures an increase of at least one unit in the flow each time the sink z is labeled. Furthermore the maximum flow is certainly bounded by the sum of all the capacities. And hence we cannot continue to increase the flow forever.

Fractional capacities on the edges are handled by multiplying all capacities by their least common denominator. The following example will illustrate this technique.

Example 5 Let us consider the network in Fig. 6.22(a). If we multiply each capacity by 6, we obtain the by now familiar network in Fig. 6.22(b) with its maximum flow. In dividing both the flows and the capacities in Fig. 6.22(b) by 6, we get the original network back again together with a maximum flow. (See Exercises 6.4-11 and 6.4-12.)

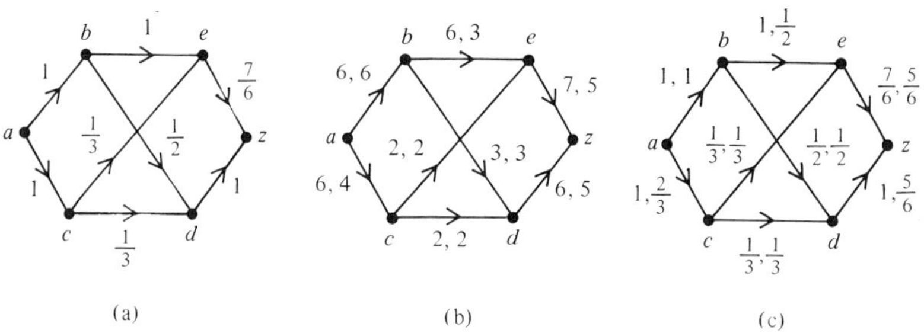

Figure 6.22 Obtaining a maximum flow with fractional capacities. □

Example 6 We now show how our algorithm applies to a transport network with a given flow, as in Fig. 6.23(a). (One can always begin with a zero flow in every edge, but we can usually do better than that by simple inspection.)

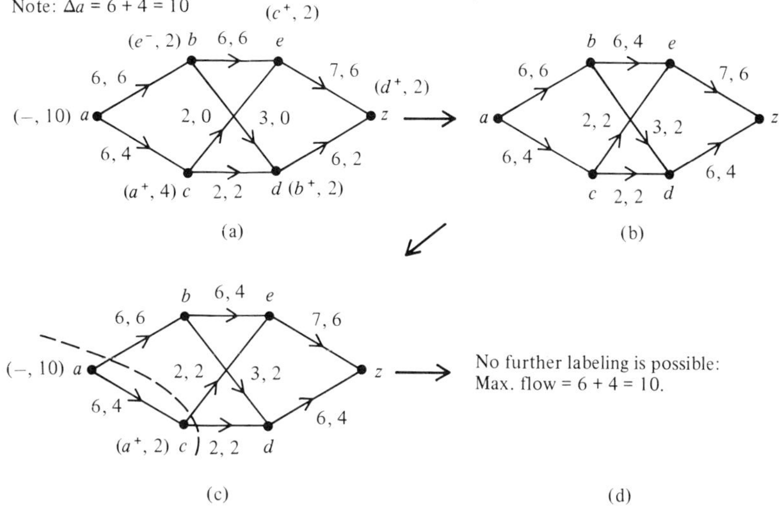

Figure 6.23 Obtaining a maximum flow with the flow augmenting algorithm.

First the source a is labeled with the pair $(-, 10)$. Vertex b cannot be labeled at this time, since edge ab is saturated. But we can increase the flow across edge ac by four units. Therefore vertex c is labeled $(a^+, 4)$.

Now d cannot be labeled yet, but the flow across edge ce can be increased by two units. Since 2 is the minimum of the numbers 2 and 4, we label vertex e with $(c^+, 2)$.

Next we have a choice between scanning vertex z or b. If we arbitrarily choose vertex b, we see that b is adjacent *to* vertex e. Since we can reduce the flow from b to e by the minimum of two or six units, we label vertex b with $(e^-, 2)$.

We now have to make another arbitrary choice, this time between vertices d and z. We choose vertex d, and it is easy to see by now (we hope) why d is given the label $(b^+, 2)$.

Finally the only unscanned vertex that remains is z. Scanning it from vertex d, we are required to give z the label $(d^+, 2)$. Notice that if it had not been possible to increase the flow across edge dz, we could have traced our way back to vertex c and labeled z from there. As things now stand, however, we back up to vertex a, increasing or decreasing the flow according to the labels on each vertex we encounter. Figure 6.23(b) is the result.

The value of our flow is now 10, and in Fig. 6.23(c) we begin the labeling process again. But we find that we can only label vertices a and c. In particular we cannot label the sink z. Hence we must have a maximum flow already.

Notice that the assignment of values flowing through each edge in Fig. 6.23(c) is different than in Fig. 6.21(b). However, the value of the flow in each network is the same—10 in each case. □

Completion Review 6.4

Complete each of the following.

1. A nonnegatively weighted digraph that has no loops is called a __________ network. Each of these networks has exactly one __________ and one __________.
2. The weight $c(x, y)$ on edge xy is called the __________ of that edge.
3. A flow in a network is a function that assigns a nonnegative number $f(x, y)$ to each __________ such that __________ and __________.
4. The value of a flow $f(x, y)$ is defined to be __________. A maximum flow is one that achieves __________ in the transport network.
5. An edge uv is saturated if __________.
6. A cut is a partition $(P, \bar{P})$ of the vertices of the network such that __________.
7. The capacity of a cut $(P, \bar{P})$, denoted __________, is equal to __________.
8. The __________ theorem states that the value of any flow is __________ the capacity of any cut. For rational capacities the maximum flow __________.
9. We can find a maximum flow by means of the __________ algorithm. A maximum flow is attained when the __________ can no longer be labeled.
10. Matching problems, supply–demand problems, and transportation problems can often be modeled as network flow problems. (*True or False?*)

Answers: **1.** transport; source; sink. **2.** capacity. **3.** edge xy; $0 \leq f(x, y) \leq c(x, y)$; the sum of the flows into any vertex equals the sum of the flows out, except for source and sink. **4.** the sum of the flows out of the source; the greatest possible flow. **5.** $f(x, y) = c(x, y)$. **6.** the source is in P and the sink is in $\bar{P}$. **7.** $c(P, \bar{P})$; the sum of the capacities of edges from P to $\bar{P}$. **8.** max flow–min cut; less than or equal to; equals the minimum cut. **9.** flow augmenting algorithm; sink. **10.** True.

Exercises 6.4

1. Consider the transport network given in the following. Find **(a)** the value of the given flow; **(b)** the capacity of the cut $(P, \bar{P})$, where $P = \{a, b, c\}$; **(c)** the capacity of the cut $(P, \bar{P})$, where $P = \{a, c, d\}$.

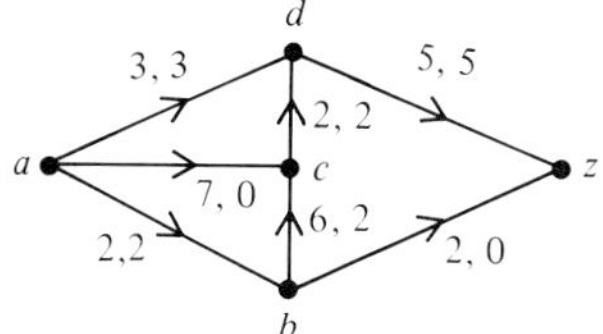

2. Use the flow augmenting algorithm to find a maximum flow in the foregoing network. You may begin with any flow. Find a minimum cut corresponding to your flow.
3. Find a maximum flow in the following network.

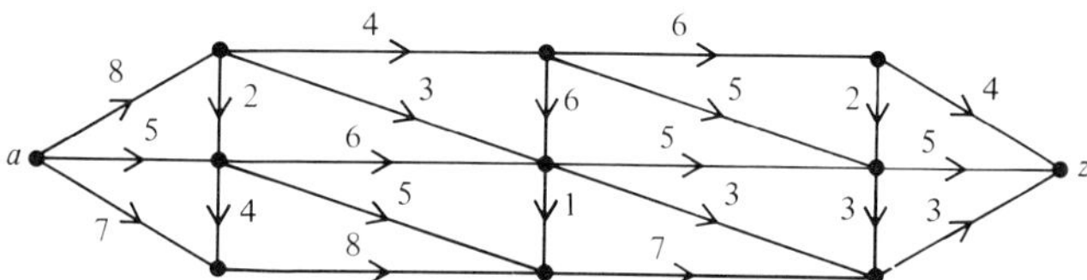

4. Determine whether the demands in Example 2 can be satisfied by determining a maximum flow in the network of Fig. 6.19(b).
5. Determine whether the supplies are adequate to satisfy the demands in the following network.

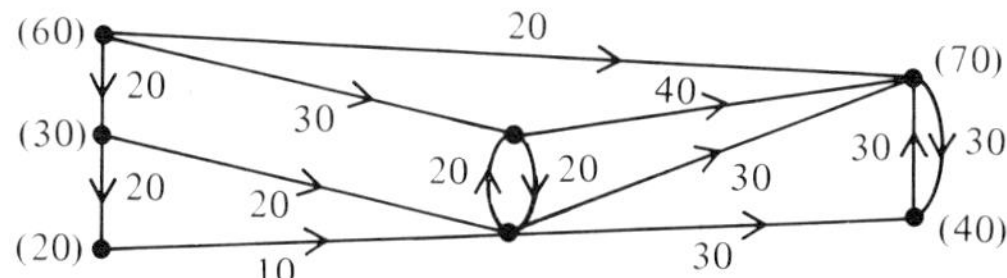

6. Determine whether the five girls in Example 3 can all find suitable dancing partners by finding a maximum flow in the network we constructed in Fig. 20(b).
7. Suppose that we are given six jobs and five men, each of whom is qualified to do some of the jobs as indicated by the following graph. Use a transport network to determine if we can find a job for each man, given that each man must do a different job.

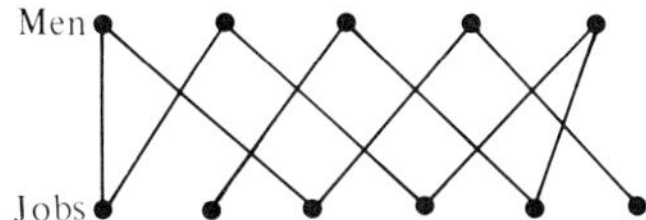

8. a) Construct a network flow model for the following problem. Five trucks are assigned to carry six kinds of household appliances to a common destination. There are three appliances of each type and the tracks can carry seven, six, five, four, and four units respectively. Can the appliances be shipped so that no two appliances of the same kind are on the same truck?
   b) Solve the problem by finding a maximum flow.

9. Modify the flow augmenting algorithm so that it can be used to find a maximum flow in the following network in which unoriented edges can have a flow in either direction, but not both. Then find the maximum flow.

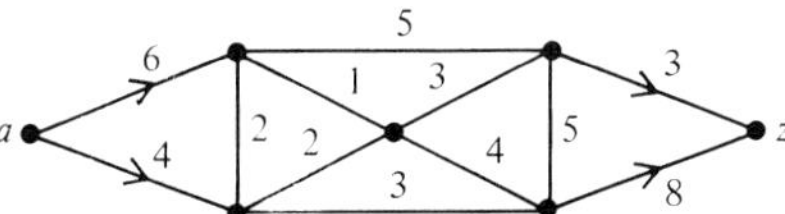

10. Find the maximum number of messengers that can be sent from vertex a to vertex z in the following network if each messenger takes a path that has no edges in common with any other messenger. (*Hint*: Use Exercise 9.)

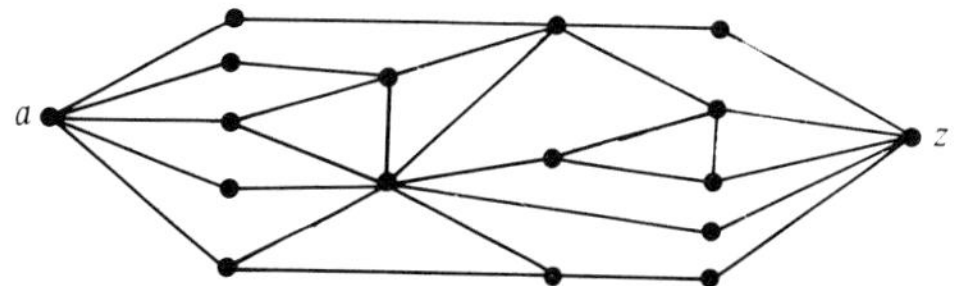

11. Verify that we did, in fact, obtain a maximum flow in Example 5.

12. Find a maximum flow in the following network having fractional capacities.

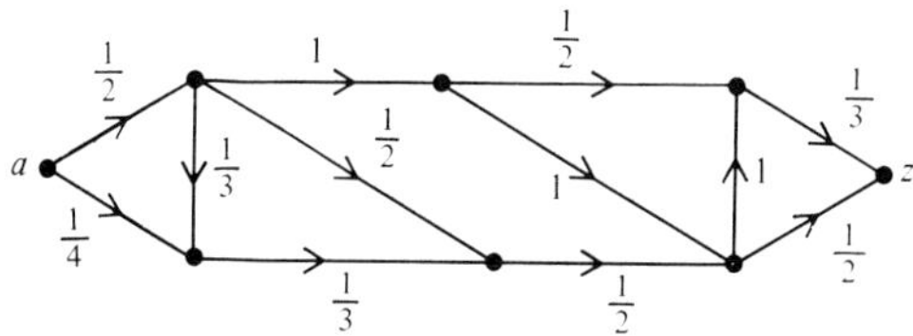

In Exercises 13–18 we sketch a proof of the max flow–min cut theorem. Let f_v denote a flow and $(P, \bar{P})$ a cut in some transport network.

13. Show that

$$f_v = \sum_{\text{all } i} f(a, v_i) - \sum_{\text{all } j} f(v_j, a).$$

14. Show that

$$f_v = \sum_{p \in P,\ \text{all } i} f(p, v_i) - \sum_{p \in P,\ \text{all } j} f(v_j, p).$$

15. Show that

$$f_v = \left[\sum_{p \in P,\ v_i \in P} f(p, v_i) + \sum_{p \in P,\ v_i \in \bar{P}} f(p, v_i)\right] - \left[\sum_{p \in P,\ v_j \in P} f(v_j, p) + \sum_{p \in P,\ v_j \in \bar{P}} f(v_j, p)\right].$$

16. Show that

$$\sum_{p\in P,\, v_i\in P} f(p, v_j) = \sum_{p\in P,\, v_j\in P} f(v_j, p).$$

17. Show that, by Exercises 15 and 16,

$$f_v = \sum_{p\in P,\, v_i\in \bar{P}} f(p, v_i) - \sum_{p\in P,\, v_j\in \bar{P}} f(v_i, p).$$

18. Show that Exercise 17 implies the following inequalities.

$$f_v \leqslant \sum_{p\in P,\, v_i\in \bar{P}} f(p, v_i) \leqslant \sum_{p\in P,\, v_i\in \bar{P}} c(p, v_i) = c(P, \bar{P}).$$

19. Use Exercise 17 to show that the flow f_v equals the sum of the flows on the edges into the sink. (*Hint*: Take $\bar{P} = \{z\}$.)

6.5 MAXIMAL MATCHINGS

"Each of five girls will accept an invitation to dance from a certain subset of the available boys at a party. Can all the girls find a desirable partner for the next dance?" We have seen in Example 3, Section 6.4, that this kind of "matching problem" can be set up in terms of network flows in transport networks. Our objective in this section is to approach these problems a little more systematically and to prove two fundamental results.

Definition 6.10 Let G be a bipartite graph in which X and Y are the disjoint sets of vertices, where no two vertices in the same set are adjacent. A (a) **matching** of X into Y is a one-to-one function f from a nonempty subset of X into Y for which $f(x) - y$ only if x is adjacent to y. A (b) **maximal matching** is one that matches the greatest possible number of elements of X with elements in Y. A (c) **complete matching** matches all the vertices of X with vertices of Y.

Example 1 a) Figure 6.20(a) is reproduced here as Fig. 6.24(a). A maximal, and also complete, matching of the set X of girls into the set Y of boys is given by the function

$$f = \{(g_1, b_4), (g_2, b_3), (g_3, b_5), (g_4, b_1), (g_5, b_6)\}.$$

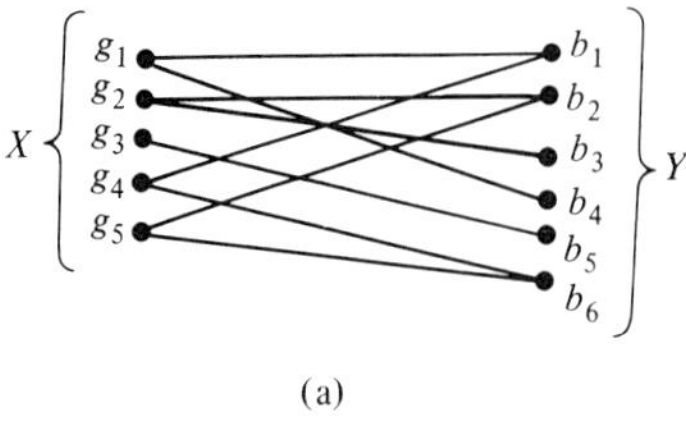

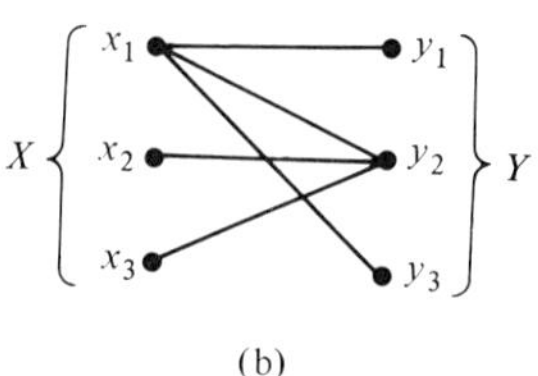

Figure 6.24 Matching problems.

Observe that the ordered pairs denote edges of the graph. Moreover, any proper nonempty subset of f is also a matching of X into Y. These are not maximal, however, since they map fewer elements of X into Y.

b) Two maximal matchings for the graph in Fig. 6.24(b) are given by the functions

$$f = \{(x_1, y_1), (x_2, y_2)\} \quad \text{and} \quad g = \{(x_1, y_3), (x_3, y_2)\}.$$

They are both maximal because the only possible match for each of the elements x_2 and x_3 is the element y_2. Neither matching is complete, however, since neither function matches every x into some y. □

The following definitions will help us to determine the size of the subset in X mapped into Y in a maximal matching. *To avoid trivial cases we henceforth assume that each vertex in X is adjacent to some vertex in Y.*

Definition 6.11 Suppose that the vertices of a bipartite graph G have been partitioned into sets X and Y, as usual, and that A is a subset of X. Then (a) $\boldsymbol{J(A)}$ will denote the vertices in Y that are adjacent to those in A. The (b) **deficiency of** $\boldsymbol{A}$ is denoted by (c) $\boldsymbol{D(A)}$ and defined to be

$$D(A) = |A| - |J(A)|,$$

while the (d) **deficiency of** $\boldsymbol{G}$ (that is, of the entire bipartite graph) is defined to be the maximum of the deficiencies of the subsets of X, and the deficiency of G is denoted by (e) $\boldsymbol{D(G)}$.

Example 2 a) In Example 1(a), $J(X) = Y$, since every vertex in X is adjacent to some vertex in Y. Hence the deficiency of X is

$$D(X) = |X| - |J(X)| = |X| - |Y| = 5 - 6 = -1.$$

It can be verified that each subset of X also has a nonpositive deficiency and that $D(G)=0$, since $D(\varnothing)=0$.

b) In Example 2(b) we also have $J(X)=Y$. So we may write

$$D(X)=|X|-|J(X)|=3-3=0.$$

However, $D(G)\neq 0$. To see this, let $A=\{x_2, x_3\}$. Then

$$D(A)=|A|-|J(A)|=|A|-|\{y_2\}|=2-1=1.$$

Therefore we conclude that $D(G)\geq 1$, since the maximum deficiency must be at least as large as the deficiency of A, namely, 1. □

Theorem 6.9 The maximum number of vertices of X that can be matched into vertices of Y in a bipartite graph G is $|X|-D(G)$.

Proof: We begin by constructing a transport network corresponding to the graph G. The capacities of edges from X to Y are each a number M, taken to be at least as large as $|X|$. The other edges, from the source a to each x_i in X or from each y_i in Y to the sink z, are each given a capacity of one. [See Fig. 6.25(a).]

It is clear that all we have to show is that there is a maximum flow of $|X|-D(G)$ in this network, since any flow will correspond to a matching of X into Y.

We will first show that the capacity of any cut $(P, \bar{P})$ in this transport network is at least $|X|-D(G)$.

Let P contain the vertices in a subset A of X, in a subset B of Y, and the source a, as in Fig. 6.25(b). If there is an edge from A to the set $Y-B$, then

$$c(P, \bar{P})\geq M\geq |X|\geq |X|-D(G),$$

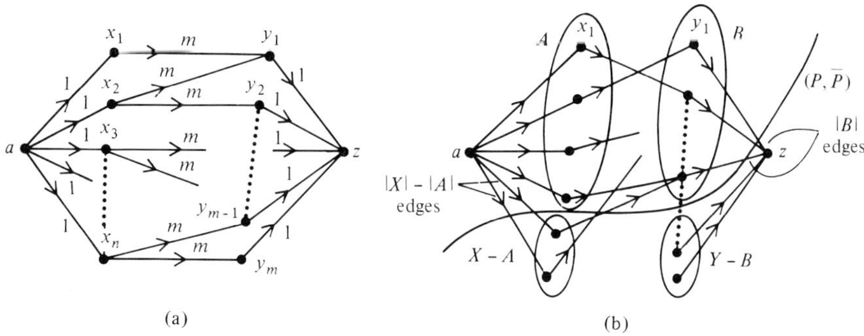

Figure 6.25 Diagrams for the proof of the maximal matching theorem.

since $D(G) \geqslant 0$. (See Exercise 6.5-7.) If no edge in G connects vertices of A with vertices of $Y - B$, then

$$c(P, \bar{P}) = |X| - |A| + |B| \geqslant |X| - |A| + J(A)$$

Since $|B| \geqslant J(A)|$. But $D(G) \geqslant |A| - |J(A)|$ for every subset A in X. This implies that

$$|X| - |A| + |J(A)| \geqslant |X| - D(G).$$

Hence

$$c(P, \bar{P}) \geqslant |X| - D(G)$$

as claimed.

Next consider the cut where P is the set of vertices in A, $J(A)$, and the source a, for some subset A of X such that $D(A) = D(G)$. Then, as in the first part of the proof, with $B = J(A)$, we see that there are no edges connecting A with $Y - J(A)$. Hence our last inequality becomes an equation, namely,

$$c(P, \bar{P}) = |X| - D(G).$$

Since no cut can attain a lower value than the one we have just found, our cut must be a minimum. The max flow–min cut theorem says that there is a flow having the value of this cut. Since this is what we had to demonstrate, the proof is complete. ■

Theorem 6.10 There is a complete matching of X into Y in a bipartite graph G if and only if $|J(A)| \geqslant |A|$ for every subset A of X.

Proof: If $|J(A)| \geqslant |A|$ for every subset A of X, then

$$D(A) = |A| - |J(A)| \leqslant 0 \quad \text{and} \quad D(G) = 0,$$

since $D(\varnothing) = |\varnothing| - |J(\varnothing)| = 0$. Hence all of X can be matched into Y by Theorem 6.9.

Conversely, if $|J(A)| < |A|$ for some subset A of X, then we cannot map A one to one into $J(A)$, which is the set of all possible images of vertices in A. Hence there cannot be a one-to-one mapping of the entire set X into Y, since its restriction to A would be a one-to-one mapping into $J(A)$. This completes the proof. ■

Although one can find a complete matching when one exists by means of our flow augmenting algorithm (Section 6.4), it is sometimes not clear that one exists until the algorithm ends. Theorem 6.10 gives us a way to show the existence of a matching without having actually to find one. Since we have to check all possible subsets of X, however, it is not very practical. Here is a test that is somewhat easier to apply.

Theorem 6.11 Suppose that in a bipartite graph G whose vertices are partitioned into sets X and Y each vertex in X is adjacent to at least k vertices in Y. Suppose that no vertex in Y is adjacent to more than k vertices in X. Then there is a complete matching of X into Y.

Proof: Let A be a subset of X. Then there are at least $k \cdot |A|$ edges from A to $J(A)$, but there are no more than $k \cdot |A|$ edges from $J(A)$ to X, some of which are attached to vertices in A. Since the last set of edges includes the first set, we can write

$$k \cdot |A| \leqslant k \cdot |J(A)| \quad \text{or} \quad |A| \leqslant |J(A)|.$$

Since this is true for each subset of X, Theorem 6.10 guarantees a complete matching of X into Y. ■

Example 3 Suppose that every girl at a dance likes at least three of the boys but no boy is liked by more than three girls. Then it is possible for each of the girls to dance with a boy whom she likes when all get up to dance. (We are assuming that each boy will dance with any girl. See Exercise 6.5-8.) □

Example 4 We would like to send a message made up of the words in the set

$$\{\text{cat, he, ate, hop, cop}\}$$

To disguise the message we want to use a single letter of each word to represent it so that each letter corresponds to a distinct word. Can this be done?

Let each word correspond to a vertex, giving us the set X, and let the individual letters a, c, e, h, o, p, and t correspond to vertices in a set Y. Vertices in X and Y are connected by an edge if the letter belongs to the corresponding word. (See Fig. 6.26.)

Since each word contains at least two of the letters, and no letter belongs to more than two words, Theorem 6.11 ensures a matching from X into Y. Hence we can indeed represent each word by a distinct letter in that word.

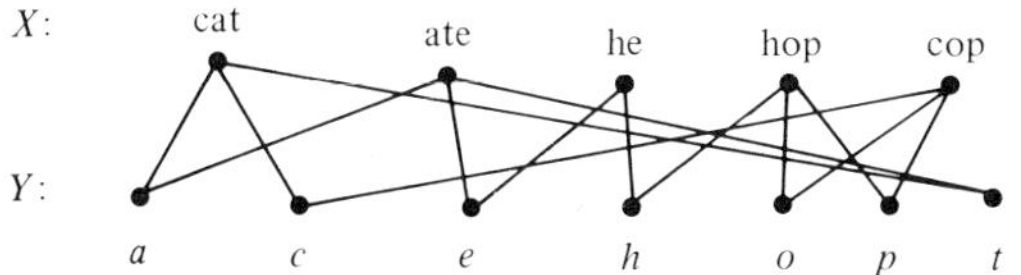

Figure 6.26 Matching and secret codes. □

Completion Review 6.5

Complete each of the following.

1. A matching in the bipartite graph G in which X and Y are the disjoint sets is a ______________ for which $f(x) = y$ only if______________.
2. A maximal matching is one that matches______________ with elements in Y.
3. A complete matching of X into Y matches______________.
4. If $J(A)$ is the set of vertices in Y that are adjacent to vertices in a subset A of X, then the deficiency of A, denoted______________, is defined to be______________.
5. The deficiency of G is defined to be______________ and is denoted by______________ ______________.
6. The maximum number of vertices of X that can be matched into vertices of Y is ______________.
7. There is a complete matching of X into Y iff______________ for every subset A of X.
8. There is a complete matching of X into Y if each vertex in X is adjacent to at least k vertices in Y and______________.

Answers: **1.** one-to-one function from a subset of X into Y; x is adjacent to y. **2.** the maximum number of vertices in X. **3.** every element in X with vertices in Y. **4.** $D(A)$; $|A| - |J(A)|$. **5.** the maximum $J(A)$ over all subsets A of X; $D(G)$. **6.** $|X| - D(G)$. **7.** $|J(A)| \geqslant |A|$. **8.** no vertex in Y is adjacent to more than k vertices in X.

Exercises 6.5

For Exercises 1 to 5, refer to graphs G_1, G_2, and G_3.

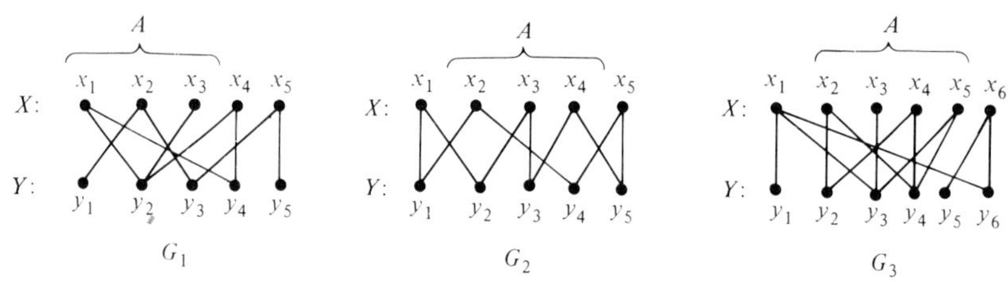

1. Find $J(A)$ for each of the graphs G_i.
2. Find $D(A)$ for each of the graphs G_i.
3. Find $D(G_i)$ for each of the graphs G_i, and explain how you obtained your answer in each case.

4. Explain on the basis of Exercise 3 which of the graphs G_i has a complete matching.

5. Find a maximal matching for each of G_i.

6. Find a maximal matching in Fig. 6.26.

7. Show that $D(G) \geqslant 0$ for any bipartite graph G. (*Hint*: Consider $A = \varnothing$.)

8. Suppose that each of the boys invited to a party has exactly five girlfriends who were invited, and each of the girls who were invited has exactly five boyfriends at the party. Can the boys and girls be paired off so that each boy goes with one of his girlfriends and vice versa? Explain.

9. Given a collection of finite sets $\{A_1, A_2, \ldots, A_n\}$, prove that one can choose a unique representative from each set if and only if every collection of m of these sets has at least m elements between them. (*Hint*: See example 4.)

10. A committee of 10 senators A, B, C, D, E, F, G, H, I, and J form the following subcommittees:

$$\{A, B, C\},\quad \{B, C, D\},\quad \{C, D, E\},\quad \{D, E, F\},\quad \{F, E, G\},\quad \{G, H, I\},$$
$$\{H, I, J\},\quad \{I, J, A\},\quad \{J, A, B\}.$$

Can a distinct senator be chosen to oversee the activities of each subcommittee so that no one reviews his own or her own subcommittee? Explain.

COMPUTER PROGRAMMING EXERCISES

(Also see the appendix and Programs A17 to A19 and A21 to A22.)

6.1. Write a program that finds the number of edges directed into and out of each vertex of a digraph, given the adjacency matrix of the digraph.

6.2. Write a program that determines if a digraph is strongly connected, given its adjacency matrix.

6.3. Write a program that determines if a digraph is unilaterally or weakly connected, given its adjacency matrix.

6.4. Write a program that finds a directed Euler circuit or trail in a digraph.

6.5. Write a program that finds a depth-first-search spanning tree in a graph or digraph by implementing Algorithm 6.1.

6.6. Write a program that orients a graph whenever it is possible to do so. When it is not possible, have your program print out the message "not connected" or indicate a bridge.

6.7. Write a program that topologically sorts the vertices in a digraph, or prints the message "not possible," by implementing Algorithm 6.2.

6.8. Write a program that determines if a digraph has a cycle.

6.9. Write a program that finds a longest path in a digraph by implementing Algorithm 6.3.

6.10. Write a program that finds a maximum flow in a transport network by implementing the flow augmenting Algorithm 6.4.

6.11. Write a program that finds a maximal matching in a bipartite graph and tells whether the matching is complete.

CHAPTER 7

APPLIED MODERN ALGEBRA

Mathematics deals in abstractions, and that is why it can be applied successfully to diverse areas of what we call "real life." For example, it might seem at first that sets, logic, and computer addition have little in common. But our discussions of these subjects in Chapters 1 and 2 revealed many similarities. These similarities can be treated systematically in terms of a mathematical structure called a "Boolean algebra," which we introduce in Section 7.5. Properties that can be deduced about this structure automatically will apply to our three previous examples if they are given the correct interpretation. To take a simple example, De Morgan's laws hold in every Boolean algebra and, of course, they are valid in our three particular examples.

Some mathematical systems are more general than others in that they include the others as special cases. It will be found that Boolean algebras, for example, are a certain kind of structure we call a "lattice." The primary concern in a lattice is that of order, that is, notions of "before" and "after," rather than operations. Order relations are of special interest to both mathematicians and computer scientists, in part because the instructions in a computer program must be executed in a certain order. However, other kinds of relations are also of great interest and have widespread applications. Hence we begin this chapter with a general discussion of relations.

7.1 RELATIONS

We frequently make comparisons between pairs of objects, saying that they are "related" or "not related" with respect to some property. Typical mathematical relations that come to mind are congruence of triangles, one number being greater than another, and one vertex or point being connected to another by a directed edge. Similarly we could say that two countries are related if they have a common border, or that two people are related if they have a common ancestor. All of these relations involve two objects at a time, and so we use the word "binary" to describe them. More precisely, we define a "binary relation" as in Definition 7.1.

Definition 7.1 Let A and B be sets. Then a (a) **binary relation** R from A to B is a subset of the Cartesian product $A \times B$. In particular a (b) **binary relation on a set** A is a subset of $A \times A = A^2$. We say that the (c) **domain** of R is the set A and its (d) **range** is the set of all second elements b of the relation's ordered pairs (a, b).

Thus we have defined a binary relations to be a set of ordered pairs, just as we previously defined a function. In fact a function $f: A \to B$ is a particular kind of binary relation. (See Definition 1.4, Section 1.2.) Observe that using ordered pairs to define relations makes sense, since certain relations, such as "greater than," are one way. That is, given two real numbers r and s, if $s < r$, then $r \not< s$.

Rather than write $(r, s) \in R$ or $(r, s) \notin R$, we usually prefer to write

$r\,R\,s$ or $r\,\not R\,s$

respectively, for "r is related to s" and "r is not related to s." Here are some examples of relations that illustrate the use of our definitions and notation.

Example 1

a) Let $A = \{$hens, cattle, sheep$\}$ and let $B = \{$eggs, wool, meat$\}$. Then $R = \{$(hens, eggs), (hens, meat), (cattle, meat), (sheep, wool), (sheep, meat)$\}$ is a relation from A to B. Notice that R is not a function from A to B because two ordered pairs have the same first element. We can write

hens R eggs, hens R meat, cattle R meat, etc.

The domain of this relation is A, and the range is B.

b) Let $A = B = \{1, 2, 3\}$ and let $R = \{(1, 2), (1, 3), (2, 3)\}$. Then

$1\,R\,2$, $1\,R\,3$, and $2\,R\,3$, but $2\,\not R\,1$, $3\,\not R\,2$, and so on.

The domain of this relation is $\{1, 2, 3\}$ and its range is $\{2, 3\}$. This relation is not a function mapping A into A because (1) there is no $x \in B$ such that $(3, x) \in R$, and (2) both $(1, 2) \in R$ and $(1, 3) \in R$. However, it is an example of a relation on A, since it is a subset of $A \times A$.

c) Let A be any set. The **identity relation** on A is the set

$$R = \{(a, a) : a \in A\}.$$

This relation is usually denoted by "$=$" and called **equality**.

d) Let us consider the relation defined on the set of real numbers by

$$\{(x, y) : x^2 + y^2 = 9, \quad x \text{ and } y \text{ are real numbers}\}.$$

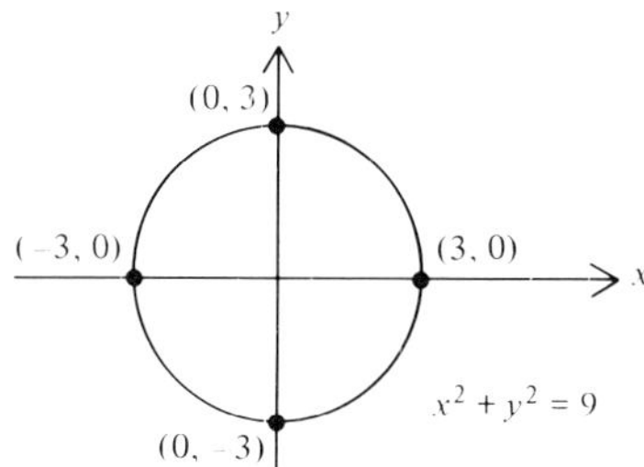

Figure 7.1 xRy iff (x, y) lies on the graph of $x^2 + y^2 = 9$, which is called the graph of the relation.

One might say that x and y are related in this example if the ordered pair (x, y) lies on a circle with center at the origin and radius three. The circle is called the graph of this relation. (See Fig. 7.1.) □

Figure 7.1 is just one way of representing relations pictorially. We assume that the reader is familiar with graphs of this type, which have, in effect, the elements of the set A listed on a horizontal axis and those of B on a vertical axis. (Do not confuse these with the "graphs" of Chapters 4 to 6.)

A pictorial scheme that is sometimes more suitable for representation of relations (or functions) on discrete sets is given by **arrow diagrams**. The elements of A and B are written in two parallel columns, as in Fig. 7.2(a), and we draw an arrow from x to y if and only if $x\ R\ y$. (Also see Section 1.2.)

In Fig. 7.2(b), however, both of our sets A and B are the same. In this case we can let each element correspond to single vertex or node and draw an arrow or directed edge from x to y whenever $x\ R\ y$. This gives us the **directed graph (or digraph) of the relation**. Thus Figs. 7.2(a) and 7.2(b) give us the arrow diagram and directed graph of the relations in Examples 1(a) and 1(b) respectively. (For those who have read Chapter 6, it will be evident that Fig. 7.2(b) is also a directed graph, but we will *not* call it the directed graph of the relation.)

Matrices also give us a convenient way to represent relations on finite sets. We form a **matrix of a given relation** from A to B by letting one row correspond to each element of A and one column to each element of B. A "1" is placed in the row corresponding to x and the column corresponding to y if and only if $x\ R\ y$; otherwise this entry is a zero. Figure 7.3 shows the matrices of the relations in Fig. 7.2.

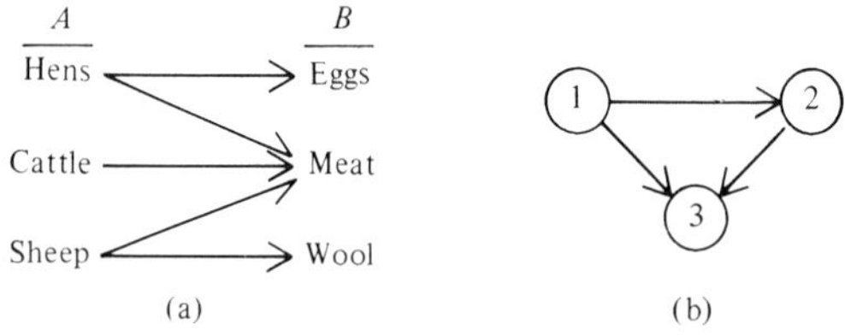

Figure 7.2 Visual representations of relations. (a) Arrow diagram. (b) Directed graph.

$$\begin{array}{l} \\ \text{Hens} \\ \text{Cattle} \\ \text{Sheep} \end{array} \begin{array}{c} \begin{array}{ccc} \text{Eggs} & \text{Meat} & \text{Wool} \end{array} \\ \begin{bmatrix} 1 & 1 & 0 \\ 0 & 1 & 0 \\ 0 & 1 & 1 \end{bmatrix} \end{array} \qquad \begin{array}{c} \\ 1 \\ 2 \\ 3 \end{array} \begin{array}{c} \begin{array}{ccc} 1 & 2 & 3 \end{array} \\ \begin{bmatrix} 0 & 1 & 1 \\ 0 & 0 & 1 \\ 0 & 0 & 0 \end{bmatrix} \end{array}$$

(a) (b)

Figure 7.3 Matrices of relations.

The following definition has its counterpart in Definition 1.6, Section 1.2, where we defined the inverse of a function.

Definition 7.2 Let R be a relation from A into B. Then the **inverse of R**, which is written R^{-1}, is the set of ordered pairs

$$R^{-1} = \{(b, a): (a, b) \in R\}.$$

Hence $b\,R^{-1}\,a$ if and only if $a\,R\,b$.

While the inverse of a function may fail to exist because of restrictions about first elements, the inverse of a relation always exists. To obtain the arrow diagram or directed graph of R^{-1}, one only has to point the arrows in the opposite direction. Moreover, to obtain the matrix of R^{-1}, one can interchange the rows and columns of the matrix of R, that is, take its transpose. Letting R be the relation on $A = \{1, 2, 3\}$ in Example 1(b), Fig. 7.2(b), and Fig. 7.3(b), we obtain the inverse relation's directed graph and matrix in Figs. 7.4(a) and 7.4(b) respectively.

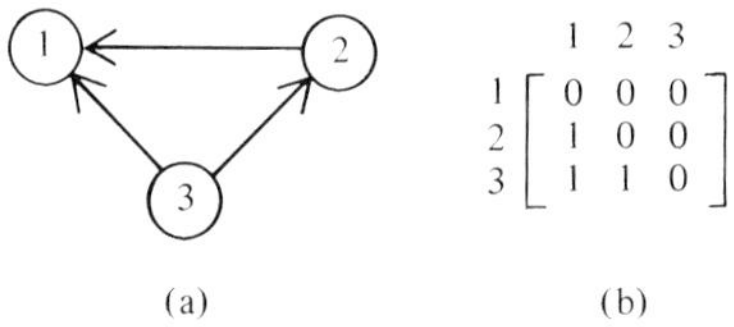

Figure 7.4 The digraph and matrix of relation R^{-1}, where the relation R is given in Fig. 7.2(b) and 7.3(b).

Let us consider some further examples of relations and their inverses before going on.

Example 2

a) Let R be the relation defined on the real numbers by "$>$," that is, "is greater than." Then R^{-1} is the relation, also defined on the real numbers, "$<$" or "is less than."

b) If R is the relation "is a descendent of" on the set of people, then R^{-1} is the relation "is an ancestor of."

c) If R is the relation "is congruent to" on the set of triangles, then R^{-1} is also the relation "is congruent to." That is, $R^{-1} = R$. □

When one considers family relationships it is natural to try to find relationships that extend through several people. Relations by marriage work this way, for instance. This idea is generalized by the following definition.

Definition 7.3 Let R be a relation from A to B and let S be a relation from B to C. The **composition** of R with S, denoted by $R \circ S$, is the set

$$R \circ S = \{(a, c)\text{: There is a } b \in B \text{ such that } (a, b) \in R \text{ and } (b, c) \in S\}.$$

In other words, $a(R \circ S)c$ iff $a\,R\,b$ and $b\,S\,c$ for some b in B.

We caution the reader that if R is a relation on A, then $R \circ R^{-1}$ is not necessarily an identity relation on A. For example, let $R = \{(b, c), (a, c)\}$. Then $R^{-1} = \{(c, b), (c, a)\}$, and $R \circ R^{-1} = \{(a, a), (a, b), (b, b), (b, a)\}$, which is not the identity relation on the set $A = \{a, b, c\}$.

Here is another example in which we compose relations.

Example 3 In Fig. 7.5 we can see how a previously defined relation R [in Example 7.1(a)] can be composed with a new relation

$$S = \{(\text{eggs, food}), (\text{meat, food}), (\text{wool, clothing})\}$$

to yield the relation

$$R \circ S = \{(\text{hens, food}), (\text{cattle, food}), (\text{sheep, food}), (\text{sheep, clothing})\}.$$

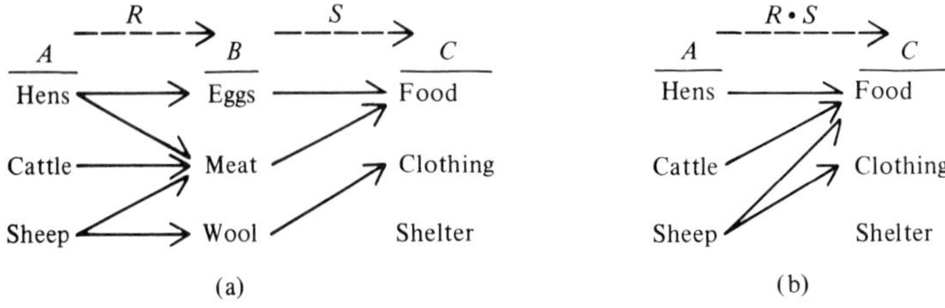

Figure 7.5 Composition of relations. □

If M_R and M_S denote the matrices of the relations R and S, we can obtain a matrix of $R \circ S$, which we denote by $M_{R \circ S}$, by first getting the product $M_R M_S$ of these matrices (see Section 4.6), and then changing every nonzero entry in this product to a 1. Thus

$$M_R M_S = \begin{array}{c} \\ h \\ c \\ s \end{array} \overset{\begin{array}{ccc} e & m & w \end{array}}{\begin{bmatrix} 1 & 1 & 0 \\ 0 & 1 & 0 \\ 0 & 1 & 1 \end{bmatrix}}, \quad \begin{array}{c} \\ e \\ m \\ w \end{array} \overset{\begin{array}{ccc} f & c & shl \end{array}}{\begin{bmatrix} 1 & 0 & 0 \\ 1 & 0 & 0 \\ 0 & 1 & 0 \end{bmatrix}} = \begin{bmatrix} 2 & 0 & 0 \\ 1 & 0 & 0 \\ 1 & 1 & 0 \end{bmatrix} \xrightarrow{\text{Change to}} \begin{array}{c} \\ h \\ c \\ s \end{array} \overset{\begin{array}{ccc} f & c & shl \end{array}}{\begin{bmatrix} 1 & 0 & 0 \\ 1 & 0 & 0 \\ 1 & 1 & 0 \end{bmatrix}}$$

The relation matrix of $R \circ S$ is sometimes called the "Boolean product" of the matrices M_R and M_S. We can define such a product more generally by means of the notation of logic gates we employed in Section 2.6.

Recall that inputs of 0's and 1's are combined by logic gates according to Table 7.1.

Table 7.1

$\vee$	0	1
0	0	1
1	1	1

$\wedge$	0	1
0	0	0
1	0	1

For example, $0 \wedge 1 = 0$, $1 \wedge 1 = 1$, $0 \vee 1 = 1$, $1 \vee 1 = 1$, and so on. Then we can formulate Definition 7.4.

Definition 7.4 If $A = (a_{ij})$ and $B = (b_{ij})$ are, respectively, m-by-r and r-by-n matrices of 0's and 1's, then their **Boolean product** is defined by the m-by-n matrix

$$A \otimes B = (c_{ij}),$$

where

$$c_{ij} = (a_{i1} \wedge b_{1j}) \vee (a_{i2} \wedge b_{2j}) \vee \dots \vee (a_{in} \wedge b_{nj})$$

for $i = 1, 2, \dots, m$ and $j = 1, 2, \dots, n$.

In other words the ijth entry in $A \otimes B$ is a 1 iff the ith row of A and the jth column of B both have a 1 in at least one of their $k = 1, 2, \dots, r$ corresponding positions. Otherwise this entry is a 0.

Example 4 Let us define a relation R on the set $A = \{1, 2, 3, 4\}$ by means of the relation matrix

$$M = \begin{bmatrix} 1 & 0 & 1 & 0 \\ 0 & 1 & 1 & 0 \\ 0 & 1 & 0 & 0 \\ 0 & 0 & 0 & 1 \end{bmatrix}$$

Then the Boolean product of M with itself, that is,

$$M \otimes M = \begin{bmatrix} 1 & 1 & 1 & 0 \\ 0 & 1 & 1 & 0 \\ 0 & 1 & 1 & 0 \\ 0 & 0 & 0 & 1 \end{bmatrix} = M^{②}$$

also gives a relation on A. In terms of the digraph, D, of R, $M \otimes M$ defines a relation R^2 in which $x\,R^2\,y$ iff x and y are connected by a sequence of two directed edges in D. In other words, $R^2 = R \circ R$. □

More generally, one can show, as in Theorem 4.14, that $M^{ⓝ}$ *defines a relation* R^n *on* A *such that* $x\,R^n\,y$ *iff* x *is connected to* y *by a sequence of* n *directed edges in* D. [R^n can be defined equivalently as the n-fold composition $R \circ R \circ \cdots \circ R$, or as $R^{n-1} \circ R$, where $R^1 = R$.]

There are other, less common, ways to define new relations, especially if both our relations R and S are from A to B, that is, subsets of $A \times B$. In this case, their union $R \cup S$, their intersection $R \cap S$, and their complements with respect to $A \times B$ such as $\bar{R}$ will also be subsets of $A \times B$ and, consequently, relations from A to B.

Example 5 Let $A = \{x, y, z\}$ be a set of students and let $B = \{183, 285, 385, 495\}$ be a set of courses. Then

$$R = \{(x, 183), (x, 285), (y, 385), (z, 285)\}$$

might designate which courses the students are now taking for each student, while

$$S = \{(x, 183), (y, 495), (z, 285), (z, 385), (z, 495)\}$$

might represent which courses each student is interested in.

a) The relation $R \cap S = \{(x, 183), (z, 285)\}$ would then show which students are interested in courses they are now taking and those courses.

b) $R \cup S = \{(x, 183), (x, 285), (y, 385), (y, 495), (z, 285), (z, 385), (z, 495)\}$ shows which students are interested in certain courses or are taking certain courses, or both.

c) $\bar{R} = \{(x, 385), (x, 495), (y, 183), (y, 285), (y, 495), (z, 183), (z, 385), (z, 495)\}$ shows which courses each student is not presently taking. □

Definition 7.5 Given sets $A_1, A_2, \ldots, A_n$, we define the (a) **Cartesian product** $A_1 \times A_2 \times \cdots \times A_n$ as the set of all ordered n-tuples

$$A_1 \times A_2 \times \cdots \times A_n = \{(a_1, a_2, \ldots, a_n): \quad a_i \in A_i\}.$$

Moreover, we define an (b) ***n*-ary relation** as a subset of $A_1 \times A_2 \times \cdots \times A_n$.

Definition 7.5 is a natural extension of the definition of a product of two sets and binary relations. Thus a ternary relation is a set of ordered triples, a quaternary relation is a set of ordered 4-tuples, and so on. An example of a **ternary relation** is

$$S = \{(x, y, z): \quad x^2 + y^2 + z^2 = 1, \quad \text{for all real } x, y, \text{ and } z\}.$$

Hence x, y, and z are related by S if they all lie on a sphere with radius 1 centered at the origin in three-dimensional space, often denoted by $\mathbb{R}^3$.

N-ary relations have important applications to organization of data in tables and data banks and are usually studied in detail in a course on data structures.

Completion Review 7.1

Complete each of the following.

1. A binary relation R from A to B is ________________. The domain of R is ________________ and the range of R is the set ________________.
2. A binary relation on A is ________________. In particular the identity relation on A is ________________.
3. An arrow diagram of a relation from A to B shows the elements of A and B ________________ with arrows going from A to B.
4. A directed graph or ________________ of a relation R on A has arrows going from a to b when ________________.
5. A matrix of a given relation has a 1 in row a and column b when ________________, and has a ________________ in this position otherwise.
6. The inverse of a relation R, written ________________, is the set ________________.
7. If A is a relation from A to B and S is a relation from B to C, then the composition of R with S, written ________________, is the set ________________.

8. The Boolean product of matrix $A = (a_{ij})$ with the matrix $B = (b_{ij})$, both matrices of zeros and ones, is the matrix $C = (c_{ij})$, where $c_{ij} =$ ________________.

9. If R has the relations matrix M, then the n-fold composition R^n of R with itself has the relations matrix ________________.

10. The Cartesian product of sets $A_1, A_2, \ldots, A_n$ is the set ________________. An n-ary relation is defined to be a ________________.

Answers: **1.** a subset of $A \times B$; A; $\{b \in B\colon (a, b) \in R\}$. **2.** a subset of $A \times A$; $\{(a, a)\colon a \in A\}$. **3.** written in columns. **4.** digraph; aRb (a is related to b). **5.** aRb; 0. **6.** R^{-1}; $\{(b, a)\colon (a, b) \in R\}$. **7.** $R \circ S$; $\{(a, c)\colon aRb$ and bRc for some b in $B\}$. **8.** 1 if at least one of $a_{ik}b_{kj}$ is 1 and 0 otherwise. **9.** $M^{\textcircled{n}}$. **10.** $\{(a_1, a_2, \ldots, a_n)\colon a_i \in A_i$ for each $i\}$; a subset of $A_1 \times A_2 \times \cdots \times A_n$.

Exercises 7.1

In Exercises 1 to 6 give the smallest possible domain, the range, an arrow diagram or a directed graph of the relation, and a matrix describing the binary relation for each.

1. $R = \{(1, a), (2, b), (3, c), (3, a)\}$.
2. $S = \{(2, a), (2, b), (2, c)\}$.
3. $T = \{1, 2, 3, 4\} \times \{a, b, c\}$.
4. $U = \{(1, 2), (2, 3), (3, 4), (4, 1)\}$.
5. $V = \{(a, a), (a, b), (a, c)\}$.
6. $W = \{(a, 1), (b, 2), (c, 3), (c, 4), (b, 3), (a, 2)\}$.
7. Find the inverses R^{-1}, S^{-1}, and T^{-1} of the binary relations in Exercises 1 to 3.
8. Find $R \circ W$, where R and W are as given in Exercises 1 and 6.
9. a) Write the binary relation R from A to B given by the following matrix as a set of ordered pairs.

$$\begin{array}{c} \\ 1 \\ 2 \\ 3 \\ 4 \end{array}\begin{array}{c} \begin{array}{ccc} a & b & c \end{array} \\ \begin{bmatrix} 1 & 0 & 1 \\ 1 & 1 & 0 \\ 0 & 1 & 1 \\ 1 & 1 & 0 \end{bmatrix} \end{array}$$

b) Is $1\,R\,c$ true?
c) Is $2 \not{R}\, c$ true?

10. a) Write the binary relation S on A given by the following directed graph as a set of ordered pairs ($A = \{1, 2, 3, 4\}$).

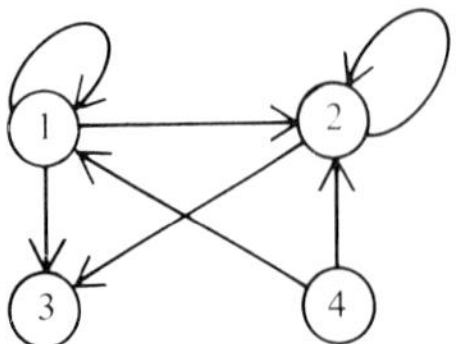

b) Find all the elements x of A such that $1\ S\ x$.
c) Find all the elements y of A such that $y\ S\ 1$.

11. What is the graph of each of the following relations?
a) $A = \{(x, y): x + y = 1, \text{ with } x, y \text{ real}\}$;
b) $B = \{(x, y): y = x^2, \text{ with } x \text{ real}\}$;
c) $C = \{(x, y): x^2 + y^2 = 1 \text{ with } x, y \text{ integers}\}$;
d) $D = \{(x, y): y = x^2, \text{ with } x \text{ and } y \text{ integers}\}$.

12. Give interpretations of the binary relations in Examples 1(a) and 1(b). Do the same with the binary relation S given in Example 3.

13. Find the binary relations $R \cup S$, $R \cap S$, and $\bar{R}$ for the binary relations R and S defined in Exercises 1 and 2.

14. Suppose that the integers x, y, and z satisfy the ternary relation T if $x^2 + y^2 + z^2 \leqslant 1$. List the set of ordered triples that satisfy T.

15. Let R be a binary relation defined on a set of integers as follows: $x\,R\,y$ if x and y have the same remainder upon division by 3. List the elements in the following sets: **(a)** $\{y: 1\,R\,y\}$; **(b)** $\{y: 2\,R\,y\}$; **(c)** $\{y: 3\,R\,y\}$.

16. Let $S = \{a, b, c, d, e, f, g, h, i, j\}$, and suppose that we partition S by writing $S = \{a, b, c\} \cup \{e, f\} \cup \{g, h, i, j\}$. Let $x\,R\,y$ mean that x is in the same partition subset as y. Find each of the following sets. **(a)** $\{x: x\,R\,a\}$; **(b)** $\{x: x\,R\,i\}$; **(c)** $\{y: e\,R\,y\}$.

17. Find the Boolean product $A \otimes B$ of the matrices

$$A = \begin{bmatrix} 0 & 1 & 0 & 0 \\ 1 & 0 & 1 & 1 \\ 1 & 1 & 0 & 0 \end{bmatrix} \quad \text{and} \quad B = \begin{bmatrix} 1 & 0 \\ 0 & 1 \\ 1 & 1 \\ 0 & 1 \end{bmatrix}.$$

18. a) Find the relation matrix M of the relation R given in the following digraph.
b) Find the matrix $M^{②}$ and the corresponding relation $R^2 = R \circ R$.

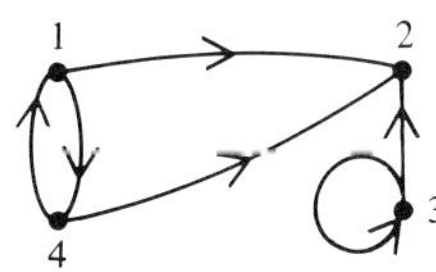

19. The **Boolean sum** of two m-by-n matrices $A = (a_{ij})$ and $B = (b_{ij})$ of 0's and 1's is given by the m-by-n matrix $A \oplus B = (c_{ij})$, where $c_{ij} = a_{ij} \vee b_{ij}$ for $i = 1, 2, \ldots, m$ and $j = 1, 2, \ldots, n$.
a) Show that if A and B are the relation matrices of relations R_1 and R_2 on $X \times Y$, then $A \oplus B$ is the relation matrix of $R_1 \cup R_2$.
b) Show that if A corresponds to the digraph D of a relation, then $I \oplus A \oplus A^{②} \oplus \cdots \oplus A^{(n-1)}$ corresponds to the path matrix of D, where I is the identity having the same dimensions as A.

20. Express each of the following tables as a ternary relation using the rows as the elements of the relation.

Part	Project	Quantity
p_1	j_1	10
p_2	j_1	15
p_3	j_2	8
p_4	j_3	12

Part	Project	Color
p_1	j_1	Red
p_2	j_2	Blue
p_4	j_3	Red
p_5	j_4	Green

7.2 EQUIVALENCE RELATIONS AND PARTITIONS

The kinds of relations one encounters most often are the binary relations on a set A, and these are the types we discuss from here on. One way to distinguish between different (binary) relations is by means of certain properties they satisfy. In this section we will discuss the following three properties.

Definition 7.6 Let R denote a binary relation on a set A.

a) We say that R is **reflexive** if $x R x$ for every x in A.

b) We say that R is **symmetric** if for every x and y in A, $x R y$ implies that $y R x$.

c) We say that R is **transitive** if for all x, y, and z in A, $x R y$ and $y R z$ implies that $x R z$.

In terms of the directed graph of the relation, R is reflexive if there is a loop at each vertex; R is symmetric if every time there is an arrow from x to y, then there is also an arrow from y to x; and R is transitive if, whenever an arrow goes from x to y, and another arrow goes from y to z, then there must be an arrow from x to z.

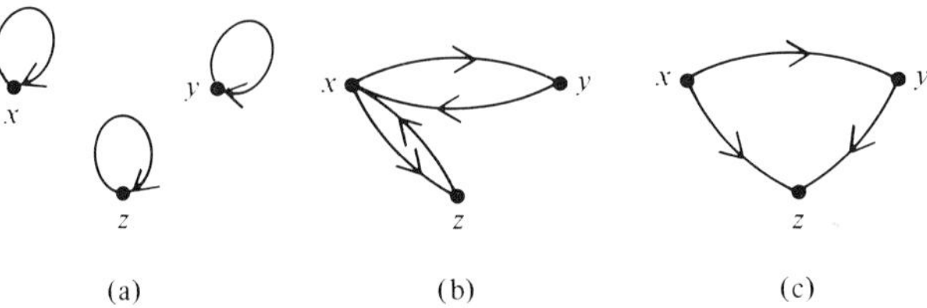

Figure 7.6 Directed graphs of (a) reflexive, (b) symmetric, and (c) transitive relations.

Example 1

a) The relation "$\leqslant$" (is less than or equal to) on the set of real numbers is reflexive, since $x \leqslant x$ is always true ($x = x$). "Less than or equal to" is also a transitive relation, since

$$x \leqslant y \quad \text{and} \quad y \leqslant z \quad \text{always implies that} \quad x \leqslant z.$$

However, this relation is not symmetric, since $1 \leqslant 2$ does not imply that $2 \leqslant 1$, for example.

b) The relation defined on the set $A = \{a, b, c, d\}$ by the matrix

$$\begin{array}{c} \\ a \\ b \\ c \\ d \end{array} \begin{array}{c} \begin{array}{cccc} a & b & c & d \end{array} \\ \begin{bmatrix} 0 & 1 & 0 & 1 \\ 1 & 0 & 1 & 1 \\ 0 & 1 & 0 & 1 \\ 1 & 1 & 1 & 1 \end{bmatrix} \end{array}$$

is clearly not reflexive, since some of the elements on the main diagonal are zeros. But it is symmetric, since the *corresponding rows and columns are the same.* (In fact, the matrix is also called symmetric.) Our relation is not transitive though because $a\,R\,b$ and $b\,R\,a$ are both true, but $a \not R\, a$.

c) The binary relation defined by the following directed graph is neither symmetric nor reflexive, since we have $a \not R\, a$, and $a\,R\,b$ does not imply that $b\,R\,a$. This relation does enjoy transitivity, however. To check this it is necessary to verify that in all cases where we have $x\,R\,y$ and $y\,R\,z$, we also have $x\,R\,z$. [This is somewhat easier to verify in Fig. 7.6(c).]

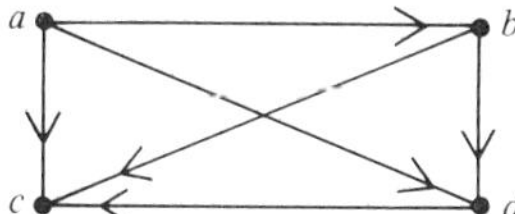

Figure 7.7 This relation is transitive, but neither symmetric nor reflexive.

d) Let $A = Z$, the set of integers. We define $a\,R\,b$ if there is an integer n such that

$$a - b = 3n.$$

This relation enjoys all three properties: It is reflexive because $a - a = 3 \cdot 0 = 0$ for every integer a. It is symmetric because $a - b = 3n$ implies that $b - a = 3(-n)$, and if n is an integer, then so is $-n$. Finally, it is transitive because

$$a - b = 3n \quad \text{and} \quad b - c = 3m \quad \text{imply that} \quad a - c = 3(n + m)$$

for any integers m and n. This relation is usually called **congruence modulo 3**, and we write $a \equiv b \pmod 3$ if $a\,R\,b$.

e) The binary relation defined by the following matrix also satisfies the reflexive, symmetric, and transitive properties.

$$\begin{array}{c} \\ a \\ b \\ c \\ d \\ e \\ f \end{array}\begin{array}{c} \begin{array}{cccccc} a & b & c & d & e & f \end{array} \\ \begin{bmatrix} 1 & 1 & 0 & 0 & 0 & 0 \\ 1 & 1 & 0 & 0 & 0 & 0 \\ 0 & 0 & 1 & 1 & 1 & 0 \\ 0 & 0 & 1 & 1 & 1 & 0 \\ 0 & 0 & 1 & 1 & 1 & 0 \\ 0 & 0 & 0 & 0 & 0 & 1 \end{bmatrix} \end{array}$$

It is easy to see that this relation is reflexive, since all the entries on the major diagonal are 1. The relation is symmetric because the ij entry equals the ji entry for every i and j. That is, the matrix is symmetric. And the transitive property, which would normally be difficult to verify, is clearly satisfied here, since the blocks of 1's indicate that the elements fall into disjoint subsets $\{a, b\}$, $\{c, d, e\}$, and $\{f\}$ such that any two elements x and y are related iff they belong to the same set! (We will say more about transitivity later.) □

Binary relations that satisfy all three of the properties we have been discussing are extremely important because they allow us to assert that elements thereby related to one another are in some sense "the same," even when they are not identical. For example, a set of three Apple computers and a set of three Commodore computers have the same number of objects, even though they are certainly not the same set. Similarly, the fractions 2/4 and 3/6 represent the same rational number, but they are not the same fraction. Perhaps it would be better to use the word "equivalent" for the kinds of comparison we are making.

Definition 7.6 If R is a binary relation on a set A, then we call R an (a) **equivalence relation** if R is reflexive, symmetric, and transitive. Moreover, if $x \in A$, then the set denoted by

$$[x] = \{y \in A : x\,R\,y\}$$

is called the (b) **equivalence class of *x*** (relative to R).

It is reassuring to note that the identity relation on any set is an equivalence relation, and the equivalence class of any element x relative to this relation includes only x. Here are some further examples.

Example 2

a) The relation given by the matrix in Example 1(e) is an equivalence relation. We have already verified this, and we only need point out again that there are exactly three equivalence classes: $\{a, b\}$, $\{c, d, e\}$, and $\{f\}$. These sets correspond to the "blocks" of 1's in the matrix that we brought to your attention earlier. Observe that, for example, $[a]=[b]=\{a, b\}$.

b) The relation "congruence modulo 3" is an equivalence relation on the set of integers. We have also verified this fact in Example 1(e). The equivalence class of 1, for example, is the set

$$[1]=\{\ldots, -2, 1, 4, 7, \ldots, 3n+1, \ldots\}.$$

In fact there are only two other equivalence classes that one can obtain from this binary relation. They are the sets

$$[2]=\{\ldots, -4, -1, 2, 5, 8, \ldots, 3n+2, \ldots\}$$

and

$$[3]=\{\ldots, -3, 0, 3, 6, 9, \ldots, 3n, \ldots\}.$$

c) None of the relations defined in Examples 1(a) to 1(c) are equivalence relations, since each of these fails to satisfy at least one of the reflexive, symmetric, or transitive properties.

d) Let $A = \cup_{i \in I} A_i$ be any partition of a set A, and define $x R y$ if x and y belong to the same partition set A_i. Then the reader can check (see Exercise 7.2-8) that R is an equivalence relation on A. If $x \in A_i$, then we have in fact $[x] = A_i$.

The collection of equivalence classes $[x]$ induced by an equivalence relation R on a set A is sometimes called the **quotient of *A* by *R***, and it is denoted by "A/R." As indicated in Example 2(a), (b), and (d), we can make the following statement about this collection of sets.

Theorem 7.1 Let R be an equivalence relation on a set A. Then $A = \cup_{x \in A} [x]$ is a partition of the set A.

Proof: We first observe that $x \in [x]$ since $x R x$ by the reflexive property of R. Hence it is true that $A = \cup_{x \in A} [x]$, since every member of A is included in this union. We need only show that for any x and y in A, the sets $[x]$ and $[y]$ are either disjoint or identical. Suppose that $[x] \cap [y] \neq \varnothing$. We will show that every element w in $[x]$ is also in $[y]$.

Now $w \in [x]$ implies that $x R w$, and by symmetry, $w R x$. Suppose that

$z \in [x] \cap [y]$. Then $x R z$ and $y R z$ are both true. But $w R x$ and $x R z$ imply, by transitivity, that $w R z$. And this, together with $y R z$, or $z R y$, gives us $w R y$. Hence $w \in [y]$, and so $[x] \subset [y]$. Similarly, we can show that $[y] \subset [x]$, giving us $[x] = [y]$ as claimed, and this completes the proof. ■

Further Discussion of Transitivity

Equivalence relations have many applications in advanced counting techniques, cryptology, and computability theory, just to take a few examples. Since these applications are beyond the scope of this text, let us address the practical question of verifying when the properties of reflexivity, symmetry, and transitivity are satisfied on finite sets. The first two properties require a relatively simple visual examination of a matrix giving the relation. [See Example 1(e).] It is somewhat simpler to determine whether a relation is transitive if we know that it is symmetric. The reason this is true is contained in the following theorem.

Theorem 7.2 Let R be a symmetric binary relation on a set A. Then R is transitive if and only if for every x, y, and z in A, $x R y$ and $x R z$ imply that $y R z$.

Proof: Let us suppose that R is transitive and that $x R y$ and $x R z$ holds for some x, y, and z in A. Since R is symmetric, we know that $y R x$, and, with $x R z$, transitivity yields $y R z$, as required.

Conversely, suppose that $x R y$ and $y R z$ are true. Symmetry gives us $y R x$. But we are assuming that for any a, b, and c in A $a R b$ and $a R c$ implies that $b R c$. Letting $a = y$, $b = x$, and $c = z$, we see that $y R x$ and $y R z$ yield $x R z$. Since $x R z$ is what we had to show, this completes the proof. ■

Algorithm 7.1 ***Transitivity Algorithm*** Given a matrix $M = (m_{ij})$ that describes a symmetric binary relation on the set $\{1, 2, 3, \ldots, n\}$, to determine if the relation is transitive:

1. [Select row i.] Repeat steps 2 and 3 for $i = 1, 2, \ldots, n$.
2. [Select column j.] Repeat step 3 for $j = 1, 2, \ldots, n-1$, $j \neq i$.
3. [Select column $k \geqslant j$.] If $m_{ij} = 1$, then do step 4 for $k = j, j+1, \ldots, n$.
4. [Apply Theorem 7.2.] If $m_{ik} = 1$ and $m_{jk} = 0$, then output "not transitive" and stop.
5. [Success.] Output "transitive."
6. End of Algorithm 7.1.

The primary value of Algorithm 7.1 is that one can use it to check visually whether a symmetric matrix defines a transitive relation. While this is not nearly as simple to achieve as our tests for symmetry and reflexivity, Fig. 7.8 shows that it is considerably easier to use this algorithm than to check all combinations of ordered triples (a, b, c).

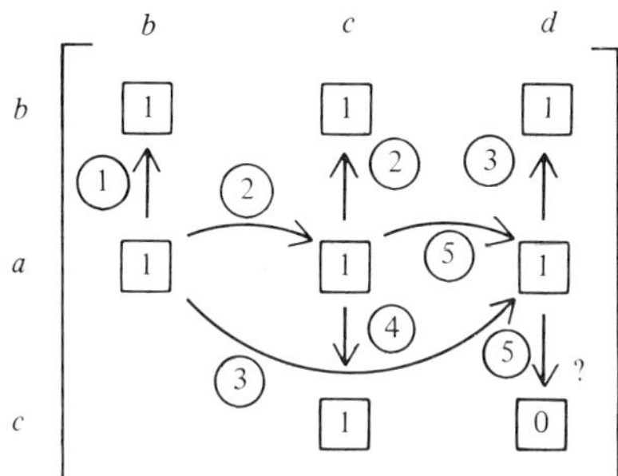

Figure 7.8 The two arrows labeled "5," for example, indicate that m_{ac} and m_{ad} are found to be 1 and 1, respectively, and we check that $m_{cd}=0$.

Examining row a, where $b<a<c<d=n$, we might encounter the following cases.

1. $m_{ab}=1$, $m_{ab}=1$, and $m_{bb}=1$.
2. $m_{ab}=1$, $m_{ac}=1$, and $m_{bc}=1$.
3. $m_{ab}=1$, $m_{ad}=1$, and $m_{bd}=1$.
4. $m_{ac}=1$, $m_{ac}=1$, and $m_{cc}=1$.
5. $m_{ac}=1$, $m_{ad}=1$, but $m_{cd}=0$.

An output "not transitive" follows case 5, and the algorithm stops.

We can also ask whether a relation R is transitive in terms of its **transitive closure**, which is defined as the set

$$R^{\infty}=\{\cup R^n : n=1, 2, \ldots\}.$$

You will recall that R^n is the n-fold composition $R \circ R \circ \cdots \circ R$. By definition of unions, $(a, b) \in R^{\infty}$ iff $(a, b) \in R^n$ for some n. Moreover, by definition of n-fold compositions, (a, b) belongs to R^n iff there are $n-1$ "intermediate" elements c_i such that (a, c_1), $(c_1, c_2), \ldots, (c_{n-1}, b)$ all belong to R. (See Exercise 7.1-10.)

We can now state the following.

Theorem 7.3 R is a transitive relation iff $R = R^\infty$.

Proof: First we claim that R^∞ is a transitive relation. To see this, observe that if we have a, b, and c such that $a R^\infty b$ and $b R^\infty c$, then, by definition of $\cup R^n$, $(a, b) \in R^i$ and $(b, c) \in R^j$ for some i and $j = 1, 2, \ldots$. But by the last observation preceding the statement of the theorem, it can be seen that $(a, c) \in R^{i+j}$. Hence $a R^\infty c$, and R^∞ is transitive.

By the first paragraph, if $R = R^\infty$, then R is transitive.

Next it is clear that we always have $R \subset R^\infty$. Suppose then that R is transitive. We will show that $R = R^\infty$.

Assume that $(a, b) \in R^\infty$. Then, as mentioned, there must be a sequence (a, c_1), $(c_1, c_2), \ldots, (c_{n-1}, b)$ of ordered pairs in R. Transitivity, however, and the first two pairs imply that (a, c_2) belongs to R. From this and the fact that the pair (c_2, c_3) belongs to R, we deduce that (a, c_3) also belongs to R. Continuing in this way, we finally obtain $(a, b) \in R$. Hence $R^\infty \subset R$, and so $R^\infty = R$, as claimed, if R is transitive. ■

If R is a relation on a finite set, then it is not difficult to compute the relation matrix of the transitive closure of R. One way is to compute the path matrix, as defined in Section 4.6 for graphs of the digraph of R.

Theorem 7.4 If a relation R is given by an n-by-n matrix M, then the relation matrix of R^∞ is the path matrix of the digraph of R, that is, the matrix obtained from $(I + M)^{n-1}$ by replacing all nonzero entries by 1's. (I denotes the n-by-n identity matrix.)

Proof: The path matrix P of the digraph D of R has an ijth entry of 1 iff there is a path in D from the ith to the jth element. By a "path" we mean a finite sequence of ordered pairs $(i, c_1), (c_1, c_2), \ldots, (c_{k,j})$ in R. There is such a sequence iff (i, j) belongs to R^k for some k and, therefore, to R^∞, completing the proof. ■

Example 3 The transitive closure of the relation given by the matrix

$$M = \begin{array}{c} \\ 1 \\ 2 \\ 3 \\ 4 \\ 5 \\ 6 \end{array} \begin{array}{c} \begin{array}{cccccc} 1 & 2 & 3 & 4 & 5 & 6 \end{array} \\ \begin{bmatrix} 1 & 0 & 1 & 0 & 1 & 1 \\ 0 & 1 & 0 & 1 & 0 & 0 \\ 1 & 0 & 1 & 0 & 1 & 1 \\ 0 & 1 & 0 & 1 & 0 & 0 \\ 1 & 0 & 1 & 0 & 1 & 1 \\ 1 & 0 & 1 & 0 & 1 & 1 \end{bmatrix} \end{array}$$

is given by M itself, since in this case M is also the path matrix of the relation. One can check this by calculating $(I + M)^{n-1} = (I + M)^5$ as in Section 4.6. Substituting 1's for the nonzero entries in this sum gives us the path matrix, but it also gives us M. Hence the relation given by M is its own transitive closure. By Theorem 7.4, the relation is transitive. □

Observe that the relation given by Example 3 is symmetric and reflexive in addition to being transitive. Hence it is an equivalence relation. Unlike Example 1(e), however, its entries are not arranged in blocks. This is merely a result of elements in the same equivalence classes not always being in adjacent rows. In fact the reader can easily verify that the equivalence classes are $\{1, 3, 5, 6\}$ and $\{2, 4\}$.

Completion Review 7.2

Complete each of the following.

1. A relation R is reflexive on A if for every a in A ________________.

2. A relation R is symmetric on A if for every a and b in A ________________.

3. A relation R is transitive on A if for every a, b, and c in A ________________.

4. A relation R is an equivalence relation on A if R is ________________.

5. If R is an equivalence relation on A, then the set of all y in A such that $x\,R\,y$ is called the ________________ of x and denoted by ________________.

6. The set of all equivalence classes determined by an equivalence relation R on a set A ________________ the set A into a set of disjoint sets called the ________________ of A by the relation R.

7. The transitive closure of a relation R is defined to be the set ________________. In fact R is a transitive relation iff $R =$ ________________.

8. If M is the relation matrix of R, then the relation matrix of the transitive closure of R is the ________________ of the digraph of R.

Answers: **1.** $a\,R\,a$. **2.** $a\,R\,b$ implies $b\,R\,a$. **3.** $a\,R\,b$ and $b\,R\,c$ implies $a\,R\,c$. **4.** reflexive, symmetric, and transitive. **5.** equivalence class; $[x]$. **6.** partitions; quotient. **7.** $R^\infty = \{\cup R^n, n = 1, 2, \ldots\}$; $R = R^\infty$. **8.** path matrix.

Exercises 7.2

1. Let $A = \{1, 2, 3\}$ and tell which of the following relations on A is **(a)** reflexive; **(b)** symmetric; **(c)** transitive.

$$R = \{(1, 1), (1, 2), (1, 3), (3, 3)\}.$$

$$S = \{(1, 1), (1, 2), (2, 2), (2, 3), (3, 2), (2, 1)\}.$$

$$T = \{(1, 1), (2, 2), (3, 3), (1, 2), (2, 3), (1, 3)\}.$$

$$U = A \times A.$$

2. Determine whether each of the following binary relations is reflexive, symmetric, or transitive. **(a)** x is greater than y, on the real numbers; **(b)** similarity of triangles; **(c)** A is a subset of B, on some power set $P(S)$; **(d)** $p \to q$ ("p implies q") on a nonempty set of propositions; **(e)** A has the same number of elements as B, for subsets of a finite set U; **(f)** x is an ancestor of y, on a set of people.
3. Determine whether the binary relations defined by the following matrices are reflexive, symmetric, or transitive.

(a) $\begin{bmatrix} 1 & 0 & 1 \\ 0 & 0 & 0 \\ 1 & 0 & 1 \end{bmatrix}$; **(b)** $\begin{bmatrix} 0 & 1 & 0 & 1 \\ 1 & 0 & 1 & 1 \\ 0 & 1 & 0 & 0 \\ 1 & 1 & 0 & 1 \end{bmatrix}$; **(c)** $\begin{bmatrix} 1 & 1 & 0 & 0 & 0 \\ 1 & 1 & 0 & 0 & 0 \\ 0 & 0 & 1 & 1 & 1 \\ 0 & 0 & 1 & 1 & 1 \\ 0 & 0 & 1 & 1 & 1 \end{bmatrix}$.

4. Which of the binary relations of those defined in the following exercises are equivalence relations? **(a)** 1; **(b)** 2; **(c)** 3.
5. Describe the equivalence classes for each of the equivalence relations in Exercises 1, 2, and 3.
6. Let R be the equivalence relation "congruence modulo 3" on the set of integers. Describe each of the following sets. **(a)** [17]; **(b)** [−4]; **[c]** [−72].
7. Let n be any positive integer. We say that "x is congruent to y modulo n" and write "$x \equiv y \pmod{n}$" if the integers x and y satisfy $x - y = mn$ for some integer m.
   a) Show that congruence modulo n is an equivalence relation.
   b) What are the equivalence classes in the quotient of the integers by congruence modulo n?
8. Let $A = \cup_{1 \leqslant i \leqslant n} A_i$ be any partition of a set A. Show that "$x R y$ if and only if x and y belong to the same set A_i" is an equivalence relation on A and that $[x] = A_i$ if and only if $x \in A_i$.
9. Let R be a symmetric and transitive binary relation on a set A, and suppose that $M = (m_{ij})$ describes this relation. Show that if $m_{ij} = 1$ for any $j \neq i$, then $m_{ii} = 1$ as well.
10. Show that (a, b) belongs to the n-fold composition $R^n = R \circ R \circ \cdots \circ R$ of a relation R with itself iff there are $n - 1$ intermediate elements c_i such that (a, c_1), (c_1, c_2), $\ldots, (c_{n-2}, c_{n-1})$, (c_{n-1}, b) all belong to the relation R.
11. **a)** Show that an intersection $\cap_{R \subset R_i} R_i$ of transitive relations R_i, all containing a relation R, is itself a transitive relation.

b) Show that the transitive closure R^∞ of a relation R is the "smallest" transitive relation containing R; that is, show that $R^\infty =$ the intersection of all transitive relations containing R.

12. Find the transitive closure of each of the following relations.

a) $R = \{(1, 2), (2, 3), (3, 3)\}$ on the set $A = \{1, 2, 3\}$.

b) R is given by the matrix

$$M = \begin{bmatrix} 0 & 1 & 0 & 1 \\ 1 & 0 & 1 & 1 \\ 0 & 1 & 0 & 0 \\ 1 & 1 & 0 & 1 \end{bmatrix}.$$

c) Let A be a set of cities, where $a\,R\,a$ means that there is a direct communications link between city a and city b.

d) Let A be the set of all people who ever lived and write $x\,R\,y$ if x is a parent of y.

e) Let R be the relation indicated by the following directed graph.

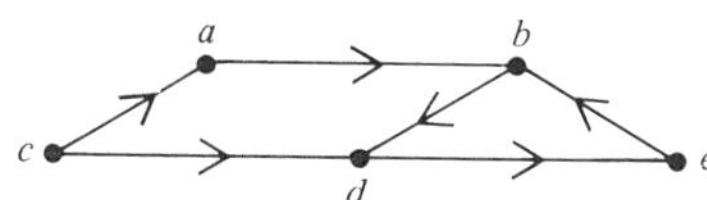

13. Warshall's algorithm for computing the transitive closure of a relation given by its n-by-n matrix $M = (m_{ij})$: (1). Repeat step 2 for $k = 1, 2, \ldots, n$. (2) For each m_{ij} such that $1 \leqslant i, j \leqslant n$, do the following. If $m_{ik} = 1$, $m_{kj} = 1$, and $m_{ij} = 0$, then change m_{ij} to a 1. (3) Output M.

Apply Warshall's algorithm to Exercise 12(b).

14. a) Algorithm 7.1 checks that $m_{ij} = 1$ and $m_{ik} = 1$ imply that $m_{jk} = 1$ if $j \leqslant k$. Why do we not also have to check the cases $j > k$?

b) Can we just check cases where $j < k$? Explain.

c) The following example shows that step 2 of Algorithm 7.1 would not be adequate if we only let $j = i, i + 1, \ldots, n$. Explain.

$$\begin{bmatrix} 1 & 1 & 0 \\ 1 & 1 & 1 \\ 0 & 1 & 1 \end{bmatrix}$$

15. Show that a reflexive binary relation R is an equivalence relation iff $a\,R\,a$ and $a\,R\,c$ imply that $b\,R\,c$ for all a, b, and c in the set in question.

16. Prove that if $|A| = n$, then there are **(a)** 2^{n^2} relations on A; **(b)** $2^{n^2 - n}$ reflexive relations on A; **(c)** $2^{n(n+1)/2}$ symmetric relations on A.

7.3 PARTIAL ORDERINGS AND DILWORTH'S THEOREM

The properties of reflexivity, symmetry, and transitivity are useful for characterizing objects that are very much alike although perhaps not identical. These properties enable us to formulate the definition of an "equivalence relation," which lumps all such objects together into a single "equivalence class." For many objects that are not alike, however, we often want to say more than $a \neq b$. Such objects may have some important order relationship, such as one being no larger than the other, or one being a "part" of the other. The properties of reflexivity and transitivity are very important in describing such relationships. But we will also need the following additional property.

Definition 7.8 Let R denote a binary relation on a set A. We say that R is **antisymmetric** if whenever $a\,R\,b$ and $b\,R\,a$ are both true, then $a = b$.

Notice that Definition 7.8 does not say that $a\,R\,b$ or $b\,R\,a$ must hold for any particular a and b in A. Nor does it say that if $a\,R\,b$ is true, then $b\,R\,a$ or $b\,\not R\,a$ must be true. Let us consider the following examples.

Example 1

a) The relation "$\leqslant$" on the integers is clearly antisymmetric, since $a \leqslant b$ and $b \leqslant a$ imply that $a = b$.

b) The relation "$<$" is also antisymmetric on the integers, even though we never have both $a < b$ and $b < a$ for any pair of integers. We sometimes say that antisymmetry is satisfied in cases like this "vacuously."

c) The relation "$\subset$" of set inclusion is antisymmetric on any collection of sets.

d) The relation "b divides a" (recall Section 1.4, Definition 1.10a) written "$b|a$" is antisymmetric on the set of positive integers, since $b|a$ and $a|b$ imply that $a = b$ for positive integers a and b.

e) Notice that the relation "$b|a$" is not antisymmetric on the entire set of integers. The property fails to hold, for example, with $2|(-2)$ and $(-2)|2$, but $-2 \neq 2$.

Definition 7.9 A relation R on a set A is called a (a) **partial ordering** (or a (b) **partial order**) if R is (1) reflexive, (2) antisymmetric, and (3) transitive. We say that A, together with R, written (A, R) is a (c) **partially ordered set or** (d) **poset**.

Example 2

a) The relation "less than or equal to" is a partial ordering of the integers Z. However, the relation "less than" is not a partial ordering of Z, since it fails to be reflexive.

b) The relation of "set inclusion" is a partial ordering of any collection of sets. Notice that for some sets in the collection we might have $A \not\subseteq B$ and $B \not\subseteq A$. In other words, these sets could turn out to be incomparable. Nevertheless, the collection of sets qualifies as a poset because all three properties are satisfied for sets that are comparable.

c) The positive integers are a partially ordered set under the relation $b|a$. We have already noted that antisymmetry is satisfied, and the reflexive property is obviously true. To show transitivity let us suppose that $b|a$ and $a|c$. Then there are positive integers m and n such that $a = mb$ and $c = na$. Hence $c = n(mb) = (nm)b$; and since nm is a positive integer, we conclude that b divides c.

d) Given a rooted tree T (see Section 5.3), then the binary relation "$v R u$ if v is an ancestor of or equal to u" is a partial ordering of the vertices of T.

The notation "$\precsim$" is often used for R in a partial ordering. We say that "a precedes b" or "b precedes a," and we write "$a \precsim b$" or "$b \precsim a$," respectively, if a and b are comparable elements. Moreover, we say that $(A, \precsim)$ is **totally ordered** and that the set A is a **chain** if every pair of elements in A are comparable.

If our set A has finitely many elements, then the concepts we have discussed thus far have a nice graphical representation. Example 2(d), that of a rooted tree, suggests a simplification of our usual relation directed graph, since it is easy to understand the "ancestor" relation directly from the tree itself without any further arrows being drawn. (See Fig. 7.9.) The only change we will make is to draw the root of the tree at the bottom of the graph, giving us what is called a "Hasse diagram" of the relation.

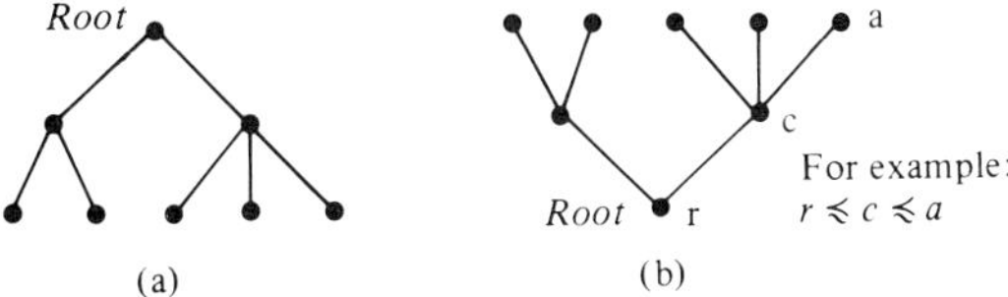

Figure 7.9 Forming a Hasse diagram. (a) Rooted tree usual picture. (b) Hasse diagram.

More generally, to form the **Hasse diagram** of a finite poset $(A, \precsim)$, we draw an edge from a up to b whenever $a \precsim b$ and there are no c's $(a \neq c \neq b)$ such that $a \precsim c \precsim b$. (We say that a is an **immediate predecessor** of b in this case.) Moreover, b is drawn higher than a in this case. No loops are drawn, and edges that would normally be included because of transitivity are omitted.

Example 3

a) In Fig. 7.10(a) we have a Hasse diagram of the elements of the set $\{1, 2, 3, 4\}$ partially ordered by the relation "less than or equal to." This set is also totally ordered and forms a chain.

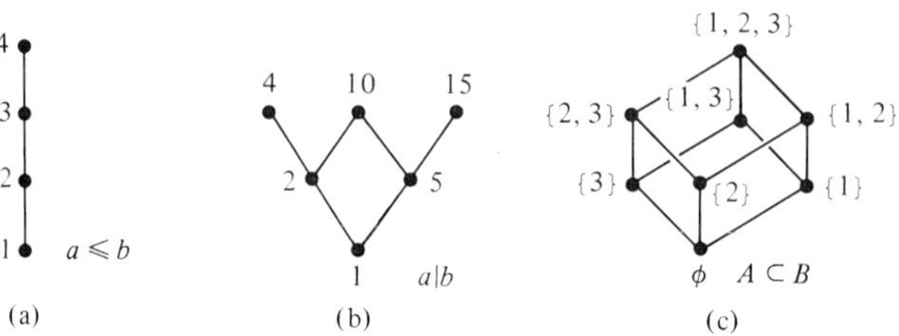

Figure 7.10 Three Hasse diagrams.

b) In Fig. 7.10(b) we have given a Hasse diagram of the set $\{1, 2, 4, 5, 10, 15\}$ partially ordered by "$a|b$." (See Example 2c.) It is clear that this is not a chain just by looking at the graph.

c) The subsets of the set $\{1, 2, 3\}$ are partially ordered by set inclusion, and the Hasse diagram of this relation is given in Fig. 7.10(c). Notice that we omit the edge connecting the set $\varnothing$ with the set $\{1, 2\}$, for example, even though we have $\varnothing \subset \{1, 2\}$, since $\varnothing \subset \{1\} \subset \{1, 2\}$. □

We say that x is **maximal** in a poset $(A, \preccurlyeq)$ if $x \preccurlyeq y$ implies that $x = y$; that is, no element y is preceded by x. Similarly we say that an element z is **minimal** in $(A, \preccurlyeq)$ if $y \preccurlyeq z$ implies that $y = z$; that is, no element y precedes z. For example, the maximal elements in Fig. 7.10 are 4 (in part a); 10, 4, and 15 (in part b); and $\{1, 2, 3\}$ (in part c). The minimal elements are 1 (in parts a and b) and $\varnothing$ (in part c).

Every finite poset has at least one minimal and one maximal element. This follows from the fact that the Hasse diagram is a directed graph with no cycles, since it is antisymmetric. (See Exercise 7.3-17.) Hence it can be topologically sorted by Algorithm 6.2, Section 6.2. In other words, we can label the vertices with the integers $1, 2, \ldots, n$ consistent with the partial ordering. The vertex labeled "1," for instance, is clearly a minimal element. To take another example, if we list the elements in Fig. 7.10(c), $\varnothing, \{1\}, \{2\}, \{3\}, \{1, 2\}, \{1, 3\}, \{2, 3\}, \{1, 2, 3\}$, and assign to each set its position on the list, this is consistent with the partial ordering by set inclusion.

Dilworth's Theorem

We define an **antichain** to be a subset of a partially ordered set in which no two distinct elements are comparable. Hence an antichain is the precise opposite of a chain, which you will recall is a subset in which every pair of elements are comparable.

Example 4 Two antichains in our poset of Fig. 7.10(c) are

$$\{\{1, 2\}, \{1, 3\}, \{2, 3\}\} \quad \text{and} \quad \{\{1\}, \{2\}, \{3\}\}.$$

The set of numbers $\{4, 10, 15\}$ gives us an antichain in Fig. 7.10(b), since none of these numbers divides the others. □

A subset of a partially ordered set is obviously partially ordered. Therefore, we can speak about a subset of a partially ordered set as being a chain. The **length** of a chain in a poset is the number of elements in it, provided that the chain is finite. Thus $\{\varnothing, \{1\}, \{1, 2\}, \{1, 2, 3\}\}$ is a chain of length 4 in Fig. 7.10(c).

There is an interesting connection between chains and antichains that is brought out in the following celebrated theorem of R. P. Dilworth.

Theorem 7.5 If $(A, \lesssim)$ is a finite poset and the size of the largest antichain in A is n, then A can be partitioned into n distinct chains.

We give suggestions for a proof of Dilworth's theorem in Exercise 7.3-9. Let us consider the following application.

Example 5 A project requiring many individual activities can be partially ordered by considering which activities cannot begin until certain other activities have been completed. (Recall our activity graphs of Section 6.3.) In other words, $a_i \lesssim a_j$ if activity $a_i = a_j$ or if a_i precedes a_j. Then an antichain of size k consists of k activities that can be assigned to k different workers or teams working at the same time, since no activity in the antichain precedes any of the others. In fact *the greatest number of activities that can go on at the same time is the size of the largest antichain.*

Let us now consider assigning activities to a single worker. We could first assign this individual a "minimal" activity, in other words, some a_1 not preceded by any other. Then assign an a_2 such that $a_1 \lesssim a_2$, and so on, until we reach a "maximal" activity. Then begin assigning activities to another worker in the same way. If we continue to assign activities in this way to all the workers, how many workers will we need? Dilworth's theorem says that if k is the size of the largest set of incomparable activities, then we can make k lists (chains) of activities, one for each of the workers.

□

The problem of actually finding a partition of the partially ordered set into disjoint chains is often solved by a modification of the network flow techniques we studied in Section 6.4. Trial and error may suffice, however, in small-scale examples. (See Exercises 7.3-5 and 7.3-8.)

The following theorem is an almost immediate consequence of Dilworth's theorem, and it is somewhat easier to apply.

Theorem 7.6 If A is a partially ordered set consisting of $mn + 1$ elements, then A has either an antichain of size $n + 1$ or a chain of size $m + 1$.

Proof: If the size of the largest antichain is less than or equal to n, then Dilworth's theorem says that A can be partitioned into n chains or fewer. If the length of the longest chain were less than $m+1$, then A would have a total of not more than nm elements, a contradiction. Thus A has some chain of length $m+1$, completing the proof. ■

Example 6 Suppose that we have been breeding white mice and we suddenly find that we have $50=7\cdot 7+1$ of them! Then there are either $8=7+1$ of these mice that are unrelated to each other or there are eight mice that form a sequence in which each one, after the first, is a descendent of the preceding one. To see this one only need order the mice by "$m_i \preccurlyeq m_j$ if $m_i=m_j$ or m_i is a descendent of m_j." It is not hard to see that this is a partial ordering of our set of mice. □

Completion Review 7.3

Complete each of the following.

1. A binary relation R on a set A is antisymmetric if for every a and b in A

   ________________.

2. A relation R on A is called a partial ordering if R is ________________,

   ________________, and ________________. In that case A (or the pair (A, R) is

   known as a ________________.

3. A set S of elements in a poset A is called a chain, or a ________________ set, if

   ________________.

4. A picture that shows the relationships in a poset is called a ________________

   diagram.

5. An element x in a poset $(A, \preccurlyeq)$ is minimal (maximal) in A if ________________.

6. An antichain in a poset A is a subset of A such that ________________.

7. Dilworth's theorem states that if in a finite poset A the largest chain is of size n, then A

   can be ________________.

8. If A is a poset of size $mn+1$, then either A has an antichain of size ________________

   or a chain of size ________________.

Answers: **1.** aRb and bRa implies that $a=b$. **2.** reflexive; antisymmetric; transitive; partially ordered set or poset. **3.** totally ordered set; every pair of elements in S is comparable. **4.** Hasse. **5.** whenever $z \preccurlyeq x$ ($x \preccurlyeq z$), then $z=x$. **6.** no two elements are comparable. **7.** partitioned into n distinct antichains. **8.** $m+1$; $n+1$.

Exercises 7.3

1. Which of the following relations on the set $A = \{1, 2, 3, 4, 5\}$ are antisymmetric? Which are partial orderings?
   a) $R_1 = \{(1, 2), (3, 4)\}$.
   b) $R_2 = A \times A$.
   c) $R_3 = \{(1, 2), (3, 4), (1, 1), (2, 2), (3, 3), (4, 4), (5, 5)\}$.
   d) $R_4 = \{(1, 1), (2, 2), (3, 3), (4, 4), (5, 5), (1, 5), (5, 1)\}$.
2. Give the Hasse diagram of the partial ordering "a divides b" for the set of integers $\{1, 2, 3, 4, 6, 8, 12, 18, 24\}$.
3. Which elements are immediate predecessors of the following numbers in the poset of Exercise 2? **(a)** 12; **(b)** 18; **(c)** 24.
4. Which elements are maximal in Exercise 2? Which are minimal?
5. Referring to the following Hasse diagram, answer each of the following questions.
   a) Which elements are maximal?
   b) Which elements are minimal?
   c) Find a chain of length 5.
   d) Find an antichain of size 4.
   e) Find all z such that $z \preccurlyeq g$.

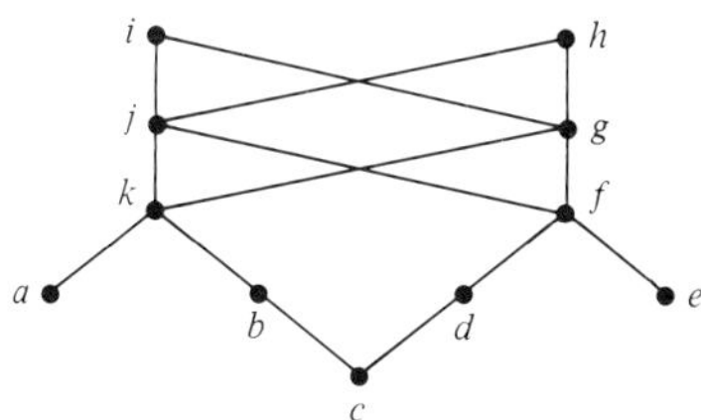

6. Suppose that we order the set of positive integers greater than 1 with the relation "a divides b." What are the minimal and maximal elements, if any?
7. If we order the power set of $A = \{1, 2, 3, 4\}$ minus the sets $\varnothing$ and A by the relation $S \preccurlyeq T$ if $S \subset T$, what are the minimal and maximal elements?
8. **a)** Find an antichain having the greatest possible size in Exercise 7.
   b) Partition the posets of Exercise 7 into the fewest possible chains.
9. Prove Dilworth's theorem (Theorem 7.5) by induction on the size n of the poset A, by showing that
   a) If $n = 1$, A is a chain.
   Then assume that the theorem is true for $n < k$. Consider two separate cases.
   b) In the first case suppose that the only maximal antichains sets with only maximal or only minimal elements.
   c) Then consider A minus a maximal element x and a minimal element y such that $y \preccurlyeq x$.
   d) In the second case there is a maximal antichain B of size j with nonminimal and nonmaximal elements.
   e) Consider sets $L = \{$elements in A no greater than some element in $B\}$ and $U = \{$elements in A no less than some element in $B\}$.

10. Suppose 37 activities are ordered as in Example 5. Explain why there are either seven activities that can be done simultaneously, or there are seven activities that can be listed in sequence such that one cannot be started until its predecessors are completed.
11. A point (x, y) in the first quadrant defines a rectangle with vertices (x, y), $(x, 0)$, $(0, y)$, $(0, 0)$, as in the following illustration. Show that the five rectangles R_i defined by any five distinct points in the first quadrant are such that there are three rectangles, one inside the other, or there are three rectangles R_i, R_j, and R_k such that no rectangle is inside any of the other two. (*Hint*: Define $(x, y) \precsim (u, v)$ if $x \precsim u$ and $y \precsim v$.)

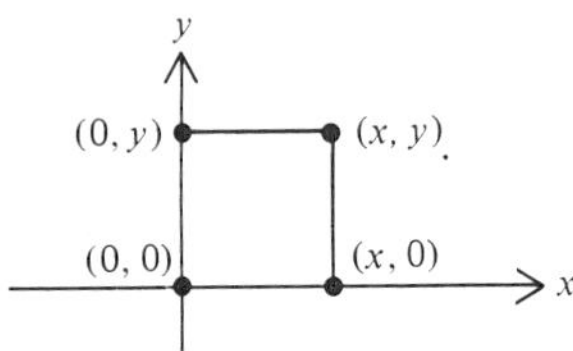

12. Show that if A_1 and A_2 are posets with relations $\precsim_1$ and $\precsim_2$, respectively, then $A_1 \times A_2$ is partially ordered by "$(x_1, x_2) \precsim (y_1, y_2)$ if both $x_1 \precsim_1 y_1$ and $x_2 \precsim_2 y_2$ are true."
13. Write an algorithm for determining whether a relation given by a matrix M is antisymmetric.
14. Show that the inverse R^{-1} of a partial order is also a partial order.
15. Let B be a digraph having no cycles. Explain why the path matrix of G describes a partial ordering of the vertices of G. (Recall that $A = (a_{ij})$ is the path matrix of G if $a_{ij} = 1$ when there is a path from i to j and $a_{ij} = 0$ otherwise.)
16. How many antisymmetric relations are there on a set having n elements? Prove your answer.
17. Prove that the directed graph of a partial ordering has no cycles of length greater than 1.

7.4 LATTICES

Among the partially ordered sets we discussed in the previous section, the set of all subsets of a set partially ordered by set inclusion is the one to which we have devoted the most space in this book. In fact Sections 1.1, 2.2, and 2.3 discussed sets and set operations in some detail, and so we are now reasonably familiar with relationships between set inclusion and the set operations of union, intersection, and complementation. For example, given sets A and B, we know that A and B are subsets of $A \cup B$, and we also know that $A \cap B$ is a subset of both A and B. Using the symbolism of partial orders we would write

$$A \precsim A \cup B, \quad B \precsim A \cup B$$

and

$$A \cap B \preccurlyeq A, \; A \cap B \preccurlyeq B.$$

Moreover, any set containing both A and B as subsets contains the set $A \cap B$. Any set contained in both A and B as subsets is also contained in the set $A \cap B$. These observations motivate the following definition.

Definition 7.10 Let $(A, \preccurlyeq)$ be a poset. For elements x, y, and z in A, we say that z is a (a) **lower bound** of x and y if $z \preccurlyeq x$ and $z \preccurlyeq y$; we say that z is an (b) **upper bound** of x and y if $x \preccurlyeq z$ and $y \preccurlyeq z$. An upper bound, z, of x and y is their (c) **least upper bound** (lub) if for every other upper bound, u, of x and y we have $z \preccurlyeq u$. A lower bound, z, of x and y is their (d) **greatest lower bound** (glb) if for every other lower bound, u, of x and y we have $u \preccurlyeq z$.

We will use the following notation.

$\boldsymbol{x} \vee \boldsymbol{y}$ = the lub of x and y, and $\boldsymbol{x} \wedge \boldsymbol{y}$ = the glb of x and y.

You have probably noticed that the symbols $\wedge$ and $\vee$ were formerly used in connection with propositions and logic. This was not an accident, as we shall soon see.

We also note that there is no danger in referring to *the* lub and *the* glb of x and y, since there can be at most one of each for a given x and y. For example, let z_1 and z_2 both be lubs of x and y. Then Definition 7.10(c) implies that z_1 and z_2 are both upper bounds of x and y, and so we have $z_1 \preccurlyeq z_2$ and $z_2 \preccurlyeq z_1$. Since our relation is antisymmetric, $z_1 = z_2$. We can show that the glb of two elements is unique in the same way.

Example 1 a) In the poset of sets ordered by set inclusion we have

$$A \vee B = A \cup B \quad \text{and} \quad A \wedge B = A \cap B.$$

b) In Fig. 7.11(a) we have the Hasse diagram of the set $\{1, 2, 3, 4, 6, 9\}$ ordered by the relation "a divides b." Then, for example,

$$4 \wedge 6 = 2, \quad 4 \wedge 3 = 1, \quad 4 \wedge 9 = 1, \quad 6 \wedge 2 = 2, \quad \text{etc.}$$

and

$$2 \vee 3 = 6, \quad 2 \vee 2 = 2, \quad 3 \vee 1 = 3, \quad \text{and so on.}$$

Notice that some pairs do not have a lub. For example, $4 \vee 6$ is not defined in the diagram.

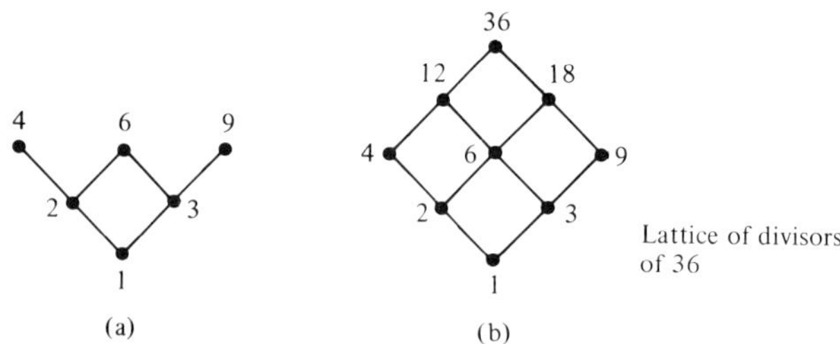

Figure 7.11 Part (b) is a lattice but part (a) is not.

c) In the poset of Fig. 7.11(b), on the other hand, the lub and the glb of every pair of elements in the set $\{1, 2, 3, 6, 9, 12, 18, 36\}$ are members of the same set. For example, $4 \vee 6 = 12$ and $4 \wedge 18 = 2$. □

Definition 7.11 A (a) **lattice** is a partially ordered set in which every pair of elements has a lub and a glb. If x and y are elements in a lattice, we say that $x \vee y$ and $x \wedge y$ are the (b) **join** and (c) **meet** of x and y respectively.

Example 2

a) The subsets of a set ordered by "$\subset$" is a lattice. Figure 7.12 gives such a lattice for the set $\{a, b, c\}$.

b) Figure 7.11(a) does not give us a lattice because the lub of two elements is not always in the set $\{1, 2, 3, 4, 6, 9\}$.

c) Figure 7.11(b) on the other hand, does give us a lattice. In fact the glb of any two elements is their greatest common divisor, and the lub of two elements is their least common multiple.

d) If we order the set $\{1, 2, 3, 4, 6, 9\}$ with the relation "less than or equal to," then this set becomes a lattice in which the lub of two elements is their maximum and the glb of two elements is their minimum. □

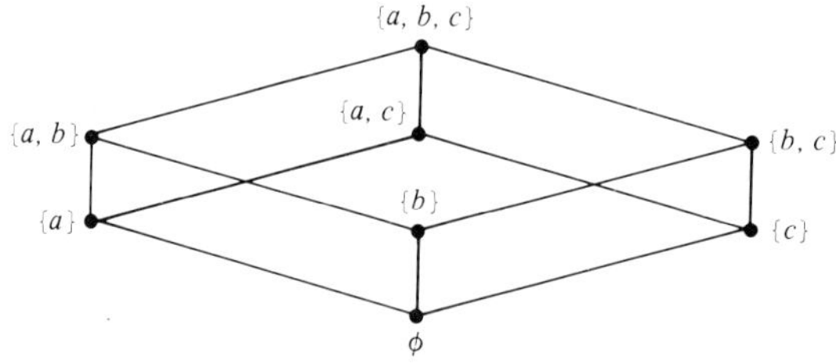

Figure 7.12 Lattice of subsets of $\{a, b, c\}$.

In the context of a lattice it is natural to consider "join" and "meet" as operations, just as we do the union and intersection of sets, for example. Indeed a lattice is sometimes defined as an algebraic system consisting of a set L and two binary operations called meet and join that satisfy the first three of the following four properties. However, each of the properties that follows is a consequence of *our* definition of a lattice as a special kind of partially ordered set.

Theorem 7.7 The following laws hold for every x, y, and z in a lattice L.

1. Commutative

1(a). $x \vee y = y \vee x$. 1(b). $x \wedge y = y \wedge x$.

2. Associative

2(a). $x \vee (y \vee z) = (x \vee y) \vee z$. 2(b). $x \wedge (y \wedge z) = (x \wedge y) \wedge z$.

3. Absorption

3(a). $x \vee (x \wedge y) = x$. 3(b). $x \wedge (x \vee y) = x$.

4. Idempotence

4(a). $x \vee x = x$. 4(b). $x \wedge x = x$.

Proof: Properties 1 and 4 follow directly from the definition of join and meet as lub and glb of x and y. The absorption property, 3(b) says that the glb of x and $x \vee y$ must be x. To prove this, we note that $x \precsim x \vee y$. If z precedes both x and $x \vee y$, then z certainly precedes x. Hence x is the glb of x and $x \vee y$, as claimed. Part 3(a) is proved similarly.

The proof of the associative law, 2(a), very much resembles arguments that we have used to show that two sets are equal. Specifically, let $w = x \vee (y \vee z)$. Then $x \precsim w$ and $y \vee z \precsim w$. The last relationship gives us $y \precsim w$ and $z \precsim w$. Hence $x \vee y \precsim w$ and $z \precsim w$. But this implies that

$$(x \vee y) \vee z \precsim w = x \vee (y \vee z).$$

It is just as easy to show that

$$x \vee (y \vee z) \precsim (x \vee y) \vee z.$$

By antisymmetry, the two elements are equal.

It should be noted that the idempotence properties are really independent of our definition of a lattice as a partially ordered set. They follow from the other three properties alone. We can prove property 4(b), for example, writing

$$\begin{aligned} x \wedge x &= x \wedge (x \vee (x \wedge y) && \text{by 3(a),} \\ &= x \wedge (x \vee z) && \text{with } z = x \wedge y, \\ &= x && \text{by 3(b),} \end{aligned}$$

completing the proof of 4(b). Property 4(a) is proved in a similar way, thus completing the proof of Theorem 7.7. ■

Observe the resemblance between Theorem 7.7 and certain parts of theorems we have stated for sets and logical propositions. Of course, there are certain omissions in our present theorem. Perhaps you would be willing to believe that we can prove "distributive laws" for lattices such as those we have proved for sets. That is, we have seen that

$$A \cap (B \cup C) = (A \cap B) \cup (A \cap C)$$

for any three sets A, B, and C. The corresponding statement

$$x \wedge (y \vee z) = (x \wedge y) \vee (x \wedge z)$$

does not, however, hold in general for lattices.

Example 3 Consider the lattice in Fig. 7.13. We can see that $a \wedge b = 0$ and $a \wedge c = 0$ from the diagram. Hence

$$(a \wedge b) \vee (a \wedge c) = 0 \vee 0 = 0.$$

However,

$$\begin{aligned} a \wedge (b \vee c) &= \\ a \wedge I \quad &= a. \end{aligned}$$

Therefore,

$$a \wedge (b \vee c) \neq (a \wedge b) \vee (a \wedge c).$$

Hence the lattice in Fig. 7.13 *does not* obey the distributive law, $x \wedge (y \vee z) = (x \wedge y) \vee (x \wedge z)$.

Definition 7.12 A lattice L will be called **distributive** if for every x, y, and z in L we have

$$x \wedge (y \vee z) = (x \wedge y) \vee (x \wedge z)$$

and

$$x \vee (y \wedge x) = (x \vee y) \wedge (x \vee z).$$

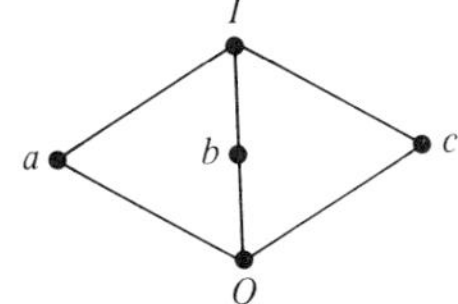

Figure 7.13 A nondistributive lattice; $(a \wedge b) \vee (a \wedge c) \neq a \wedge (b \vee c)$.

□

Thus Fig. 7.13 is not a distributive lattice, but the lattice of subsets of a set is distributive.

The empty set, the universal set, and complements play special roles in the algebra of subsets. The following definition characterizes these roles for lattices.

Definition 7.13 A poset has a (a) **lower bound**, O (respectively, an (b) **upper bound**, I) if for all x in the poset we have $0 \preccurlyeq x$ (respectively, $x \preccurlyeq I$). Let L be a lattice with a lower bound and an upper bound. We say that L is (c) **complemented** if for every element x in L there is an element y in L (called a (d) **complement** of x) such that

$$x \wedge y = 0 \quad \text{and} \quad x \vee y = I.$$

The elements O and I can be shown to give us analogs of the properties that we normally associate with the empty set $\varnothing$ and the universe U, namely,

$$x \vee I = I, \quad x \wedge I = x, \quad x \vee O = x, \quad \text{and} \quad x \wedge O = O.$$

To show that the first of these is true, for example, suppose that $x \vee I = y$. Then $x \preccurlyeq y$ and $I \preccurlyeq y$. But $y \preccurlyeq I$ by the definition of I. Hence antisymmetry implies that $y = I$ and, consequently, $x \vee I = I$. (See Exercise 7.4-11 where you are asked to prove the others.)

Although we have defined "complements" in lattices so that they imitate the complement $\bar{A}$ of a set A, there is one important difference. Complements of elements in a lattice need not be unique. In other words, an element can have two different complements. Moreover, it is possible that some elements will not have any complement. Let us consider some examples.

Example 4

a) Consider the lattice in Fig. 7.13, our nondistributive lattice. Then I and O function as the upper and lower bound respectively. It is easy to see that the element b has the elements a and c as two distinct complements.

b) The lattice in Fig. 7.11(b) has the numbers 36 and 1 as its upper and lower bounds respectively. Moreover, the numbers 4 and 9, for example, have each other as unique complements, since $4 \vee 9 = 36$ and $4 \wedge 9 = 1$. However, there is no complement for the number 2, since $2 \vee x = 36$ only if $x = 36$, and $2 \wedge y = 1$ only if $y = 3$, $y = 9$, or $y = 1$.

c) Consider the lattice of propositions relative to some universal truth set U. This lattice has the tautology $\mathcal{T}$ as an upper bound and the contradiction $\mathcal{F}$ as a lower bound. For every proposition p, the negation of p, which you will recall is written $\neg p$, is its unique complement. □

If a lattice has both an O and an I, we say that it is **bounded**. In a bounded distributive lattice we can speak of *the* complement of an element due to the following theorem.

Theorem 7.8 In a bounded distributive lattice L, complements of elements are unique when they exist.

Proof: Let us suppose that an element x has two complements a and b. Then a and b have the following properties.

$$x \vee a = I, \quad x \vee b = I, \quad x \wedge a = O, \quad \text{and} \quad x \wedge b = O.$$

The distributive laws enable us to write

$$a = a \vee O = a \vee (x \wedge b) = (a \vee x) \wedge (a \vee b) = I \wedge (a \vee b) = a \vee b.$$

Similarly we have

$$b = b \vee O = b \vee (x \wedge a) = (b \vee x) \wedge (b \vee a) = I \wedge (b \vee a) = b \vee a.$$

Since $a \vee b = b \vee a$, we conclude that $a = b$, as required. ■

When the complement of an element x is unique, we denote it by the symbol x'. This allows us to state *De Morgan's laws* as Theorem 7.9.

Theorem 7.9 If x and y are any two elements in a bounded distributive complemented lattice L, then

$$(x \wedge y)' = x' \vee y' \quad \text{and} \quad (x \vee y)' = x' \wedge y'.$$

Proof: To show that $x' \wedge y'$ is the complement of $x \vee y$, we write

$$(x \vee y) \vee (x' \wedge y') = ((x \vee y) \vee x') \wedge ((x \vee y) \vee y') = I \wedge I = I$$

and

$$(x \vee y) \wedge (x' \wedge y') = ((x \wedge x') \wedge y') \vee ((y \wedge x') \wedge y') = O \vee O = O.$$

The proof that $x' \vee y'$ is the complement of $x \wedge y$ is similar and is left as an exercise. ∎

Completion Review 7.4

Complete each of the following.

1. Let $(A, \preccurlyeq)$ be a poset. Then z is a lower (upper) bound of x and y in A if __________. Furthermore z is the greatest lower bound (least upper bound) of x and y if __________. The greatest lower (upper) bound of x and y is written symbolically as __________.
2. A lattice is defined to be a poset in which every pair of elements has a __________ and a __________, which are called the __________ and the __________ of x and y respectively.
3. The operations of join and meet in a lattice are both commutative and associative. (*True or False*?)
4. The absorption laws state that __________ and __________.
5. The idempotence laws state that __________ and __________.
6. The set of subsets of a given set form a lattice under the operations of union and intersection only if the set is finite. (*True or False*?)
7. Distributive laws of join over meet and meet over join hold in every lattice. (*True or False*?)
8. A lattice L has a lower bound O (upper bound, I) if for all x in L we have __________. If L has both an O and an I, we then say that L is __________.
9. A bounded lattice is complemented if for each x in L there is an element y in L such that __________.

10. De Morgan's laws, stated ________________ and ________________ in terms of joins and meets, hold for all bounded distributive complemented lattices. (*True or False*?)

Answers: **1.** $z \lesssim x$ and $z \lesssim y$ ($x \lesssim z$ and $y \lesssim z$) for all x and y in A; z is a lower (upper) bound of x and y, and when w is also a lower (upper) bound, then $w \lesssim z$ ($z \lesssim w$); $x \wedge y$ ($x \vee y$). **2.** lub; glb; join; meet. **3.** True. **4.** $x \vee (x \wedge y) = x$; $x \wedge (x \vee y) = x$. **5.** $x \vee x = x$; $x \wedge x = x$. **6.** False: It is always a lattice. **7.** False. **8.** $O \lesssim x$ ($x \lesssim I$); bounded. **9.** $x \wedge y = O$ and $x \vee y = I$. **10.** $(x \vee y)' = x' \wedge y'$; $(x \wedge y)' = x' \vee y'$; True.

Exercises 7.4

1. Referring to Fig. 7.11(b), find the **(a)** lub of 12 and 18; **(b)** lub of 12 and 9; **(c)** glb of 12 and 18; **(d)** glb of 3 and 4.
2. Consider the poset of the positive integers ordered by the relation $a \lesssim b$ if a divides b.
   a) Why is this a lattice? In particular, what is $x \wedge y$ and $x \vee y$ for each pair of elements?
   b) Find all upper and lower bounds of this lattice.
3. Prove that the glb in a poset must be unique.
4. Consider the lattice consisting of the set of all subsets of the set $\{1, 2, 3, 4\}$ ordered by the subset relation.
   a) What is the lower bound of this lattice?
   b) What is the upper bound?
   c) Draw the Hasse diagram of this lattice as a poset.
5. Show that a lattice can have at most one upper bound and at most one lower bound.
6. Give an example of a lattice that has **(a)** a lower bound but no upper bound; **(b)** an upper bound but no lower bound.
7. Consider Fig. 7.13. Show that $a \vee (b \wedge c) \neq (a \vee b) \wedge (a \vee c)$.
8. Which of the posets shown in the following figure are lattices?

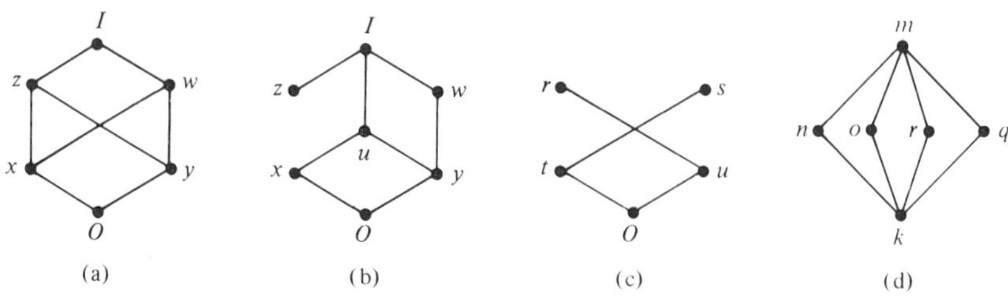

9. Consider the set $L = \{1, 2, 3, \ldots, 10\}$ ordered by "less than or equal to."
   a) Show that this yields a lattice. In particular, what are $x \wedge y$ and $x \vee y$ for each pair of elements?

b) What are the elements "I" and "O"?
c) Give an example to show that this lattice is not complemented.

10. Suppose that we have a mathematical system consisting of a set $S \neq \varnothing$ and two operations, $\wedge$ and $\vee$, that satisfy rules 1 to 4 of Theorem 7.7.
a) Show that this system is a poset if we define the partial ordering "$x \precsim y$ if and only if $x = x \wedge y$."
b) Show that this poset is a lattice with $x \vee y$ and $x \wedge y$ as the lub and glb of x and y respectively.

11. If L is a bounded lattice, show that **(a)** $x \wedge I = x$; **(b)** $x \vee O = x$; **(c)** $x \wedge O = O$.

12. Show that every finite lattice has a lower bound O and an upper bound I. (*Hint*: Consider $a_1 \wedge a_2 \wedge \ldots \wedge a_n$ and $a_1 \vee a_2 \vee \ldots \vee a_n$, where $a_1, a_2, \ldots, a_n$ are all the elements in the lattice.)

13. Show that any chain is a distributive lattice. (*Hint*: Using maximums and minimums for $\vee$ and $\wedge$, show that the distributive laws hold in each of the six cases $a \precsim b \precsim c$, $a \precsim c \precsim b$, $b \precsim a \precsim c$, $b \precsim c \precsim a$, $c \precsim a \precsim b$, and $c \precsim b \precsim a$.)

14. Show that the lattice of Fig. 7.11(b) is distributive. (*Hint*: The meet and join are the least common multiple and greatest common divisor respectively. Count the number of times each different prime factor from a, b, and c appears in each and use Exercise 13.)

15. Show, using Exercise 10, that in the lattice of propositions $p \precsim q$ if and only if $p \rightarrow q = \mathcal{T}$. (*Hint*: Use $p \rightarrow q = \neg p \vee q$.)

7.5 BOOLEAN ALGEBRAS

In the preceding section we defined the properties of complements, boundedness, and distributivity for lattices. These give us all the ingredients we need to define the following algebraic structure.

Definition 7.14 A **Boolean algebra** is a bounded, distributive, complemented lattice having at least two distinct elements.

Boolean algebras are sometimes defined a little differently, as we shall discuss later in this section. Our definition has the immediate advantage of letting us present some examples in terms of lattices, which we have already studied.

Example 1

a) The power set of $A = \{2, 3, 5\}$ with intersection and union of sets taken to be the operations of meet and join, respectively, is a Boolean algebra. The complement of an element of $P(A)$ is just the set complement. The sets A and $\varnothing$ are the bounds I and O, respectively, in this lattice. And, of course, sets obey the set operation distributive laws. [See Fig. 7.14(a).]

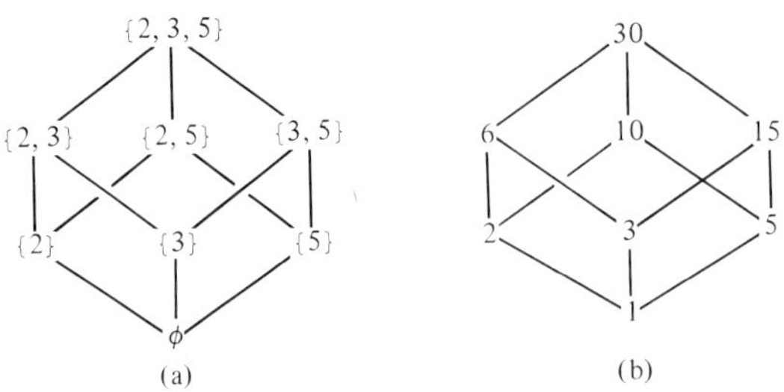

Figure 7.14 Isomorphic Boolean algebras. (a) Subsets of {2, 3, 5}. (b) Divisors of 30.

b) The divisors of 30, $B = \{1, 2, 3, 5, 6, 10, 15, 30\}$ give us another Boolean algebra, where join and meet are indicated as in Fig. 7.14(b). The meet of two numbers is their greatest common divisor, and their least common multiple is their join. The I and the O of this Boolean algebra are the numbers 30 and 1 respectively. One can check that each element has a complement, and the distributive laws can be verified as in Exercise 7.4-14.

It can be seen that the two diagrams of Fig. 7.14 have the same form. We say that the two Boolean algebras are **isomorphic** because one can give a one-to-one correspondence between their elements that preserves joins, meets, and complements. In our examples the correspondence is obvious. For example, $f(\{2, 3\}) = 6 = (2)(3)$ and $f(\{3, 5\}) = 15 = (3)(5)$. To illustrate what is meant by the word "preserved," notice that

$$f(\{2, 3\} \cap \{3, 5\}) = f(\{3\}) = 3$$

and that

$$f(\{2, 3\}) \wedge f(\{3, 5\}) = 6 \wedge 15 = 3.$$

Similarly, $f(\{2, 3\} \cup \{3, 5\}) = f(\{2, 3\}) \vee f(\{3, 5\}) = 30$, and $f(\overline{\{2, 3\}}) = f(\{5\}) = 5$, which is the complement of $6 = f(\{2, 3\})$. □

It turns out that every finite Boolean algebra is isomorphic to a Boolean algebra of the subsets of a finite set, in a manner very much like that which we have illustrated. Although we shall not give a proof of this remarkable fact, we can deduce the following corollary. (However, see Exercises 7.5-18 through 7.5-21.)

Theorem 7.10 A finite Boolean algebra has 2^n elements for some positive integer n.

Proof:

$$|P(A)| = 2^n \quad \text{if } |A| = n.$$ ■

Example 2

a) The lattice of divisors of 50, that is, the set $\{1, 2, 5, 10, 25, 50\}$ is not a Boolean algebra under the join and meet operations as greatest common divisor and least common multiple because this set has six elements, and 6 is not a power of 2.

b) The set of all subsets of $\{1, 2, 5, 10, 25, 50\}$ is a Boolean algebra having $2^6 = 64$ elements, taking meet and join as intersection and union respectively.

c) Infinite Boolean algebras also exist. As an example, take the set of all subsets of the integers, with join and meet as in part (b). □

The operations of join and meet in a Boolean algebra obey a variety of laws. Some of these, such as the *distributive laws*, are explicitly stated in its definition. Others, such as the *associative, commutative, absorptive, and idempotence laws*, are implied by the fact that a Boolean algebra is a lattice. Still others, such as *De Morgan's laws*, were deduced from a combination of both of the previous kinds. (See Theorem 7.9, Section 7.4, for example.)

Boolean algebras are often defined without direct reference to lattices or even posets, however, by stating a small number of laws from which all the others can be deduced. We include this definition for the sake of completeness.

Definition 7.15 ***(Alternate to 7.14)*** A **Boolean algebra** is a nonempty set B that is closed under two binary operations $+$ and $*$, and a unary operation $'$. In addition, B has two distinct elements denoted by 0 and 1, which satisfy the following laws for every a, b, and c in B.

1. Commutative

1(a). $a + b = b + a$. 1(b). $a * b = b * a$.

2. Distributive

2(a). $a + (b * c) = (a + b) * (a + c)$. 2(b). $a * (b + c) = (a * b) + (a * c)$.

3. Complements

3(a). $a + a' = 1$. 3(b). $a * a' = 0$.

4. Identity

4(a). $a + 0 = a$. 4(b). $a * 1 = a$.

One usually identifies the $+$ operation with $\vee$ and the $*$ operation (read "times") with $\wedge$. Moreover, just as in ordinary multiplication, the operation symbol $*$ is often omitted, and we simply write expressions such as $a*b$ as "ab." We will usually follow this practice.

When we say that B is **closed** with respect to the three operations, we mean that the elements resulting when these operations are applied to elements in B remain in B. For example, in Fig. 7.14(b) the least common multiple of any two numbers in the diagram is always a number in the diagram. Hence the set of numbers in this diagram is closed under the operation of taking their least common multiples. Likewise, this set is closed under the operation of taking greatest common divisors. These operations may be identified with the Boolean operations of $*$ and $+$ respectively. Similarly, the complement of each element with respect to the numbers which play the roles of "1" and "0," namely, 30 and 1, respectively, are always found in the figure. Thus the set of divisors of 30 is closed under the operation of complementation.

The 0 and the 1 are identified with what we have called the lower and upper bounds of the lattice respectively. If we interchange the 0 and the 1, and interchange the $+$ and the $*$ operations in any statement of a Boolean algebra, we obtain what is called the *dual* of that statement. For example,

$$(a1)+(0+b)=a+b \quad \text{is the dual of} \quad (a+0)(1b)=ab.$$

And, as with the special case of a Boolean algebra of sets, we have the following **principle of duality**.

Theorem 7.11 Let S and S^* be a statement and its dual in a Boolean algebra. Then S is true if and only if S^* is true.

One can now show that all of the properties we associated with our first definition of a Boolean algebra (7.14) are implied by our alternate definition.

If we can show that Definition 7.15 defines a lattice, with $a+b$ replaced by $a \vee b$, ab replaced by $a \wedge b$, and the complement of a, a', behaving as it should in a lattice, then the alternate definition immediately tells us that our lattice is distributive and complemented. Hence we have to show that Definition 7.15 implies that the absorption and associative laws hold and that 0 and 1 are the lower and upper bounds of the lattice. (Also see Exercise 7.4-10.)

Theorem 7.12 Let a, b, and c be any three elements in a system defined in Definition 7.15. Then the following are always true.

1. Idempotent laws

1(a). $a+a=a$. 1(b). $aa=a$.

2. Boundedness laws

2(a). $a+1=1$. 2(b). $a0=0$.

3. Absorption laws

3(a). $a + (ab) = a.$ 3(b). $a(a + b) = a.$

4. Associative laws

4(a). $a + (b + c) = (a + b) + c.$ 4(b). $a(bc) = (ab)c.$

Proof: To prove 1(a) we can write

$$a = a + 0 = a + (aa') = (a + a)(a + a') = (a + a)1 = a + a.$$

Then 1(b) follows by 1(a) and duality (Theorem 7.11).

To prove 2(a) consider the following string of equalities.

$$a + 1 = (a + 1)(a + a') = a + (1a') = a + (a'1) = a + a' = 1.$$

As before, 2(b) then follows by 2(a) and duality.

To show that 3(a) is true, write

$$a + (ab) = (a1) + (ab) = a(1 + b) = a(b + 1) = a1 = a.$$

Then 3(b) follows by 3(a) and duality, as before. (*Note.* The reader should actually dualize each of the proofs as an exercise, although, by Theorem 7.11 this is not required.)

We now sketch a proof of 4(a), leaving some of the details to be filled in by the reader in Exercise 7.5-8.

Let $x = a + (b + c)$ and let $y = (a + b) + c$. Then one can show that $ax = a$ and that $ay = a$; hence $ax = ay$. Similarly, the reader is asked to show that $a'x = a'(b + c)$ and that $a'y = a'(b + c)$, giving us $a'x = a'y$. Therefore, we can write

$$x = 0 + x = (aa') + x = (ax) + (a'x) = (ay) + (a'y) = (aa') + y = 0 + y = y,$$

as required. The other associative law follows from 4(a) and duality, completing the proof of Theorem 7.12. ■

Now that we have verified that our two definitions of a Boolean algebra are equivalent (given Exercises 7.4-10 and 7.5-8), let us point out that there are important laws, such as De Morgan's laws, that hold given either definition. Moreover, De Morgan's laws can be written more generally as

$$(x_1 + x_2 + \cdots + x_n)' = x_1' * x_2' * \cdots * x_n'$$

and

$$(x_1 * x_2 * \cdots * x_n)' = x_1' + x_2' + \cdots + x_n'$$

and proved by induction.

Meaningful expressions in the variables $x_1, x_2, \ldots, x_n$, the symbols 0 and 1, and the operations of $+$, $*$, and complementation are known as **Boolean expressions**. These are functions of n variables we can evaluate by substituting a 0 or a 1 for each x_i. Then we evaluate expressions such as $0 \wedge 1$ as we did circuits in Section 2.6.

Recall that the meet, join, and complement operations correspond to AND-, OR-, and NOT-gates respectively. The symbols used for these gates and the tables defining the corresponding Boolean arithmetic are shown in Fig. 7.15.

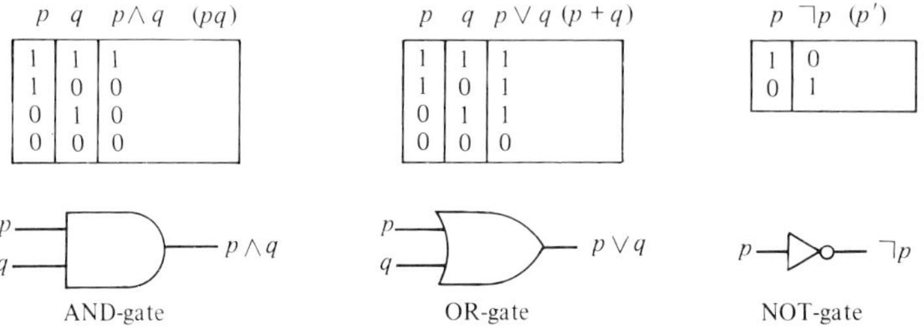

p	q	$p \wedge q$ (pq)
1	1	1
1	0	0
0	1	0
0	0	0

p	q	$p \vee q$ $(p+q)$
1	1	1
1	0	1
0	1	1
0	0	0

p	$\neg p$ (p')
1	0
0	1

Figure 7.15 Logic gates and tables.

The circuits constructed from AND-gates, OR-gates, and NOT-gates with inputs from sources $x_1, x_2, \ldots,$ and x_n give us another example of a Boolean algebra. You will recall that one of the problems we considered in Section 2.6 was the simplification of such circuits, a problem that is equivalent to the simplification of logical circuits as in, for example, $\neg((\neg p) \vee (\neg q)) = p \wedge q$. We will reconsider this problem in the next section, but first we must discuss the problem of writing such expressions in a standardized form, one that corresponds to the polynomial notation of a sum of products.

Definition 7.16 A nonzero Boolean expression in the variables $x_1, x_2, \ldots, x_n$ is said to be in **disjunctive normal form** if it is a sum of products of the form $y_1 y_2 \ldots y_n$, where each y_i is equal to either x_i or x_i'; moreover, "0" is also in disjunctive normal form.

Thus $xyz + x'yz + xy'z'$ is a Boolean expression in the variables x, y, and z that is in disjunctive normal form. But $xy + y'z'$ is not in disjunctive normal form since, for example, xy does not contain the factor z or z'. Also, $xy + x'y'$ is not in disjunctive normal form in the three variables x, y, and z, but it is in disjunctive normal form with respect to x and y alone.

Notice that in a Boolean algebra we can perform simplifications such as "$2x$" $= x + x = x$ and "x^2" $= xx = x$. Moreover, *the following steps can be used to change any Boolean expression of sums and products into disjunctive normal form.*

1. Use De Morgan's laws and the involution laws ($1' = 0$, $0' = 1$, and $(x')' = x$) to move the complement operation into parentheses so that it applies only to the variables x_i.

2. Then use the distributive laws to express the result as a sum of products.
3. Next we eliminate any repeated products or sums, as described, using the idempotence and absorption laws, and arrange the variables in the correct order with the commutative laws.
4. Finally, if the variable x_i is missing from a product, multiply that product by $(x_i + x_i') = 1$, and repeat steps 2 and 3.

Example 3 Let us transform the Boolean expression

$$(xy)'z + x'z + xy(y' + z')'$$

into disjunctive normal form, using the aforesaid four rules.

$$\begin{aligned}
&(xy)'z + x'z + xy(y' + z')' && = \text{(by step 1)} \\
&(x' + y')z + x'z + xy(yz) && = \text{(by step 2)} \\
&x'z + y'z + x'z + xyyz && = \\
&2x'z + y'z + xy^2z && = \text{(by step 3)} \\
&x'z + y'z + xyz && = \text{(by step 4)} \\
&x'z(y + y') + (x + x')y'z + xyz && = \text{(by step 2)} \\
&x'yz + x'y'z + xy'z + x'y'z + xyz.
\end{aligned}$$

The final expression is in disjunctive normal form in the variables x, y, and z. □

Could we have obtained a different disjunctive normal form for the expression in Example 3? The answer is contained in the following theorem.

Theorem 7.13 A nonzero Boolean expression in the variables $x_1, x_2, \ldots, x_n$ has precisely one disjunctive normal form, disregarding order of summands.

Proof: The four steps given in the foregoing show that we can transform a Boolean expression into at least one disjunctive normal form.

Now suppose that there are at least two forms for some Boolean expression, E. That is, suppose that we have two different sums of products P_i and Q_i for E:

$$E = P_1 + P_2 + \cdots + P_n = Q_1 + Q_2 + \cdots + Q_m.$$

Then one of the P's is different from all the Q_i's, or one of the Q's is different from all the P_j's. Suppose that P_1 is different from the Q's. Observe that each product includes all the variables or their complements. Therefore, if $P_1 \neq Q$, then x_i is in P_1 and x_i' is in Q, or the reverse. Hence $P_1Q_i = 0$ for all $i = 1, 2, \ldots, m$, and so

$$P_1E = P_1(Q_1 + Q_2 + \cdots + Q_m) = 0 + 0 + \cdots + 0 = 0.$$

But P_1 is also different from all the other P_i and $P_1P_1 = P_1$, by the idempotence laws. Hence

$$P_1E = P_1(P_1 + P_2 + \cdots + P_n) = P_1 + 0 + 0 + \cdots + 0 = P_1.$$

We conclude that $P_1 = 0$, which contradicts the fact that P_1 is a product of different variables or their complements. This contradiction implies that our disjunctive normal form must be unique. ■

Completion Review 7.5

Complete each of the following.

1. Defined as a lattice, a Boolean algebra is a ______________ having at least ______________ elements.
2. Every finite Boolean algebra has ______________ elements, for some positive integer n.
3. An isomorphism between Boolean algebras is a ______________ that preserves ______________. Every finite Boolean algebra is isomorphic to a Boolean algebra of ______________.
4. A Boolean algebra may be defined in terms of two binary operations, usually denoted ______________, and a unary operation with the four laws known as ______________, ______________, ______________, and ______________.
5. The principle of duality is valid in a Boolean algebra. (*True or False*?)
6. Meaningful expressions in the variables $x_1, x_2, \ldots, x_n$, the symbols 0 and 1, and the operations of $+$, $*$, and complementation are known as ______________.
7. A sum of products of the form $y_1y_2 \ldots y_n$, where each y_i is either x_i or x_i' is known as a ______________.
8. A nonzero Boolean expression in the variables $x_1, x_2, \ldots, x_n$ has precisely ______________ disjunctive normal form, disregarding the ______________.

Answers: **1.** bounded distributive complemented lattice; two distinct. **2.** 2^n. **3.** one-to-one correspondence; sums, products, and complements; subsets of a set. **4.** + and $*$; commutativity; distributivity; complements; identities. **5.** True. **6.** Boolean expressions. **7.** disjunctive normal form. **8.** one; order of the summands.

Exercises 7.5

1. For each of the following lattices, explain why it either is or is not a Boolean algebra.

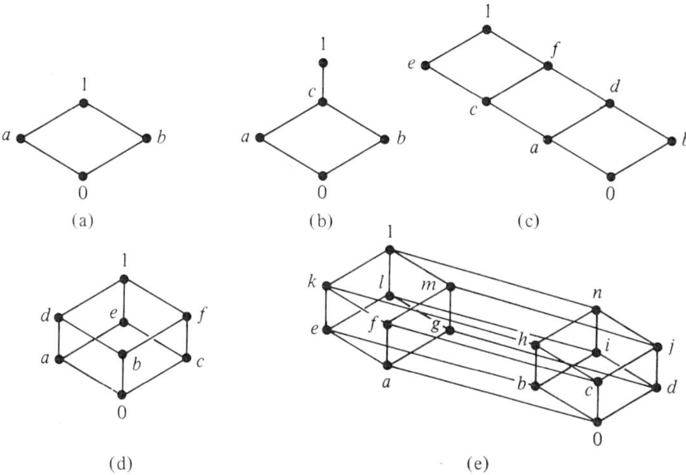

2. Show that the set $D = \{1, 3, 5, 7, 15, 21, 35, 105\}$ of divisors of 105 is a Boolean algebra under the operations of $a + b =$ the least common multiple of a and b, and $ab =$ the greatest common divisor of a and b. (*Hint*: Consider the one-to-one correspondence between D and the subsets of the set $\{3, 5, 7\}$.)

3. Show that the set $S = \{1, 2, 3, 5, 6, 7, 10, 14, 15, 21, 30, 35, 42, 70, 105, 210\}$ of divisors of 210 is a Boolean algebra under the operations of Exercise 2. (*Hint*: Consider subsets of the set $\{2, 3, 5, 7\}$.)

4. Show that the set of positive divisors of 70 is a Boolean algebra under the operations of Exercise 2.

5. Which of the following numbers of elements are permissible for a Boolean algebra? **(a)** 9; **(b)** 16; **(c)** 25; **(d)** 32; **(e)** 128.

6. Write the duals of the following statements: **(a)** $x'z' = x'yz' + yz'$; **(b)** $yz'(x + x') = xyz' + x'yz'$; **(c)** $x'yzt + x'yz't = x'yt$.

7. Dualize the proofs given of parts 1 to 3 of Theorem 7.14, thereby obtaining direct proofs of parts 1(b) to 3(b).

8. Complete the proof of Theorem 7.14 (4a), showing that if $x = a + (b + c)$ and $y = (a + b) + c$, then $ax = a = ay$ and $a'x = a'(b + c) = a'y$.

9. Prove the generalized De Morgan's laws.

10. Draw a lattice diagram for the Boolean algebra of all the subsets of the set $\{2, 3, 5, 7\}$ under the operations of union, intersection, and set complementation. (*Hint*: See Exercise 1.)

11. Give an isomorphism between the lattices in Exercises 10 and 3.

12. Evaluate the Boolean expression $xy' + yz' + x + y + z$ when $x = 1$, $y = 0$, and $z = 0$.

13. Which of the following are in disjunctive normal form? **(a)** x^2; **(b)** $xy'z + x(x + y)$; **(c)** $xyzx$; **(d)** $x + x + y$; **(e)** $xyz + xyz'$. (*Note*: x, y, and z are all the variables.)

14. Put each of the following expressions into disjunctive normal form. **(a)** xy; **(b)** x; **(c)** $xy + (x + y)'z$; **(d)** $xy(x' + z)'$; **(e)** $(x + y + z)(xyz)$; (*Note*: Assume that x, y, and z are all the variables.)

15. Find the disjunctive normal form of **(a)** the set $(A \cap B) \cup (C \cap \bar{B})$; **(b)** the logical expression $(p \wedge q) \vee (r \wedge (\neg q))$; **(c)** the circuit diagram.

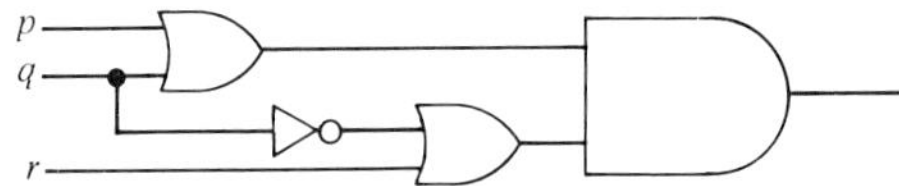

16. Show that there are 2^{2^n} different disjunctive normal forms on n distinct variables.

For the remaining exercises we will need the following definition: An **atom** in a lattice is an element x such that $0 \leqslant x \neq 0$ and whenever $0 \leqslant y \leqslant x$ then $y = 0$ or $y = x$.

17. Find the atoms in the Boolean algebra of **(a)** the subsets of the set $\{3, 5, 7\}$; **(b)** Exercise 2; **(c)** the subsets of the set $\{2, 3, 5, 7\}$; **(d)** Exercise 3.

18. Let $a_1, a_2, \ldots, a_n$ denote all the atoms below an element $x \neq 0$ in a finite Boolean algebra B, and let $y = a_1 \vee a_2 \vee \cdots \vee a_n$. Show that **(a)** $x \vee y' = 1$; **(b)** $x \wedge y' = 0$; **(c)** $x = x \wedge (y \vee y')$; **(d)** and hence $x = y$.

19. Show that a nonzero element of a finite Boolean algebra can be written as a join of atoms below it in one and only one way. (See Exercise 19.)

20. Show that the mapping $f(x) =$ "the set of all atoms below x" is an isomorphism from a finite Boolean algebra A onto the lattice of subsets of A.

7.6 MINIMIZATION AND SWITCHING CIRCUITS

In the preceding section we obtained a standard form for Boolean expressions called the disjunctive normal form. This is, you will recall, a (Boolean) sum of products, where each product includes every variable x_i under discussion or its complement x_i'. We will say that these products are **fundamental products** hereafter. (They are also known as "miniterms.")

The disjunctive normal form gives us a way to account for all possible non-equivalent Boolean expressions in a given set of variables. It also gives us a way to determine whether two Boolean expressions are equivalent, since the disjunctive normal form (which we will write "dnf" from now on) is unique. However, we were first motivated to discuss dfn's by the question of "simplification" of Boolean expressions. We discuss this question in the first part of this section. Then we show how to apply our techniques to the simplification of electrical switching circuits, which can be conceived of as yet another example of a Boolean algebra.

Let us first reconsider what we mean by simplifying logical circuits.

Example 1 The logical circuit in Fig. 7.16(a) is considerably more complicated than the one given in Fig. 7.16(b), and yet the two are equivalent. This is because the first can be simplified symbolically by writing

$$xyz + xyz' = xy(z + z') = xy(1) = xy.$$

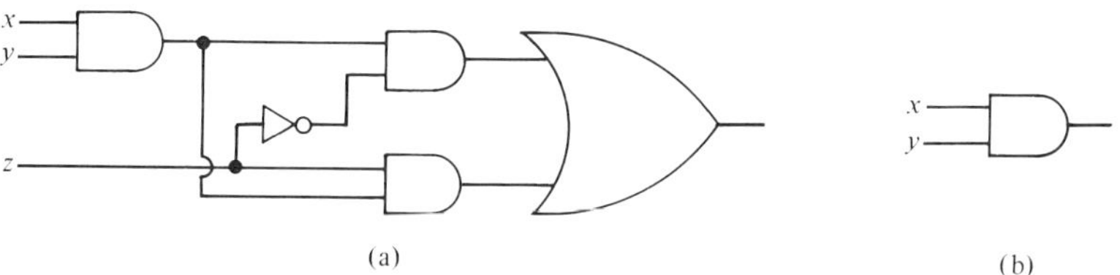

Figure 7.16 Circuit (a) $xyz+xyz'$ or $(x \wedge y \wedge z) \vee (x \wedge y \wedge (7z))$ is simplified into circuit (b) xy or $x \wedge y$ by elimination of $(z+z')$. □

As the foregoing example illustrates, *when two fundamental products differ by a single variable and its complement, we can eliminate the variable and the complement from these products.*

We can also simplify sums of products as in the following example.

Example 2 The logical circuit in Fig. 7.17(a) can be pared down to that given in Fig. 7.17(b). The two logical circuits are equivalent because

$$xyz + xy = xyz + xy1 = xy(z + 1) = xy1 = xy.$$

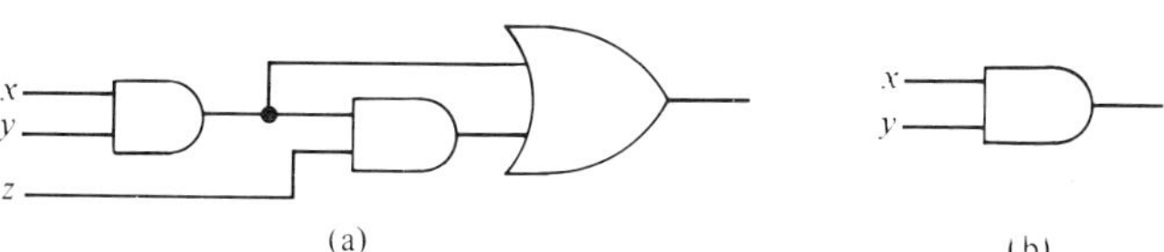

Figure 7.17 Circuit (a) $xyz+xy$ (or $(x \wedge y \wedge z) \vee (x \wedge y)$ is simplified into circuit (b) xy (or $x \wedge y$) by absorption of xyz by the term xy. □

The general principle at work in Example 2 is this: *If, in a sum of products of the variables and their complements, one product is contained in another, then the larger product may be eliminated.*

One can systematically apply our two rules to simplify a sum of products, obtaining at last what is called a **minimal sum of products form**. What we mean by the word "minimal" is given in the following definition.

Definition 7.17 Let E be a Boolean expression in the form of a sum of products, where each product contains some or all of the variables $x_1, x_2, \ldots,$ and x_n or their complements. Let $E(f)$ denote the number of factors x_i or x_i' and let $E(s)$ denote the number of summands in E. If H is an equivalent Boolean expression, we will say that (a) ***H* is simpler than *E*** when $H(f) \leq E(f)$, $H(s) \leq E(s)$, and at least one of these inequalities is strict. We will say that H is (b) **minimal** if no equivalent sum of products is simpler than H.

We now discuss a geometrical method for obtaining a minimal sum of products from the dnf of a Boolean expression. The method consists of making charts called "Karnaugh maps" of the dnf when the number of variables is not too large. We first consider the method for three variables, and then the four-variable case.

A **two-by-four Karnaugh map** is a rectangular checkerboard in which the eight squares represent all possible fundamental products of a dnf on three variables, say, x, y, and z. A square is checked whenever one's dnf has the corresponding fundamental product. [See Fig. 7.18(a).] Two squares will be considered "adjacent" if they have an edge in common, where, in the two-by-three case, the rightmost and leftmost edges in the table are considered to be identical, as if they wrapped around to form a cylinder. [See Fig. 7.18(b).] Moreover, one must label rows and columns so that adjacent rows or columns differ in exactly one variable and its complement.

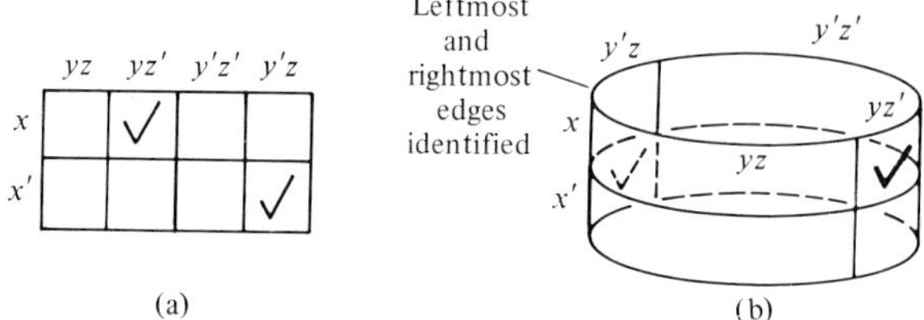

Figure 7.18 Fundamental products xyz' and $x'y'z$ are checked in the Karnaugh map (a) $E = xyz' + x'y'z$. Part (b) gives a three-dimensional version.

One can identify the simplified products in a Karnaugh map as maximal (checked) rectangles of area one, two, four, or eight square units, where a "maximal" reactangle is *one that cannot be contained in a larger rectangle.* These four cases correspond to products having exactly three, two, one, or zero variables respectively. A variable is eliminated from a product if the corresponding maximal rectangle includes both the variable and its complement. *To form a minimal sum of products, we choose the least possible number of maximal rectangles that cover all the checks.*

Example 3

a) Let $E_1 = xyz' + xy'z' + x'yz' + x'y'z$. This Boolean expression is represented by the Karnaugh map in Fig. 7.19(a). We see that three maximal rectangles

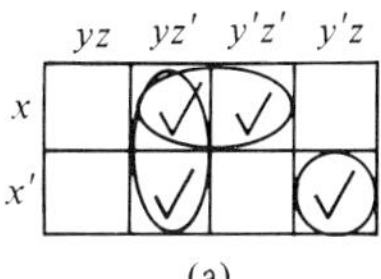

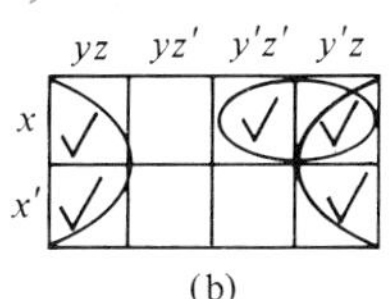

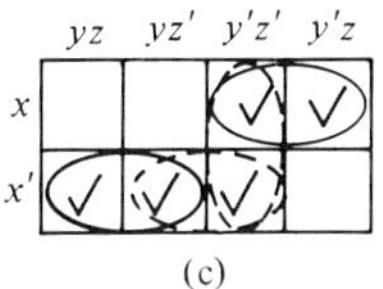

Figure 7.19 Three Karnaugh maps. (a) $E_1 = yz' + xz' + x'y'z$. (b) $E_2 = z + y'x$. (c) $E_3 = x'y + xy' + x'z'$ and $E_3 = x'y + xy' + y'z'$.

are needed to cover all the checks. The isolated square is $x'y'z$. The horizontal one-by-two rectangle is $xz'(y + y') = xz'$. And the vertical one-by-two rectangle is $yz'(x + x') = yz'$. Hence $E_1 = yz'' + xz' + x'y'z$.

b) Let $E_2 = xyz + x'yz + xy'z' + xy'z + x'y'z$. We represent this in Fig. 7.19(b). The two-by-two square is $(x + x')(y + y')z = z$, and the one-by-two rectangle is $xy'(z + z') = xy'$. Hence $E_2 = z + xy'$.

c) Let $E_3 = xy'z' + xy'z + x'yz + x'yz' + x'y'z'$. Then E_3 is represented in Fig. 7.19(c). Notice that there are two different ways that we can cover the checks with three maximal rectangles. Hence $E_3 = x'y + xy' + x'z' = x'y + y'z' + xy'$ gives us two different ways to express E_3 as minimum sums of products. □

Let us reconsider our construction of the carry bit circuit of Section 2.6 within the context of minimization.

Example 4 When adding two binary digits x and y and a digit z carried over from the previous stage of an addition problem, the new carry digit, c, can be determined by Table 7.2.

Table 7.2

x:	1	1	1	1	0	0	0	0
y:	1	1	0	0	1	1	0	0
z:	1	0	1	0	1	0	1	0
c:	1	1	1	0	1	0	0	0

A circuit diagram corresponding to the Boolean expression

$$E = xyz + xyz' + xy'z + x'yz$$

will give the output required by the last row of Table 7.2. To minimize E we simply construct the Karnaugh map in Fig. 7.20. This gives us $E = xy + yz + xz = (x + z)y + xz$. In the notation of Section 2.6, this gives us $((x \vee z) \wedge y) \vee (x \wedge z)$, which is precisely the expression we used in Example 5, Section 2.6.

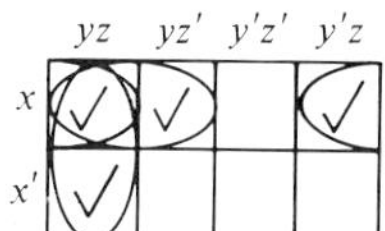

Figure 7.20 Simplifying the carry bit circuit by means of a Karnaugh map. $E = xy + yz + xz$.

□

Four-by-four Karnaugh maps can be used to simplify Boolean expressions in four variables. As in the two-by-four maps, the left and right edges of the table are identified with each other. We also identify the top and bottom edges in order to determine "adjacent" squares. The final terms in the minimum sum of products are given by maximal rectangles having areas of one, two, four, eight, or 16 square units, which correspond to products having four, three, two, one, or zero factors respectively.

Example 5 We can use four-by-four Karnaugh maps, as in Fig. 7.21, to simplify Boolean expressions in four variables such as

$$E = xyzw + xyz'w + xy'zw + xy'zw' + xy'z'w' + xy'z'w + x'y'zw + x'y'zw'$$
$$+ x'y'z'w' + x'y'z'w + x'yzw + x'yz'w.$$

Notice how adjacent rows and adjacent columns in Fig. 7.21 differ by exactly one variable and its complement. The reader will also observe that we have included three nonmaximal rectangles in Fig. 7.21 (corresponding to $xy'w'$, yw, and $x'y'$) merely to show how other terms with two or three variables correspond to rectangles. The terms $xy'w'$, yw and $x'y'$ do not appear in the final minimized expression $E = y' + w$.

□

Karnaugh maps for five variables, x, y, z, w, and u, consist of two four-variable maps, as in Fig. 7.21, one for u and the other for u'. In addition to the usual adjacencies,

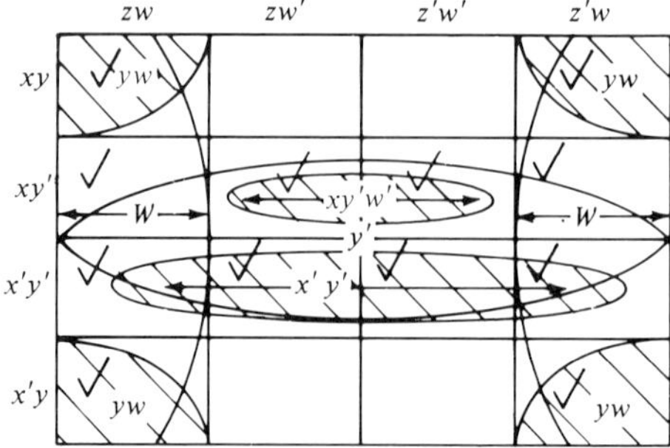

Figure 7.21 A four-by-four Karnaugh map. $E = y' + w$. (Note: The three shaded rectangles are *not* maximal.

squares in corresponding positions in the two maps are considered adjacent. For example, the $xyzw$ squares in the u and u' maps are adjacent. For six variables, x, y, z, w, u, and v, one can use four four-by-four maps, one for each of uv, $u'v$, $u'v'$, and uv', with appropriate descriptions of adjacencies. However, we shall omit the details in the five- and six-variable cases.

Switching Circuits

Suppose that we have a number of electrical switches $A, B, C, \ldots$, which are connected to some utility, such as the light bulb in Fig. 7.22, and to a power source. Each switch can be put into an "on" or an "off" position, allowing the circuit to conduct or not conduct electric current respectively. We will say that the circuit is "on" or "off" when it can either conduct or not conduct.

When two switches are connected **in parallel**, as in Fig. 7.23(a), the circuit will be on if at least one switch is on. This type of connection can be used when, for example, one wants to turn on a light from two different locations. On the other hand, two switches are connected **in series**, as in Fig. 7.23(b), if both switches are required to be on for the circuit to be on. This type of circuit design might be used where one of the switches is a safety device such as a circuit breaker or a fuse. Finally, we will denote by A' a switch that is on when switch A is off and off when A is on. Then switch A' might be used with an automatic monitoring device that switches on whenever the lights are turned off in an office, for example.

Apart from pairs such as A and A', different capital letters will denote different switches that can be turned off and on independently of one another. We use the notation

$$A \wedge B \quad \text{and} \quad A \vee B$$

for series and parallel connections of switches A and B respectively. A switch can also be connected in series or in parallel with itself, as in $A \wedge A$, and this is indicated by the

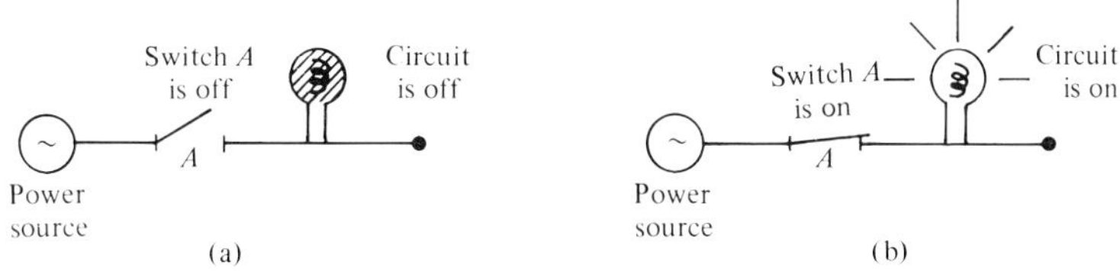

Figure 7.22 Switching circuit conventions.

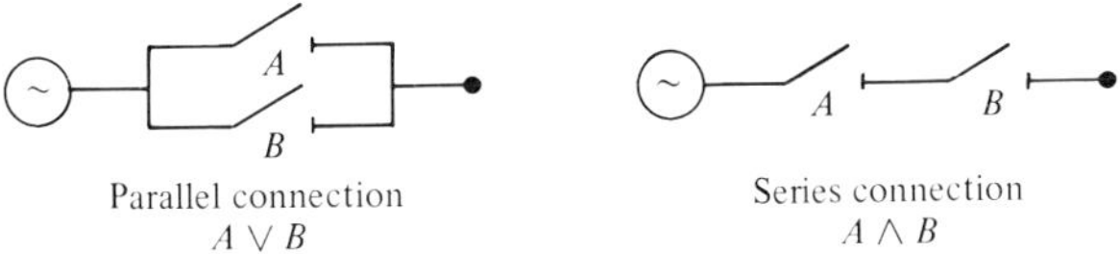

Figure 7.23 (a) Parallel connection, $A \vee B$. (b) Series convention, $A \wedge B$.

same letter being used for "both" switches. More generally, two switches are **equivalent** if they are on together and off together.

It is useful to think of the switches as giving off–on input and the circuits as giving the corresponding off–on output. These inputs and outputs can be summarized as in Table 7.3.

Table 7.3

A	B	$A \vee B$
On	On	On
On	Off	On
Off	On	On
Off	Off	Off

A	B	$A \wedge B$
On	On	On
On	Off	Off
Off	On	Off
Off	Off	Off

A	A'
On	Off
Off	On

We can use these tables to give us the outputs of more complicated **switching circuits**, as in the following example.

Example 6 The switching circuit given in Fig. 7.24 can be written symbolically as $((A' \vee B) \wedge C) \vee (A \wedge C')$. Table 7.4 shows when it is on or off. The table is constructed in the same way we previously devised our truth tables.

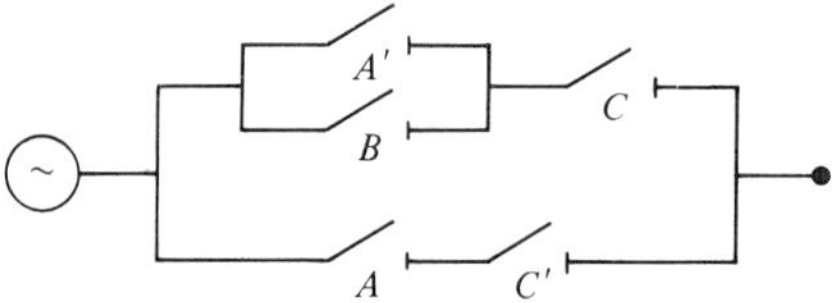

Figure 7.24 The switching circuit $((A' \vee B) \wedge C) \vee (A \wedge C')$.

Table 7.4

A	B	C	A'	C'	$A' \vee B$	$(A' \vee B) \wedge C$	$A \wedge C'$	$((A' \vee B) \wedge C) \vee (A \wedge C')$
On	On	On	Off	Off	On	On	Off	On
On	On	Off	Off	On	On	Off	On	On
On	Off	On	Off	Off	Off	Off	Off	Off
On	Off	Off	Off	On	Off	Off	On	On
Off	On	On	On	Off	On	On	Off	On
Off	On	Off	On	On	On	Off	Off	Off
Off	Off	On	On	Off	On	On	Off	On
Off	Off	Off	On	On	On	Off	Off	Off

□

It should be obvious that the system of switching circuits is much like previous systems we have studied, namely, the Boolean algebras of logical circuits, propositions, and sets. For example, the symbols we are using for series and parallel switching circuits are identical to the symbols we used for conjunction and disjunction of propositions respectively. The relationship between a proposition p and its negation $\neg p$ is equivalent to that between switches A and A'. And, of course, the truth values T and F correspond to "on" and "off" respectively. ($\mathcal{T}$ and $\mathcal{F}$ might be used for switching circuits that are always on or always off.) Hence we can state the following theorem.

Theorem 7.14 The algebra of switching circuits is a Boolean algebra.

Let us now apply what we have learned about disjunctive normal forms and the minimization of Boolean expressions to our switching circuits.

Example 7 The switching circuit in Fig. 7.25(b) is a simplification of the one in Fig. 7.25(a). To see this we express the circuit in Fig. 7.25(a) as

$$(A \wedge B \wedge C) \vee (A' \wedge B \wedge C') \vee (A \wedge B' \wedge C) \vee (A' \wedge B' \wedge C')$$

Since this is a dnf, we can immediately use a Karnaugh map to obtain a minimum sum of products form, as in Fig. 7.26 of $(A \wedge C) \vee (A' \wedge C')$. This Boolean expression corresponds to Fig. 7.25(b).

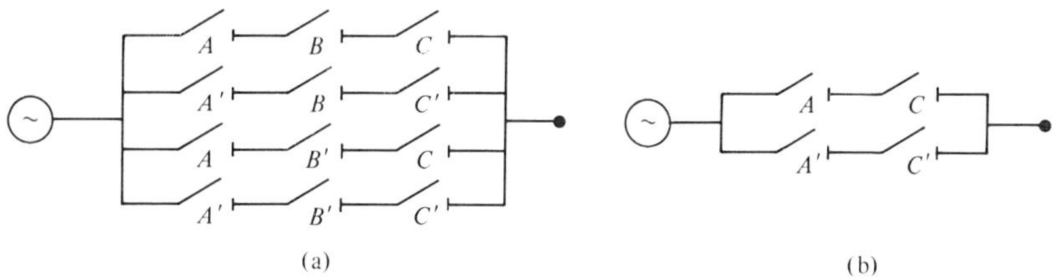

Figure 7.25 Switching circuit (a) is simplified to switching circuit (b).

	$B \wedge C$	$B \wedge C'$	$B' \wedge C'$	$B' \wedge C$
A	√			√
A'		√	√	

Figure 7.26 $(A \wedge B \wedge C) \vee (A' \wedge B \wedge C') \vee (A \wedge B' \wedge C) \vee (A' \wedge B' \wedge C')$ is simplified into $(A \wedge C) \vee (A' \wedge C')$. □

Example 8 Let us return to Example 6 and try to simplify Fig. 7.24. We first write it out symbolically, obtaining

$$((A' \vee B) \wedge C) \vee (A \wedge C') = (A' \wedge C) \vee (B \wedge C) \vee (A \wedge C').$$

Now we write this in disjunctive normal form, obtaining

$$(A' \wedge B \wedge C) \vee (A' \wedge B' \wedge C) \vee (A \wedge B \wedge C) \vee (A \wedge B \wedge C') \vee (A \wedge B' \wedge C').$$

We can find a minimum sum of products form for this expression from its Karnaugh map, as in Fig. 7.27.

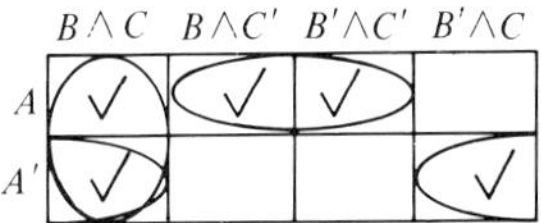

Figure 7.27 Simplifying Fig. 7.24 with a Karnaugh map.

The rectangles we have found tell us that

$$(A' \wedge C) \vee (B \wedge C) \vee (A \wedge C')$$

is a minimum sum of products, and this gives us a switching circuit with six switches and the five logical connectives $\wedge$ and $\vee$. □

Completion Review 7.6

Complete each of the following.

1. If E and H are equivalent Boolean sums of products, then E is simpler than H if ________, ________, and at least ________.
2. A two-by-four Karnaugh map is a ________ in which the eight squares represent all ________.
3. One must label adjacent rows or columns in a Karnaugh map so that they differ by ________.
4. One considers rectangles of dimensions ________ in a Karnaugh map. These correspond to products having ________ variables. A maximal rectangle is one that cannot be ________. Maximal rectangles correspond to ________.

5. In a two-by-four Karnaugh map, in addition to the usual adjacencies, the ______________ are also considered to be adjacent. In a four-by-four Karnaugh map we also consider the ______________ to be adjacent.

6. When two switches are connected in parallel, the circuit will be on if ______________.

7. When two switches are connected in series, the circuit will be on if ______________.

8. When a switch is on, then its complement is ______________, and vice versa.

9. The algebra of switching circuits is a ______________ algebra.

Answers: **1.** $E(f) \leq H(f)$; $E(s) \leq H(s)$; one of these inequalities is strict. **2.** rectangular checkerboard; fundamental products on three variables. **3.** a complement in exactly one variable. **4.** 1, 2, 4, 8, . . . ; all $n, n-1, n-2, n-3, \ldots$; contained in another rectangle; simplified products in the final expression. **5.** left- and right-hand edges; top and bottom edges. **6.** at least one of the switches is on. **7.** both switches are on. **8.** off. **9.** Boolean.

Exercises 7.6

In Exercises 1 to 10, give the Karnaugh map for each and a minimum sum of products form of the Boolean expression. Write the expression in disjunctive normal form first, if necessary. (Assume three variables for Exercises 1 to 6 and four variables for Exercises 7 to 10.)

1. $xyz' + x'yz + x'y'z'$.
2. $xy'z + x'yz'$.
3. $xy + xyz' + xy'z'$.
4. $x + y + x'yz$.
5. $x(y + xz')$.
6. $(x'y + z)'$.
7. $x'yzw + xyzw + xy'zw' + x'y'zw' + xyzw'$.
8. $x'yzw + xyzw + x'y'z'w' + x'y'z'w + x'z'w'$.
9. $(x + y)(z + w) + (x + y')zw$.
10. $xy + yz + (xz)' + (xy)'zw$.

In Exercises 11 and 12 find a minimum sum of products form for the switching circuit for each. Draw the switching circuit corresponding to your answer.

11. $(A \wedge B \wedge C') \vee (A \wedge B' \wedge C) \vee (A' \wedge B \wedge C)$.
12. $(A \wedge B \wedge C) \vee (A' \wedge B \wedge C') \vee A$.

In Exercises 13 to 16, write a Boolean expression for the switching circuit in each. Then obtain an "on–off" table for the circuit, such as Table 7.4.

13.

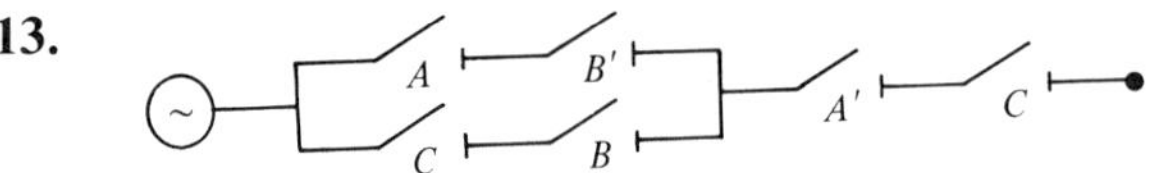

14.

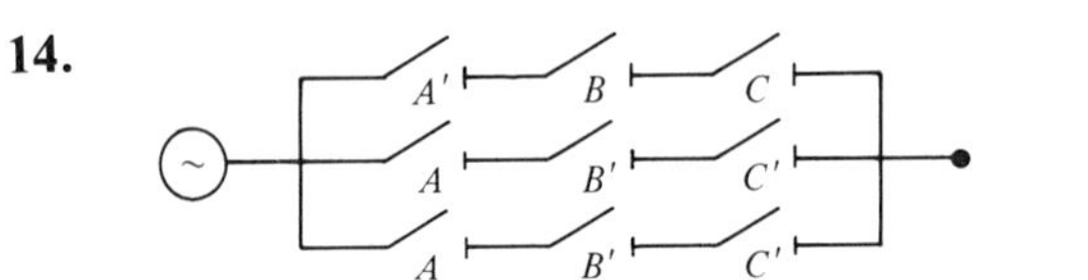

15.

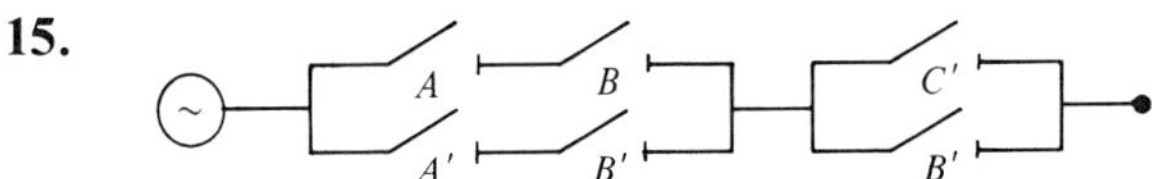

16.

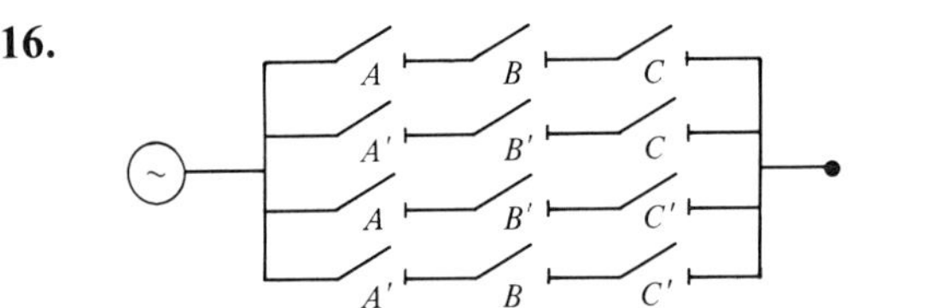

17. Find a minimum sum of products form for the switching circuits in Exercises 13 and 15. Draw the circuits.

18. Find a minimum sum of products form for the switching circuits in Exercises 14 and 16. Draw the circuits.

19. Verify that the logical circuit

$$((\neg x) \wedge (\neg y) \wedge c) \vee (x \wedge (\neg y) \wedge (\neg c)) \vee ((\neg x) \wedge (y) \wedge (\neg c)) \vee (x \wedge y \wedge c)$$

for calculating the sum bit in a full-adder is presently in minimum sum of products form.

7.7 GROUPS

In this section we study an algebraic structure that has only a single binary operation. This algebraic structure, called a "group," has applications in many areas of mathematics, physics, chemistry, and the social science. It has even been used to find solutions to popular puzzles such as Rubik's cube. In the next section we show how groups may be applied to binary coding theory. For the present, however, we discuss some of the essential definitions concerned with groups and related structures.

Definition 7.18 A **group** consists of a nonempty set G and a binary operation$*$ under which G is closed and such that the following laws hold.

1. (Associative law) For every a, b, and c in G, $a*(b*c)=(a*b)*c$.
2. (Identity) There exists an element e in G (called an **identity**) such that $e*a=a*e=a$ for every a in G.
3. (Inverses) For every a in G, there exists an element a' in G (called an **inverse** of a) such that $a*a'=a'*a=e$.

Observe that the identity element of a group is unique. If there were two identities, e and f, then we would have

$$e=ef=f.$$

Similarly, each element in a group has a unique inverse.

We sometimes refer to the group as the pair $(G, *)$, or more simply as G, where the operation is understood. Moreover, since $*$ is a binary operation, it follows that $a*b$ is an element of G for every a and b in G. This property, you may recall, is called "closure," and we say that G is "closed" with respect to $*$.

Example 1 Is the set of integers Z a group with respect to addition?

The integers are closed with respect to addition, where $*$ is taken to be "$+$". The associative law holds, and 0 is the identity for $+$. Moreover, for every integer a, the integer $-a$ fulfills the role of an inverse, since $a*a'$ amounts to $a+(-a)=(-a)+a=0$ in $(Z, +)$. Hence $(Z, +)$ is a group. □

Example 2 Is the set Q of rational numbers a group with respect to multiplication?

The product of two rational numbers is always a rational number, and multiplication of rational numbers is certainly associative. Furthermore, 1 is the identity for the multiplication of rational numbers. However, there is no rational number that acts as a multiplicative inverse for 0, since one cannot divide by zero. Hence (Q, x) is not a group.

On the other hand, if r is a nonzero rational number, then so is $1/r$. And $1/r$ is the multiplicative inverse of r. So $(Q-\{0\}, \times)$ is a group. □

Example 3 Is the set $Z_n=\{0, 1, 2, \ldots, n-1\}$ of remainders upon division by the integer n, with the binary operation of addition modulo n, a group?

Recall from Section 1.8 that Z_n is closed under addition modulo n. We will leave it to the reader to show that Z_n is associative, that 0 is its identity, and that each element in Z_n has an inverse. Hence $(Z_n, +(\text{mod } n))$ is a group for every positive integer n. □

Example 4 Let G consist of all strings of the binary digits 0 and 1 of length n, and for two such strings x and y, let $x*y$ be the string z of length n obtained as follows: The ith digit of z is a 0 if x and y have the same digit in the ith place, but it is a 1 if x and y differ in the ith place. (For example, $0110*1011=1101$.) Is $(G, *)$ a group?

It is clear that G is closed under operation $*$, since two strings x and y of length n yield a string $x*y$ of length n. The identity is just the string of length n having all zeros, and each string is its own inverse. (For example, check that $1011*0000=1011$ and that $1011*1011=0000$.) The associative law is verified by observing that for both $a*(b*c)$ and $(a*b)*c$ the ith entry is a 0 if and only if all of the ith entries of a, b, and c are the same. Hence $(G, *)$ is a group. We will have more to say about this group and its application to coding theory in the next section. □

Groups can have infinitely many or finitely many elements, as the previous four examples show. Moreover, in each of these examples we have $a*b=b*a$ for every element in the group. In other words, the operation in each of these groups is commutative. Such groups are called **Abelian**. Not all groups, however, are Abelian groups.

Example 5 Consider the set S_3 of all one-to-one correspondences of the set $\{1, 2, 3\}$ onto itself. If f and g are two such correspondences, let $f*g$ be the composition (formerly written "$g \circ f$") of these two functions, where f maps first, followed by g. [For example, if $f(1)=2$, $f(2)=1$, and $f(3)=3$, and if $g(1)=3$, $g(2)=2$, and $g(3)=1$, then $f*g(1)=g(f(1))=g(2)=2$, $f*g(2)=g(f(2))=g(1)=3$, and $f*g(3)=g(f(3))=g(3)=1$.] Is $(S_3, *)$ a group? Is it Abelian?

First of all, composition of functions is an associative operation (see Exercise 7.7-18) and so $*$ is associative. The identity mapping, I, from $\{1, 2, 3\}$ onto itself is the identity element of S_3. Moreover, the composition of two one-to-one correspondences from $\{1, 2, 3\}$ onto itself is also a one-to-one correspondence, and if f is a one-to-one correspondence, then so is the function f^{-1}. Hence $f^{-1}=f'$. Thus S_3 is a group.

There is a useful way to visualize the group S_3 using an equilateral triangle, as in Fig. 7.28. Let r_1 denote the reflection of the triangle about the axis of symmetry L. Then $r_1(1)=2$, $r_1(2)=1$, and $r_1(3)=3$. Similarly, let r_2 and r_3 denote reflections of the

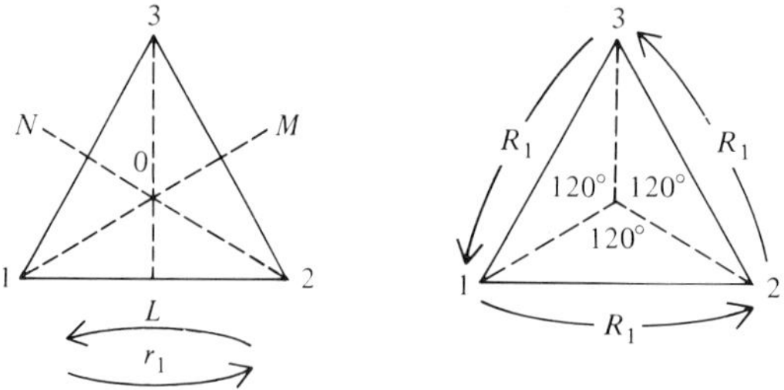

Figure 7.28 Two symmetries of the equilateral triangle, including a 120°-rotation and a reflection.

triangle about the axes of symmetry M and N respectively. Furthermore, let R_1 and R_2 denote counterclockwise rotations of 120 and 240 degrees, respectively, about the point 0. Then the six symmetries I, r_1, r_2, r_3, R_1, and R_2 give us all possible one-to-one correspondences between the vertices of the triangle.

In this group of **symmetries of the equilateral triangle** (or **symmetric group**), S_3, we note that $r_1 * r_2 = R_2$, while $r_2 * r_1 = R_1$. Hence $r_1 * r_2 \neq r_2 * r_1$, and so S_3 is not an Abelian group. □

We define a **semigroup** to be a nonempty set S that is closed under an associative binary operation $*$. A **monoid** is defined to be a semigroup $(S, *)$ that has an identity element. Thus every group is both a semigroup and a monoid, and every monoid is a semigroup. This means that every example of a group that we have given so far is also an example of a semigroup and of a monoid. However, $(N, +)$, the set of positive integers under addition, gives us an example of a semigroup that is neither a group nor a monoid, since the additive identity, 0, does not belong to N. Likewise, the set of all strings that can be formed from a finite set $\{a, b, \ldots, z\}$ gives us a monoid under the operation of concatenation (following one string by another). The identity is the "empty string." Since there are no inverses, however, this system is not a group.

Subgroups, Cosets, and Lagrange's Theorem

A **subgroup**, H, of a group $(G, *)$ is a subset of G that is also a group under the binary operation $*$. Specifically, H is closed under $*$; the associative law holds in H; the identity of G is in H; and if x belongs to H, then x' also belongs to H. As a practical matter, though, *to show that a nonempty subset H of G is a subgroup of G, it is sufficient to show that*

1. *H is closed under $*$, and that*
2. *whenever x is in H, then so is x'.*

The sufficiency of (1) and (2) follows from the observation that, first, if the associative law holds in G, then it must also hold in H, since products in H are also products in G. Second, if x being in H implies that x' is in H, then $x * x' = e$ is also in H by closure. Hence H is a group under $*$ if (1) and (2) can be verified.

Example 6

a) The set of even integers E is a subgroup of $(Z, +)$. This follows from the fact that the sum of two even integers is an even integer, and if n is even, then so is $-n$.

b) The set of rotations $\{I, R_1, R_2\}$ is a subgroup of the group of symmetries of the triangle, S_3. (The mapping I may be considered to be a rotation of 360 degrees.) Moreover, the composition of two rotations is evidently a rotation; and R_1 is the inverse of R_2, since $R_2 * R_1 = R_1 * R_2 = I$.

c) The set of binary strings $\{0000, 1110, 0011, 1101\}$ is a subgroup of the group of binary strings of length $n = 4$ that we defined in Example 4. The reader can check that this set is closed under the operation $*$ that we defined previously. Moreover, we recall that each element is its own inverse. □

We may define an equivalence relation on a group G by means of one of its subgroups, H. We will say that aRb if there is an h in H such that $a = b * h$.

To check that R is an equivalence relation on G, we first observe that $a = a * e$ for every a in G and that e is an element of H. Hence aRa, and R is reflexive. Furthermore, if aRb, then we can write $a = b * h$ for some h in H. But h' is also in H, and it follows that $b = a * h'$, since $a * h' = (b * h) * h' = b * (h * h') = b * e = b$. Hence bRa, and R is symmetric. Finally, if aRb and bRc are true because of $a = b * x$ and $b = c * y$ for some x and y in H, then it follows that $a = c * (y * x)$. Since $y * x$ is in H by closure, we conclude that aRc. Hence R is also transitive, and we have shown that R is an equivalence relation as claimed.

Now given an element g in G, we call the set of elements $g * H = \{g * h\colon h \text{ in } H\}$ a (left) **coset** of the subgroup H. The coset $g * H$ is the equivalence class of the element g. Thus the equivalence relation R partitions the group G into cosets of the form $g * H$. In other words, our equivalence relation has divided up G into disjoint sets $g * H$ whose union is G. Notice that one of the cosets is merely H itself, since $e * H$ consists of the elements $e * h = h$.

There is a natural one-to-one correspondence f between the elements of any two cosets $x * H$ and $y * H$ given by $f(x * h) = y * h$, for each h in H. This mapping from $x * H$ to $y * H$ is one to one because if $y * h_1 = y * h_2$, then multiplying each expression on the left by y' yields $h_1 = h_2$. The mapping is onto $y * H$, since every element in $y * H$ is of the form $y * h$ for some h in H.

Now, if G is a group having finitely many elements, so is any subgroup H of G. Moreover, the discussion given above shows that every coset $g * H$ has the same number of elements as H. Suppose that G is partitioned into exactly r cosets $g * H$. If the number of elements in G—called the **order** of G—is n and if the order of H is s, it follows that $n = rs$. In other words, the discussion we have given may be summed up in the following theorem.

Theorem 7.15 ***Lagrange's Theorem*** If G is a group and H is a subgroup of G, then the left cosets of H partition G into a union of disjoint subsets. Moreover, if G is a group of finite order, then any two cosets of H have same number of elements, and the order of H divides the order of G.

Example 7 What are the cosets of the subgroup $H = \{I, R_1, R_2\}$ of rotations of the group of symmetries S_3?

Let us first observe that $g * H = H$ for all elements g in H, essentially because H is closed. Hence, $I * H = R_1 * H = R_2 * H = H$. Moreover, the reader can check that each of the reflections in S_3 give us the same coset, that is, $r_1 * H = r_2 * H = r_3 * H$. For example, $r_1 * I = r_1$, $r_1 * R_1 = r_3$, and $r_1 * R_2 = r_2$. Hence, $r_1 * H = \{r_1, r_2, r_3\}$.

Thus, the two sets H and $r_1 * H$ are disjoint and their union is clearly S_3.

We also notice that every coset of H has three elements and that 3 divides 6, the order of S_3. □

Example 8 What are the cosets of the subgroup $H = \{I, r_1\}$ of the group S_3?

Since $R_1 * r_1 = r_2$, and $r_2 * r_1 = R_1$, the cosets $R_1 * H$ and $r_2 * H$ both equal the set $\{R_1, r_2\}$. The other two distinct cosets of H are $H = I * H = r_1 * H$ and $R_2 * H = r_3 * H = \{R_2, r_3\}$. Thus the cosets H, $R_1 * H$, and $R_2 * H$ are disjoint and their union is S_3.

Moreover, every coset of H has two elements, and 2 divides 6, the order of S_3. □

We observe that the identity of a group G always gives us a subgroup of G and that G is a subgroup of itself. We call $\{e\}$ and G "trivial" subgroups of G. Lagrange's theorem is often useful in telling us when a finite group does not have a subgroup of a certain order. For example, if G is a group of order 6, then G cannot have a subgroup of order 4, since 4 does not divide 6. More generally, a group of prime order cannot have any nontrivial subgroups. (Why not?) We pursue similar questions in the exercises.

Completion Review 7.7

Complete each of the following.

1. A group is a nonempty set that is ____________ under a binary operation $*$ and satisfies the ____________, ____________, and ____________ laws.
2. A group may have more than one identity element. (*True or False*?)
3. A each element y of a group has exactly one element x in the group, called its ____________, such that $y * x = x * y =$ ____________.
4. All groups obey the commutative law. (*True or False*?)
5. An example of a non-Abelian group is ____________.
6. An algebraic structure consisting of a set and a binary operation that obeys the laws of closure and associativity is called a ____________. If the structure has an identity element as well, then it is called a ____________.
7. A subset H of a group G that is itself a group under the operation of G is called a ____________ of G. If g is an element of G, then the set $g * H =$ ____________ is called a ____________.

8. The set of left cosets of a group ______________ the group. Moreover, ______________ states that there is a one-to-one correspondence between any two left cosets of the group.

9. If G is a group of finite order and H is subgroup of G, then the order of H ______________ the order of G.

10. A group of prime order has no nontrivial subgroups. (*True or False*?)

Answers: **1.** closed; associative; identity; inverse. **2.** False. **3.** inverse; the identity, e. **4.** False. **5.** S_3, the symmetries of the equilateral triangle. **6.** semigroup; monoid. **7.** subgroup; $\{g*h$: for all h in $H\}$; left coset of G. **8.** partitions; Lagrange's theorem. **9.** divides. **10.** True.

Exercises 7.7

1. Find the inverse of each element in Z_4, the group of integers modulo 4 under addition.

2. Give a table showing how to form $a*b$ for each pair of elements in the group of binary strings {0000, 1110, 0011, 1101} under the $*$ operation that we defined in Example 4.

3. Write an operations table for the group S_3 of symmetries of the equilateral triangle.

For Exercises 4 to 10 determine whether the operation defined on the given set defines a group, an Abelian group, a semigroup, or a monoid.

4. The set nZ = the set of integral multiples of the integer n, under ordinary addition.

5. The set $-N$ = the negative integers, under ordinary addition.

6. The set of integers under multiplication.

7. The set of positive real numbers under multiplication.

8. The set of two-by-two matrices with integer entries under **(a)** addition of matrices; **(b)** multiplication of matrices.

9. The set of binary strings {0000, 1110, 0011} under the operation $*$ defined in Example 4.

10. The subsets of a set S under **(a)** union; **(b)** intersection.

11. Explain why a group of prime order cannot have nontrivial subgroups.

12. Show that the set of integral multiples of 3, $3Z$, is a subgroup of $(Z, +)$.

13. Find all the cosets of the subgroup 3Z in Z.

14. Can a set of order 10 have subgroups of orders 3, 4, 6, 7, 8, or 9? Explain.

15. Find all the subgroups of S_3.

16. Prove that if H and K are subgroups of G, then $H \cap K$ is a subgroup of G.

17. Give an example to show that if H and K are subgroups of G, then $H \cup K$ need not be a subgroup of G.

18. Show that when all the compositions are well defined, then composition of functions is an associative operation. That is, show that $f\circ(g\circ h)=(f\circ g)\circ h$ for the functions f, g, and h.
19. Show that if e is the identity of a group G and f is an identity of a subgroup of G, then $f=e$.

7.8 GROUP CODES

In this section we consider the problem of transmitting and receiving coded messages in the form of binary strings, that is, sequences of 0's and 1's. Each string will represent a **word**, and we will say that a **code** is a collection of words that are used to represent distinct messages. We will confine our attention to **block codes**, which are codes in which all the words have the same length, n.

One often thinks of coding in terms of keeping data secret. But an equally important problem is that of transmitting data so that, in spite of some information being lost along the way, the intended message can still be made out at the point of destination. For example, interference might cause some of the 1's in a word to become 0's and some of the 0's to become 1's. A process by which we recover the transmitted word is known as **error correction**.

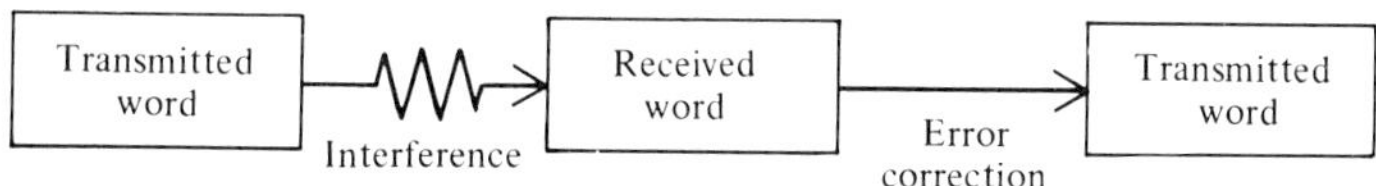

Figure 7.29 The error-correction process.

To take a simple example, suppose that we encode the instructions "buy" and "sell" some stocks as the words 0000 and 1111 respectively. If the instruction to "buy" is somehow garbled in transmission and is received as 1000, what should the receiver do? Since 1000 is, intuitively speaking, more like 0000 than 1111 (three 0's versus one 1), a reasonable course of action might be to buy rather than sell.

Now let us make the intuitive idea of our little example more precise. We will denote by S the set of all binary sequences of length n. If x and y are words in S, we define $x*y$ to be that word in S that has a 1 in the ith position iff x and y differ in that position. For example, $(1000)*(1110)=0110$, since 1000 and 1110 differ in the second and third positions. In the preceding section we showed that *$(S, *)$ is a group in which each word is its own inverse and having the string of n zeros as its identity element*. (See Example 4, Section 7.7.)

We define the **weight, $w(x)$**, of a word x to be the number of 1's in x. For example, $w(0111)=3$ and $w(1000)=1$. The weight $w(x*y)$ of $x*y$ is evidently the *number of places in which the words x and y differ*. It is natural to use this quantity as a measure of how "far apart" x and y are. Therefore, we will define the **distance, $d(x, y)$**, between x and y by the formula

$$d(x, y)=w(x*y).$$

For example, the distance between 1000 and 0000 is 1, and the distance between 1000 and 1111 is 3.

Now suppose that we want to decode a received word y; that is, we want to determine which transmitted word among all the possible transmitted words $x_1, \ldots, x_n$ corresponds to y. The **minimum-distance decoding criterion** states that if one computes $d(x_i, y)$ for $i = 1, 2, \ldots, n$, then one should choose an x_i for which $d(x_i, y)$ is a minimum. In our last example we chose 0000 rather than 1111 as the transmitted word because the weight of $0000 * 1000$ was smaller than the weight of $1111 * 1000$. Of course, the minimum-distance criterion may lead to ambiguous choices, as when the received word is 1001, a word that is just as close to 0000 as it is to 1111.

An important property of the distance measure that sometimes allows us to avoid such ambiguities is contained in the following theorem. It is called the "triangle inequality."

Theorem 7.15 For every x, y, and z in G we have $d(x, y) \leqslant d(x, z) + d(z, y)$.

Proof: If we compare the contributions in the ith places of x, y and $x * y$ with the weights $w(x)$, $w(y)$, and $w(x * y)$, we quickly realize that

$$w(x * y) \leqslant w(x) + w(y).$$

Moreover, since, for every z in G, $z * z = e$, the identity of G, we can write $x * y = x * (z * z) * y = (x * z) * (z * y)$. Hence

$$w(x * y) = w((x * z) * (z * y)) \leqslant w(x * z) + w(z * y)$$

or

$$d(x, y) \leqslant d(x, z) + d(z, y),$$

as claimed. ∎

Let us now define the **minimum distance of a code C** to be the minimum distance between any two distinct codewords in C. Then the preceding theorem enables us to prove the following.

Theorem 7.16 If one follows the minimum-distance decoding criterion for a code of minimum distance $2k + 1$, then one can correct k or fewer transmission errors in each word of the code.

Proof: Suppose that code word x were transmitted and a word y received such that

$$d(x, y) \leqq k.$$

We have to show that if z is any other word in our code, then x rather than z will be selected. That is, we have to show that z will not be selected.

By our assumption about the minimum distance of the code, we may write

$$d(x, z) \geqq 2k + 1.$$

But our previous theorem tells us that

$$d(x, z) \leqq d(x, y) + d(y, z),$$

or

$$d(x, z) - d(x, y) \leqq d(y, z).$$

Since

$$k + 1 = (2k + 1) - k \leqq d(x, z) - d(x, y)$$

implies $k + 1 \leqq d(z, y)$, we conclude that $d(x, y) < d(z, y)$, and so x will be selected rather than z. ∎

Example 1 What is the minimum distance of the code (a) {0000, 1111}? (b) Of the code {00000, 11111}? What are the error-correcting capacities of each of these codes?

a) The minimum distance of the code {0000, 1111} is 4. As noted, however, this code can only correct a single transmission error per word.

b) The minimum distance of the code {00000, 11111} is 5. Since $5 = (2)(2) + 1$, the error-correcting capacity of this code is 2. For example, each of 11000, 10100, and 00101 would be decoded as 00000, while each of 11011, 10111, and 00111 would be decoded as 11111. □

We will say that a subset G of the set S of strings of length n is a **group code** if $(G, *)$ is a subgroup of $(S, *)$. One reason why group codes are convenient to work with is contained in the following theorem.

Theorem 7.17 The minimum distance of a group code G is equal to the minimum weight of the nonzero words in G.

Proof: First of all, given any nonzero word x in G,

$$w(x) = d(x, e) \geqq \underset{\substack{y, z \in G \\ y \neq z}}{\text{minimum}} [d(y, z)].$$

because the zero word, e, is in G. But for any y and z in G, $y * z$ is also in G. Since $w(y * z) = d(y, z)$, we also can write

$$d(y, z) \geqq \underset{\substack{x \in G \\ x \neq e}}{\text{minimum}} [w(x)].$$

Combining our two inequalities tells us that

$$\underset{\substack{x \in G \\ x \neq e}}{\text{minimum}} [w(x)] = \underset{\substack{y, z \in G \\ y \neq z}}{\text{minimum}} [d(y, z)].$$ ■

Example 2 What is the minimum distance in the subgroup $G = \{00000, 11111, 11000, 00111\}$ of the group of strings of length 5?

Rather than calculating the distances between the $C(4, 2) =$ six pairs of words, it suffices to note that 11000 is the nonzero element of minimum weight in G, namely, 2. Hence 2 is the minimum distance of G. □

Let $(G, *)$ be a group code and suppose that y is a received word. If we let $x = x_0$, then we observe that $d(y, x_0)$ will be the minimum of the distances $d(y, x)$ for all x in G if and only if $y * x_0$ is a word of minimum weight in the coset $y * G$. Therefore, the minimum-distance decoding criterion tells us we may take $x_0 = y * (y * x_0)$ as the transmitted word. These observations lead to the following procedure for decoding a received word when we have a group code.

Algorithm 7.2 If $(G, *)$ is a group code and y is a received word, to find the corresponding transmitted word by means of the minimum-distance criterion:

1. Find the coset $y * G$ in G.
2. Find an element g of minimum weight in $y * G$. (This element is called the "leader of $y * G$.")
3. Output $y * g$ as the transmitted word and stop.
4. End of Algorithm 7.2.

Example 3 Suppose that $G = \{00000, 10101, 01010, 11111\}$ and that we have received the words $y = 11010$ and $z = 11101$. What should we take as the transmitted words?

The relevant cosets are

$$y*G=\{10000, 00101, 11010, 01111\}$$

and

$$z*G=\{01000, 11101, 0010, 10111\},$$

and we may take the leader of each coset to be the first element listed in each case. (Notice, however, that the third element in $z*G$ could also have been chosen as the leader in the second coset.)

Therefore, the transmitted word corresponding to y is given by our algorithm as

$$y*10000=11010*10000=01010,$$

which is the third element listed in G.

The transmitted word corresponding to z is given by our algorithm as

$$z*01000=11101*01000=10101,$$

which is the second element listed in G.

Observe that if we had chosen the element 00010 to be the leader in $z*G$, we would have obtained

$$z*00010=11101*00010=11111$$

as the transmitted word. This ambiguity arises because the distance between z and each of 10101 and 11111 is only 1. The minimum distance of the group G, however, is 2. By Theorem 7.17 the error-correcting capacity of the group is, therefore, zero. To be able to correct a single error in transmission, our theorem requires that the minimum distance of a group be at least $(2)(1)+1$. □

Completion Review 7.8

Complete each of the following.

1. A code, as defined in this section, is a collection of words, or________________. In block codes all the words have the same________________.

2. The process by which we recover a transmitted word from a received word is known as________________.

3. If x and y are words in a block code S, we defined $x*y=$________________. Moreover, we defined $w(e)=$________________ and $d(x, y)=$________________.

4. If $S = \{x_1, x_2, \ldots, x_n\}$ and y is a received word from S, then the minimum-distance decoding criterion states that one should choose x_i as the transmitted word if ________________.

5. The triangle inequality in S states that________________.

6. The minimum distance of a code G is the________________.

7. If the minimum distance of a code is $2k+1$ and one uses the minimum-distance decoding criterion, then one can correct________________ in each word.

8. The minimum distance of a group code is equal to the________________ of the non-zero words in G.

9. If $(G, *)$ is a group code and y is a received word, then our algorithm instructs us to first find the________________ of G. Then we find an element________________ called the leader of the coset. Finally, we output________________ as the transmitted word.

Answers: **1.** binary strings; length. **2.** error correction. **3.** that word in S that has a 1 in the ith position iff x and y differ in that position; the number of 1's in x; $w(x*y)$. **4.** $d(x_i, y)$ is a minimum for $i = 1, 2, \ldots, n$. **5.** $d(x, y) \leqslant d(x, z) + d(z, y)$. **6.** minimum $d(x, y)$ over all distinct x and y in G. **7.** k or fewer errors. **8.** minimum weight. **9.** coset $y*G$; g of minimum weight; $y*g$.

Exercise 7.8

1. What are all the words of length 4 having a weight of **(a)** 1; **(b)** 2; **(c)** 3; **(d)** 4?
2. What are all the words of length 5 having a weight of **(a)** 0; **(b)** 2; **(c)** 4?
3. Find all the words y of length 4 such that $d(x, y) = 2$ if $x = 0011$.
4. Find all the words y of length 5 such that $d(x, y) = 1$ if $x = 01011$.
5. What is the minimum distance of the group code $\{0000, 0011, 1101, 1110\}$?
6. What is the minimum distance of the group code $\{000000, 001111, 010011, 011100, 100110, 101001, 110101, 111010\}$?
7. What is the maximum number of transmission errors that can be corrected with the group of Exercise 5? Why?
8. What is the maximum number of transmission errors that can be corrected with the group of Exercise 6? Why?
9. What are the cosets of each of the following elements in the group of Exercise 5, and what are the leaders in each coset? **(a)** 0011; **(b)** 1011; **(c)** 1100.
10. What are the cosets of each of the following elements in the group of Exercise 6, and what are the leaders of each coset? **(a)** 001111; **(b)** 100000; **(c)** 110110.

11. Use the code of Exercise 5 to determine the transmitted word, if the received word is **(a)** 1101; **(b)** 0101; **(c)** 1100.

12. Use the code of Exercise 6 to determine the transmitted word if the received word is **(a)** 100110; **(b)** 010101; **(c)** 110110.

COMPUTER PROGRAMMING EXERCISES

(Also see the appendix and Programs A23 to A24.)

7.1. Write a program that finds the Boolean product of two matrices.

7.2. Write a program that finds the Boolean sum of two matrices.

In Exercises 3 to 9 we assume that the relation R is on finite set A and is given by a relation matrix.

7.3. Write a program that finds the relation matrix of the n-fold composition of R with itself, R^n.

7.4. Write a program that determines if a relation R is reflexive.

7.5. Write a program that determines if a relation R is symmetric.

7.6. Write a program that determines if a relation R is transitive. Have your program also give the transitive closure of R.

7.7. Write a program that determines if a relation R is an equivalence relation. In the event that R is an equivalence relation, have your program list the equivalence classes of R.

7.8. Write a program that determines if a relation R is a partial ordering of A.

7.9. Write a program that finds a maximal antichain in a poset (A, R).

7.10. Write a program that inputs a Boolean expression in x, y, and z and prints out its truth table.

7.11. Write a program that finds the disjunctive normal form of a Boolean expression in x, y, and z.

7.12. Write a program that finds a minimum sum of products form for a Boolean expression in disjunctive normal form.

CHAPTER 8

FURTHER TOPICS IN COUNTING AND RECURSION

The techniques we discussed in Chapter 3 for solving counting problems required no more than simple elementary algebra. In this chapter, however, we discuss methods that demand somewhat more from the reader in the way of mathematical background. Beginning with Section 8.3, for example, we require some familiarity with rules for manipulating polynomials and power series, which are used in the powerful counting techniques of "generating functions." Indeed we use these "generating functions" to find closed-form solutions to recurrence relations. Sections 8.1 and 8.2 also discuss closed-form solutions to certain recurrence relations. An acquaintance with methods for finding roots of polynomials would be advisable for these two sections.

8.1 HOMOGENEOUS LINEAR RECURRENCE RELATIONS

We have seen that one can write certain recurrence relations in closed form. For example, recall that the solution to the recurrence relation with initial condition

$$a_n = r \cdot a_{n-1} \quad \text{for } n \geqslant 1,$$
$$a_0 = k$$

for constants r and k, can be written in the closed form

$$a_n = k \cdot r^n \quad \text{for all } n \geqslant 0.$$

That is, the general term, a_n, can be given directly as a function of the variable n instead of in terms of the previous term a_{n-1}. Some methods for obtaining closed-form solutions, such as the method of backward substitution that we discussed in Section 3.7, rely heavily on insight, or, in other words, educated guessing at a solution. Mechanical methods for obtaining closed-form solutions, however, are only known for certain classes of recurrence relations. The following type occurs quite frequently in applications, and so we include it here.

Definition 8.1 An equation of the form

$$k_0 a_n + k_1 a_{n-1} + \cdots + k_m a_{n-m} = f(n) \tag{8.1}$$

where the k_i are constants and $f(n)$ is a function of the variable n is called a (a) **linear recurrence relation with constant coefficients**. The recurrence relation is of (b) **order m** provided that k_0 and k_m are not zero. If $f(n)$ is identically zero, then we say that the recurrence relation is (c) **homogeneous**. Otherwise it is (d) **non-homogeneous**. We say that the left-hand side of Eq. (8.1) set equal to zero is the (e) **homogeneous part** of the recurrence relation.

Equation (8.1) defines the nth term, a_n, of the recurrence relation in terms of the preceding m terms $a_{n-1}, a_{n-2}, \ldots, a_{n-m}$, and the function f is evaluated for the integers n. The recurrence relation is called "linear" by analogy with linear equations such as $3x+2y=1$, where the variables x and y only occur in the first degree.

Example 1

a) $4a_n+3a_{n-1}-2a_{n-2}=3n+1$ is a second-order linear recurrence relation with constant coefficients 4, 3, and -2. The general term a_n is defined in terms of the two preceding terms a_{n-1} and a_{n-2}, and by the function $f(n)=3n+1$. Since this function is not identically zero for all values of n, the recurrence relation is nonhomogeneous.

b) $4a_n+3a_{n-1}-2a_{n-2}=0$ is a homogeneous second-order linear recurrence relation. In fact it is the homogeneous part of the relation in Example 1(a).

c) The recurrence relation $a_n+n\cdot a_{n-1}=1$ has variable as well as constant coefficients, since the coefficient of a_{n-1} is the variable n. (However, we call this relation **linear** because all the a_i terms are of the first degree.)

d) $a_n=(a_{n-1})^2$ is not linear because a_{n-1} occurs in the second degree. Similarly, $a_n=a_{n-1}\cdot a_{n-2}$ is not linear.

e) $a_n=a_{n-1}+a_{n-3}$ is a linear recurrence relation with constant coefficients of order 3, since it can be put into the form

$$a_n-a_{n-1}+0\cdot a_{n-2}-a_{n-3}=0. \qquad \square$$

To see how to solve linear recurrence relations with constant coefficients, let us consider the recurrence relation that gives us the Fibonacci numbers. (See Example 3, Section 3.6.)

Example 2 Find the closed-form solution to the recurrence relation

$$a_n=a_{n-1}+a_{n-2}, \quad \text{for } n\geqslant 2,$$

with the initial conditions $a_0=1$ and $a_1=1$. (From now on the word "solution" will mean "closed-form solution.")

We first observe that our recurrence relation $a_n-a_{n-1}-a_{n-2}=0$ is similar in form to a second-degree polynomial. Moreover, if we recall our solution to the recurrence relation $a_n=r\cdot a_{n-1}$, a natural guess at a solution in the present example is $a_n=r^n$, where r is the root of the polynomial

$$x^2-x-1=0.$$

Indeed if we substitute $a_n=r^n$ on the left-hand side of our recurrence relation, we obtain

$$a_n-a_{n-1}-a_{n-2}=r^n-r^{n-1}-r^{n-2}=r^{n-2}(r^2-r-1)=0,$$

for any r such that $r^2-r-1=0$.

Now the polynomial $x^2 - x - 1$ has the two distinct roots,

$$r = \frac{1+\sqrt{5}}{2} \quad \text{and} \quad s = \frac{1-\sqrt{5}}{2},$$

neither of which satisfy both of our initial conditions $a_0 = 1$ and $a_1 = 1$. However, if $a_n = r^n$ and $a_n = s^n$ are both solutions of $a_n - a_{n-1} - a_{n-2} = 0$, then for any constants A and B, we have

$$(Ar^n + Bs^n) - (Ar^{n-1} + Bs^{n-1}) - (Ar^{n-2} + Bs^{n-2}) =$$
$$A(r^n - r^{n-1} - r^{n-2}) + B(s^n - s^{n-1} - s^{n-2}) =$$
$$0 + 0 = 0.$$

Hence $a_n = Ar^n + Bs^n$ is also a solution. In fact we can find a solution of the form

$$a_n = A\left(\frac{1+\sqrt{5}}{2}\right)^n + B\left(\frac{1-\sqrt{5}}{2}\right)^n \tag{8.2}$$

that satisfies our initial conditions.

Substituting $n = 0$ and $n = 1$ in Eq. (8.2), we obtain the system of equations

$$1 = A\left(\frac{1+\sqrt{5}}{2}\right)^0 + B\left(\frac{1-\sqrt{5}}{2}\right)^0$$
$$1 = A\left(\frac{1+\sqrt{5}}{2}\right)^1 + B\left(\frac{1-\sqrt{5}}{2}\right)^1.$$

Upon solving this system of equations for A and B, we find that

$$A = (1 + \sqrt{5})/(2\cdot\sqrt{5}) \quad \text{and} \quad B = -(1 - \sqrt{5})/(2\cdot\sqrt{5}).$$

Substituting these values of A and B in Eq. (8.2), we obtain

$$a_n = \frac{1}{\sqrt{5}}\left(\frac{1+\sqrt{5}}{2}\right)^{n+1} - \frac{1}{\sqrt{5}}\left(\frac{1-\sqrt{5}}{2}\right)^{n+1}$$

as the closed form of the nth Fibonacci number. □

The equation $x^2 - x - 1 = 0$ that we obtained by looking at the recurrence relation $a_n - a_{n-1} - a_{n-2} = 0$ is called the "characteristic equation" of the relation. Finding its roots was the key to solving the recurrence relation. More generally, one can show that the following result is true.

Theorem 8.1 ***Homogeneous Case*** Given the relation $k_0a_n + k_1a_{n-1} + \cdots + k_ma_{n-m} = 0$, if the roots $r_1, r_2, \ldots, r_m$ of its **characteristic equation**

$$k_0x^m + k_1x^{m-1} + k_2x^{m-2} + \cdots + k_m = 0$$

are all *distinct*, then the general form of the solutions of the recurrence relation is given by

$$a_n = A_1(r_1)^n + A_2(r_2)^n + \cdots + A_m(r_m)^n,$$

where the constants A_i are determined by m initial conditions.

A proof that Theorem 8.1 gives us the correct form of solutions of the recurrence relation directly parallels the discussion of the closed form of the Fibonacci numbers, and so we omit it. Instead let us illustrate how the theorem is used in the following examples.

Example 3 Find the solution to $(1/2)a_n - a_{n-1} = 0$ for $n > 0$, and $a_0 = 2$.

The characteristic equation of this recurrence relation, as defined in the statement of Theorem 8.1, is easily found by writing $a_n = x^n$ and substituting in the recurrence relation. This gives us

$$(1/2)x^n - x^{n-1} = 0.$$

Dividing by x^{n-1}, which we assume is not zero, we get the characteristic equation

$$(1/2)x - 1 = 0.$$

The only root of this equation is $r = 2$. Hence the general form of the solutions we seek is

$$a_n = A \cdot (2)^n.$$

Since $a_0 = 2$, we can write

$$2 = A \cdot 2^{\circ} = A.$$

Therefore, the closed form of the solution is

$$a_n = 2 \cdot 2^n = 2^{n+1}, \quad \text{for } n \geqslant 0.$$

□

Example 4 Solve the recurrence relation $a_n + 2a_{n-1} - a_{n-2} - 2a_{n-3} = 0$, given the initial conditions $a_0 = a_1 = 0$ and $a_2 = 1$.

The characteristic equation is found by writing

$$x^n + 2x^{n-1} - x^{n-2} - 2x^{n-3} = 0,$$

or

$$x^3 + 2x^2 - x - 2 = 0.$$

Using the rational root test (see Exercise 8.1-8), for example, we quickly find the roots to be $r_1 = 1$, $r_2 = -1$, and $r_3 = -2$. The general form of our solution is, by Theorem 8.1,

$$a_n = A_1(1)^n + A_2(-1)^n + A_3(-2)^n.$$

Substituting the initial conditions such as $a_0 = 0$ [so that $0 = A_1(1)^0 + A_2(-1)^0 + A_3(2)^0$] in the foregoing gives us the system of equations

$$0 = A_1 + A_2 + A_3,$$

$$0 = A_1 - A_2 - 2A_3,$$

$$1 = A_1 + A_2 + 4A_3.$$

Solving these equations simultaneously, we obtain $A_1 = 1/6$, $A_2 = -1/2$, and $A_3 = 1/3$. Hence our closed-form solution is

$$a_n = (1/6)(1)^n + (-1/2)(-1)^n + (1/3)(-2)^n.$$ □

If the roots of the characteristic equation of a recurrence relation are not distinct, then the general form of the solution is somewhat different than that given in Theorem 8.1. Recall that "r is a root of multiplicity k of a polynomial $p(x)$" if $(x - r)^k$ is a factor of $p(x)$. Then we can state the following theorem.

Theorem 8.2 If r is a root of multiplicity k of the characteristic equation in Theorem 8.1, then

$$(B_1 n^{k-1} + B_2 n^{k-2} + \cdots + B_{k-1} n + B_k) r^n$$

is a solution of the corresponding homogeneous recurrence relation, for some constants $B_1, B_2, \ldots, B_k$ determined by the initial conditions.

Theorem 8.2 says that we must determine k constants for each of the roots of the characteristic equation that are repeated exactly k times. We illustrate how this is done in the following example.

Example 5 Solve the recurrence relation $a_n - 2a_{n-2} + a_{n-4} = 0$, given the initial conditions $a_0 = 2$, $a_1 = a_3 = 0$, and $a_2 = 6$.

We first obtain the characteristic equation

$$x^4 - 2x^2 + 1 = 0$$

and notice that the polynomial factors into

$$x^4 - 2x^2 + 1 = (x+1)^2(x-1)^2.$$

Hence our characteristic equation has the roots $+1$ and -1, each of multiplicity 2. By Theorem 8.2, the general form of our solution is, therefore,

$$a_n = (B_1 n^1 + B_2)(1)^n + (B_3 n^1 + B_4)(-1)^n.$$

We now determine the four constants B_i by means of our four initial conditions. Substituting as before, we obtain a system of four equations

$$2 = B_1 \cdot 0 + B_2 + B_3 \cdot 0 + B_4,$$

$$0 = B_1 \cdot 1 + B_2 - B_3 \cdot 1 - B_4,$$

$$6 = B_2 \cdot 2 + B_2 + B_3 \cdot 2 + B_4,$$

$$0 = B_2 \cdot 3 + B_2 - B_3 \cdot 2 - B_4.$$

This system has the solution $B_1 = B_2 = B_3 = B_4 = 1$. Hencc

$$a_n = (1 \cdot n + 1)(1)^n + (n \cdot 1 + 1)(-1)^n$$

is the solution to our recurrence relation. □

Completion Review 8.1

Complete each of the following.

1. A linear recurrence relation with constant coefficients has the form________________. It is called linear because all the________________.
2. The relation is of order m if________________. It is homogeneous if________________ ________________. Otherwise it is called________________. The homogeneous part is________________.

3. The characteristic equation of the relation in the first question is ________.

4. If the characteristic equation of a homogeneous linear recurrence relation with constant coefficients has distinct roots r and s, then its general solution has the form ________.

5. If r is root of multiplicity three in the fourth question, then ________ is a solution.

Answers: **1.** $k_0a_n + k_1a_{n-1} + \cdots + k_ma_{n-m} = f(n)$; terms are of degree 1 in the a_i's. **2.** k_0 and k_m are not zero; $f(n)$ is identically zero; nonhomogeneous; $k_0a_n + k_1a_{n-1} + \cdots k_ma_{n-m} = 0$. **3.** $k_0x^m + k_1x^{m-1} + \cdots + k_m = 0$. **4.** $Ar^n + Bs^n$. **5.** $(An^2 + Bn + C)r^n$.

Exercises 8.1

1. Which of the following recurrence relations are linear?
   a) $a_n + a_{n-1} = a_{n-2} + n^2$.
   b) $(a_n)^2 - a_{n-1} = 1$.
   c) $a_na_{n-1} = 1$.
   d) $na_n + (n-1)a_{n-1} = a_{n-2}$.
   e) $(a_n + a_{n-1})^2 = a_{n-2}$.
2. Which of the following is a linear recurrence relation with constant coefficients? And which of the latter are homogeneous?
   a) $a_n + a_{n-2} = a_{n-1} + a_{n-3}$.
   b) $a_n = n$.
   c) $a_n + na_{n-1} = 1$.
   d) $a_n + 2a_{n-1} + 3a_{n-2} = n - 3$.
   e) $a_n = 2^n$.
   f) $a_n = 2^n + n$.
3. Find the characteristic equation and the form of solutions of each of the following recurrence relations.
   a) $a_n - a_{n-1} = 0$.
   b) $a_n = 3a_{n-2} - 2a_{n-1}$.
   c) $a_n = 8a_{n-2} - 16a_{n-4}$.
4. Find a homogeneous linear recurrence relation with constant coefficients having as its characteristic equation:
   a) $r^2 + 5r + 6 = 0$.
   b) $r^{10} + r^5 + r = 0$.
   c) $(r+1)^3 = 1$.
5. Find the particular solutions in Exercise 3 that have the corresponding initial conditions:
   a) $a_0 = 2$.
   b) $a_0 = 1$ and $a_1 = 2$.
   c) $a_0 = 1$, $a_1 = 2$, $a_2 = 0$, and $a_3 = 2$.

6. Find the particular solutions in Exercise 3 that have the corresponding initial conditions:
   a) $a_0 = -2$.
   b) $a_0 = 2$ and $a_1 = 1$.
   c) $a_0 = 2$, $a_2 = 24$, and $a_1 = a_3 = 0$.
7. Verify that neither $[(1+\sqrt{5})/2]^n$ nor $[(1-\sqrt{5})/2]^n$ satisfies both of the initial conditions giving the Fibonacci numbers.
8. **(Rational root test)** Prove that the polynomial $a_0 + a_1x + a_1x^2 + \cdots + a_nx^n$ has a rational root of the form p/q in lowest terms, where a_i, p, and q are integers, only if p divides a_0 and q divides a_n. [*Hint*: Substitute p/q for x and then multiply by q^n.]
9. Use Exercise 8 to find the roots of the following.
   a) $x^3 - 7x - 6 = 0$.
   b) $16x^4 - 72x^2 + 81 = 0$.
   c) $8x^3 + 4x^2 + 2x + 1 = 0$.
10. Find the solution to the recurrence relation $a_n - 7a_{n-2} - 6 = 0$, where $a_0 = 3$, $a_1 = 4$, $a_2 = 14$.
11. Find the solution to $a_n + 6a_{n-1} - 7a_{n-2} - 6a_{n-3} = 0$, where $a_0 = a_1 = a_2 = 1$.
12. The number of ways to give away a \$1 bill or a \$5 bill on successive days until we have given a total of \$$n$ is given by $a_n = a_{n-1} + a_{n-5}$, for $n \geqslant 5$, with $a_i = 1$ for $n = 1, 2, 3, 4$, and $a_5 = 2$.
   a) Find the characteristic equation of this recurrence relation.
   b) Show that the characteristic equation does not have any rational roots.
13. Show that when n is large, F_n, the nth Fibonacci number, is approximately equal to $(1/\sqrt{5})[(1+\sqrt{5})/2)]^n$. [*Hint*: See Example 2 and notice that $|(1-\sqrt{5})/2| < 1$.]
14. Show that when n is large, F_{n+1}/F_n is approximately equal to $(1+\sqrt{5})/2$, the so-called "golden mean," also called the "divine proportion." (*Hint*: See Exercise 13.)

8.2 THE NONHOMOGENEOUS CASE

In this section we continue our discussion of linear recurrence relations with constant coefficients. Having discussed the homogeneous case in the previous section, we now turn to the general situation. The following rule allows us to use our discussion of the homogeneous case as a stepping stone to the solution of the general case.

Theorem 8.3 ***Nonhomogeneous Case*** The general solution of a linear recurrence relation with constant coefficients is the sum of a particular solution (denoted $a_n^{(p)}$) and the general solution (denoted $a_n^{(h)}$) of the homogeneous part. In other words,

$$a_n = a_n^{(p)} + a_n^{(h)}.$$

Partial Proof: To see why $a_n^{(p)}+a_n^{(h)}$ is a solution to the nonhomogeneous recurrence relation $k_0a_n+k_1a_{n-1}+\cdots+k_ma_{n-m}=f(n)$, we substitute $a_i^{(p)}+a_i^{(h)}$ for a_i, $i=n, n-1, \ldots, n-m$, obtaining

$$\begin{aligned}
&k_0(a_n^{(p)}+a_n^{(h)})+k_1(a_{n-1}^{(p)}+a_{n-1}^{(h)})+\cdots+k_m(a_{n-m}^{(p)}+a_{n-m}^{(h)})\\
&\quad=(k_0a_n^{(p)}+k_1a_{n-1}^{(p)}+\cdots+k_ma_{n-m}^{(p)})+(k_0a_n^{(h)}+k_1a_{n-1}^{(h)}+\cdots+k_ma_{n-m}^{(h)})\\
&\quad=(k_0a_n^{(p)}+k_1a_{n-1}^{(p)}+\cdots+k_ma_{n-m}^{(p)})+0=f(n).
\end{aligned}$$

Since we obtain $f(n)$, $a_n^{(p)}+a_n^{(h)}$ is a solution, as claimed. (We will not show, however, that it is the only solution.) ■

While there is no completely systematic way to find a particular solution of a linear recurrence equation with constant coefficients, for certain types of functions $f(n)$ we can, at least, suggest the correct form. Table 8.1 gives a few of these functions.

Table 8.1

$f(n)$	Particular Solution	Condition
A polynomial in n of degree d	A polynomial in n of degree d	None
$A\cdot k^n$	$B\cdot k^n$	k not a root of the characteristic polynomial
$A\cdot k^n$	$B\cdot n^m\cdot k^n$	k a root of the characteristic polynomial of multiplicity m
$g(n)+h(n)$	Particular solution to $\sum_{i=0}^{m} k_ia_{n-i}=g(n)$ + Particular solution to $\sum_{i=0}^{m} k_ia_{n-i}=h(n)$	None

A, B, k, and d are constants.

Now we illustrate each of these types.

Example 1 Let us find the general solution to the recurrence relation $a_n+3a_{n-1}+2a_{n-2}=n^2$.

First we notice that the homogeneous part of our relation is

$$a_n+3a_{n-1}+2a_{n-2}=0,$$

whose characteristic equation is

$$x^2 + 3x + 2 = 0.$$

The roots of the characteristic equation are $r_1 = -1$ and $r_2 = -2$, giving us the solution to the homogeneous part in the form

$$a_n^{(h)} = A(-1)^n + B(-2)^n.$$

Next $f(n)$ is a polynomial of degree $d = 2$, namely, $f(n) = n^2$. Therefore, Table 8.1 tells us that our particular solution will also be a polynomial in n of degree 2. That is, it will be of the form $a_n^{(p)} = Cn^2 + Dn + E$, where we must determine C, D, and E in order to obtain a particular solution. We do this by substituting $a_i^{(p)}$ in our recurrence relation, $i = n, n-1, n-2$:

$$(Cn^2 + Dn + E) + 3(C(n-1)^2 + D(n-1) + E) + 2(C(n-2) + D(n-2) + E) = n^2$$

or

$$6Cn^2 + (-14C + 6D)n + (11C - 7D + 6E) = 1n^2 + 0n + 0.$$

Equating coefficients of terms of like degree, we get

$$6C = 1, \quad -14C + 6D = 0, \quad \text{and} \quad 11C - 7D + 6E = 0.$$

Solving this system of three equations we get $C = 1/6$, $D = 14/36$, and $E = 32/216$. Hence the general solution is

$$a_n = (1/6)n^2 + (14/36)n + 32/216 + A(-1)^n + B(-2)^n. \qquad \square$$

Example 2 Let us find the solution to the recurrence relation $a_n + 3a_{n-1} + 2a_{n-2} = 3^n$.

The homogeneous part of our recurrence relation is the same as in Example 1, and its solutions are of the form

$$a_n^{(h)} = A(-1)^n + B(-2)^n.$$

Since the function $f(n) = 3^n$, Table 8.1 tells us to look for a particular solution of the form

$$a_n^{(p)} = C(3)^n.$$

Substituting in our recurrence relation, we get

$$C(3)^n + 3C(3)^{n-1} + 2C(3)^{n-2} = 3^n,$$

which we can rewrite as

$$C\cdot(3)^n + C\cdot(3)^n + \frac{2}{9}C\cdot(3)^n = \frac{20}{9}C\cdot(3)^n = 1\cdot(3)^n.$$

Equating coefficients of 3^n, we obtain $C=9/20$. Hence our particular solution is $a_n^{(p)}=(9/20)(3)^n$, and our general solution is

$$a_n = \frac{9}{20}(3)^n + A(-1)^n + B(-2)^n. \qquad \square$$

Example 3 Now let us solve the recurrence relation $a_n + 3a_{n-1} + 2a_{n-2} = (-2)^n$.

The difference between this example and the previous one is that -2 is a root of multiplicity 1 in the characteristic equation of the homogeneous part. Table 8.1 tells us that we should look for a particular solution of the form

$$a_n^{(p)} = Cn^1(-2)^n.$$

Substituting in the recurrence relation, we eventually obtain

$$0Cn(-2)^n + 2C(-2)^n = (-2)^n,$$

and so $C=1/2$. Therefore, the general solution is

$$a_n = \tfrac{1}{2}n(-2)^n + A(-1)^n + B(-2)^n. \qquad \square$$

Example 4 To solve the recurrence relation $a_n + 3a_{n-1} + 2 = n^2 + (-2)^n$, we notice that the homogeneous part is the same as in the last three examples, while the function $f(n)$ is the sum of the functions in Examples 1 and 3. Hence Table 8.1 directs us to take the sum of the particular solutions in those examples as a particular solution to our present recurrence relation. Consequently our general solution is of the form

$$a_n = (1/6)n^2 + (14/36)n + (32/216) + \tfrac{1}{2}n(-2)^n + A(-1) + B(-2)^n. \qquad \square$$

Example 5 Finally, let us solve the recurrence relation $a_n + 3a_{n-1} + 2a_{n-2} = (-2)^n$, $n>1$, $a_0=0$, and $a_1=1$. Observe that in this example we want more than just the form of the solution. In fact we already know the general form of the solution from Example 3. It is

$$a_n = \tfrac{1}{2}(-2)^n\cdot n + A(-1)^n + B(-2)^n.$$

Substituting the values obtained from the initial conditions, we obtain the two equations

$$0 = \tfrac{1}{2}(-2)^0 \cdot 0 + A(-1)^0 + B(-2)^0,$$
$$1 = \tfrac{1}{2}(-2)^1 \cdot 1 + A(-1)^1 + B(-2)^1.$$

The solution to this system of equations is $A = 2$ and $B = -2$. Therefore, the solution to our recurrence relation is

$$a_n = \tfrac{1}{2}(-2)^n \cdot n + 2(-1)^n - 2(-2)^n, \quad n \geq 2.$$ □

Completion Review 8.2

Complete each of the following.

1. The general solution of a linear recurrence relation with constant coefficients is the sum of a ________________ solution and the ________________ of the homogeneous part.
2. If the function $f(n)$ in the recurrence relation is a polynomial of degree n, then a particular solution is ________________. It may be found by means of the ________________.
3. If the function $f(n)$ is of the form $A \cdot k^n$, then there is a particular solution of the form ________________, where k is a root of the characteristic equation of multiplicity m.
4. If $f(n)$ is of the form $g(n) + h(n)$, then there is a particular solution that is a ________________ of the solutions of the equations in which $f(n) = g(n)$ and $f(n) = h(n)$.

Answers: **1.** particular; general. **2.** polynomial of degree n; initial conditions. **3.** $B \cdot n^m \cdot k^n$. **4.** sum.

Exercises 8.2

1. Find the homogeneous part of each of the following recurrence relations.
(a) $a_n - a_{n-1} = n^2$; **(b)** $a_n = 3a_{n-2} - 2a_{n-1} + 2^n$;
(c) $a_n + n = 8a_{n-2} - 16a_{n-4}$.

2. Find a particular solution to each of the recurrence relations in Exercise 1.

3. Find the general solution to each of the recurrence relations in Exercise 1.

4. Find the solutions to Exercise 1 satisfying the corresponding initial conditions:
(a) $a_0 = 2$; **(b)** $a_0 = 1$ and $a_1 = 2$; **(c)** $a_0 = 1$, $a_1 = 2$, $a_2 = 0$, and $a_3 = 2$.

5. Jane Doe deposits \$$P$ into a savings account at the beginning of each year, and the account pays her $100r$ percent interest compounded annually. The recurrence model for the amount a_n that she has on deposit after she makes her payment at the beginning of the nth year is $a_n = (1+r)a_{n-1} + P$, where $a_0 = P$. Solve this recurrence model using the methods of this section.

6. The number of regions a_n into which the plane is divided by n lines, such that no two are parallel and no three intersect in a point, is $a_n = a_{n-1} + n$, where $a_0 = 1$. Solve this recurrence relation by the methods of this section.

7. The number of regions a_n formed by n intersecting circles in the plane, each pair of which intersects at exactly two points, and such that no three circles intersect in a point, is given by the recurrence relation $a_n = a_{n-1} + 2(n-1)$, where $a_1 = 2$. Solve this recurrence relation by the methods of this section.

8. In the Tower of Hanoi puzzle (see Exercise 3.6-17), the number of moves a_n required to transfer n disks from one disk to another is given by the recurrence relation $a_n = 2a_{n-1} + 1$, where $a_1 = 1$. Solve this recurrence relation by the methods of this section.

9. The number of comparison steps b_n required to sort 2^n registers by means of the Bose-Nelson procedure (see Exercises 3.7-14 and 3.7-15) is given by the recurrence relation $b_n = 2b_{n-1} + 3^n$, where $b_1 = 1$. Solve this recurrence relation by the methods of this section.

10. Solve the simultaneous system of recurrence relations

$$\begin{aligned} a_n &= 3a_{n-1} + 2b_{n-1} \\ b_n &= a_{n-1} + 2b_{n-1}. \end{aligned} \qquad a_0 = 1 \quad b_0 = 0$$

(*Hint*: Eliminate b_{n-1} and solve for b_n. Then use substitution in the first equation.)

11. Solve the simultaneous system of recurrence relations

$$\begin{aligned} a_n &= 2^n \\ b_n &= b_{n-1} + a_{n-1}. \end{aligned} \qquad a_0 = 1, \quad b_0 = 0$$

12. Let a_n and b_n be the number of low-energy and high-energy nuclear particles in a reactor at time n. Suppose that when a high-energy particle (HP) strikes an atomic nucleus it is absorbed and the nucleus emits one low-energy particle (LP) and three HPs. But when an LP is absorbed by a nucleus, two LPs and two HPs are emitted. Assume there is exactly one HP at time $n = 0$ and no LPs and at every n each particle reacts with a nucleus. Give recurrence relations modeling this system and solve, given the initial condition(s).

8.3 GENERATING FUNCTIONS AND RECURRENCE RELATIONS

Given a sequence of numbers $a_0, a_1, a_2, \ldots, a_r, \ldots$, we say that the formal sum

$$A(x) = \sum_{r=0}^{\infty} a_r x^r = a_0 + a_1 x + a_2 x^2 + \cdots + a_r x^r + \cdots$$

is the **generating function** of the sequence. In particular, if the sequence is defined by a recurrence relation, then we say that the sum is the generating function of the recurrence relation. We will not be concerned with the evaluation of such sums for particular values of x. We will, however, want to manipulate these sums as we do polynomials.

For example, given two such sums $A(x)=\sum_{r=0}^{\infty} a_r x^r$ and $B(x)=\sum_{r=0}^{\infty} b_r x^r$, we can combine coefficients of terms of like degree and write

$$A(x)+B(x)=\sum_{r=0}^{\infty}(a_r+b_r)x^r.$$

Another technique that we will soon find useful is factoring out a common factor from each term of the sum, as, for example, in

$$(ka_0x^m+ka_1x^{m+1}+ka_2x^{m+2}+\cdots)=kx^m(a_0+a_1x+a_2x^2+\cdots).$$

It will also be useful to be able to recognize alternate forms of certain formal sums, such as

$$1/(1-x)=1+x+x^2+\cdots+x^r+\cdots. \tag{8.3}$$

To justify Eq. (8.3), one can multiply each term on the right-hand side by $(1-x)$, that is, by 1 and by $-x$, and add the resulting terms as follows:

$$\begin{array}{l} 1+x+x^2+x^3+\cdots+x^r+\cdots \\ \quad -x-x^2-x^3-\cdots-x^r+\cdots \\ \hline 1+0+\ 0\ +\ 0\ +\cdots+0+\cdots \end{array}$$

Since we obtain a 1 when we add corresponding terms, we conclude that Eq. (8.3) is valid.

Now let us see how the ideas we have discussed can be used to solve recurrence relations.

Example 1 How can we use generating functions to obtain a closed-form solution for the recurrence relation $a_r=2a_{r-1}$, $r\geqslant 1$ when a_0 is given?

Our first step is to find the generating function of the recurrence relation. We do this by multiplying each side of our recurrence relation by x^r, obtaining

$$a_rx^r=2a_{r-1}x^r.$$

We then sum each side of the resulting equation for $r=1, 2, 3, \ldots$. We obtain

$$a_1x+a_2x^2+a_3x^3+\cdots=2a_0x+2a_1x^2+2a_2x^3+\cdots \tag{8.4}$$

Now if $A(x) = a_0 + a_1x + a_2x^2 + \cdots$ is the required generating function, then the left-hand side of Eq. (8.4) is just $A(x) - a_0$, since it is missing the a_0 term. Moreover, we can factor $2x$ out of every term on the right-hand side of Eq. (8.4), giving us $2xA(x)$. Hence Eq. (8.4) becomes

$$A(x) - a_0 = 2xA(x),$$

or

$$A(x) - 2xA(x) = a_0,$$

or

$$A(x)[1 - 2x] = a_0,$$

or

$$A(x) = a_0/[1 - 2x].$$

The right-hand side of our last equation closely resembles the left-hand side of Eq. (8.3), our formula for $1/(1 - x)$. In fact if we substitute $2x$ for x in Eq. (8.3), we find that we can also write $A(x)$ in the form $A(x) = a_0 + (2x)a_0 + (2x)^2a_0 + (2x)^3a_0 + \cdots$, or

$$A(x) = a_0 + 2a_0x + 2^2a_0x^2 + \cdots + 2^ra_0x^r + \cdots \tag{8.5}$$

Since the coefficient of x^r in Eq. (8.5) is 2^ra_0, we conclude that $a_r = 2^ra_0$ is the closed-form solution of our original recurrence relation. □

A key step in Example 1 was changing the form of $A(x)$ from a fraction to a summation. In more complicated examples, this is often done by means of the method of "partial fractions," which one usually learns in a calculus course. The following example will serve to remind the reader how the method of partial fractions can be used to reduce certain fractions involving more than one linear factor into a number of fractions, each of which has exactly one linear factor. The general method can be found in almost any elementary calculus text, such as Thomas and Finney's *Calculus with Analytic Geometry* (fifth edition, Addison-Wesley, 1984).

Example 2 Suppose that a generating function of a sequence has been reduced to the fractional form $A(x) = (3 - 7x)/(1 - 5x + 6x^2)$. Write $A(x)$ in summation form by means of partial fractions, and then find the rth term a_r of the sequence.

We may begin by factoring the denominator of $A(x)$:

$$1 - 5x + 6x^2 = (1 - 2x)(1 - 3x).$$

Hence our partial fraction decomposition of $A(x)$ will be of the form

$$\frac{3x - 7x}{(1 - 5x + 6x^2)} = \frac{R}{(1 - 2x)} + \frac{S}{(1 - 3x)}$$

for two constants R and S. A popular method of finding these constants is to multiply both sides of the previous equation by $1 - 5x + 6x^2$. Then we equate coefficients of corresponding terms on each side of the resulting equation. Thus

$$3 - 7x = R(1 - 3x) + S(1 - 2x) = (R + S) + (-3R - 2S)x.$$

Equating corresponding coefficients gives us the system of equations

$$\begin{aligned} R + S &= 3, \\ -3R - 2S &= -7. \end{aligned}$$

Solving this system of equations, we obtain $R = 1$ and $S = 2$. Therefore, we may write

$$A(x) = 1/(1 - 2x) + 2/(1 - 3x).$$

Expanding each of the fractions with the aid of Eq. (8.3), we obtain

$$A(x) = \sum_{r=0}^{\infty} 2^r x^r + 2 \sum_{r=0}^{\infty} 3^r x^r = \sum_{r=0}^{\infty} (2^r + 2 \cdot 3^r) x^r.$$

We may now read the coefficient of x^r directly, and so $a_r = 2^r + 2 \cdot 3^r$ is the rth term. □

Example 3 Find the closed form of the recurrence relation $a_r = a_{r-1} + a_{r-2}$, $a_0 = a_1 = 1$, using generating functions.

As in Example 1, we begin by finding the generating function of a_r. We first multiply each side of the recurrence relation by x^r. Then we sum both sides of the resulting equation for $r = 2, 3, 4, \ldots$, since our recurrence formula is only valid for these values. We obtain

$$\sum_{r=2}^{\infty} a_r x^r = a_2 x^2 + a_3 x^3 + \cdots = A(x) - a_0 - a_1 x$$

on the left-hand side. On the right-hand side we get

$$\begin{aligned}\sum_{r=2}^{\infty} (a_{r-1}x^r + a_{r-2}x^r) \\ &= x \sum_{r=2}^{\infty} a_{r-1}x^r + x^2 \sum_{r=2}^{\infty} a_{r-2}x^{r-2} \\ &= x(a_1x + a_2x^2 + \cdots) + x^2(a_0 + a_1x + a_2x^2 + \cdots) \\ &= x[A(x) - a_0] + x^2[A(x)].\end{aligned}$$

Using the conditions $a_0 = a_1 = 1$, we obtain the equation

$$A(x) - 1 - x = x[A(x) - 1] + x^2A(x).$$

We solve this equation for $A(x)$, obtaining $A(x) = 1/(1 - x - x^2)$.

Now we cannot use Eq. (8.3) to express $A(x)$ until we decompose $1/(1 - x - x^2)$ into partial fractions. This can be done by first writing the denominator as

$$1 - x - x^2 = [x + (1 + \sqrt{5})/2][x - (1 - \sqrt{5})/2].$$

Letting $k_1 = (1 + \sqrt{5})/2$ and $k_2 = (1 - \sqrt{5})/2$, we eventually find that

$$A(x) = (k_1/\sqrt{5})/(1 - k_1x) - (k_2/\sqrt{5})/(1 - k_2x). \tag{8.6}$$

Now we may use Eq. (8.3), and we see that each fraction in Eq. (8.6) has an expansion of the form $(k_i/\sqrt{5}) \sum_{r=0}^{\infty} (k_i)^r x^r$. Hence we may add these sums term by term and obtain the rth term of the resulting expression for $A(x)$. We conclude that the closed form of our recurrence relation is given by

$$a_r = (1/\sqrt{5})(k_1)^{r+1} - (1/\sqrt{5})(k_2)^{r+1},$$

or

$$a_r = \frac{1}{\sqrt{5}}\left(\frac{1+\sqrt{5}}{2}\right)^{r+1} - \frac{1}{\sqrt{5}}\left(\frac{1-\sqrt{5}}{2}\right)^{r+1},$$

just as in Example 2, Section 8.1, that is, the closed form of the Fibonacci numbers. □

Our method for solving recurrence relations by means of generating functions can be summarized as follows:

Given a recurrence relation defining a_r for $r \geqslant m$ and m initial conditions, do the following.

1. Multiply each side of the mth-order recurrence relation by x^r and sum both sides of the result for $r = m, m+1, m+2, \ldots$.
2. Manipulate the sums until you can solve for $A(x)$ in terms of an algebraic expression in the variable x.
3. Change the algebraic form of $A(x)$ to a summation form $\sum_{r=0}^{\infty} a_r x^r$, and read off the coefficient a_r of x^r.

The partial fractions technique is very useful for carrying out step 3 of our procedure, especially with the aid of Eq. (8.3). But Eq. (8.3) can only take us so far. Here is a generalization of Eq. (8.3) that will increase our ability to convert algebraic forms into summation forms and vice versa. For positive integers n we have

$$1/(1-x)^n = \sum_{r=0}^{\infty} C(r+n-1, r)x^r, \quad n \geqslant 1. \tag{8.7}$$

The expressions "$C(r+n-1, r)$" are the binomial coefficients. Indeed we may establish Eq. (8.7) by means of a counting argument, as follows. By Eq. (8.3) we have

$$1/(1-x)^n = [1/(1-x)]^n = (1 + x + x^2 + \cdots)^n.$$

To find the coefficient of x^r in the last expression, we notice that x^r consists of r indistinguishable factors of x. The number of ways to form x^r in the last product is equal to the number of ways to distribute these r indistinguishable factors of x among the n factors $(1 + x + x^2 + \cdots)$. According to Section 3.3, this comes to $C(r+n-1, r)$ ways. Hence $C(r+n-1, r)$ is the coefficient of x^r in the expansion of $1/(1-x)^n$.

Example 4 Find the coefficient of x^{13} in the expansion of

$$(x^2 + x^3 + x^4 + \cdots)^5.$$

We first extract x^2 from each of the terms in $x^2 + x^3 + x^4 + \cdots$, obtaining

$$\begin{aligned}&[x^2(1 + x + x^2 + x^3 + \cdots)]^5 \\ &\quad = x^{10}(1 + x + x^2 + x^3 + \cdots)^5 \\ &\quad = x^{10}[1/(1-x)^5].\end{aligned}$$

The coefficient of x^{13} in the original expression is the coefficient of x^3 in $1/(1-x)^5$, since $x^{13}=x^3x^{10}$. By Eq. (8.7), the coefficient of x^3 in $1/(1-x)^5$ is $C(3+5-1, 3)=C(7, 3)=35$, and this is the number we seek. □

Example 5 Use generating functions to solve the recurrence relation $a_r=a_{r-1}+1$, where $a_0=1$.

Multiplying both sides of our equation by x^r and summing over $r=1, 2, 3, \ldots$, we obtain

$$\sum_{r=1}^{\infty} a_r x^r = \sum_{r=1}^{\infty} a_{r-1} x^r + \sum_{r=1}^{\infty} 1 x^r.$$

Hence

$$A(x)-1=xA(x)+[1/(1-x)]-1$$

Solving for $A(x)$, we obtain

$$A(x)=1/(1-x)^2.$$

By Eq. (8.7), with $n=2$, we conclude that $a_r=C(r+2-1, r)=r+1$ for $r=0, 1, 2, \ldots$. □

It is often useful to be able to recognize special cases of Eq. (8.7) when they are written out in summation form. For example, with $n=2$ we obtain

$$1/(1-x)^2=\sum_{r=0}^{\infty} C(r+2-1, r)x^r=\sum_{r=0}^{\infty}(r+1)x^r.$$

In other words, we obtain the interesting little formula

$$1/(1-x)^2=1+2x+3x^2+4x^3+\cdots. \tag{8.8}$$

Example 6 Use Eq. (8.8) and the method of generating functions to solve the recurrence relation $a_r=a_{r-1}+(r-1)$, where $a_0=1$.

Multiplying both sides of our recurrence relation by x^r and summing over $r=1, 2, 3, \ldots$, we obtain

$$\sum_{r=1}^{\infty} a_r x^r = \sum_{r=1}^{\infty} a_{r-1} x^r + \sum_{r=1}^{\infty} (r-1)x^r. \tag{8.9}$$

The left-hand side of Eq. (8.9) is merely $A(x)-a_0=A(x)-1$. Moreover, you should now recognize the first sum on the right-hand side as $xA(x)$. The second sum on the right-hand side can be rewritten as follows:

$$\sum_{r=1}^{\infty} (r-1)x^r$$

$$= x^2 \sum_{r=1}^{\infty} (r-1)x^{r-2}$$

$$= x^2(1 + 2x + 3x^2 + \cdots) = x^2/(1-x)^2$$

by Eq. (8.8). Hence Eq. (8.9) becomes

$$A(x) - 1 = xA(x) + x^2/(1-x)^2.$$

Solving this last equation for $A(x)$, we get

$$A(x) = 1/(1-x) + x^2/(1-x)^3.$$

To calculate the rth coefficient a_r in the summation form of $A(x)$, we see that the first fraction's contribution to a_r is 1, while the second fraction's contribution is $C((r-2)+2-1,\ r-2) = r(r-1)/2$. Hence $a_r = 1 + r(r-1)/2$ is the closed-form solution. □

Generalized Binomial Coefficients

Returning to Eq. (8.7), we notice that $1/(1-x)^n$ can also be written as $(1-x)^{-n}$. Now $(1-x)^n$ can be expanded by the binomial theorem when n is a positive integer. Indeed we can also expand $(1-x)^{-n}$ by a generalization of the binomial theorem that agrees with Eq. (8.7) if we simply formulate the following definition.

Definition 8.2 For all real numbers z and nonnegative integers r, we define the **generalized binomial coefficients** $\mathbf{C(z,\ 0) = 1}$ and

$$C(z, r) = z(z-1)(z-2)\ldots(z-r+1)/r!,\ r > 0.$$

It is clear that if z is a positive integer, then our generalized definition of $C(z, r)$ agrees with our former definition, namely, $C(z, r) = z!/[r!(z-r)!]$. Moreover, it can be shown by elementary algebra that when z is a negative integer we have

$$C(-z, r) = C(r + z - 1, r) \tag{8.10}$$

Hence our expansion of $1/(1-x)^n = (1-x)^{-n}$ can be written in a form that resembles the usual binomial theorem

$$(1-x)^{-n} = \sum_{r=0}^{\infty} C(-n, r)(-x)^r \tag{8.11}$$

which, as the reader can check, agrees with Eq. (8.8).

More generally it can be shown (by methods of advanced calculus) that the formula

$$(1+x)^z = \sum_{r=0}^{\infty} C(z, r)x^r \tag{8.12}$$

is valid for all real x between -1 and $+1$ and all real numbers z. For example,

$$(1+x)^{1/2} = 1 + [(-1/2)/1!]x + [(-1/2)(-3/2)/2!]x^2 + \cdots.$$

We will soon give an application of Eq. (8.12) to solutions of recursive relations, but first we will need to known how to multiply two generating functions. The correct rule is much like the way we multiply polynomials. For example, by the distributive law of ordinary arithmetic we find that

$$\begin{aligned}(a_0 + a_1x + a_2x^2)(b_0 + b_1x + b_2x^2) \\ = a_0b_0 + (a_0b_2 + a_1b_1 + a_2b_0)x^2 + (a_1b_2 + a_2b_1)x^3 + (a_2b_0)x^4.\end{aligned}$$

More generally we can state that

$$\left(\sum_{r=0}^{\infty} a_rx^r\right)\left(\sum_{r=0}^{\infty} b_rx^r\right) = \left(\sum_{r=0}^{\infty} c_rx^r\right),$$

where, for each r, (8.13)

$$c_r = \sum_{k=0}^{r} a_kb_{r-k}.$$

Example 7 What is the number of ways, a_r, to place parentheses in the expression $x_1 + x_2 + x_3 + \cdots + x_r$ so that exactly two numbers are added at each stage, as, for example, when the r numbers are added on a hand calculator?

To clarify what is meant by our problem, $a_3 = 2$, since we can only write the sum $x + y + z$ in the two distinct ways $x + (y + z)$ and $(x + y) + z$ by inserting parentheses. Moreover, it is clear that $a_2 = a_1 = 1$, and we will arbitrarily set $a_0 = 0$.

Next for each positive integer n, $0 \leqslant n \leqslant r$, divide the sum $x_1 + x_2 + \cdots + x_r$ into two smaller sums

$$x_1 + x_2 + \cdots + x_n \quad \text{and} \quad x_{n+1} + x_{n+2} + \cdots + x_r.$$

Now for each n there are a_n ways to add the first sum and a_{r-n} ways to add the second sum by placing parentheses as specified. Hence there are $a_n a_{r-n}$ ways to add and then combine the two smaller sums. Therefore, for each r we have

$$a_r = a_1 a_{r-1} + a_2 a_{r-2} + \cdots + a_{r-1} a_1, \tag{8.14}$$

which we notice is just the coefficient of x^r in the expansion of $A(x)A(x) = (0 + a_1 x + a_2 x^2 + \cdots + a_r x^r + \cdots)^2$.

Multiplying each side of Eq. (8.14) by x^r and summing over $r = 1, 2, 3, \ldots$, we obtain $A(x) - x$ on the left-hand side. On the right-hand side we obtain $[A(x)]^2$, as mentioned in the previous paragraph. In other words, we can write

$$A(x) - x = [A(x)]^2.$$

Solving this quadratic equation for the root that gives us $A(0) = a_0 = 0$, we have

$$A(x) = (1/2) - (1/2)(1 - 4x)^{1/2}.$$

If we now expand $(1 - 4x)^{1/2}$ by Eq. (8.12), we find that the coefficient of x^r is

$$\begin{aligned} (-4)^r C(1/2, r) &= (-4)^r \frac{(\frac{1}{2})(\frac{1}{2} - 1)(\frac{1}{2} - 2) \cdots (\frac{1}{2} - r + 1)}{r!} \\ &= 2^r \frac{(-1)(1)(3)(5) \cdots (2r - 3)}{r!} \\ &= (-2/r)C(2r - 2, r - 1), \end{aligned}$$

where the last step can be verified by induction. Since the coefficient of x^r in the expansion of $(-1/2)(1 - 4x)^{1/2}$ is the coefficient a_r in $A(x)$ for $n = 1, 2, 3, \ldots$, we may write

$$a_r = (1/r)C(2r - 2, r - 1), \quad \text{for } r \geqslant 1.$$

□

Completion Review 8.3

Complete each of the following.

1. The formal sum $\sum_{r=0}^{\infty} a_r x^r$ is called the ________________ of the sequence $a_0, a_1, a_2, \ldots, a_r, \ldots$.

2. Two such sums $A(x) = \sum_{r=0}^{\infty} a_r x^r$ and $B(x) = \sum_{r=0}^{\infty} b_r x^r$ can be added, with the result ________________. Moreover, $ka_0x^m + ka_1x^{m+1} + ka_2x^{m+2} + \cdots$ can be written ________________.

3. Written as a formal sum, $1/(1-x) =$ ________________. More generally, $1/(1-x)^n =$ ________________ for a positive integer n.

4. The generalized binomial coefficient $C(z, r) =$ ________________ for all real numbers z and nonnegative integers r. It occurs in the formula $(1+x)^z =$ ________________, which is valid for all real x between ________________ and all real numbers z.

5. Multiplying $A(x)B(x)$ for the $A(x)$ and $B(x)$ in the second question, yields $C(x) = \sum_{r=0}^{\infty} c_r x^r$, where $c_r =$ ________________.

6. Given a recurrence relation defining a_r for $r \geqslant m$ and m initial conditions, we can apply the method of generating functions by first ________________ each side by ________________ and then ________________ the result for $r =$ ________________.

7. Continuing where the preceding question left off, we next solve for ________________.

8. Finally, we change the algebraic form of $A(x)$ to ________________ and read off the coefficient ________________.

Answers: **1.** generating function. **2.** $\sum_{r=0}^{\infty} (a_r + b_r)x^r$; $kx^m A(x)$. **3.** $1 + x + x^2 + x^3 + \cdots$; $\sum_{r=0}^{\infty} C(r+n-1, r)x^r$. **4.** $z(z-1)(z-2)\ldots(z-r+1)/r!$; $\sum_{r=0}^{\infty} C(z, r)x^r$; -1 and $+1$. **5.** $\sum_{r=0}^{k} a_k b_{r-k}$. **6.** multiplying; x^r; summing; $m, m+1, m+2, \ldots$. **7.** $A(x)$ as an algebraic expression. **8.** summation form; of x^r.

Exercises 8.3

For Exercises 1 to 6 write the first five terms of the generating function of the given sequence in each. Then write the generating function itself using sum notation.

1. $1, -1, 1, -1, \ldots, (-1)^n, \ldots$.

2. $0, 2, 4, 6, 8, \ldots, 2n, \ldots$.

3. $1, 2, 4, 8, 16, \ldots, 2^n, \ldots$.

4. $1, 1/2, 1/3, 1/4, \ldots, 1/n, \ldots$.

5. $1, 1/2, 1/8, 1/16, \ldots, 1/2^n, \ldots.$

6. $1, -2, 3, -4, 5, \ldots, n(-1)^n, \ldots.$

For each of Exercises 7 to 12 write the first five terms of the generating function of the given recurrence relation.

7. $a_r = 3a_{r-1}$, and $a_0 = 1$.

8. $a_r = (-1)a_{r-1}$, and $a_0 = 2$.

9. $a_r = a_{r-1} - a_{r-2}$, $a_0 = 1$, and $a_2 = -1$.

10. $a_r = a_{r-1} + 2a_{r-2}$, $a_0 = 1$, and $a_2 = 2$.

11. $a_r = ra_{r-1}$, and $a_0 = 1$.

12. $a_r = a_{r-1} + r$.

For each of Exercises 13 to 20 write the given fraction in summation form.

13. $1/(1-4x)$.

14. $1/(1-x^2)$.

15. $3/(1-2x)$.

16. $1/(1+x)$. [*Hint*: Use $(1+x) = (1-(-x))$.]

17. $x/(1-x)$.

18. $1/(1-x) + 1/(1-3x)$.

19. $1/(1-3x+2x^2)$. [*Hint*: Factor the denominator.]

20. $x/(x-1)$.

For each of Exercises 21 to 26 use a generating function to solve the given recurrence relation.

21. $a_r = 4a_{r-1}$, $a_0 = 1$.

22. $a_r = -2a_{r-1}$, $a_0 = 1$.

23. $a_r = 4a_{r-2}$, $a_0 = a_1 = 1$.

24. $a_r - a_{r-2} = 0$, $a_0 = 0$, $a_1 = 1$.

25. $6a_r = 5a_{r-1} - a_{r-2}$, $a_0 = 1$, $a_1 = 0$.

26. $3a_r - 4a_{r-1} + a_{r-2} = 0$, $a_0 = a_1 = 1$.

27. Express $1/(1-x)^3$ as an infinite sum in the variable x.

28. Express $x/(1+x)^4$ as an infinite sum in the variable x.

29. Find the coefficient of x^{10} in the infinite sum expansion of $(1+x+x^2+x^3+\cdots)^4$.

30. Find the coefficient of x^{10} in the infinite sum expansion of $(x+x^3+x^5+\cdots)^3$.

Solve each of the recurrence relations in Exercises 31 to 34 using a generating function.

31. $a_r = a_{r-1} + 2r, a_0 = 0$.

32. $a_r = a_{r-2} + r$, $a_0 = a_1 = 1$.

33. $a_r = a_{r-1} + 2(r-1)$, $a_0 = 1$.

34. $a_r = a_{r-1} + r(r+1)$, $a_0 = 1$.

Expand each of the expressions in Exercises 35 and 36 using the generalized binomial theorem, Eq. (8.12), giving the first four terms.

35. $(1+2x)^{1/2}$.

36. $(1-4x)^{-1/2}$.

37. Write the first four terms in the expansion of
$(1+x+x^2+x^3+\cdots)(1-x+x^2-x^3+-\cdots)$.

38. Write the first four terms in the expansion of $(x+x^2+x^3+\cdots)^2$.

8.4 GENERATING FUNCTIONS AND COUNTING

In the previous section we used generating functions to help us find closed-form solutions to recurrence relations. The recurrence relations themselves were used to describe solutions to counting problems. In this section we show how some counting problems can be modeled and solved directly with the aid of generating functions.

Let us begin by considering the binomial expansion

$$(1+x)^n = C(n, 0) + C(n, 1)x + C(n, 2)x^2 + \cdots + C(n, n)x^n$$

for a positive integer n. This expansion is the generating function of the finite sequence whose terms, $a_r = C(n, r)$, are the binomial coefficients. Each of these coefficients counts the number of ways to form the term x^r by taking x's from r factors $(1+x)$ and 1's from $n-r$ factors $(1+x)$.

As we discussed in Chapter 3, $C(n, r)$ tells us the number of ways to form a subset from a set of n distinct objects by selecting r of these objects. Another important way to interpret $C(n, r)$ is as the number of integer solutions to the equation

$$e_1 + e_2 + \cdots + e_n = r,$$

where $0 \leqslant e_i \leqslant 1$ for each i. The constraints upon the variables e_i mean that each of the e_i in our equation may be used at most once. In the following examples we see how we can use generating functions to solve both the subset selection problem when the objects are not all distinct and, equivalently, the integer solution problem when some of the e_i's are allowed to be greater than 1.

Let us first consider the product $(1+x)^2(1+x+x^2)^3$. Each term in the expansion of this product will be formed by taking either a 1 or an x from each of the two factors in $(1+x)^2$ and a 1, or an x, or an x^2 from each of the three factors in $(1+x+x^2)^3$. For example, $(x)(1)(x^2)(1)(1)$ is just one way to form the term x^3. The coefficient of x^3 in our product $(1+x)^2(1+x+x^2)^3$ is the number of ways to make all such selections. Equivalently, the coefficient of x^3 is the number of integer solutions to the equation

$$e_1 + e_2 + e_3 + e_4 + e_5 = 3,$$

where $0 \leqslant e_1, e_2 \leqslant 1$, and $0 \leqslant e_3, e_4, e_5 \leqslant 2$. The variables e_1 and e_2 correspond to selections from the two factors in $(1+x)^2$ while the variables e_3, e_4, and e_5 correspond to selections from the three factors in $(1+x+x^2)^3$. Setting $e_3 = 2$, to take an example, is equivalent to selecting x^2 from the first quadratic factor.

When we ask more generally about the coefficient of x^r in this expansion or about the integer solutions of $e_1 + \cdots + e_5 = r$, we say that the product $(1+x)^2 \times (1+x+x^2)^3$ is the "generating function of the coefficient of x^r" or the "generating function of the number of such solutions."

Still another useful interpretation of the coefficient of x^3 in the expansion of $(1+x)^2(1+x+x^2)^3$ is this: It is the number of ways to select three objects (corresponding to the x's) from five kinds of objects (corresponding to the five factors), where there are one of each of the first two kinds of objects and two (indistinguishable) objects in each of the remaining three categories.

Example 1 Find the number of ways to select three balls from one red, one blue, two green, two white, and two black balls.

Our problem is equivalent to finding the coefficient of x^3 in the previous paragraph. In other words, we can solve this problem by considering the generating function $(1+x)^2(1+x+x^2)^3$. We need not multiply out $(1+x)^2(1+x+x^2)^3$, however, to find that the coefficient of x^3 is 15. Merely note that if x's are chosen from both of the first two factors, then we can choose the last x in three ways, one from any of the last three factors. But if x^2 is chosen from one of the last three factors, then we can choose the last x from one of the remaining four factors in four ways. Hence $3+(3)(4)=15$ is the number of ways to select the three balls. □

In the next two examples we focus our attention on how to obtain the generating function that solves a counting problem. In particular, we show how to change the given counting problem into a problem asking for the number of integer solutions to an equation of the type mentioned prior to Example 1.

Example 2 Find the generating function for the number of ways to select r balls from a collection of four red balls, four green balls, and four blue balls.

Our problem is equivalent to finding the number of integer solutions to the equation

$$e_1+e_2+e_3=r,$$

where $0 \leq e_i \leq 4$ and each e_i corresponds to a different color ball. Hence a formal sum $1+x+x^2+x^3+x^4$ will correspond, in turn, to each e_i in the generating function for selecting the r balls, which is

$$(1+x+x^2+x^3+x^4)^3.$$

□

Example 3 Find the generating function for the number of ways to distribute 20 identical balls among six distinct boxes so that each box gets at least two balls.

We first restate the problem as the number of integer solutions to the equation

$$e_1+e_2+e_3+e_4+e_5+e_6=20,$$

where $2 \leq e_i$, $i=1,2,3,4,5,6$. Now we let each e_i correspond to a factor $x^2+x^3+x^4+\cdots$, and the generating function is easily seen to be

$$(x^2+x^3+x^4+\cdots)^6.$$

□

The solutions to the problems whose generating functions we found in the last two examples require us to calculate certain coefficients of the generating functions. Calculation of the coefficients of terms of generating functions can often be simplified with the aid of expansion formulas such as the binomial theorem, in addition to others of the kind that we discussed in the last section. For the sake of convenience, we list these equations as Eqs. (8.15) to (8.17), in addition to Eq. (8.19), which is just the result of substituting $-x^m$ for x in Eq. (8.15), and Eq. (8.18), which can be verified by multiplying both sides by $1-x$ and subtracting like terms.

$$(1+x)^n = 1 + C(n, 1)x + C(n, 1)x^2 + \cdots + C(n, n)x^n. \tag{8.15}$$

$$1/(1-x) = 1 + x + x^2 + x^2 + \cdots + x^r + \cdots. \tag{8.16}$$

$$1/(1-x)^n = 1 + C(1+n-1, 1)x + C(2+n-1, 2)x^2 + \cdots + C(r+n-1, r)x^r + \cdots. \tag{8.17}$$

$$(1-x^{n+1})/(1-x) = 1 + x + x^2 + \cdots + x^n. \tag{8.18}$$

$$(1-x^m)^n = 1 - C(n, 1)x^m + C(n, 2)x^{2m} - \cdots + (-1)^n x^{mn}. \tag{8.19}$$

Example 4 Find the number of ways to distribute 20 identical balls among six distinct boxes so that each box gets at least two balls.

We have already calculated the generating function for this problem as

$$(x^2 + x^3 + x^4 + \cdots)^6.$$

The solution to our problem is the coefficient of x^{20} in the generating function. To find this coefficient, we begin by factoring out x^2, and we obtain

$$[x^2(1 + x + x^2 + \cdots)]^6 = x^{12}(1 + x + x^2 + \cdots)^6.$$

Applying Eq. (8.16) then gives us

$$x^{12}[1/(1-x)]^6.$$

Hence the coefficient of x^{20} in this product is the coefficient of x^8 in $[1/(1-x)]^6$, which, by Eq. (8.17), is $C(8+6-1, 8) = C(13, 8)$. Therefore, there are $C(13, 8)$ ways to distribute the balls. □

Example 5 Find the number of ways to distribute 20 identical balls among six distinct boxes if the first box can have no more than three balls, but the remaining boxes can have any number of balls.

As in Example 4, we can let a factor of $(1+x+x^2+\cdots)$ correspond to each box that has an unlimited capacity, and we let $(1+x+x^2+x^3)$ correspond to the box that can hold no more than three balls. Then the generating function for the distribution problem is

$$(1+x+x^2+x^3)(1+x+x^2+\cdots)^5$$

where we wish to find the coefficient of the x^{20} term, as in the previous example. The second factor in the generating function can be rewritten with the aid of Eq. (8.16) as $1/(1-x)^5$. Since we are multiplying this by $(1+x+x^2+x^3)$, the coefficient of x^{20} in the generating function is the sum of the coefficients of x^{20}, x^{19}, x^{18}, and x^{17}, in $1/(1-x)^5$. Hence by Eq. (8.17), the sum of these coefficients, and the answer to our distribution problem, is

$$C(5+20-1,\,20)+C(5+19-1,\,19)+C(5+18-1,\,18)+C(5+17-1,\,17)$$

or

$$C(24,\,20)+C(23,\,19)+C(22,\,18)+C(21,\,17). \qquad \square$$

Example 6 How many ways are there to collect \$12 from 18 people, if the first 17 people can each give either a dollar or nothing and the last person can give \$1, \$10, or nothing?

The generating function of this problem is $(1+x)^{17}(1+x+x^{10})$, where we seek the coefficient of x^{12}. We can expand the factor $(1+x)^{17}$ by Eq. (8.15). Since we are multiplying this factor by $(1+x+x^{10})$, the coefficient of x^{12} in the product is

$$C(17,\,12)+C(17,\,11)+C(17,\,2),$$

and this is the number of ways to collect the \$12. $\square$

For our next example, we show how the idea of finding coefficients of generating functions can be used to discover and to check combinatorial identities.

Example 7 Verify the identity

$$C(m+n,\,r)=\sum_{k=0}^{r} C(m,\,k)C(n,\,r-k).$$

We observe that $C(m+n,\,r)$ is the coefficient c_r of x^r in the expansion of $f(x)=(1+x)^{m+n}$. Let $g(x)=(1+x)^m=\sum_{k=0}^{m} a_k x^k$ and let $h(x)=(1+x)^n=\sum_{k=0}^{n} b_k x^k$.

Then $f(x)=g(x)h(x)$, and by Eq. (8.13) of Section 8.3, the coefficient c_r is

$$a_0b_r + a_1b_{r-1} + a_2b_{r-2} + \cdots + a_rb_0.$$

But for each k, $a_k = C(m, k)$ and $b_{r-k} = C(n, r-k)$. Hence $C(m+n, r) = C(m, 0)C(n, r) + C(m, 1)C(n, r-1) + \cdots + C(m, r)C(n, 0)$ as required. □

Exponential Generating Functions

The generating functions we have considered up to this point are sometimes called "ordinary generating functions" or "enumerators for combinations." Indeed a look at the solutions to the examples in this section will convince the reader that we have been entirely concerned with combinations. Notice, however, that we can rewrite Eq. (8.15) in the form

$$(1+x)^n = P(n, 0)/0! + [P(n, 1)/1!]x + [P(n, 2)/2!]x^2 + \cdots + [P(n, r)/r!]x^r + \cdots + [P(n, n)/n!]x^n, \quad (8.20)$$

since $C(n, r) = P(n, r)/r!$. Thus $(1+x)^n$ also tells us something about permutations, as long we we are willing to overlook the divisors of $r!$. In fact we call $(1+x)^n$ the "exponential generating function" of the sequence of $P(n, r)$'s. More generally we have the following definition.

Definition 8.3 Given a sequence $a_0, a_1, a_2, \ldots, a_r, \ldots$, we say that

$$F(x) = a_0/0! + a_1x/1! + a_2x^2/2! + \cdots + a_rx^r/r! + \cdots$$

is the **exponential generating function** of the sequence.

The adjective "exponential" is derived from the power series expansions of the exponential functions

$$e^x = 1 + x/1! + x^2/2! + \cdots + x^r/r! + \cdots \quad (8.21)$$

and, more generally,

$$e^{nx} = 1 + nx/1! + n^2x^2/2! + \cdots + n^rx^r/r! + \cdots \quad (8.22)$$

where $e = 2.718281\ldots$ is the base of natural logarithms. (These formulas are derived in most elementary calculus books.) Thus e^x is the exponential generating function of the sequence $1, 1, 1, \ldots, 1, \ldots$ and e^{nx} is the exponential generating function of the sequence $1, n, n^2, \ldots, n^r, \ldots$.

Exponential generating functions are very useful for solving problems concerning arrangements. For example, the exponential generating function of the number of arrangements of r objects selected from unlimited supplies of n different kinds of objects is

$$(1+x+x^2/2!+x^3/3!+\cdots)^n=(e^x)^n=e^{nx}.$$

But the coefficient of x^r in e^{nx}, by Eq. (8.22), is n^r. Hence there are n^r such arrangements. Notice that this answer agrees with the answer we obtain by applying the multiplication rule of Section 3.1: Since there are n choices for each of r positions, we simply multiply r factors of n to obtain n^r, as before.

The two formulas,

$$(1/2)(e^x+e^{-x})=1+x^2/2!+x^4/4!+\cdots, \tag{8.23}$$

and

$$(1/2)(e^x-e^{-x})=x+x^3/3!+x^5/5!+\cdots, \tag{8.24}$$

are sometimes of value in the simplification of exponential generating functions. This will be illustrated in the following example.

Example 8 How many r-digit quaternary sequences (those whose digits are 0, 1, 2, and 3) are there with an odd number of 0's and an even number of 1's?

Letting the factor $(x+x^3/3!+x^5/5!+\cdots)$ correspond to the choice of 0's, and letting $(1+x^2/2!+x^4/4!+\cdots)$ correspond to the choice of 1's, we find that the exponential generating function for this problem is

$$(x+x^3/3!+\cdots)(1+x^2/2!+\cdots)(1+x+x^2/2!+x^3/3!+\cdots)^2.$$

Using Eqs. (8.21), (8.23), and (8.24), we can write our generating function more simply as

$$(1/2)(e^x-e^{-x})(1/2)(e^x+e^{-x})e^xe^x=(1/4)(e^{4x}-1).$$

The coefficient of $x^r/r!$ in the last expression is $(1/4)(4^r)=4^{r-1}$ if $r>0$. Hence there are 4^{r-1} such sequences. □

In our last example we see how one might use the techniques of the last section to combine exponential generating functions.

Example 9 How many ways can we put 20 people into three rooms with at least one person in each room?

To see that this is a problem concerning permutations, think of forming 20-tuples of room assignments. We form a generating function modeling the problem by letting each room correspond to a factor $(x+x^2/2!+x^3/3!+\cdots)=e^x-1$. Notice that the term $x^0=1$ is missing from the expansion because each room must be assigned to at least one person. Hence the exponential generating function of this problem is

$$(x+x^2/2!+x^3/3!+\cdots)^3=(e^x-1)^3=e^{3x}-3e^{2x}+3e^x-1.$$

When we expand each power of e with Eq. (8.22) and combine like powers of x^r, we obtain

$$\begin{aligned} e^{3x}-3e^{2x}+3e^x-1 \\ &= \sum_{r=0}^{\infty}(3^r x^r/r!)-3\sum_{r=0}^{\infty}(2^r x^r/r!)+3\sum_{r=0}^{\infty}(x^r/r!)-1 \\ &= \sum_{r=0}^{\infty}(3^r-3\cdot 2^r+3)x^r/r!-1. \end{aligned}$$

Hence the required number of ways to arrange the people in the 20 rooms is

$$3^{20}-3\cdot 2^{20}+3,$$

since this is the coefficient of $x^r/r!$ in the exponential generating function when $r=20$. □

Completion Review 8.4

Complete each of the following.

1. The generating function of the finite sequence of binomial coefficients $C(n, r)$ is ________________ in closed form. This is also the generating function of the number of integer solutions of the equation $e_1+e_2+\cdots+e_n=r$, where each $e_i=$ ________________.

2. The generating function of the number of ways to select three objects from five kinds of objects, where there are one of each of the first two kinds of objects and two (indistinguishable) objects in each of the remaining three categories, is ________________.

3. The generating function for the number of ways to distribute 20 identical balls among

six distinct boxes so that each box gets at least two balls is______________. The number of ways is the coefficient of______________.

4. $1+x+x^2+\cdots+x^n=$______________ in closed form.

5. $(x^2+x^3+x^4+\cdots)^6=$______________ in closed form.

6. The formal sum $a_0/0!+a_1/1!+a_2/2!+\cdots+a_n/n!+\cdots$ is called the ______________ of the sequence $a_0, a_1, a_2, \ldots$.

7. The exponential generating function of the number of permutations of r objects selected from unlimited supplies of n different kinds of objects is______________. In closed form this amounts to______________.

8. A closed form of the exponential generating function of 1, 1, 1, . . . is ______________.

9. A closed form of the sum $1+x^2/2!+x^4/4!+\cdots$ is______________.

10. A closed form of the sum $x+x^3/3!+x^5/5!+\cdots$ is______________.

Answers: **1.** $(1+x)^n$; 0 or 1. **2.** $(1+x)^2(1+x+x^2)^3$. **3.** $(x^2+x^3+\cdots)^6$; x^2. **4.** $(1-x^{n+1})/(1-x)$. **5.** $x^{12}[1/(1-x)]^6$. **6.** exponential generating function. **7.** $(1+x+x^2/2!+x^3/3!+\cdots)^n$; e^{nx}. **8.** e^x. **9.** $(1/2)(e^x+e^{-x})$. **10.** $(1/2)(e^x-e^{-x})$.

Exercises 8.4

1. Find a generating function to model the number of ways to select r balls from four red balls, three green balls, and five white balls.
3. Find a generating function to model the number of ways to select r balls from an unlimited number of red balls, green balls, and white balls.
3. Use your generating function in Exercise 1 to find the number of balls when $r=8$.
4. Use your generating function in Exercise 2 to find the number of balls when $r=6$.
5. Find a generating function to model the number of integral solutions to the equation $e_1+e_2+e_3+e_4=r$, where $0 \leqslant e_i$, $i=1, 2, 3, 4$.
6. Do Exercise 4, substituting the constraints $1 \leqslant e_1$, $e_2 \leqslant 3$, $0 \leqslant e_3$, and $5 \leqslant e_4$.
7. Use your generating function in Exercise 4 to find the number of solutions when $r=5$.
8. Use your generating function in Exercise 5 to find the number of solutions when $r=7$.
9. Do Exercise 5, substituting the constraints that the e_i are positive and odd.
10. **a)** Find a generating function to model the number of ways to distribute 10 identical balls among five distinct boxes if the first and second boxes can have no more than three balls, but the remaining boxes can have an unlimited number of balls.

    b) Use your generating function to solve the problem.

11. **a)** Find a generating function to model the number of ways to collect \$25 from 10 people if the first nine people can each give \$1 or nothing and the last person can give \$1, \$5, or \$10 or nothing.
b) Use your generating function to solve the problem.

12. Find the number of ways to select 25 donuts from plain, jelly-filled, custard-filled, lemon-filled, vanilla-glazed, chocolate-glazed, and coconut surprise donuts, with at least two of each kind selected.

13. How many different committees of 40 senators can be formed, with no two senators from the same state on the same committee?

14. How many ways are there to get a sum of 23 when 10 dice are thrown?

15. Verify the identity $\sum_{r=0}^{n} [C(n, r)]^2 = C(2n, n)$.

16. Find the exponential generating function of the sequence 1, 2, 3, 4, . . . , n, . . .

17. Find the exponential generating function of the sequence 1, 0, -1, 0, 1, 0, -1, 0, 1, ...

18. Find the exponential generating function of the sequence 0, 1, 0, -1, 0, 1, 0, -1, 0, ...

19. Find the exponential generating function for the permutations of none, 1, 2, . . . , n of n identical objects.

20. Find the exponential generating function of all $p+q$ of $p+q$ objects, where p are of one type and q are of another type.

21. Find the exponential generating function of the permutations of two objects of one kind and three objects of another kind if the objects are taken 0, 1, 2, 3, 4, or 5 at a time.

22. **a)** Find a generating function for the number a_r of r-digit quaternary sequences that contain an even number of 0s.
b) Use your generating function to find a_r.

23. **a)** Find a generating function for the number a_r of r-digit quaternary sequences in which each of the digits 0, 1, and 2 all appear at least once.
b) Use your generating function to find a_r.

24. **a)** Find a generating function for the number of ways to put r people in two different rooms with at least one person in each room.
b) Use your generating function to find a_r.

25. **a)** Find a generating function for the number of ways to put r distinct objects into n distinct cells so that no cell is empty and the order within a cell does not matter.
b) Show that the coefficient a_r of $x^r/r!$ in your generating function is equal to

$$\sum_{i=0}^{r} (-1)^i C(n, i)(n-i)^r.$$

[*Note*: The numbers $S(n, r) = (1/n!)a_r$ are called "Stirling numbers of the second kind."]

c) Find the number of ways to put r distinct objects into n nondistinct cells with no cell left empty.
d) Find the number of ways of distributing r distinct objects into n distinct cells with empty cells allowed.

COMPUTER PROGRAMMING EXERCISES

8.1. Write a program that calculates a_n for a given n when a_r is given in terms of a homogeneous linear recurrence relation of order r together with r initial conditions.

8.2. Write a program that will find the nth Fibonacci number, given our closed-form solution in Section 8.1. Compare the time it takes your program to find F_{50} with the time taken by the program in Exercise 1 with $a_n - a_{n-1} - a_{n-2} = 0$, $a_0 = a_1 = 1$.

8.3. Write a program that will find the nth term of the product of two generating functions, given the first n terms of each.

8.4. Write a program that will find the first r coefficients in the expansion of $(1-x)^{-n}$.

8.5. Write a program that will calculate the generalized binomial coefficient $C(z, r)$ for any real number z and nonnegative integer r.

8.6. Write a program to calculate the first 20 terms in the expansion of e^1, the exponential generating function of 1, 1, 1, . . . , given by Eq. (8.21) of Section 8.4.

APPENDIX

PROGRAMS IN BASIC AND EXERCISES

In this appendix we present several computer programs illustrating the concepts discussed in Chapters 1 through 7. These programs were written for and tested on a Commodore 64K microcomputer. With very little modification, they can be loaded and run on most popular microcomputers. (See your user's manual.) They are presented here for self-study or classroom discussion. Modifications of the programs are suggested in the accompanying exercise sets. The reader is strongly encouraged to try to think of other modifications and/or improvements. These are not presented as state-of-the-art programs: They merely illustrate the algorithms we previously discussed. If they stimulate further thought and discussion about discrete mathematics and, in particular, about algorithms, then they will have accomplished their goal.

Commonly Used BASIC Statements: A Primer/Refresher

Our programs are written in the computer language BASIC, the dialect "spoken" by the 64K Commodore. This is not the most powerful version of BASIC, but it is quite sophisticated and, at the same time, easily understood, even by those unfamiliar with the language. Our programs use only a handful of what are called "statements" or "system commands," the commands that instruct the computer to input or output data, assign values to variables, execute a command out of the sequence given by line numbers, and so on. We illustrate these in two preliminary programs before going on to the "discrete mathematics programs."

Program A1: BASIC Statements

```
05 REM THIS PROGRAM ILLUSTRATES THE STATEMENTS 'REM',
10 REM 'LET','READ','PRINT','DATA','END', and 'RUN'
20 LET A=3          The number 3 is assigned to the variable A.
40 READ B
30 READ C                        The values -1 and 5 are assigned to variables C and B, respectively
50 PRINT                         from line 80. Line 30 executed before line 40.
60 PRINT "A+B+C="     Message 'A+B+C=' printed after a line is skipped.
70 PRINT A+B+C      Sum of values assigned to A, B, and C is printed on next line.
80 DATA -1,5

90 END      Execution of program stops.
RUN      Computer told to execute the program in memory.

A+B+C=                      These two lines are the program's output.

7
```

The italicized notes at the right tell you what the program's statement is supposed to accomplish. They are not to be loaded with the program. Here is a more detailed explanation of these commands.

1. REM (message) A commentary on the program that is listed among the commands. It has, however, no effect on the program's execution. See lines 05 and 10.

2. LET X = (a value) This assigns a value to the variable X. (The notation "X ← (value)" is *not* used in BASIC.) The word "LET" is optional. See line 20.

3. DATA This initializes a sequence of values that can be used by a READ statement. The values are read (assigned) sequentially until all data lists are exhausted. See line 80.

4. READ X X is assigned the value of the first term in a data sequence that has not been already read. See lines 30 and 40.

5. (a) PRINT If nothing follows this statement, then a line is skipped in output. See line 50 and the output.

5. (b) PRINT "(message)" This causes whatever is within the quotation marks to be printed. See line 60 and the output.

5. (c) PRINT X The *value* of the variable X is printed. See line 70 and the output.

6. END This indicates the last statement in the program. See line 90.

7. RUN This statement is not part of the program. Rather it tells the computer to run the program that is in memory. The statements in the program are executed in order of their line numbers unless there is a "branching statement" to the contrary. A new line can also be indicated with colons.

The following program includes a variety of branching statements as well as different ways to output data.

Program A2: More BASIC Statements

05 REM THIS PROGRAM ILLUSTRATES THE STATEMENTS 'GOTO', 'READ'	
06 REM 'A(K)', 'IF . . . THEN', AND 'PRINT' WITH COMMAS AND SEMICOLONS.	
07 GOTO 9	*The computer executes line 9 after line 7, skipping 8. Hence X is not*
08 LET X = 1	*assigned the value 1. The value printed will be 0 (assigned instead).*
09 PRINT X	
10 LET K = 0	
15 LET K = K + 1	*The preceding value of K is raised by 1*
20 READ A(K)	*A(K) is called a subscripted variable.*
30 IF K < 5 THEN GOTO 15	
40 FOR K = 1 TO 5	*K is set equal to 1 until statement 60 is reached. Then K is raised to 2 and we return to line 50 . . . .*
50 PRINT A(K)	
60 NEXT K	*K is raised by 1 each time line 60 is reached, unless K ⩾ 5. In that case we proceed to line 70.*
70 FOR K = 1 TO 5	

```
80   PRINT A(K),       The comma divides the screen into four columns for printing output.
90   NEXT K
95   PRINT
100  FOR K = 1 TO 5
110  PRINT A(K);       The semicolon suppresses all spacing.
120  NEXT K
130  DATA 2, 4, 6, 8, 10        These are the values assigned in line 20.
140  END

RUN

0      The computer assigns X the value 0 because line 8 was skipped. This is what is printed
2  }   by line 9.
4  }
6  }   This is what is printed by lines 40–60;
8  }
10 }
2           4           6           8      lines 70–90;
10
2 4 6 8 10                                 lines 95–120
```

Now let us discuss the new statements in Program A2.

8. GOTO (line number) The next line to be executed is the one following the word GOTO. See line 07.

9. READ A(K) The only thing new here is the notation A(K), which you can think of as a_k. Thus lines 15–30 read in the values of a_1, a_2, a_3, a_4, and a_5. Up to 10 such variables can be read in for a_k without additional commands.

10. IF (condition) THEN (command) The "command" following the word "then" is executed if the "condition" is true. Otherwise the next line is scanned. See line 30.

11. FOR K = 1 to N . . . NEXT K The sections of the program between the FOR and NEXT portions are executed in turn for K = 1, 2, 3, . . . , N. (K can also be incremented by numbers S other than 1 if we follow the "N" with the formula STEP S.) See lines 40–60, 70–90, and 100–120 for "loops" of this form.

12(a). PRINT . . . . , The comma causes whatever is printed directly after ". . . ." to be placed in the next of four columns. See line 80 and the corresponding output.

12(b). PRINT . . . . ; The semicolon causes the next thing to be printed after ". . . ." to follow it without spacing. See line 110 and the corresponding output.

There you have nearly all the commands we will use for our programs. For those who would like to learn more about BASIC, there are several excellent introductions we could name, for example, Donald Spencer's *A Guide to BASIC Programming* (second edition), which is published by Addison-Wesley. But we recommend that the reader simply plunge right into the programs that follow after trying the exercises.

Exercises for Programs A1 and A2

1. Explain what would happen if we were to make each of the following changes in Program A1.
  a) Add a line "55 PRINT."
  b) Make line 40 line 30 and make line 30 line 40.
  c) Put a comma after the second quotation mark in line 60.
  d) Put a semicolon after the quotation mark in line 60.
  e) Write "5, −1" after the word DATA in line 80.
  f) Write 2, 3, 4 after the word DATA in line 80.
  g) Delete lines 05 and 10.

2. How could we change the value of A to be 6 in Program A1?

3. Write a program that assigns values of 3 and 4 to variables X and Y, respectively, using a LET for the first and READ for the second. Then have the program print out the value of $X^2 + XY$ (written "X↑2 + X∗Y" on the 64K, for example).

4. Modify your program of Exercise 3 with FOR .... NEXT loops so that it processes the pairs of values (X, Y): (3, 4), (5, 6), (7, 8), (9, 10), . . . , (19, 20). Let it print out the values of $X^2 + XY$ **(a)** on different lines; **(b)** in widely spaced columns; **(c)** one right after the other.

Program A3: FIBONACCI (See Algorithm 1.3, Section 1.4.)

```
80   PRINT
90   PRINT "THE FIBONACCI NUMBERS"
95   PRINT " FROM 1 to 17711"
100  LET N1=1        N1 is the first Fibonacci number.
105  PRINT N1,
110  LET N2=1        N2 is the second Fibonacci number.
115  PRINT N2,
120  LET N=N1+N2     Line 120 defines the remaining Fibonacci numbers recursively.
130  PRINT N,
140  LET N1=N2       Lines 140 and 150 ensure that each Fibonacci number is the sum of
150  LET N2=N        of the previous two.
160  IF N<13000 THEN GOTO 120  The next-to-last number will be less than 13000.
170  END
RUN
```

```
THE FIBONACCI NUMBERS
FROM 1 TO 17711                              The output begins here.
1         1         2         3
5         8         13        21
34        55        89        144
233       377       610       987
1597      2584      4181      6765
10946     17711
```

Program A4: Nested POLY 1 (See Algorithm 1.2, Section 1.3.)

```
100 REM THIS PROGRAM FINDS THE VALUE OF P(X)=3X↑4+2X↑3+4X↑2-5X+1
105 REM WHEN X=-3 BY MEANS OF NESTING.
110 LET P=3        P is initially set equal to the coeffiecient of X↑4.
115 LET X=-3
120 FOR K=1 to 4     Lines 120-140 calculate (((3x+2)x+4)x-5)x+1.
125 READ A
130 LET P=P*X+A
140 NEXT K
150 PRINT P
200 DATA 2, 4, -5, 1
205 END
RUN
241     This is the output P(-3).
```

Program A5: NESTED POLY 2

Note: This is a more elaborate version of Program A3.

```
90  PRINT
92  REM EVALUATES P(-3) IN NESTED FORM
93  REM THE INTERMEDIATE CALCULATIONS
94  REM YIELD THE QUOTIENT AND REMAINDER
95  REM OF P(X)/(X+3)
100 PRINT "P(X)/(X+3)="
105 PRINT "X↑3", "X↑2", "X↑1", "X↑0"
110 LET P=3: LET X=-3: PRINT P,       A sequence of more than one instruction can be
                                      written on a single line if they are separated by
                                      colons as on line 110.
120 FOR K=1 to 4
125 READ A
130 LET P=P*X+A
140 PRINT P,     Line 140 prints the coefficients of the quotient.
150 NEXT K
155 PRINT "IS THE REMAINDER AND THE"
160 PRINT "                VALUE OF P(-3)"
170 PRINT "WHEN P(X)="
175 PRINT "3*X↑4+2*X↑3+4*X↑2-5*X+1"
200 DATA 2, 4, -5, 1
210 END
RUN
```

```
P(X)/(X + 3) =
X↑3        X↑2        X↑1        X↑0
3          -7         25         -80
241        IS THE REMAINDER AND THE
           VALUE OF P(-3)
WHEN P(X) =
3*X↑4+2*X↑3+4*X↑2-5*X+1
```

The coefficients are printed below the terms.

Exercises for Programs A3, A4, and A5

1. What change in Program A3 will get it to evaluate $P(2)$?

2. What changes in Program A3 will get it to evaluate $p(-3)$ when $p(x) =$
(a) $x^4 + x^3 + x^2 + x + 1$; **(b)** $x^3 + x^2 + x + 1$;
(c) $x^5 + x^4 + x^3 + x^2 + x + 1$?

3. What changes must be made in Programs A4 and A5 corresponding to those you made in Exercise 2?

4. Why is "241" printed on the fourth line of output in Program A4?

Program A6: PRIMETEST (See Algorithm 1.4, Section 1.4.)

This is an example of an interactive program. It prompts the user to type in a number from the keyboard after the question mark. The number, N, is then tested for primality by determining if the quotient of any integer from 2 through the square root of N, written SQR(N) in the program, yields an integer, namely, the integer part (INT) of the square root of N.

```
100  INPUT "PRINT A NUMBER";N
110  FOR K=2 to INT(SQR(N))
120  LET P=(N/K)-INT(N/K)
130  IF P=0 THEN GOTO 200
140  NEXT K
150  PRINT N;"     IS A PRIME     "
160  STOP
200  PRINT N;"     IS DIVISABLE by  ";K
210  END
RUN
PRINT A NUMBER? 7
7      IS A PRIME
RUN

PRINT A NUMBER? 8
8    IS DIVISABLE BY 2
```

The INPUT command prints the message in quotes followed by a question mark. User types in data, which are assigned to N.

Line 110 gives trial divisors.

If none of the K's divide N, then line 150 is executed and no further lines are processed.

Otherwise lines 150–160 are skipped, and lines 200–210 are processed instead.

The question is output; the first 7 is input; the next line is output. The program is then run again with the number 8.

Our programs hereafter will *not* be interactive. However, the reader may want to modify them so that data can be inputted directly from the keyboard. This example can serve as a model of how that can be done. (*Note*: STOP halts the program.)

Program A7: SIEVE (See Exercise 1.4-14.)

```
5  PRINT
10 REM LISTS PRIMES FROM 2 THROUGH P
12 REM USING SIEVE OF ERATOSTHENES
13 REM WHERE P IS IN DATA (LINE 76)
14 READ P
15 DIM A(P)
```

The DIM statement in line 15 is needed because A(P) has more than 10 subscripts, that is, P > 10.

```
20 FOR N=2 to P
25 LET A(N)=N
30 NEXT N
35 FOR N=2 to P/2
40 IF A(N)=0 then GOTO 60
45 FOR M=(2*N) to P step N
```

Line 40 locates the next nonzero N remaining on the list.

Lines 45–60 erase every Nth number beginning with 2N, that is, set them=0.

```
50 LET A(M)=0
55 NEXT M
60 NEXT N
65 FOR N=2 to P
70 IF A(N)<>0 THEN PRINT A(N);
```

Line 70 prints the numbers remaining on the list. The symbol < > is read "≠".

```
75 NEXT N
76 DATA 200
77 PRINT
78 PRINT "ARE THE PRIMES FROM 2–";P
80 END
RUN
```

The output begins here.

```
2   3   5   7   11   13   17   19   23   29   31
37   41   43   47   53   59   61   67   71   73
79   83   89   97   101   103   107   109   113
127   131   137   139   149   151   157   163
167   173   179   181   191   193   197   199
ARE THE PRIMES FROM 2–200
```

Exercises for Programs A6 and A7

1. Run Program A6, inputting the numbers **(a)** 53; **(b)** 153; **(c)** 1111111; **(d)** 111111111111. What is your output in each case?

2. Why is it only necessary to try divisors up to INT(SQR(N))?

3. **a)** Run Program A6 with 131071. What was the last divisor tested?

b) Change line 110 in Program A6 to read FOR K = 2 TO N/2, and run the new program with 131071 again. What change do you notice in how the program runs?

c) What was the last divisor tested in the new program with 131071?

4. Find all the primes from 2 through 500 by changing the DATA statement in Program A7, line 76, to read 'DATA 500.'

5. Make Program A7 interactive by replacing lines 14 and 76 with an appropriate input statement.

Program A8: PARTITION (See Example 4, Section 2.2.)

```
05   REM HOW THE SUBSETS A, B, AND C PARTITION U INTO 8 CELLS
06   PRINT "WHEN THE NUMBER OF ELEMENTS IN SETS"
11   PRINT "A AND B AND C,":PRINT "A AND B."
12   PRINT "A AND C,": PRINT "B AND C,"
13   PRINT "A," PRINT "B,": PRINT "C,"
14   PRINT "AND THE UNIVERSE U"
15   PRINT "ARE, RESPECTIVELY,"
20   FOR K = 1 TO 8
30   READ N(K):PRINT N(K);",";
40   NEXT K
50   PRINT :PRINT"                    THEN"
60   S(1) = N(1)
70   S(2) = N(2) - S(1): S(3) = N(3) - S(1)
72   S(4) = N(4) - S(1)
75   S(5) = N(5) - (S(1) + S(2) + S(3))
80   S(6) = N(6) - (S(1) + S(2) + S(4))
85   S(7) = N(7) - (S(1) + S(3) + S(4))
90   S(8) = N(8) - (S(1) + S(2) + S(3) + S(4) + S(5) + S(6) + S(7))
95   PRINT S(2); "  ARE IN A AND B, BUT NOT C"
100  PRINT S(3); "  ARE IN A AND C, BUT NOT B"
110  PRINT S(4); "  ARE IN B AND C, BUT NOT A"
120  PRINT S(5); "  ARE IN A, BUT NOT B AND NOT C"
130  PRINT S(6); "  ARE IN B, BUT NOT A AND NOT C"
140  PRINT S(7); "  ARE IN C, BUT NOT A AND NOT B"
150  PRINT S(8); "  ARE NOT IN A, NOR IN B, NOR IN C"
160  PRINT S(1); "  ARE IN A AND B AND C"
170  PRINT "THESE SETS PARTITION U"
200  DATA 5, 8, 11, 12, 40, 24, 28, 80
205  END
```

Lines 60–90 mimic the calculations discussed in Example 4, Section 2.2.

On each of lines 95–160 the value of S(K) is printed just before the message.

```
RUN
WHEN THE NUMBER OF ELEMENTS IN SETS
A AND B AND C
A AND B
A AND C
B AND C
A
B
C
AND THE UNIVERSE U
ARE, RESPECTIVELY
5, 6, 11, 12, 40, 24, 28, 80,
       THEN
3      ARE IN A AND B BUT NOT C
6      ARE IN A AND C BUT NOT B
7      ARE IN B AND C BUT NOT A
26     ARE IN A BUT NOT B AND NOT C
9      ARE IN B BUT NOT A AND NOT C
10     ARE IN C BUT NOT A AND NOT B
14     ARE NOT IN A NOR IN B NOR IN C
5      ARE IN A AND B AND C
THESE SETS PARTITION U
```

This is the output from Program A8.

Program A9: SETS AND BINARY (See Theorem 2.6, Section 2.2.)

```
05  REM THIS PROGRAM LISTS A INTERSECT B AND
06  REM A UNION B-COMPLEMENT WHEN A, B, AND U
07  REM ARE GIVEN IN DATA
10  READ N
15  PRINT "U =              ";
20  FOR K = 1 TO N
30  READ N(K)
40  PRINT N(K);
50  NEXT K
55  PRINT
58  PRINT "A =             ";
60  FOR K = 1 to N
70  READ A(K)
80  IF A(K) = 1 THEN PRINT N(K);       N(K) is the kth element of A if A(K)=1.
90  NEXT K
95  PRINT
98  PRINT "B =             ";
100 FOR K = 1 to N
110 READ B(K)
120 IF B(K) = 1 THEN PRINT N(K);       The kth element is printed if B(k)=1 is true.
130 NEXT K
135 PRINT
138 PRINT "A INTSCT B =          ";
140 FOR K = 1 TO N
150 IF A(K) = 1 AND B(K) = 1 THEN PRINT N(K);    The kth element is printed if it is in
                                                 both A and B.
160 NEXT K
170 PRINT
178 PRINT "A UNION B*CMPL =";

180 FOR K = 1 TO N
190 IF A(K) = 1 OR B(K) = 0 THEN PRINT N(K);  The kth element is printed if it is in at
200 NEXT K                                    least one of A or B complement.
210 DATA 5        Line 210 gives |U|.
220 DATA 1, 3, 5, 6, 7    Line 220 gives the list of elements in U.
230 DATA 0, 1, 0, 0, 1    Lines 230 and 240 define A and B by indicating the presence or
240 DATA 1, 0, 1, 0, 1    absence of an element with the binary digits 1 and 0.
250 END
RUN
U =                1 3 5 6 7  }
A =                3 7        }
B =                1 5 7      }  This is the output of Program A8.
A INTSCT B =       7          }
A UNION B-CMPL =   3 6 7      }
```

Exercises for Programs A8 and A9

1. Run Program A8 with the data in line 200 given as 4, 7, 9, 10, 14, 20, 23, 44. Which of these is the number of elements in A intersect B?
2. What instructions, placed after line 170, would get the computer to output the number of elements in the set **(a)** $A \cup B$; **(b)** $\overline{(A \cup B)}$; **(c)** $\overline{(A \cap B \cap C)}$?
3. What instructions, placed after line 200 in Program A9, would get the computer to output the elements in the set **(a)** $A \cup B$; **(b)** $\bar{A}$; **(c)** $\bar{B}$; **(d)** $\overline{A \cap B}$; **(e)** $\overline{A \cup B}$?
4. Run Program A9 assuming $U = \{3, 5, 6, 8, 14, 17, 20, 25\}$, $A = \{5, 8, 20\}$, and $B = \{5, 14, 20, 25\}$ after making the appropriate changes in the data statements.
5. Write a computer program that will define operations on three subsets A, B, and C of a universe U, as in Program A9.

Program A10: N FACTORIAL (See Algorithm 3.1, Section 3.1)

```
100  REM THIS PROGRAM FINDS THE VALUE
101  REM OF N FACTORIAL, N IN DATA
105  READ N:PRINT
108  PRINT " N      ","N FACTORIAL"
110  LET F=1          Line 110 defines 0!=1.
120  FOR K=1 TO N     Lines 120–140 calculate N! recursively.
130  LET F=F*K
140  NEXT K
145  PRINT N,F
150  DATA 12          Line 150 contains the present value of N.
155  END
RUN

N             N FACTORIAL
12            479001600
```

Program A11: BINOMIAL (See Example 4, Section 3.2)

```
90   REM THIS PROGRAM USES THE RECURSIVE DEFINITION
91   REM OF PASCAL'S TRIANGLE TO CALCULATE BINOMIAL
92   REM COEFFICIENTS
93   REM WE ALSO INTRODUCE DOUBLY SUBSCRIPTED VARIABLES
100  DIM C(25, 25)     The DIM statement in line 100 prepares space for 676=(26) (26)
                       doubly subscripted variables, C0,0–C25,25.
110  FOR N=0 TO 25     Lines 110–140 initialize certain of these variables at 1.
120  LET C(N, 0)=1
130  LET C(N, N)=1
140  NEXT N
150  FOR N=1 TO 25     Lines 150–190 form a nested loop, that is, a large N loop containing
160  FOR R=1 TO N-1    a smaller R loop in lines 160–180.
170  LET C(N, R)=C(N-1, R)+C(N-1, R-1)
```

```
180 NEXT R
190 NEXT N
200 PRINT " N"," R","C(N,R)"
210 FOR K=1 TO 3
220 READ N
230 READ R
240 PRINT N,R,C(N,R)
250 NEXT K
260 DATA 5,3,10,7,25,15
270 END
RUN
N          R          C(N,R)
5          3          10
10         7          120
25         15         3268760
```

Three of the values calculated for C(N, R) will be printed by lines 210–250. The variable K merely serves as a counter.

The output is printed in tabular form.

Program A12: BIRTHDAY (See Example 3, Section 3.5.)

```
100 REM MODEL CALCULATES THE PROBABILITY THAT
101 REM AT LEAST TWO OF N RANDOMLY CHOSEN
102 REM PEOPLE HAVE THE SAME BIRTHDAY
103 PRINT "       N","       PR(N)"
120 READ N
130 LET Q=1
140 FOR K=1 TO N
150 LET Q=Q*(366-K)/365
160 NEXT K
170 LET P=1-Q
180 PRINT N,
185 PRINT P
190 GOTO 120
200 DATA 5,10,15,20,25,50,100
210 END
RUN
N          PR(N)
5          .0271355736
10         .116948177
15         .252901319
20         .411438383
25         .568699703
50         .97037358
100        .999999693
?OUT OF DATA ERROR IN 120
```

Lines 140–160 calculate the probability that no two of the people have the same birthday, assuming 365 possible days.

Line 170 calculates the required probability.

The commas in the PRINT statements give us the output in tabular form.

After 100 is read, there are no more data available for line 120 to read.

Exercises for Programs A10, A11, and A12

1. Run Program A10 with the following numbers in data. **(a)** 0; **(b)** 1; **(c)** 5; **(d)** 10; **(e)** 20; **(f)** 30. Use a GOTO statement as in Program A12, line 190, so that (a) to (f) can be done at once.

2. Modify Program A10 so that it will calculate **(a)** $(2N)!$; **(b)** $2(N!)$.
3. Use Program A11 to evaluate **(a)** $C(8, 6)$; **(b)** $C(15, 10)$; **(c)** $C(25, 20)$.
4. Modify Program A11 so that it will evaluate $C(30, 15)$.
5. Modify Program A11 so that it will print all the values of $C(\mathrm{N}, R)$ for N and R less than or equal to 5.
6. Write a program for calculating $C(N, R)$ based on Algorithm 3.2.
7. Use Program A12 to find the probability that at least two people from 28 randomly selected people have the same birthday.
8. Modify Program A12 so that we do not get an OUT OF DATA message after all the data have been used up.

Program A13: BUBLSORT (See Algorithm 3.4, Section 3.7.)

```
101  REM THIS PROGRAM APPLIES THE BUBBLE SORT ALGORITHM
102  REM TO THE LAST N NUMBERS IN DATA WHERE N IS THE
103  REM FIRST NUMBER IN THE DATA STATEMENT
105  READ N
110  DIM A(N)
120  FOR K=1 TO N
130  READ A(K)
140  NEXT K
150  FOR B=N TO 2 STEP -1      The statement STEP -1 reduces B by 1 each time.
160  FOR I=1 TO B-1
170  IF A(I)<=A(I+1) THEN GOTO 200   We go to 200 if a_i and a_i+1 are in the correct order.
180  LET E=A(I)
185  LET A(I)=A(I+1)      Otherwise lines 180–190 make an exchange in the positions of
190  LET A(I+1)=E         a_i and a_i+1.
200  NEXT I
210  NEXT B
220  FOR I=1 to N
230  PRINT A(I);
240  NEXT I
250  DATA 9      The number of items sorted is nine.
251  DATA 63, 72, 9, 6, 82, 18, 104, 2, 1
RUN
1  2  6  9  18  63  72  82  104     This is the output, the sorted sequence of line 251.
```

Exercises for Program A13

1. Use Program A13 to sort the following list of numbers in ascending order: 13, 45, 6, 65, 7, 19, 20, 111, 46, 675, 12, 4.
2. Insert the following statements into Program A13: 165 LET C=C+1; 215 PRINT C. What is the significance of the number that is printed?
3. Change statement 170 so that your program will print a list of numbers in descending order.
4. Modify Program A13 so that it will print only the minimum and maximum numbers on your list of data.

Program A14: MATRIX ARRAY (See Section 4.5.)

```
100  REM THIS PROGRAM READS DATA INTO A
101  REM 4 by 3 MATRIX ARRAY BY ROWS AND
102  REM PRINTS IT IN A COMPACT FORMAT
105  DIM A(4, 3)
110  FOR I=1 TO 4      The I loop controls the rows and the J loop controls the columns.
115  FOR J=1 TO 3
120  READ A(I,J)
125  NEXT J
130  NEXT I
135  FOR I=1 TO 4      Lines 135–160 print the matrix.
140  FOR J=1 TO 3
145  PRINT TAB(J*4-4);A(I, J);   The TAB statement will print each column four spaces apart.
150  NEXT J
155  PRINT    Line 155 ensures that a line is skipped after each row is printed.
160  NEXT I
165  DATA 1, 2, 3, 4, 5, 6, 7, 8, 9, 10, 11, 12
RUN
1     2     3
4     5     6
7     8     9        This is the output.
10    11    12
```

Note: It is much easier to read and print a matrix on large computers using more powerful versions of BASIC than are usually available on a microcomputer. We will, however, continue to use the method illustrated in our programs, since we expect that the user usually will be limited to a "personal computer."

Matrix operations can also be handled with brief BASIC commands on a large mainframe computer. Smaller microcomputers can carry out matrix operations as shown in the following programs.

Program A15: MATRIX ADD (See Algorithm 4.3, Section 4.5.)

```
10   REM THIS PROGRAM ADDS MATRICES A AND B
11   REM WHEN THEIR DIMENSIONS AND ENTRIES
12   REM ARE GIVEN IN DATA
13   PRINT "A="
20   READ N: READ M     Lines 20–30 define the dimensions of the matrices as N and M.
30   DIM A(N,M), B(N,M)
40   FOR I=1 TO N      Lines 40–100 read in and print A.
50   FOR J=1 TO M
60   READ A(I,J)
70   PRINT TAB(J*4-4);A(I,J);
80   NEXT J
90   PRINT
100  NEXT I
105  PRINT:PRINT "B="
```

Lines 110–170 read in and print B.

```
110 FOR I=1 TO N
120 FOR J=1 TO M
130 READ B(I,J)
140 PRINT TAB(J*4-4);B(I,J);
150 NEXT J
160 PRINT
170 NEXT I
175 PRINT:PRINT "A+B="
```

Lines 180–230 add corresponding entries of A and B and print them by rows.

```
180 FOR I=1 TO N
190 FOR J=1 TO M
200 PRINT TAB(J*4-4);A(I,J)+B(I,J);
210 NEXT J
220 PRINT
230 NEXT I
```

Line 240 contains N and M.

```
240 DATA 2, 3
```

Lines 251–252 contain the entries of A by rows.

```
251 DATA 2, 6, 3
252 DATA 7, 8, 2
```

Lines 351–352 contain the entries of B by rows.

```
351 DATA -1, -9, -3
352 DATA 32, 15, 25
RUN
```

This is the output.

```
A=
 2    6    3
 7    8    2
B=
-1   -9   -3
 32   15   25
A+B=
 1   -3    0
 39   23   27
```

Program A16: MATMULT (See Algorithm 4.4, Section 4.6.)

```
01 REM THIS PROGRAM MULTIPLIES TWO MATRICES
02 REM IN DATA * SEE 240-250
05 READ M: READ N
06 READ P: READ Q
08 DIM A(M,N): DIM B(P, Q)
```

Matrix A is M by N.

Matrix B is P by Q, P=N.

Lines 10–35 read in A.

```
10 FOR I=1 TO M
15 FOR J=1 TO N
20 READ A(I, J)
30 NEXT J
35 NEXT I
38 FOR I=1 TO P
40 FOR J=1 TO Q
45 READ B(I, J)
50 NEXT J
55 NEXT I
```

Lines 60–90 compute the product of A and B. The entries are computed by rows.

```
60 FOR I=1 TO M
65 FOR J=1 TO Q
```

```
68  LET P(I, J)=0
70  FOR K=1 TO N
75  LET P(I, J)=P(I, J)+A(I, K)*B(K, J)
80  NEXT K
85  NEXT J
90  NEXT I
94  PRINT
95  PRINT "THE PRODUCT OF A AND B IS "
96  PRINT
97  FOR I=1 TO M        Lines 97–140 print the product matrix.
100 FOR J=1 TO Q
110 PRINT TAB(J*4-4);P(I,J);
120 NEXT J
130 PRINT
140 NEXT I
150 PRINT:PRINT "WHEN A="
160 PRINT
170 FOR I=1 TO M        Lines 170–179 print A.
173 FOR J=1 TO N
175 PRINT TAB(J*4-4);A(I,J);
177 NEXT J
178 PRINT
179 NEXT I
180 PRINT:PRINT "AND B=":PRINT
190 FOR I=1 TO P        Lines 190–199 print B.
193 FOR J=1 TO Q
195 PRINT TAB(J*4-4);B(I,J);
197 NEXT J
198 PRINT
199 NEXT I
242 DATA 3, 2                   Line 242: M, N.
243 DATA 2, 2                   Line 243: P, Q.
251 DATA 6, 9, 15, 7, 42, 1     Line 251: A by rows.
252 DATA 0, -5, 17, 3           Line 252: B by rows.
RUN
THE PRODUCT OF A AND B IS
153     -3
119     -54
17      -207
WHEN A=
6       9
15      7
42      1
AND B=
0       -5
17       3
```

This is the output.

Exercises for Programs A14, A15, and A16

1. Run Program A14 with the data 2, 4, 6, 8, 10, 12, 14, 16, 18, 20, 22, 24, 26, 28. What is the output? Why? What if the data are 2, 4, 6?
2. Change line 120 in Program A14 to read "120 READ A(J,I)." What is the output? Why?
3. Modify Program A14 so that it will accept data for an N-by-M matrix, where N and M are given in a data statement. Design it so that the reading and printing are both done in one nested loop.
4. Use Program A15 to add the matrices

$$A=\begin{bmatrix} 1 & 2 & 3 & 4 & 5 \\ 6 & 7 & 8 & 9 & 10 \\ 4 & 3 & 2 & 1 & 0 \end{bmatrix} \quad \text{and} \quad B=\begin{bmatrix} 3 & 6 & 7 & 5 & 9 \\ -1 & 0 & 1 & 2 & 6 \\ 3 & 5 & 7 & 9 & 11 \end{bmatrix}.$$

   How can one input the data correctly?

5. Use Program A16 to multiply the matrices

$$A=\begin{bmatrix} 5 & 3 & 7 \\ 4 & 8 & 9 \\ 1 & 0 & 2 \\ 0 & 1 & 7 \end{bmatrix} \quad \text{and} \quad B=\begin{bmatrix} 3 & 5 \\ 4 & 0 \\ 1 & 6 \end{bmatrix}.$$

   How can one input the data correctly?

6. What happens if, in Exercise 5, we try to multiply matrix A by the matrix $B=[3 \quad 5]$? Why?

Program A17: VERTEX DEG

```
3000 REM THIS PROGRAM FINDS DEGREES OF VERTICES
3001 REM OF A GRAPH WHOSE ADJACENCY MATRIX IS
3002 REM IN DATA. SEE LINES 3071-3100.
3010 READ N                              The dimension of the adjacency matrix is N.
3015 PRINT:PRINT
3020 FOR I=1 TO N                        Lines 3020–3060 include the reading of A from data: line 3030.
3025 FOR J=1 TO N
3030 READ A(I,J)
3035 LET S=S+A(I,J)                      The sums of the entries in each row are added in line 3035.
3040 NEXT J
3045 PRINT "DEG(";I;")=";S               This sum is printed in line 3045.
3050 IF S/2<>INT(S/2) THEN D=D+1         Odd vertices are tallied.
3055 S=0                                 A new row sum is initialized in line 3055.
3060 NEXT I
3065 PRINT "THERE ARE ";D;" ODD VERTICES"
3070 DATA 5                              There are N=5 vertices.
3071 DATA 0, 1, 1, 0, 1
```

```
3072 DATA 1, 0, 1, 0, 1        Lines 3071–3075 give the entries of A by rows.
3073 DATA 1, 1, 0, 1, 0
3074 DATA 0, 0, 1, 0, 1
3075 DATA 1, 1, 0, 1, 0
RUN
DEG(1) = 3  ⎫
DEG(2) = 3  ⎪
DEG(3) = 3  ⎬ This is the output.
DEG(4) = 2  ⎪
DEG(5) = 3  ⎭
THERE ARE 4 ODD VERTICES
```

Program A18: PATH MATRIX (See Definition 4.20, Section 4.6.)

```
5000 REM THIS PROGRAM FINDS THE PATH MATRIX
5001 REM OF A GRAPH GIVEN ITS ADJACENCY MATRIX
5002 REM IN DATA. SEE LINES 5300–5350.
5010 READ N
5011 PRINT: PRINT "THE ADJACENCY MATRIX IS"
5015 FOR I = 1 TO N
5020 FOR J = 1 TO N
5021 LET I(J,J) = 1
5022 READ A(I,J)              I is the N-by-N identity matrix.
5024 PRINT TAB(J * 4 – 4);A(I,J);
5025 LET P(I,J) = A(I,J) + I(I,J)     Line 5025 begins the nesting procedure continued in 5070,
5030 NEXT J                           defining P=A+I.
5033 PRINT
5035 NEXT I
5036 PRINT
5038 IF N = 2 THEN GOTO 5162
5040 FOR L = 1 TO N – 2       Lines 5040–5160 define P=
5045 FOR I = 1 TO N           I+A+A^2+ · · · +A^(N-1) in nested form
5050 FOR J = 1 TO N           (. . . ((A+I)A+I)A . . . )A+I.
5060 LET M(I,J) = 0
5065 FOR K = 1 to N
5070 LET M(I,J) = M(I,J) + P(I,K) * A(K,J)
5075 NEXT K
5080 NEXT J
5085 NEXT I
5130 FOR I = 1 TO N
5135 FOR J = 1 TO N
5138 LET P(I,J) = M(I,J)
5140 LET P(J,J) = P(J,J) + I(J,J)
5150 NEXT J
5155 NEXT I
5160 NEXT L
5162 PRINT "THE PATH MATRIX IS"
5165 FOR I = 1 TO N
5170 FOR J = 1 TO N
```

```
5175  IF P(I,J)< >0 THEN LET P(I,J)=1
5177  LET K=K+P(I,J)
5180  PRINT TAB(J*4-4);P(I,J);
5190  NEXT J
5195  PRINT
5200  NEXT I
5205  IF K<N↑2 THEN GOTO 5220
5210  PRINT "THE GRAPH IS CONNECTED"
5215  STOP
5220  PRINT "THE GRAPH IS NOT CONNECTED"
5320  DATA 5
5321  DATA 0, 1, 0, 1, 0
5322  DATA 1, 0, 0, 0, 0
5323  DATA 0, 0, 0, 1, 1
5324  DATA 1, 0, 1, 0, 0
5325  DATA 0, 0, 1, 0, 0
RUN
```

Line 5175 converts the nonzero entries of P to 1's.

The path matrix must have all N-square entries equal to 1 for the graph to be connected.

N=5 is the number of vertices.

The entries of A are listed in data by rows: lines 5321–5325.

```
THE ADJACENCY MATRIX IS
0    1    0    1    0
1    0    0    0    0
0    0    0    1    1
1    0    1    0    0
0    0    1    0    0
THE PATH MATRIX IS
1    1    1    1    1
1    1    1    1    1
1    1    1    1    1
1    1    1    1    1
1    1    1    1    1
THE GRAPH IS CONNECTED
```

This is the output.

Exercises for Programs A17 and A18

1. Use Program A17 to find the vertex degrees of the following graph.

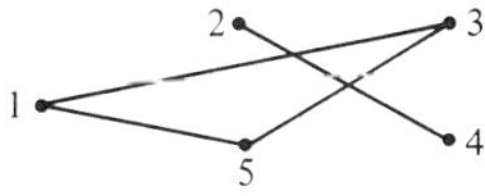

2. Can Program A18 be used to find the vertex degrees in a multigraph? What about a digraph?
3. The number D calculated in Program A17 will always be even. Why?
4. Why does Program A18 go to line 5162 if $N=2$? (See line 5038.)
5. Test Program A18 with the graph in Exercise 1.
6. Will Program A18 work on a multigraph? On a digraph?
7. What would we get if we left lines 5175 and 5177 out of Program A18?
8. Write a program that calculates the nth power of a matrix A.

Program A19: EULER T/C (See Algorithm 4.1, Section 4.1.)

```
400 REM THIS PROGRAM FINDS AN EULER TRAIL OR CIRCUIT
401 REM IN A MULTIGRAPH GIVEN AN APPROPRIATE ADJACENCY
402 REM MATRIX IN DATA
403 DIM T(30),Q(30),A(15,15)
404 READ N
```

The multigraph must be connected and have at most two vertices of odd degree. If there are odd vertices, then one of them must be taken as vertex 1.

```
405 PRINT
406 PRINT "IF THE ADJACENCY MATRIX IS"
407 FOR I=1 TO N
408 FOR J=1 TO N
410 READ A(I,J)
411 PRINT TAB(J*4-4);A(I,J);
412 NEXT J
413 PRINT
414 NEXT I
420 PRINT
425 PRINT "THEN AN EULER TRAIL OR CIRCUIT IS"
430 LET I=1
```

Lines 430–520 find the initial circuit or trail.

```
450 LET T(I)=1
```

For a trail we let T(1) equal a vertex of odd degree. K counts the number of edges and new subscripts used.

```
455 LET K=1
460 FOR J=1 TO N
470 IF A(I,J)=0 THEN GOTO 520
480 LET A(I,J)=A(I,J)-1
```

480–490 update the matrix when edge is used.

```
490 LET A(J,I)=A(J,I)-1
500 LET I=J
510 GOTO 620
520 NEXT J
530 FOR I=1 TO N
```

Lines 530–560 find an unused edge, if any.

```
540 FOR J=1 TO N
550 IF A(I,J)=0 THEN GOTO 590
560 LET T(K+1)=0
570 LET K=K+1
575 LET B1=1
580 GOTO 641
```

This marker B1 is set when there is a subcircuit that must be added to the original trail (see line 575).

```
590 NEXT J
600 NEXT I
605 IF B1=1 THEN GOTO 680
```

A special printing method is needed to add subcircuits to the first trail.

```
610 GOTO 645
```

Otherwise (line 610) the printing of the trail may proceed.

```
620 LET K=K+1
630 LET T(K)=J
640 GOTO 460
641 LET K=K+1
642 LET T(K)=I
```

Lines 642 defines where to begin a subcircuit.

```
643 GOTO 460
645 PRINT
650 FOR I=1 TO K
```

```
660  PRINT T(I);            Lines 650–670 print the trail when no t(k)=0, that is, when there is
670  NEXT I                 no subcircuit to be added to the main trail.
675  STOP
680  PRINT
710  LET U=0          Lines 710–764 concatenate the main trail with the subcircuits.
712  LET U=U+1
714  IF U=K+1 THEN GOTO 780       Line 714: If all subcircuits are concatenated, then
                                  give output.
716  IF T(U)< >0 THEN GOTO 712
718  FOR I=1 TO U-1    If T(U)=0, then relabel.
720  IF T(U+1)< >T(I) THEN GOTO 726     Find end of subcircuit.
722  LET R=I          In line 722 end of subcircuit has been found; relabeling begins.
724  GOTO 728
726  NEXT I
728  FOR I=1 TO R-1       Relabeling takes place in lines 728–763.
730  LET Q(I)=T(I)
732  NEXT I
734  FOR J=U+1 TO K+1
736  IF T(J)=0 THEN GOTO 742
738  LET Q(R-U+J-1)=T(J)
740  NEXT J
742  LET T=J
744  FOR J=R+1 TO U
746  LET Q(T+J-U-2)=T(J)
748  NEXT J
750  FOR I=1 TO T-3
752  LET T(I)=Q(I)
754  NEXT I
756  FOR I=T TO K+1
758  LET T(I-2)=T(I)
760  NEXT I
762  LET K=K-2
763  LET T(K+1)=0
764  GOTO 710
780  FOR I=1 TO K         Lines 780–784 print the trail when there were
782  PRINT T(I);          subcircuits to be added.
784  NEXT I
800  DATA 4          The number of vertices is N=4.
801  DATA 0,1,1,1         Lines 801–804 contain the adjacency
804  DATA 1,0,2,1         matrix of the multigraph by rows.
803  DATA 1,2,0,1
804  DATA 1,1,1,0
RUN

IF THE ADJACENCY MATRIX IS
0   1   1   1
1   0   2   1
1   2   0   1
1   1   1   0
```

```
THEN AN EULER TRAIL OR CIRCUIT IS
1 2 3 1 4 2 3 4
```

The output contains an Euler trail that is not a circuit.

Exercises for Program A19

1. Input the adjacency matrix of the following graph in Program A19 and run it.

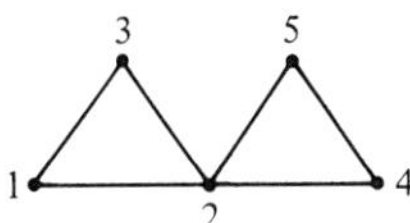

2. Input the adjacency matrix of the following graph in such a way that it will find an Euler trail.

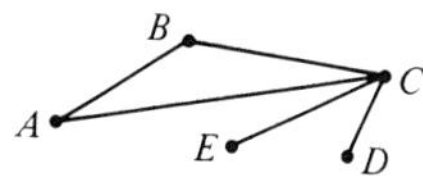

3. Follow the instructions in Exercise 3 for the following multigraph.

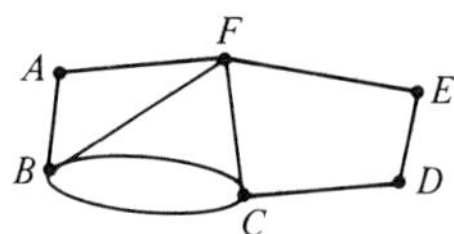

4. Add instructions to the beginning of Program A19 that will **(a)** check if your multigraph is connected; **(b)** check that there are no more than two vertices of odd degree; **(c)** print a message that there is no Euler trail when (a) and/or (b) is not satisfied.

5. Add further instructions to Program A19 such that when Exercises 4(a) and 4(b) are satisfied the computer finds an odd vertex v, if any, and sets T(1)=v in line 450.

6. Test the enhanced version of the program you obtained in Exercises 4 and 5 with the following matrices.

(a) $\begin{bmatrix} 0 & 1 & 1 & 1 \\ 1 & 0 & 1 & 1 \\ 1 & 1 & 0 & 1 \\ 1 & 1 & 1 & 0 \end{bmatrix}$; **(b)** $\begin{bmatrix} 0 & 1 & 0 & 0 \\ 1 & 0 & 0 & 0 \\ 0 & 0 & 0 & 1 \\ 0 & 0 & 1 & 0 \end{bmatrix}$; **(c)** $\begin{bmatrix} 0 & 1 & 0 & 0 \\ 1 & 0 & 1 & 1 \\ 0 & 1 & 0 & 1 \\ 1 & 1 & 1 & 0 \end{bmatrix}$.

Program A20: MIN COST TREE (See Algorithm 5.1, Section 5.1.)

```
4000 REM THIS PROGRAM FINDS A MINIMUM SPANNING TREE
4001 REM OF A GRAPH WHOSE COST MATRIX IS IN DATA—
4002 REM SEE 5330
4010 READ N: PRINT "IF THE COST MATRIX IS" :DIM C(N,N),A(N,N),P(N,N)
```

```
4020 FOR I=1 to N
4030 FOR J=1 TO N
4040 READ C(I,J): PRINT TAB(J*4-4);C(I,J);
4050 LET A(I,J)=0:LET P(I,J)=0
4060 NEXT J
```

C(I, J) is cost matrix.
A(I, J) is tree matrix.
P(I, J) checks for cycles.

```
4070 PRINT
4075 NEXT I
4080 FOR E=1 TO N-1
```

Line 4080 begins a loop that looks for the N−1 edges of the tree.

```
4100 LET B=1.E 20
```

B is set=to the highest possible cost of any "edge."

```
4110 FOR I=1 TO N-1
4120 FOR J=I+1 TO N
```

Lines 4110–4180 find least cost not yet considered.

```
4130 IF C(I,J)>=B THEN GOTO 4170
4140 LET B=C(I,J)
4150 LET I0=I
4160 LET J0=J
4170 NEXT J
4180 NEXT I
4190 IF B=1.E20 THEN GOTO 4300
```

We branch to output statement when there are no longer "finite" edges.

```
4200 LET C=C(I0,J0)
4215 LET C(I0,J0)=1.E30
4220 IF P(I0,J0)=0 THEN GOTO 4260
4230 REM
```

Otherwise (IO, JO) is added to the tree, its cost C is added to the sum M, and matrices A, C, and P are updated. The cost of (IO, JO) is raised so that it is no longer considered.

```
4240 GOTO 4100
4250 REM
```

However, if a cycle would be formed with (IO, JO), then we go back to find a different edge in line 4100.

```
4260 LET A(I0,J0)=1:LET A(J0,I0)=1
4265 LET M=M+C
4270 GOSUB 5000
4280 REM
```

In this "GOSUB" statement, we continue from line 5000 until reaching a RETURN statement. This returns us to the next line, 4280.

```
4290 NEXT E
4300 REM
```

Lines 4300–4390 give us the output: the tree matrix and the minimum cost.

```
4310 PRINT "THE ADJACENCY MATRIX OF THE MINIMUM COST TREE IS"
4320 FOR I=1 TO N
4330 FOR J=1 TO N
4340 PRINT TAB(J*4-4);A(I,J);
4350 NEXT J
4360 PRINT
4370 NEXT I
4380 PRINT "THE MINIMUM COST IS ";M
4390 END
5000 REM
```

Line 5000 begins the P(I, J) matric updating subroutine. You will recognize it from Program A16.

```
5010 FOR I=1 TO N
5020 FOR J=1 TO N
5030 LET I(J,J)=1
5040 LET P(I,J)=A(I,J)+I(I,J)
5050 NEXT J
5060 NEXT I
5065 IF N=2 THEN GOTO 5220
5070 FOR L=1 TO N-2
5080 FOR I=1 TO N
```

```
5090 FOR J=1 TO N
5100 LET M(I,J)=0
5110 FOR K=1 TO N
5120 LET M(I,J)=M(I,J)+P(I,K)*A(K,J)
5130 NEXT K
5140 NEXT J
5150 NEXT I
5155 FOR I=1 TO N
5160 FOR J=1 TO N
5165 LET P(I,J)=M(I,J)
5170 LET P(J,J)=P(J,J)+I(J,J)
5180 NEXT J
5190 NEXT L
5220 RETURN
5320 REM
```

The RETURN statement brings us to one line after GOSUB 5000, namely, line 4280.

```
5330 DATA 5
```

The number of vertices N=5.

```
5331 DATA 0,8,7,10,5
5332 DATA 8,0,3,12,9
5333 DATA 7,3,0,1,7
5334 DATA 10,12,1,0,1.E30
5335 DATA 5,9,7,1.E30,0
```

Lines 5331–5336 give the cost matrix by rows.

An entry of 1.E30=10^{30} means that there is no edge in that position.

```
RUN
THE ADJACENCY MATRIX OF THE MINIMUM COST TREE IS
0    0    1    0    1
0    0    1    0    0
1    1    0    1    0
1    0    0    0    0
THE MINIMUM COST IS 16
```

This is the adjacency matrix of a minimum cost (or minimum spanning) tree.

Exercises for Program A20

1. Run Program A20 with the cost matrix

$$C=\begin{bmatrix} 0 & 2 & 4 & 6 & 8 \\ 2 & 0 & 1 & 3 & 5 \\ 4 & 1 & 0 & 7 & 9 \\ 6 & 3 & 7 & 0 & 1 \\ 8 & 5 & 9 & 1 & 0 \end{bmatrix}.$$

2. Use Program A20 to find a spanning tree of the following graph. Use the adjacency matrix modified with entries 1.E30 as the cost matrix.

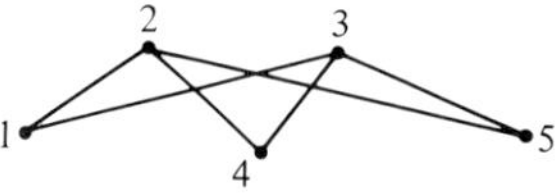

3. Which part of Program A20 seems to use the most time?

Program A21 : DFS SPAN TREE (See Algorithm 6.1, Section 6.2.)

```
1000 REM THIS PROGRAM FINDS A DEPTH-FIRST SPANNING TREE
1001 REM ROOTED AT VERTEX 1 OF A DIGRAPH GIVEN IN DATA
1002 REM OR PRINTS "NOT CONNECTED"
1010 READ N
1020 PRINT "IF THE ADJACENCY MATRIX IS"
1030 FOR I=1 TO N
1040 FOR J=1 TO N
1050 READ A(I,J)
1060 PRINT TAB(J*4-4); A(I,J);
1070 NEXT J
1080 PRINT
1090 NEXT I
1100 LET B(1)=1
1110 LET K=1                          K denotes the DFN of the vertices.
1115 IF K=N THEN GOTO 1238            We branch to the output subroutine.
1120 FOR I=K TO 1 STEP -1             Lines 1120–1200 backtrack to the first vertex
1130 FOR J=1 TO N                     where there is an outgoing unused edge that
1140 IF A(B(I),J)=0 THEN GOTO 1190    leads to a new vertex.
1150 FOR R=1 TO K
1160 IF B(R)=J THEN GOTO 1190
1170 NEXT R
1180 GOTO 1230        The chosen edge does not lead to any vertices already selected.
1190 NEXT J
1200 NEXT I
1210 PRINT "NOT CONNECTED"            If we reach line 1210, then the root cannot reach
1220 STOP                             all the other vertices, since K<N.
1230 LET K=K+1
1235 LET D(B(I),J)=1      An edge is included in the tree.
1236 LET B(K)=J           Its terminal vertex gets the DFN J.
1237 GOTO 1115            We check to see if we are done.
1238 PRINT
1239 PRINT "THEN A DFS SPANNING TREE IS"     The next few lines give our output.
1240 FOR I=1 TO N
1250 FOR J=1 TO N
1260 PRINT TAB(J*4-4);D(I,J);
1270 NEXT J
1280 PRINT
1290 NEXT I
1295 PRINT
1300 PRINT "THE DEPTH-FIRST NUMBERING OF"
1310 PRINT "THE VERTICES IS"
1320 FOR I=1 TO N
1330 PRINT B(I);
1340 NEXT I
1400 REM
1410 DATA 5              Line 1410 carries N. Lines 1411–1415 hold the adjacency matrix
1411 DATA 0,0,1,1,1      entries by rows.
```

```
1412 DATA 0,0,1,1,0
1413 DATA 1,1,0,0,0
1414 DATA 1,1,0,0,0
1415 DATA 1,0,0,0,0
```

This is adjacency matrix of a digraph in which each edge is bidirectional.

```
RUN
IF THE ADJACENCY MATRIX IS
0   0   1   1   1
0   0   1   1   0
1   1   0   0   0
1   1   0   0   0
1   0   0   0   0
THEN A DFS SPANNING TREE IS
0   0   1   0   1
0   0   0   1   0
0   1   0   0   0
0   0   0   0   0
0   0   0   0   0
```

The directed nature of the tree is evident from the lack of symmetry in this matrix.

```
THE DEPTH-FIRST NUMBERING OF
THE VERTICES IS
1   3   2   4   5
```

This numbering tells us the order in which the vertices were found.

Program A22: TOPL SORT (See Algorithm 6.2, Section 6.2.)

```
100 REM THIS PROGRAM TOPOLOGICALLY SORTS THE VERTICES OF
101 REM AN ACYCLIC DIGRAPH WITH ONE SOURCE AND ONE SINK-SEE DATA.
104 PRINT "IF THE ADJACENCY MATRIX IS"
105 READ N:DIM A(N,N), B(N), P(N), T(N)
106 FOR I=1 TO N
107 FOR J=1 TO N
108 READ A(I,J)
109 PRINT TAB(J*4-4);A(I,J);
110 NEXT J
111 PRINT
```

Begin constructing a DFS tree at the source and give it DFN 1.

```
112 NEXT I
113 LET B(1)=1
114 LET M=N
115 LET V=1
116 LET K=1
130 FOR J=1 TO N
140 IF A(V,J)=0 THEN GOTO 190
150 FOR R=1 TO K
160 IF B(R)=J THEN GOTO 190
170 NEXT R
```

Lines 130–190 continue giving DFNs until the sink is found. If A(V, J)=1, then V is not the sink, and we check if J has backtrack number.

```
180 GOTO 350
```

If J has no backtrack number, we go to 350.

```
190 NEXT J
200 LET T(M)=V
```

The sink, V, is given the sorting number M.

```
210 FOR S=1 TO N
220 LET A(S,V)=0
```

All edges terminating in the sink are removed from the digraph.

```
230  NEXT S
240  LET M=M-1
250  IF M=0 THEN GOTO 380      Unless we have finished sorting, we begin to look for a
260  LET Q=V                   new sink by checking the predecessor of the old sink.
262  LET V=P(Q)
265  GOTO 130
350  LET K=K+1
360  LET B(K)=J         J is given the backtrack number K+1.
365  LET P(J)=V         We record the predecessor of J as V, and make J the new test sink.
370  LET V=B(K)
375  GOTO 130
380  PRINT
385  PRINT "THEN THE VERTICES ARE TOPOLOGICALLY SORTED"
390  PRINT "IN THE ASCENDING SEQUENCE:"
400  FOR K=1 TO N
410  PRINT TAB(K*4-4);T(K);
420  NEXT K
440  REM
441  REM     N is read from line 450.
450  DATA 5       The adjacency matrix is read from subsequent rows.
451  DATA 0,0,0,1,1     Vertex 1 must be the source.
452  DATA 0,0,0,0,0     Vertex 2 happens to be the sink.
453  DATA 0,1,0,0,0
454  DATA 0,1,1,0,0
455  DATA 0,0,1,0,0
RUN
IF THE ADJACENCY MATRIX IS
0  0  0  1  1
0  0  0  0  0
0  1  0  0  0                                  This is the output.
0  1  1  0  0
0  0  1  0  0
THEN THE VERTICES ARE TOPOLOGICALLY SORTED
IN THE ASCENDING SEQUENCE:
1  5  4  3  2
```

Exercises for Programs A21 and A22

1. Run Program A21, using the following digraphs as data.

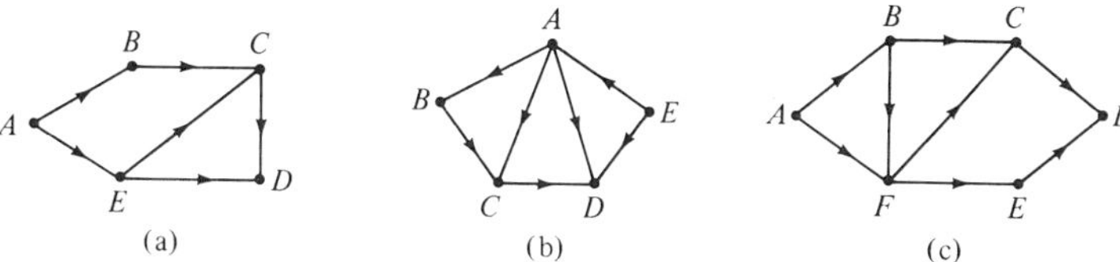

2. Run Program A21 using the following graphs as data. (*Hint*: Replace each edge with two directed edges, one in each direction.)

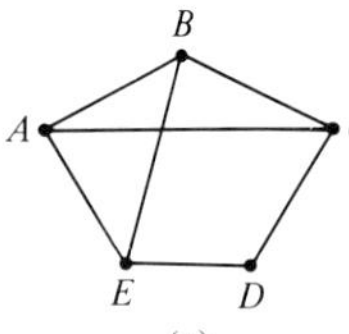

(a)

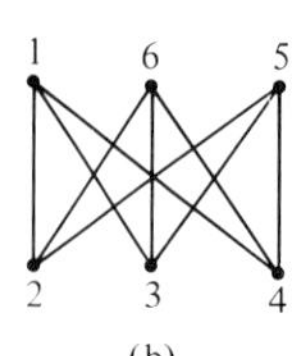

(b)

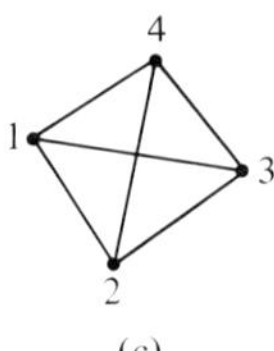

(c)

3. Modify Program A21 so that it finds the source when it is not necessarily placed at vertex 1. (*Hint*: Look for a column of zeros.)
4. Use the output of Program A21 to orient the given graph if possible.
5. Use Program A22 to topologically sort the vertices of the following digraph.

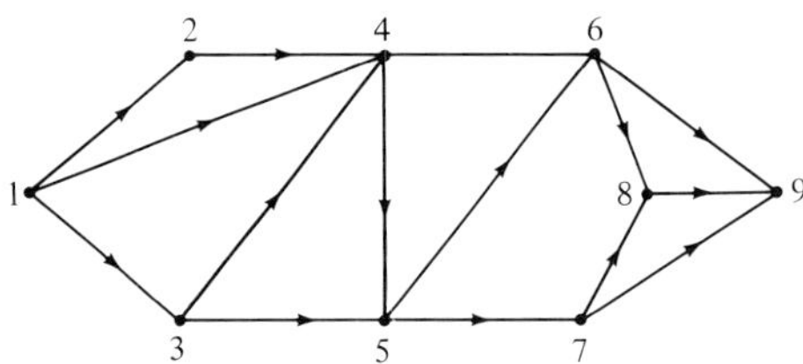

Program A23: TRANS TEST (See Section 7.2.)

```
500 REM THIS PROGRAM CHECKS TRANSITIVITY FOR A RELATION
501 REM MATRIX IN DATA. NO PRIOR CONDITIONS ASSUMED.
510 READ N: DIM A(N,N)
515 PRINT "IF THE RELATION MATRIX IS"
520 FOR I=1 TO N
530 FOR J=1 TO N
540 READ A(I,J)
550 PRINT TAB(J*4-4);A(I,J);
560 NEXT J
570 PRINT
580 NEXT I
590 PRINT "THEN THE RELATION IS"
600 FOR I=1 TO N
```

Transitivity is checked in lines 600–670. We verify for all x, y, and z that xRy and yRz implies that xRz.

```
610 FOR J=1 TO N
620 FOR K=1 TO N
630 IF A(I,J)=0 OR A(J,K)=0 THEN GOTO 650
640 IF A(I,K)=0 THEN GOTO 700
```

In line 640, if iRj and jRk do not imply that iRk, then the relation is not transitive.

```
650 NEXT K
660 NEXT J
670 NEXT I
680 PRINT "TRANSITIVE"
690 STOP
700 PRINT "NOT TRANSITIVE"
710 STOP
```

```
740 REM
750 DATA 5
751 DATA 1,1,1,1,1
752 DATA 0,1,1,1,0
753 DATA 0,0,1,0,0
754 DATA 0,0,1,1,0
755 DATA 0,1,1,1,1
RUN
IF THE RELATION MATRIX IS
1   1   1   1   1
0   1   1   1   0
0   0   1   0   0
0   0   1   1   0
0   1   1   1   1
THEN THE RELATION IS
TRANSITIVE
```

Line 750 is read as N.

Lines 751–755 are read as the entries of the relation matrix.

Notice that this relation is not symmetric.

Program A24: SYMTRANSTEST (See Algorithm 7.1, Section 7.2.)

```
500 REM THIS PROGRAM CHECKS WHETHER A
501 REM SYMMETRIC RELATION IN DATA IS TRANSITIVE.
510 READ N:DIM A(N,N)
515 PRINT "IF THE RELATION MATRIX IS"
520 FOR I=1 TO N
530 FOR J=1 TO N
540 READ A(I,J)
550 PRINT TAB(J*4-4);A(I,J);
560 NEXT J
570 PRINT
580 NEXT I
590 PRINT "THEN THE RELATION IS"
600 FOR I=1 TO N
610 FOR J=1 TO N-1
620 IF A(I,J)=0 THEN GOTO 660
630 FOR K=J TO N
640 IF A(I,K)=1 AND A(J,K)=0 THEN GOTO 700
650 NEXT K
660 NEXT J
670 NEXT I
680 PRINT "TRANSITIVE"
690 STOP
700 PRINT "NOT TRANSITIVE"
710 STOP
720 REM
750 DATA 5
751 DATA 1,0,0,1,0
752 DATA 0,1,0,0,1
753 DATA 0,0,1,0,0
```

Transitivity is checked in lines 600–670. We proceed as in Algorithm 7.1.

If iRj and iRk but $j \not R k$, then the relation is not transitive, since it is symmetric.

Line 750 contains N.

Subsequent lines contain the entries of the relation matrix.

```
754  DATA 1,0,0,1,0
755  DATA 0,1,0,0,1
RUN
IF THE RELATION MATRIX IS
1    0    0    1    0
0    1    0    0    1
0    0    1    0    0
1    0    0    1    0
0    1    0    0    1
THEN THE RELATION IS
TRANSITIVE
```

The relation matrix must be symmetric.

Exercises for Programs A23 and A24

1. Run Program A23 with the matrix from Program A24. Do you get the same answer?
2. Run Program A23 with the matrix from Program A24. Do you get the same answer? Can Program A24 be used to check transitivity on relations that are not symmetric?
3. Write a program that will check whether a relation given by a matrix is symmetric.
4. Write a program that will check whether a relation given by a matrix is reflexive.
5. Write a program that will check whether a relation given by a matrix is an equivalence relation.

Program A25: LOGIC TABLE (See Section 2.4 and 2.5.)

```
10   REM THIS PROGRAM PRINTS A LOGIC TABLE
15   REM FOR THE STATEMENT IN LINE 110
20   LET T=(1<2): LET F=(1>2)
25   PRINT " P, " Q", " S"
30   LET P=T: Q=T: GOSUB 100
40   LET P=T: Q=F: GOSUB 100
50   LET P=F: Q=T: GOSUB 100
60   LET P=F: Q=F: GOSUB 100
65   PRINT "NOTE -1=T AND 0=F"
70   STOP
100  PRINT P, Q,
110  S=NOT(P AND NOT(Q))
120  PRINT S
130  RETURN
RUN
S=NOT(P AND NOT (Q))
P          Q          S
-1         -1         -1
-1          0          0
 0         -1         -1
 0          0         -1
NOTE -1=T AND 0=F
```

T is a tautology and F is a contradiction.

For each pair of truth values the "GOSUB" directs the program to line 100, where the values of P and Q are printed, S is evaluated and printed, and we return to the next instruction.

$S = \neg(p \wedge (\wedge q))$.

This is the output.

Exercises for Program A25

1. Modify the program so that it will print a logic table for $\neg(p \vee ((\neg q) \wedge p)) \vee ((\neg p) \wedge q)$.
2. Modify the program so that it will print a logic table for $[p \vee (q \wedge (\neg r))] \vee (r \wedge p)$.

BIBLIOGRAPHY

1. Aho, A., J. Hopcroft, and J. Ullman, *The Design and Analysis of Computer Algorithms*, Addison-Wesley, Reading, Mass., 1974.
2. Appel, K., and W. Haken, "Every planar map is 4-colorable," *Bulletin of the American Mathematics Society*, 82 (1976), 711–712.
3. Baase, S., *Computer Algorithms: Introduction to Design and Analysis*, Addison-Wesley, Reading, Mass., 1978.
4. Bellman, R., K. L. Cooke, and J. A. Lockette, *Algorithms, Graphs, and Computers*, Academic Press, New York, 1970.
5. Bellmore, M., and G. L. Nemhauser, "The traveling salesman problem," *Operations Research*, 16(1968), 538–558.
6. Berge, C., *The Theory of Graphs and Its Applications*, Wiley, New York, 1962.
7. Birkhoff, G., and T. C. Bartee, *Modern Applied Algebra*, McGraw-Hill, New York, 1970.
8. Bogart, K. P., *Introductory Combinatorics*, Pitman, Marshfield, Mass., 1983.
9. Bondy, J. A., and U. S. R. Murty, *Graph Theory with Applications*, North Holland, New York, 1976.
10. Boole, G., *Laws of Thought*, reprinted by Dover, New York, 1951.
11. Busacker, R. G., and T. L. Saaty, *Finite Graphs and Networks: An Introduction with Applications*, McGraw-Hill, New York, 1965.
12. Capobianco, M., and J. C. Mollezzo, *Examples and Counter-examples in Graph Theory*, North Holland, New York, 1978.
13. Even, M., *Algorithmic Combinatorics*, Macmillan, New York, 1973.
14. Feller, W., *An Introduction to Probability Theory and Its Applications*, Wiley, New York, 1950.
15. Ford, L. R., and D. R. Fulkerson, *Flows in Networks*, Princeton University Press, Princeton, N.J., 1962.

16. Golumb, S., and L. Baumert, "Backtrack programming," *Journal of the ACM*, 12(1965), 516–524.
17. Halmos, P. R., *Naive Set Theory*, Van Nostrand, New York, 1967.
18. Henkin, L., "On Mathematical induction," *American Mathematical Monthly*, 67(1960), 323–337.
19. Hohn, F. E., *Applied Boolean Algebra, 2nd ed.*, Macmillan, New York, 1966.
20. Hu, T. C., *Combinatorial Algorithms*, Addison-Wesley, Reading, Mass., 1982.
21. Kline, M., *Mathematical Thought from Ancient to Modern Times*, Oxford University Press, New York, 1972.
22. Knuth, D. E., *The Art of Computer Programming, Vol. 1: Fundamental Algorithms*, Addison-Wesley, Reading, Mass., 1973.
23. Knuth, D. E., *The Art of Computer Programming, Vol. 3: Sorting and Searching*, Addison-Wesley, Reading, Mass., 1973.
24. Lipschutz, S., *Theory and Problems of Set Theory and Related Topics*, Schaum, New York, 1964.
25. Lipschutz, S., *Discrete Mathematics*, Schaum, New York, 1976.
26. Lipschutz, S., *Essential Computer Mathematics*, Schaum, New York, 1982.
27. Liu, C. L., *Introduction to Combinatorial Mathematics*, McGraw-Hill, New York, 1968.
28. Mendelson, E., *Boolean Algebra and Switching Circuits*, Schaum, New York, 1970.
29. Niven, I., *Mathematics of Choice*, Random House, New York, 1965.
30. Ore, O., *Graphs and Their Uses*, Random House, New York, 1963.
31. Riordan, J., *An Introduction to Combinatorial Analysis*, Wiley, New York, 1958.
32. Stanat, D. F., and D. F. McAllister, *Discrete Mathematics in Computer Science*, Prentice-Hall, Englewood Cliffs, N.J., 1977.
33. Tremblay, J. P., and R. Manohar, *Discrete Mathematical Structures with Applications to Computer Science*, McGraw-Hill, New York, 1975.
34. Tucker, A., *Applied Combinatorics*, Wiley, New York, 1980.
35. Wand, M., *Induction, Recursion, and Programming*, North Holland, New York, 1980.
36. Welsh, D. J. A., and M. B. Powell, "An upper bound to the chromatic number of a graph and its applications to time tabling problems," *The Computer Journal*, 10(1967), 85–86.

ANSWERS TO SELECTED EXERCISES

Section 1.1

1. No; "large" is not precise. **3.** No; "fast" is not precise. **5.** Yes; it is $\{1, 2, 3, 4, 5, 6, 7, 8, 9, 10\}$. **6.** Yes; it is $\emptyset$. **7.** Yes; colleges keep a record of these things. **8.** No; leads to logical paradox. **9.** $\{1, 2, 3, 4\}$. **11.** $\{1, 2, 3, 4, 5, 6, 7, 8, 9, 11, 12, 13, 14, 15, 16, 17, 18, 19, 20\}$. **13.** If A and B are both empty sets, then $A \subset B$ and $B \subset A$ imply that $A = B$.

15. 

17. (a) and (c). **18.** $P(\{0, 1, 2\}) = \{0, \{0\}, \{1\}, \{2\}, \{0, 1\}, \{0, 2\}, \{1, 2\}, \{1, 2, 0\}\}$. **19.** For both sets there is no element that is not in one but is in the other, so they have the same elements. **21.** $x_5 = 2$; $x_{10} = 1$; $x_{100} = 1$. **22.** Finite. It is $\{1, 2, 3\}$. **23.** The same d_x that worked for each x in the original set works for each x in the subset. **24.** 8. **25.** $|m - n| - 1$.

Section 1.2

1. $\{(a, d), (a, e), (a, f), (a, g), (b, d), (b, e), (b, f), (b, g), (c, d), (c, e), (c, f), (c, g)\}$. **3.** $\{(a, a), (a, b), (a, c), (b, a), (b, b), (b, c), (c, a), (c, b), (c, c)\}$. **5.** mn. **6.** Domain $= \{a, b, c\}$; codomain $= \{d, e, f, g\}$; range $= \{d, e, g\}$. **7a.** $h \notin B$. **7b.** a is the first element of two pairs.

7c. not a subset of $A \times B$. **9.** $f(1)=2$, $f(2)=4, \ldots, f(n)=2n, \ldots$; domain $= N$; range = set of positive even integers. **11.** $0 \leftarrow 1$, $1 \leftarrow 2$, $-1 \leftarrow 3$, $2 \leftarrow 4$, $-2 \leftarrow 5$, etc.; domain $= N$; range $= Z$. **13.** $\{(2, 1), (3, 2), (5, 4)\}$. **15.** $\{(x, x^3)$: x is real$\}$. **17.** $x=6$, $y=11$. **19a.** For each y in B there corresponds a unique x in A. **19b.** f^{-1} is $1-1$ and onto iff f is $1-1$ and onto. **19c.** Arrange A in a finite sequence $a_1, a_2, a_3, \ldots$ and B in a finite sequence $b_1, b_2, b_3, \ldots$. Then the correspondence is $b_1 \leftarrow a_1$, $b_2 \leftarrow a_2$, etc. **20.** $\{(x, x): x \in A\}$.

Section 1.3

1-1. Set $S(1) \leftarrow 0$. **1-2.** Repeat step 3 for $k=2, 3, \ldots, n-1$. **1-3.** Set $S(k) \leftarrow S(k-1) + f(k)$. **1-4.** Output $S(n-1)$ and stop. **1-5.** End.
3. The process would never terminate because k never equals 3.5. No, not if we apply the definition strictly. n must be an integer. **5.** $\sum_{k=2}^{n-1} f(k)$; $\sum_{k=1}^{n} f(2k)$. **7.** $10+11+12+\cdots=5005$. **9.** $2+3+4+\cdots=[n(n+1)]/2+n$. **11.** $5+8+11+\cdots=15{,}350$. **13.** $2^{101}-1$. **15.** $2+2+2+\cdots=200$. **17.** $0+0+0+\cdots=0$. **19a.** \$465. **19b.** Precisely \$0.10 $(1-2^{30})/(1-2)$, or about \$107,370,000. **20a.** $(((x+1)x+1)x+1)x+1$. **20c.** $((2x+0)x-1)x+3$. **21.** In part (c), $n=3$, $a_0=2$, $a_1=0$, $a_2=-1$, $a_3=3$. In part (d), $n=5$, $a_0=1$, $a_1=0$, $a_2=0$, $a_3=0$, $a_4=2$, and $a_5=0$. **23.** $p(-2)=59 \neq 0$. Hence $p(x)$ is not divisible by $x+2$.

Section 1.4

1. For each element x in the set of integers, let $d_x=1$. **3.** If the even numbers were finite, there would be a largest one, say m. But $2m$ is even and $2m>m$. **5.** 1, 2, 3, 5, 8, 13, 21, 34, 55, 89, 144, 233, 377, 610, 987, 1597, 2584, 4181, 6765, 10,946, and 17,711 are the first 21 Fibonacci numbers and $17{,}711 > 12{,}345$. **7-1.** Set $a_1 \leftarrow 1$, $a_2 \leftarrow 1$, $n \leftarrow 3$. **7-2.** Repeat 7-3(a)–7-3(c) until $a_n > x$. **7.3a.** Set $a_n \leftarrow a_{n-1} + 2a_{n-2}$. **7-3b.** If $a_n = x$, output yes and stop. **7-3c.** Else set $n \leftarrow n+1$; output no and stop. **7.5.** End; a_{10}. **9.** Use $d \geqslant \sqrt{p}$ in place of $d \geqslant p/2$. **11a.** $2^3 \cdot 3$. **11b.** $2^2 \cdot 3^2$. **11c.** Prime: $37 \cdot 1$. **11d.** $2^4 \cdot 3$. **11e.** $5 \cdot 13$. **11f.** $3^2 \cdot 11$. **12a.** 1, 2, 4, 6, 8, 3, 12, 24. **13a.** r is decreasing and r is a positive integer. **13b.** Rounded off, r may look like zero while $r \neq 0$. **13c.** 1. **15.** $a=rp$, $b=sp$ with integers r and s. Then $a+b=p(r+s)$, and $r+s$ is an integer. Hence $p/(a+b)$. **17.** If $n=p_1p_2 \ldots p_n$, then $n^2=(p_1p_2p_3 \ldots)(p_1p_2p_3 \ldots)$ uniquely in terms of primes. Hence if a prime p/n^2, it must divide some p_i and, therefore, n as well.

Section 1.5

1. b, d, and e are in lowest terms. **2a.** 0.375. **2f.** $0.\overline{142857}$. **3.** a, b, and d are terminating. **4a.** 23/99. **4e.** 230/99. **4f.** $-13/9$. **5.** change $-$ to $+$; apply algorithm; change $+$ to $-$. Similarly, multiply and divide by an appropriate power of 10. **7.** Neither repeating nor terminating. **9.** If $\sqrt{2}+1$ were rational, then $(\sqrt{2}+1-1)=\sqrt{2}$ would also be rational. Contradiction. **10a.** Truncation. **10d.** Rounding. **11a.** 3, 4. **11b.** 3, 5. **11c.** 3, 5. **11d.** 2, 5. **12a.** 3.5×10^2, 0.35×10^3. **13a.** 3.4500×10^2, 0.34500×10^3. **13b.** 0.00×10^0, 0.000×10^{-1}. **13c.** 3.50×10^0, 0.350×10^{-1}. **15.** In Exercise 14 we have all positive rationals $r_1, r_2, r_3, \ldots$. To list all rationals, write $0, r_1, -r_1, r_2, -r_2, \ldots$. **17a.** If we disallow repeating 9s, it is unique. **17b.** If $b=b_1b_2 \ldots$. where $b_n \neq a_{nn}$ for all n, then $b \neq a_n$ for all n, since b and a_n differ in their nth digit.

Section 1.6

1a. $2^2+2^0=5$. **1b.** $2^3=8$. **1c.** $2^3+2^2+2+2^0=15$. **1d.** $2^0=1$. **1e.** $2^9+2^8+2^7+\cdots+2^0=1023$. **2a.** $2^{-1}=0.5$. **3.** If $x=.\bar{1}$, then $2x=1.\bar{1}$. Hence $2x-x=x=1$. **5a.** 2 bits. **5b.** 2 bits. **5c.** 6 bits. **5d.** 8 bits, or 1 byte on a 8-bit = 1-byte machine. **6a.** $(1010000100100011)_2=41{,}251$. **7a.** $(001010011000)_2=664$. **7b.** $(111000111)_2=455$. **7c.** $(0.111)_2=0.875$. **9a.** $(11101)_2$. **9b.** $(1100100)_2$. **9c.** $(11001000)_2$. **11a.** $(2322)_8$. **11b.** $(402)_{16}$. **13a.** $(0.10011)_2$. **13b.** $(11011.000011)_2$. **15.** Nested: three additions and three multiplications versus three additions and five multiplications without nesting. **17.** Given $h_1h_2h_3h_4h_5h_6$. **17-1.** Set $i\leftarrow 6$, $x\leftarrow h_6$. **17-2.** Set $x\leftarrow 16x+h_{i-1}$. **17-3.** If $i=2$, stop. Output x. Else. **17-4.** Set $i\leftarrow i-1$. **17-5.** Return to step 2.

Section 1.7

1. 11000. **3.** 100110. **5.** 1000.101. **7.** 00101. **9.** 1.101. **11.** 10011010. **13.** 10000.01. **15.** 10110.1. **17.** One's complement = 010010; two's complement = 010011. **19.** One's complement = 110100; two's complement = 110101. **21.** 011011 + 101010 = (1)000101 yielding 101. **23.** 010.110 + 110.111 = (1)001.101, yielding 1.101. **25.** 000101 + 10000 = 100110, yielding −11010.

Section 1.8

1. True; $3-5=2$. **3.** True; $33\text{-}100{,}033=(-1)(100{,}000)$. **5.** False; -25 is not divisible by 12. **7.** False; $3-17$ is not divisible by 5. **9.** $q=0$, $r=4$, $4=(9)(0)+4$, $0<4<9$. **11.** $q=-1$, $r=15$, $-13=(28)(-1)+15$, $0<15<28$. **13.** 0, 1, 2, 3, 4, 5, 6, 7, 8. **15.** 0, 1, 2, 3, 4. **17.**

+	0	1	2	3	4
0	0	1	2	3	4
1	1	2	3	4	0
2	2	3	4	0	1
3	3	4	0	1	2
4	4	0	1	2	3,

×	0	1	2	3	4
0	0	0	0	0	0
1	0	1	2	3	4
2	0	2	4	1	3
3	0	3	1	4	2
4	0	4	3	2	1.

19a. $x=1$. **19b.** $x=4$. **19c.** $x=3$. **19d.** $x=2$.

21a.

Z_{10}:	0	1	2	3	4	5	6	7	8	9
$Z_2 * Z_5$:	(0, 0)	(1, 1)	(0, 2)	(1, 3)	(0, 4)	(1, 0)	(0, 1)	(1, 2)	(0. 3)	(1, 4)

21b. $f(8+9)=f(8)+f(9)=(0+1(\text{mod } 2), 3+4(\text{mod } 5))=(1, 2)$: 7. **21c.** $f((8)(9))=f(8)\cdot f(9)=(0\cdot 1(\text{mod } 2), 3\cdot 4(\text{mod } 5))=(0, 2)$: 2. **23.** $b=xa$ and $c=ya$ imply $(b+c)=(x+y)a$ and $bc=(xya)a$. **26-1.** List the elements of Z_n. **26-2.** List the elements of Z_r s times in succession below your first list. **26-3.** List the elements of Z_s r times in succession, pairing them with those in the lists of Z_r, giving a single list of the elements of $Z_r \times Z_s$. **26-4.** Let the kth elements in the lists of Z_n and $Z_r \times Z_s$ correspond. **26-5.** End.

Section 2.1

1a. $1+1=2$. **1b.** $2\cdot 3=6$. **1c.** Take a rhombus with a 45-degree angle. **1d.** $-7<5$ but $49>25$. **1e.** 1 cannot be represented as a product of primes. **1f.** $1^2+2^2=5\neq(2(3)/2)^2$. **2a.** $2k+2n=2(k+n)$, where k and n are integers. But $k+n$ is an integer. So $2(k+n)$ is even. **3a.** $S(1)$: $1=1$; $S(4)$: $1+3+5+7=16$. **3b.** $S(1)$: $1=[1(1+1)]/2$;

$S(4)$: $1+2+3+4=[4(4+1)]/2$. **3c.** $S(1)$: The angle sum of a convex polygon having one side is -180 degrees; $S(4)$: The angle sum of a convex polygon having four sides is 2(180) degrees. **3d.** $S(1)$: It is possible to make up postage of exactly 1 cent by using combinations of 2-cent and 5-cent postage stamps; $S(4)$: It is possible to make up postage of exactly 4 cents by using combinations of 2-cent and 5-cent postage stamps. **5a.** $S(1)$: $1=1$ is verified. Assume that $1+3+\cdots+2k-1=k^2$ is true for some $k\geqq 1$. Then $(1+3+\cdots+2k-1)+(2(k+1)-1)=k^2+(2(k+1)-1)=k^2+2k+1=(k+1)^2$, as in $S(k+1)$. Hence $S(k+1)$ is true. Thus $S(n)$ is true for all $n\geqslant 1$. **6d.** *Hint*: Use $(1+2+\cdots+k)^2=(k(k+1)/2)^2$. **7.** Let $S(n)$ be a statement concerning every positive integer $n\geqslant n_1$. Suppose we verify that (a) $S(n_1)$ is true, and (b) whenever $S(k)$ is assumed true for some $k\geqslant n_1$ we can show that $S(k+1)$ is true. Then $S(n)$ is true for all $n\geqslant n_1$. **9b-1.** Set SUM$\leftarrow$26. **9b-2.** Repeat step 3 for $n=28, 30, 32, \ldots, 2{,}500{,}000$. **9b-3.** Set SUM$\leftarrow$SUM $+\,n$. **9b-4.** Output SUM and stop. **9b-5.** End. **11.** The incorrect answer may suffice for some purposes but not for others, for example, if we wanted to know if the sum were a prime number.

Section 2.2

1a. $\{1, 2, 3, 6, 10\}$. **1b.** $\{1, 2, 4, 6, 7, 8\}$. **1c.** $\{5, 9\}$. **1d.** $\{3, 4, 5, 7, 8, 9, 10\}$. **1e.** $\{1, 2, 3, 4, 6, 7, 8, 10\}$. **1f.** $\{1, 2, 6\}$. **1g.** See (1f). **1h.** See (1e). **1i.** U. **1j.** $\{4, 7, 8\}$. **1k.** $\{3, 10\}$. **1l.** $\varnothing$. **3.** $\{3, 4, 8, 7, 10\}$.

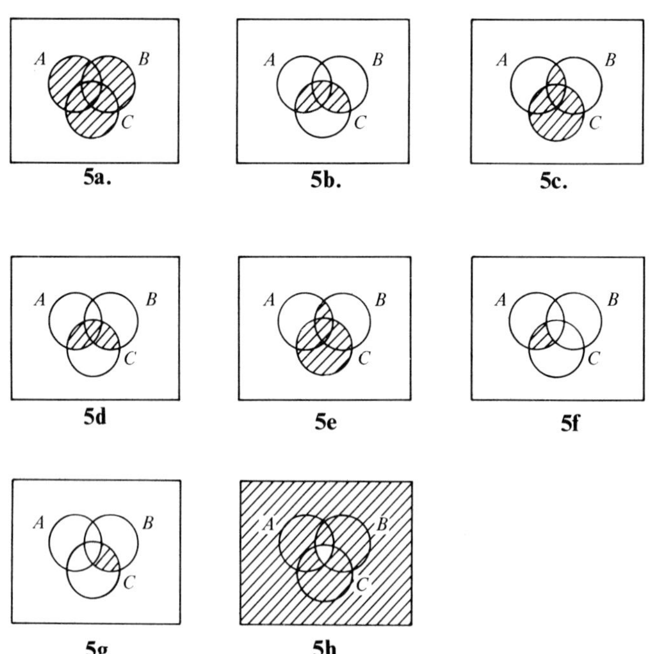

5a. 5b. 5c. 5d 5e 5f 5g 5h

7. 2^5. **9.** $A=S_{0001101110}$, $B=S_{0010100011}$. **11a.** 197. **11b.** 97. **13a.** 2. **13b.** 2. **13c.** 106. **15.** $|A|+|B|$ counts the elements in $A\cap B$ twice and all others in $A\cup B$ once. Hence $|A|+|B|-|A\cap B|$ counts each element in $A\cup B$ exactly once.

Section 2.3

1. $1-2 \neq 2-1$. **3.** $1-(2-3) \neq (1-2)-3$. **5.** $A \cup B = B \cup A = \{1, 3, 4, 5\}$; $A \cap B = B \cap A = \{1, 4\}$. In 3a, take $C = A$. Then $A \cup (B \cap A) = (A \cup B) \cap (A \cup A) = \{1, 4, 5\}$; $\overline{A \cup B} = \bar{A} \cap \bar{B} = \{2\}$. The rest are done similarly.

7.

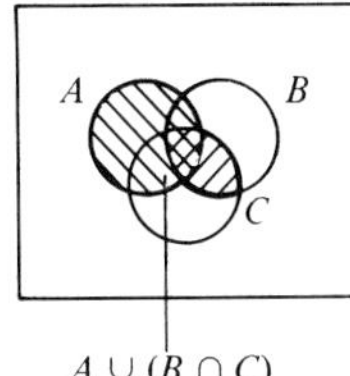

$A \cup (B \cap C)$

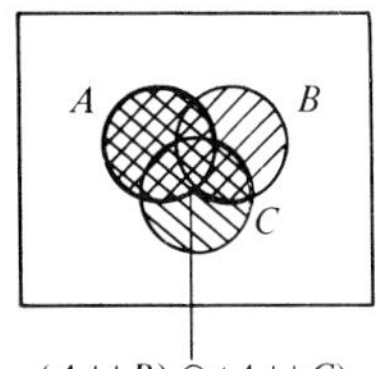

$(A \cup B) \cap (A \cup C)$

9a. If x is in $A \cup B$, then x is in A or x is in B. In other words, x is in B or x is in A. Hence x is in $B \cup A$ and $A \cup B \subset B \cup A$. Similarly, $B \cup A \subset A \cup B$. Hence $A \cup B = B \cup A$. **9b.** Similar to (9a). **11.** For Exercise (6a), if x is in $A \cup A$, then x is in A or A. Thus $A \cup A \subset A$. Since $A \subset A \cup A$, we have $A = A \cup A$. The other involution law is proved similarly. **13.** x is in $\bar{\bar{A}}$ if and only if x is not in $\bar{A}$. This is true if and only if x is in A. This shows that both $\bar{\bar{A}} \subset A$ and $A \subset \bar{\bar{A}}$. Hence $A = \bar{\bar{A}}$. **15.** x is in $A \cup (B \cup C)$ if and only if x is in A or x is in $B \cup C$. This is true iff x is in A or x is in B or x is in C. The first two "or's" are true iff x is in $A \cup B$. With the third this is true iff x is in $(A \cup B) \cup C$. Hence the two sets are equal. The other associative law is done similarly. **17a.** $\bar{A} \cup \bar{B} = A \cap B$. **17b.** $(A \cup \varnothing) \cap (B \cup A) = A$. **17c.** $(\bar{A} \cup B) \cap (A \cup B) = B$. **19a.** $\overline{\bar{A} \cup \bar{B}} = \bar{\bar{A}} \cap \bar{\bar{B}} = A \cap B$. **19b.** $(A \cup \varnothing) \cap (B \cup A) = A \cap (B \cup A) = A$, since $A \subset A \cup B = B \cup A$. **19c.** $(\bar{A} \cup B) \cap (A \cup B) = (\bar{A} \cap A) \cup B = \varnothing \cup B = B$. For each part of Exercise (19) we can also invoke Exercise (18) and the principle of duality.

Section 2.4

1a. Not a proposition. **1b.** Proposition and tautology. **1c.** Proposition. **1d.** Proposition and contradiction. **2b.** $\neg(p \wedge \neg q)$. **3a.** The course is enjoyable or the presentation is stimulating. **3b.** The course is not enjoyable and the presentation is not stimulating. **3c.** It is not the case that the course is enjoyable or that the presentation is not stimulating. **3d.** The material is significant and it is the case that either the course is enjoyable or that the presentation is stimulating. **3e.** It is either the case that the material is significant and the course is enjoyable or that the material is significant and the presentation is stimulating.

4b.

p	q	$\neg p$	$\neg q$	$\neg p \wedge \neg q$
T	T	F	F	F
T	F	F	T	F
F	T	T	F	F
F	F	T	T	T

5a.

$\mathscr{T}$	$\mathscr{F}$	$\mathscr{T} \wedge \mathscr{F}$
T	F	F

5b.

p	$\mathscr{T}$	$\mathscr{T} \vee p$
T	T	T
F	T	T

5c.

$\mathscr{F}$	p	$\mathscr{F} \wedge p$
F	T	F
F	F	F

5d.

$\mathscr{T}$	$\mathscr{F}$	$\mathscr{T} \vee \mathscr{F}$
T	F	T

6a. 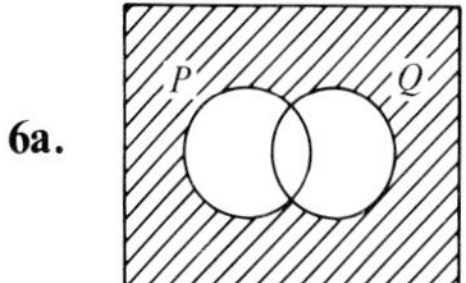

7.

p	q	$\neg(p \wedge q)$	$\neg p \vee \neg q$
T	T	F	F
T	F	T	T
F	T	T	T
F	F	T	T

Corresponding truth values are equal.

9. Its truth values are all F for every combination of values of p and q. **11.** Let $p = I$ will go to the races; $q =$ there is no exam tomorrow; $r =$ Peacemaker is running. Then $q \vee (\neg q \wedge r)$ and $q \vee r$ are equivalent.

q	r	$q \vee (\neg q \wedge r)$	$q \vee r$
T	T	T	T
T	F	T	T
F	T	T	T
F	F	F	F

Section 2.5

1a. $r \rightarrow p$. **1b.** $\neg p \rightarrow (\neg q \vee \neg r)$. **3.** $\neg p \rightarrow \neg r$: If the course is not enjoyable, then the presentation is not stimulating. **4a.** See Exercise (1a). **5a.** $q \rightarrow p$. **5c.** $r \leftrightarrow (p \vee q)$. **7.** If the professor is not delighted, then the class did not perform admirably.

8a.

p	q	$\neg p \rightarrow q$
T	T	T
T	F	T
F	T	T
F	F	F

9. c and d are tautologies and e is a contradiction.

11.

p	q	$q \rightarrow p$	$\neg p \rightarrow \neg q$
T	T	T	T
T	F	T	T
F	T	F	F
F	F	T	T

Corresponding truth values are equal.

13. One obtains all T's with the second "$\rightarrow$."
14a. A rectangle is not necessarily a square. **15.** If a program is completely debugged, then it runs perfectly. My program was completely debugged. Therefore, it ran perfectly.
17. Let p = the automatic monitoring system will be in operation; q = there is a large payroll in the office; r = there is nobody in the office; s = electric power should be turned on. Then the power should be turned on if $\neg(r \wedge \neg(q \vee r)) = \neg(r \wedge (\neg q \wedge \neg r)) = \neg((r \wedge \neg r) \wedge \neg q)) = \neg(\mathscr{F} \wedge \neg q) = \neg\mathscr{F} = \mathscr{T}$, in other words, in every case. **19a.** $p \wedge q = \neg(\neg p \vee \neg q)$.
19b. $\neg(\neg p \vee q) \vee (\neg(\neg p \vee \neg q))$. **19c.** $\neg(\neg(\neg p \vee q) \vee \neg(\neg q \vee p))$.

Section 2.6

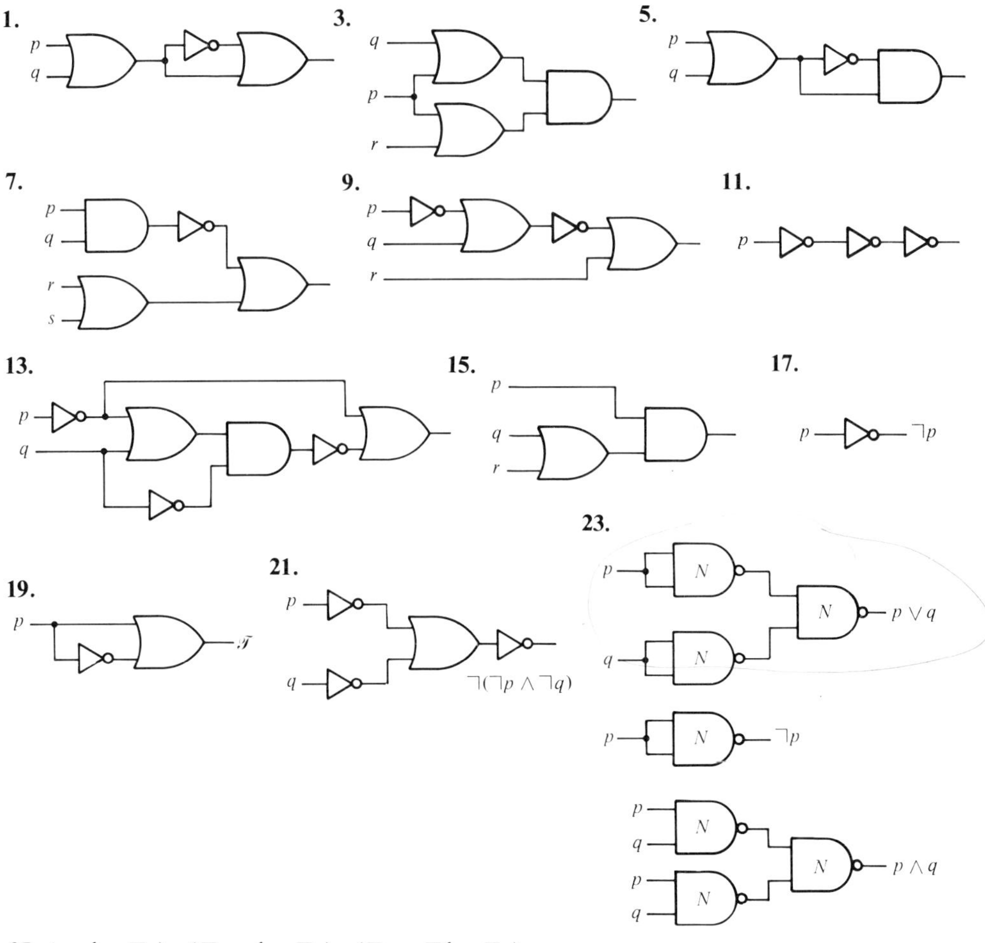

25. $(a \wedge b \wedge \neg c) \vee (\neg a \wedge b \wedge \neg c) \vee (\neg a \wedge \neg b \wedge \neg c)$.

Section 3.1

1a. 6. **1b.** 720. **1c.** 42,320. **1d.** 479,001,600. **1e.** 1. **2d.** $20/1 = 20$.
3a. n. **3b.** $n+1$. **3c.** $n(n-1)$. **4g.** About 3.56×10^{14}. **5.** 72. **7.** 8. **9.** 6.

11. 360. **13.** 28,800. **14a.** $5!/2$. **15.** 60. **17-1.** 7. Set $F \leftarrow 1$. **17-2.** (Ignored.) **17-3.** Repeat step 4 for $k = 1, 2, 3, 4, 5$. **17-4.** Set $F \leftarrow Fk$. **17-5.** Output $F(=5!)$ and stop. **17-6.** End. **19.** Given n and r. **19-1.** Set $P \leftarrow 1$, $k \leftarrow 0$. **17-3.** Set $P \leftarrow P(n-k+1)$. **17-2.** Set $k \leftarrow k+1$. **17.4.** If $k = r$, output P and stop. Else. **17-5.** Return to step 2. **21.** There are m ways to choose the first coordinate, followed by n ways to choose the second coordinate for each ordered pair. Hence there are mn ways to form elements of $A \times B$.

Section 3.2

1a. 1. **1b.** 6. **1c.** 15. **1d.** 20. **1e.** 15. **1f.** 6. **1g.** 1. **2b.** 56.

3a. $\dfrac{n!}{0!(n-0)!} = 1$. **4.** $\dfrac{n!}{r!(n-r)!} = \dfrac{n!}{(n-r)!(n-(n-r))!}$. **3b.** $\dfrac{n!}{(n-1)!1} = n$.

5.

M	100	99	98	97	96
D	5	4	3	2	1
C	20	495	16,170	784,245	75,287,520

7. 35. **9.** $x^7y^0 + 7x^6y + 21x^5y^2 + 35x^4y^3 + 35x^3y^4 + 21x^2y^5 + 7xy^6 + y^7$. **11.** $a^4 + 4a^3b + 6a^2b^2 + 4ab^3 + b^4$. **13.** $x^6 + 6x^5(2y) + 15x^4(2y)^2 + 20x^3(2y)^3 + 15x^2(2y)^4 + 6x(2y)^5 + (2y)^6$.

15.

```
                1
              1   1
            1   2   1
          1   3   3   1
        1   4   6   4   1
      .....................
1  10  45  120  210  252  210  120  45  10  1
```

17. $C(3, 2) = C(2, 2) + C(2, 1) = 1 + 2 = 3$. $C(4, 3) = C(3, 3) + C(3, 2) = 1 + 3 = 4$. $C(4, 2) = C(3, 2) + C(3, 1) = 3 + 3 = 6$. $C(4, 3) = C(3, 3) + C(3, 2) = 1 + 3 = 4$. Hence $C(5, 3) = C(4, 3) + C(4, 2) = 6 + 4 = 10$. **19.** $0 = (1-1)^n = \sum_{k=0}^{n} (1)^{n-k}(-1)^k C(n, k)$. **21.** 66. **23.** 5148. **25.** $100!/((25!)^2(50!))$.

Section 3.3

1. 8. **3a.** 5040. **3b.** 120. **3c.** 90,770. **5.** 2,522,520. **7.** 91. **9.** 126. **11.** 28. **13.** For each r, $0 \leqslant r \leqslant k$, select r elements from the m set. Then select the remaining $k-r$ elements from the n set. This can be done in $C(m, r)C(n, k-r)$ ways for each r. **15-1.** If $n_k = 0$ or $n_k = n$ for any $k = 1, 2, \ldots, r$, then output 1 and stop. Else. **15-2.** Set $M \leftarrow n$, $k \leftarrow 1$, $D \leftarrow n_k$, and $C \leftarrow M/D$. **15-3.** If $D = 1$ and $k = r$, output C and stop. Else. **15-4.** If $D = 1$ and $k < r$, set $k \leftarrow k+1$, $D \leftarrow n_k + 1$. **15-5.** Set $M \leftarrow M-1$, $D \leftarrow D-1$, $C \leftarrow C \cdot (M/D)$. **15-6.** Return to step 3. **17.** $a^4 + b^4 + c^4 + 4a^3b + 4a^3c + 4b^3a + 4b^3c + 6a^2b^2 + 6a^2c^2 + 6b^2c^2 + 4a^2bc + 4b^2ac + 4c^2ab$. **19.** $123{,}552 = [C(13, 1)C(4, 2)C(12, 1)C(4, 2)C(44, 1)]/2$.

Section 3.4

1a. 9. **1b.** 1. **3.** b and d. **5.** 24. **6a.** 24. **7.** 48. **9.** $16^r - C(4, 1)\cdot 15^r + C(4, 2)\cdot 14^r - C(4, 3)\cdot 13^r + C(4, 4)\cdot 12^r$. **11.** 455. **13.** $C(4, 1)\cdot 3^{20} - C(4, 2)\cdot 2^{20} + + C(4, 3)\cdot 1^{20} - 0$.

15. $45!\left(\frac{1}{0!}-\frac{1}{1!}+\frac{1}{2!}-\cdots-\frac{1}{45!}\right)$.

17. $n^4-5n^3+8n^2-4n$. **19.** $\overline{(A\cup B)\cup C}=\overline{(A\cup B)}\cap\bar{C}=(\bar{A}\cap\bar{B})\cap\bar{C}$.

Section 3.5

1a.

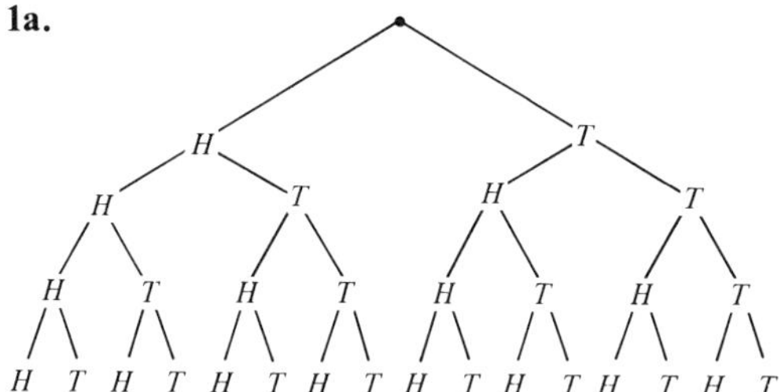

1b. The set of sequences of H's and T's of length 4. **1c.** $\{HHHT, HTHH, HHTH, THHH\}$. **1d.** 1/4. **2a.** 6/16. **3a.** 0. **3b.** 0. **3c.** 1/36. **3d.** 4/36. **3e.** 1/36. **5.** 2/106. **7a.** 1/8!. **7b.** 2/8!. **9a.** $1/(P(36, 6))$. **9b.** $1/(C(36, 6))$. **10b.** $C(13, 1)C(4, 3)C(12, 2)\cdot 4^2/C(52, 5)$. **11.** See Appendix program A11. **13.** $(366)(2)+1$. **15.** 101. **17.** No. For a counterexample, let the dots (people) be connected in the following diagram iff the people are friends.

19a. Where two people are in the set C. **19c.** As in Exercise 17,

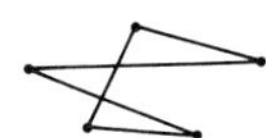

Section 3.6

1. 56. **3.** 100. **5.** 44. **7.** 15. **9.** $a_n=a_{n-1}+2a_{n-2}$, $a_0=a_1=1$; $a_{12}=2731$; $a_{24}=11{,}184{,}811$. **11.** $a_n=a_{n-1}+a_{n-2}$, $a_1=1$, $a_2=2$. **13.** $a_n=a_{n-1}+a_{n-2}$, $a_1=2$, $a_2=3$. **15.** $a_n=(1.01)a_{n-1}-600$, $a_0=50{,}000$. **17.** $a_r=2a_{r-1}+1$, $a_0=1$. **19.** In the first case there are d_{n-2} ways to permute the remaining $n-2$ integers. The first case occurs once for every $i=1, 2, \ldots, n-1$, giving $(n-1)d_{n-2}$ ways so far. In the second case first interchange i and n. Then permute only the integers in the first $n-1$ positions. This can be done $(n-1)d_{n-1}$ ways.

Section 3.7

1a. $a_n=n!$. **1b.** $a_n=3^{n-1}\cdot\frac{1}{2}$. **1c.** $a_n=2n+1$. **1d.** $a_{n,r}=C(n, r)$. **3.** $a_n=1^2+2^2+\cdots+n^2=n(n+1)(2n+1)/6$. **5.** $a_n=2+2+3+4+\cdots+n=1+[n(n+1)]/2$. **7.** $a_n=2^n-1$. **9.** 32,640; 6305. **11-1** to **5.** These are the same. **11-6.** Same as 3. **11-7.** Set $j\leftarrow j-1$. **11-8.** If $j=2$, return to step 6. **11-9.** Else output $r_1=\min$ and $r_n=\max$. **13a.** $C(r_1, r_2)C(r_3, r_4)C(r_1, r_3)C(r_2, r_4)$. **13c.** 3, 6, 7, 5 . . . 3, 6, 7, 5 . . . 3, 6, 5, 7 . . . 3, 6, 5, 7 . . . 3, 6, 5, 7 . . . (3, 7) = output. **14c.** $a_k=3a_{k-1}$, $a_0=1$. **15a.** $C(r_1, r_2)C(r_3, r_4)C(r_1, r_3)C(r_2, r_4)C(r_2, r_3)$. **15c.** $b_k=2b_{k-1}+a_{k-1}=2b_{k-1}+3^{k-1}$, $b_0=0$.

Section 3.8

1. $a_n = 2n - 3$. **2.** $a_n = (2/3)n - 1$. **3.** $a_n = \log_4 n + 1$. **4.** $An + B$. **5.** $0(n \log_4 n)$. **6.** $0(n \log_2 n)$. **7.** $0(n \log_4 n)$. **8.** Exercise (3) yields $5 = \log_4 256 + 1$ for both; Exercise (4) yields 341 for both; Exercise (5) yields 1284 versus $1280 = (256)\log_4(256)$. **9.** 49 comparisons. **11.** $[n(n-1)/2]/n^2$ gets close to 1/2 as n becomes very large. **12.** $a_n = a_{n/2} + 1$ and $a_1 = 0$ yields $a_n = \log_2 n$.

Section 4.1

1. *c, d, e, f,* and *h.* **2.** *a, b, d, e, f,* and *h.* **3.** *a, b,* and *f.* **4a.** *ACE, ACBDE, ACDE, ACBE, ACDBE, ABE, ABCE, ABDE, ABCDE,* and *ABDCE.* **5.** $\sum_{i=1}^{2m+1} (2n_i + 1) = 2(\sum_{i=1}^{2m+1} n_i) + 2m + 1 = 2(m + \sum_{i=1}^{2m+1} n_i) + 1$. **7.** There are $C(n, 2)$ pairs of vertices to join with an edge. **9.** Each vertex, except for v_1 and v_2, has as many entrances as exits. For v_1 there is one more exit; for v_2 there is one more entrance. **13.** The first vertex is chosen arbitrarily. Then permute the remaining vertices $(n-1)!$ ways. Divide by 2 since clockwise and counterclockwise circuits are equivalent. **15.** Replace every street by two edges, where edges connect intersections (points). Apply Euler's theorem. **17.** Let the square in the ith row and jth column be s_{ij}. Then $s_{15}s_{27}s_{48}s_{36}s_{57}s_{78}s_{86}s_{65}s_{84}s_{72}s_{51}s_{63}s_{42}s_{21}s_{13}s_{34}s_{26}s_{18}s_{37}s_{45}$ $s_{24}s_{12}s_{33}s_{41}s_{53}s_{61}s_{82}s_{74}s_{55}s_{67}s_{88}s_{76}s_{64}s_{83}s_{71}s_{52}s_{31}s_{43}s_{22}s_{14}s_{35}s_{16}s_{28}s_{47}s_{68}s_{56}s_{77}s_{85}s_{73}s_{81}$ $s_{62}s_{54}s_{75}s_{87}s_{66}s_{58}s_{46}s_{38}s_{17}s_{25}s_{44}s_{32}s_{11}s_{23}$ is a Hamiltonian path. **19.** Yes. Each room has an even number of doors and all rooms are connected. **20.** Yes. Traverse the fictitious edge first.

Section 4.2

1. *a, e, f.* **3.** *c, f.* **5.** *a, d, f.* **7.** *d, c, g.* **9.** 0, 7, 1, 3. **11a.** $m = n$; must alternate between disjoint sets. **11b.** m and n both even; either odd gives a vertex of odd degree. **13.** Each of the m vertices is attached to every one of the n vertices in the other. Apply the fundamental multiplication principle. **15a.** 13. **15b.** $n - 1$. **17.** c is the only graph with exactly six edges. **19.** e is connected while h is not. **21.** The following correspondence is an isomorphism: $(B_1, A_1), (B_2, A_3), (B_3, A_5), (B_4, A_7), (B_5, A_2), (B_6, A_4), (B_7, A_6)$. **23.** We can color the vertices as in the hint because there is a unique path in a tree from some vertex to all other vertices. Since no red vertex will be connected to a green vertex and vice versa, this partitions the vertices as required. **25.** Yes. M_1J_3, M_2J_2, M_3J_1, M_4J_5, M_5J_6, M_6J_4 gives such an assignment.

Section 4.3

1a. 2. **1b.** 1. **1c.** 4. **1d.** 3. **1e.** 6.

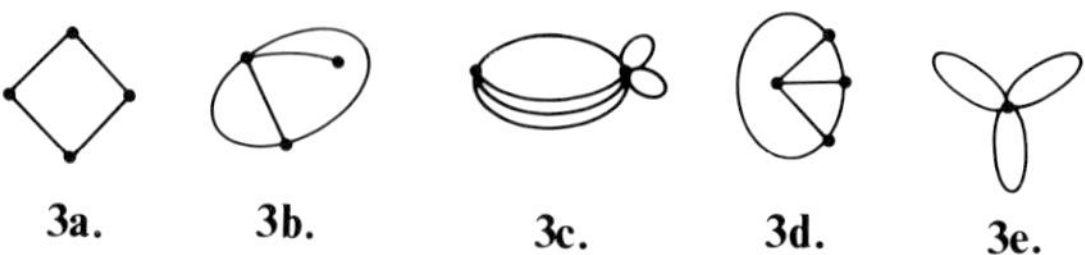

3a. **3b.** **3c.** **3d.** **3e.**

5. *a, c,* and *d.* **7a.** Yes, Neither K_5 or $K_{3,3}$ is a subgraph, or see (7b).

7b. Yes.

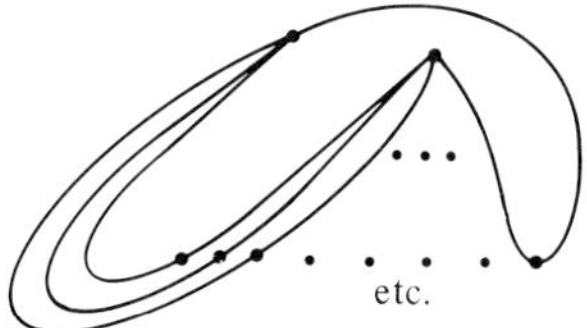

7c. No. $K_{3,3}$ is a subgraph.
9. See (7b).

11a.

11b.

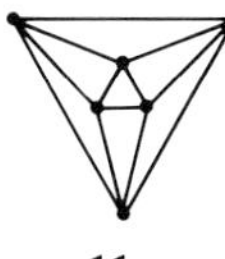

11c.

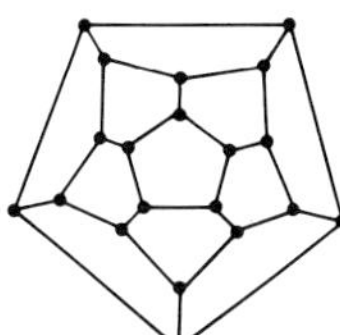

11d.

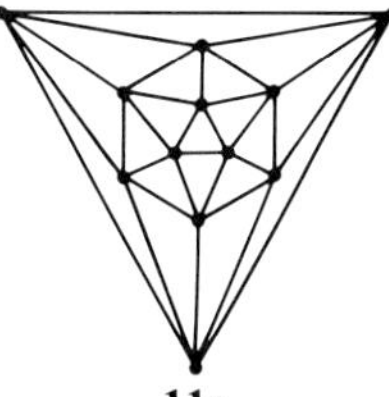

11e.

15. Suppose $3R \leqslant 2E$. But $V - E + R = 2$ implies $R = 2 - V + E$. Hence $3(2 - V + E) \leqslant 2E$ and so $E \leqslant 3V - 6$. **17.** Notice that K_5 is a subgraph, $E = 12$, and $V = 6$. Hence $12 = 3(6) - 6$. **19.** If the edge is a loop, there is one vertex and two regions. So $1 - 1 + 2 = 2$ is true. If it is not a loop, then there are two vertices and one region. So $2 - 1 + 1 = 2$ is again verified.

Section 4.4

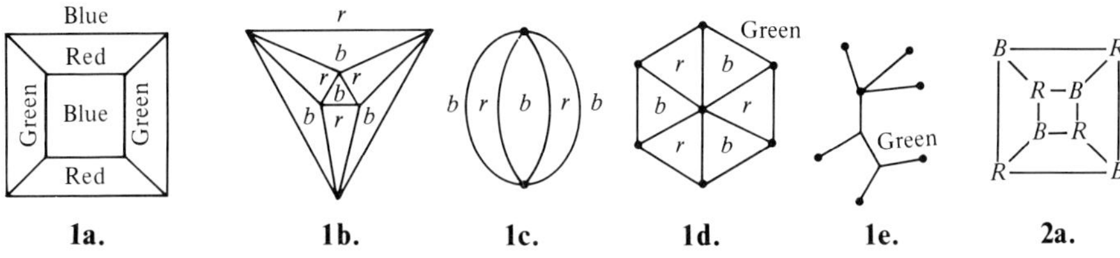

1a. **1b.** **1c.** **1d.** **1e.** **2a.**

3a. 100. **3b.** 2. **3c.** 2. **5.** Each vertex is adjacent to $n - 1$ others. Hence $\chi(K_n) \geqslant n$. But n colors suffice for any graph with n vertices. **7.** The maximum of m and n. **8.** Two-colorable if n is even. Otherwise three-colorable only. **11c.** List the vertices $v_1, v_2, v_3, v_4, v_5, v_6$. Then c_1 is assigned to v_1, v_2, and v_5; c_2 is assigned to v_3 and v_4; and c_3 is assigned to v_6. But the chromatic number of the graph is only 2. **12.** Label the intervals as they appear by rows a, b, c, d, e, f, g, h, and let each letter correspond to a vertex. Connect two vertices with an edge if the corresponding intervals overlap. We obtain the following graph for which $\chi(G) = 4$. Hence four registers suffice.

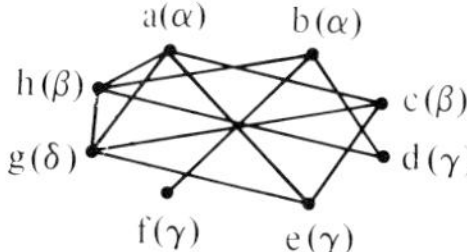

Section 4.5

1a. $\begin{bmatrix} 1 & 0 & 0 & 0 & 0 & 0 \\ 0 & 1 & 1 & 0 & 0 & 0 \\ 1 & 1 & 0 & 1 & 0 & 0 \\ 0 & 0 & 0 & 1 & 1 & 0 \\ 0 & 0 & 1 & 0 & 1 & 1 \end{bmatrix}.$ **1c.** $\begin{bmatrix} 1 & 3 \\ 2 & 3 \\ 2 & 5 \\ 3 & 4 \\ 4 & 5 \\ 5 & 5 \end{bmatrix}.$ **1c.** $\begin{bmatrix} 0 & 1 & 0 & 0 & 0 & 0 \\ 1 & 0 & 1 & 0 & 0 & 0 \\ 1 & 1 & 0 & 1 & 0 & 0 \\ 0 & 0 & 0 & 1 & 1 & 0 \\ 0 & 0 & 1 & 0 & 1 & 1 \end{bmatrix}.$

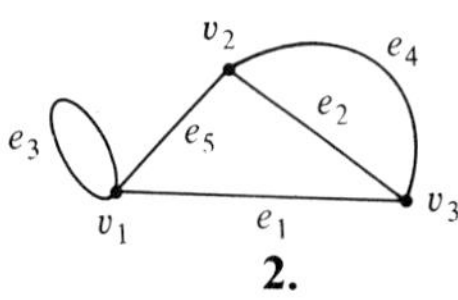

2.

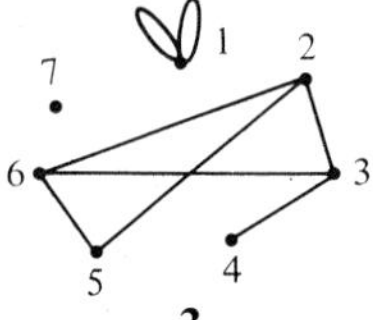

3.

5. Sum the number of times the vertex appears in the matrix.

7.

9. 2 by 2. **10a.** $\begin{bmatrix} 4 & 1 \\ 2 & 3 \end{bmatrix}.$ **11e.** $\begin{bmatrix} 4 & 1 \\ 2 & 3 \end{bmatrix}^t = \begin{bmatrix} 1 & 2 \\ 0 & -1 \end{bmatrix} + \begin{bmatrix} 3 & 0 \\ 1 & 4 \end{bmatrix} = \begin{bmatrix} 4 & 2 \\ 1 & 3 \end{bmatrix}.$

13. Interchange steps 1 and 2; interchange steps 4 and 5.

15.

	A	B	C	D
A	0	2	0	1
B	2	0	2	1
C	0	2	0	1
D	1	1	1	0

17a. Map a vertex corresponding to row i in the first matrix into the vertex corresponding to row i in the second matrix. Edge adjacencies are preserved because the matrices are equal.

17b.

$$\begin{array}{c|cc} & e_1 & e_2 \\ \hline v_1 & 1 & 1 \\ v_2 & 1 & 0 \\ v_3 & 0 & 1 \end{array} \neq \begin{array}{c|cc} & f_1 & f_2 \\ \hline u_1 & 1 & 0 \\ u_2 & 1 & 1 \\ u_3 & 0 & 1 \end{array}$$

but

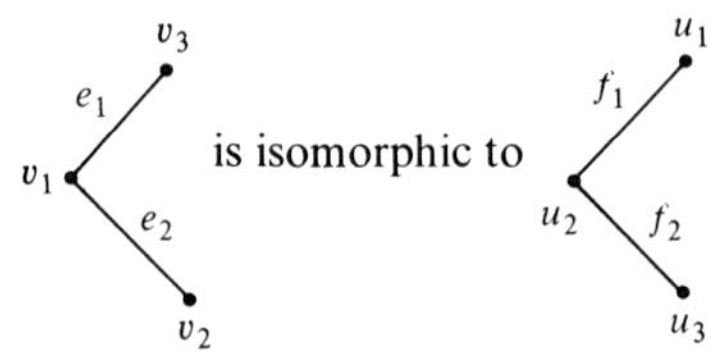

Section 4.6

1. $AB = \begin{bmatrix} -1 & 2 \\ 5 & 9 \end{bmatrix}; BA = \begin{bmatrix} 0 & 3 & 6 \\ 1 & 0 & 2 \\ 2 & 2 & 8 \end{bmatrix}.$

2. Neither AB nor BA exist. **3.** $AB = BA = A$.

4. $AB=[5]$;
$$BA=\begin{bmatrix}1&1&1&1&1\\1&1&1&1&1\\1&1&1&1&1\\1&1&1&1&1\\1&1&1&1&1\end{bmatrix}.$$

5. Given $i=2$ and $j=1$ in steps 1 and 2, we have the following sequence. **5-3.** Set $c_{12}\leftarrow 0$. **5-4.** Set $k\leftarrow 0+1=1$. **5-5.** Set $c_{21}\leftarrow c_{21}+a_{21}b_{11}=0+(-1)(2)=-2$. **5-4.** Set $k\leftarrow 1+1=2$. **5-5.** Set $c_{21}\leftarrow c_{21}+a_{22}b_{21}=-2+(0)(4)=-2$. **5-4.** Set $k\leftarrow 2+1=3$. **5-5.** Set $c_{21}\leftarrow c_{21}+a_{23}b_{31}=-2+(3)(1)=1$. (Since $k=3$, this is c_{21}.)
7a-1. Set $m\leftarrow 0$, $P\leftarrow I_{n,n}$. **7a-2.** If $m=k$, output $A^k=P$ and stop. Otherwise: **7a-3.** Set $P\leftarrow P\cdot A$, $m\leftarrow m+1$. **7a-4.** Return to step 2. **7a-5.** End.

8a.
$$A(BC)=\begin{bmatrix}2&-1\\0&1\end{bmatrix}\begin{bmatrix}6&9\\0&-2\end{bmatrix}=\begin{bmatrix}12&20\\0&-2\end{bmatrix}=\begin{bmatrix}4&4\\-2&0\end{bmatrix}\begin{bmatrix}0&1\\3&4\end{bmatrix}=(AB)C.$$

9a.
$$\begin{bmatrix}a&b\\c&d\end{bmatrix}\left(\begin{bmatrix}e&f\\g&h\end{bmatrix}\right)\begin{bmatrix}i&j\\k&m\end{bmatrix}=\begin{bmatrix}aei+afk+bgi+bhk & aej+afm+bgj+bhm\\cei+cfk+dgi+dhk & cej+cfm+dgj+dhm\end{bmatrix}$$
$$=\left(\begin{bmatrix}a&b\\c&d\end{bmatrix}\begin{bmatrix}e&f\\g&h\end{bmatrix}\right)\begin{bmatrix}i&j\\k&m\end{bmatrix}.$$

11. $A^0=I_{n,n}$. If A^r is n by n and A is n by n, then their product is n by n, since the first and last dimensions are both n. Hence A^k is n by n for all integers $k\geqslant 0$.

13. **13**
$$A_4^2=\begin{bmatrix}2&0&1&0&1&0\\0&2&0&1&0&1\\1&0&2&0&1&0\\0&1&0&2&0&1\\1&0&1&0&2&0\\0&1&0&1&0&2\end{bmatrix}$$
and the 1, 2 element is a 0. Hence there are 0 walks from v_1 to v_2 of length 2.

15.
$$(I+A_4)^5=\begin{bmatrix}51&46&35&30&35&46\\46&51&46&35&30&35\\35&46&51&46&35&30\\30&35&46&51&46&35\\35&30&35&46&51&46\\46&35&30&35&46&51\end{bmatrix}.$$
Hence the path matrix is 6 by 6 and all its entries are 1's.

17. For any other vertex v_i, v_i is adjacent to some v_j and hence is connected to itself by the path $v_iv_jv_i$ of length 2. If $n\geqslant 2$, the matrix I is unnecessary in this calculation of the path matrix.
19. Every walk containing more than $n-1$ edges contains some vertex more than once. Removing redundancies leaves us with $n-1$ or fewer vertices. **21.** The vertices that v_i can reach are precisely those that can reach v_j.

Section 5.1

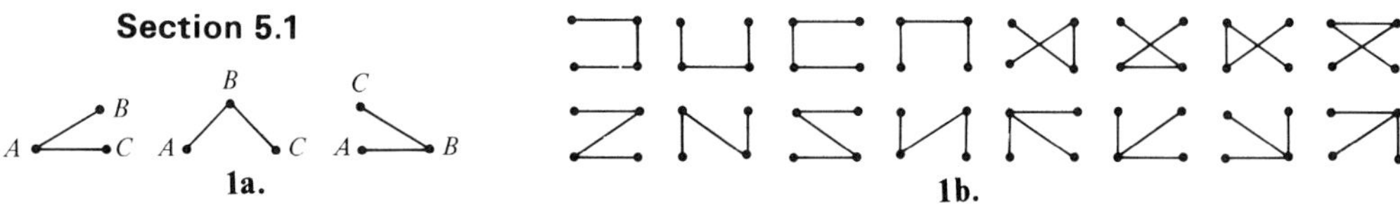

1a. **1b.**

3. Stop when you obtain $n-1$ edges. **5.** $n^{n-2}m^{m-2}$. **7.** Include edges
$\begin{bmatrix} \text{NY} & \text{L} & \text{Pa} & \text{Pe} & \text{NY} \\ \text{L} & \text{Pa} & \text{Pe} & \text{T} & \text{MC} \end{bmatrix}$ giving a total weight of 12,200.
9. Give every edge a weight of 1 and indicate the absence of an edge by M = twice the number of edges. **11.** Let a, b, c, and d be vertices of K_4, and given each edge a weight of 1. Order the edges ab, ac, cb, ... The algorithm produces a cycle. **13.** Edge set = $\{ae, af, df, bd, cd\}$ and weight = 29. **15.** Only K_1 and K_2. **17a.** 4, 4, 3. **17b.** 5, 5, 5. **17c.** 3, 2, 1. **17d.** 4, 2, 2. **18a.** $\{v_5v_2, v_1v_3, v_1v_4, v_4v_5\}$.

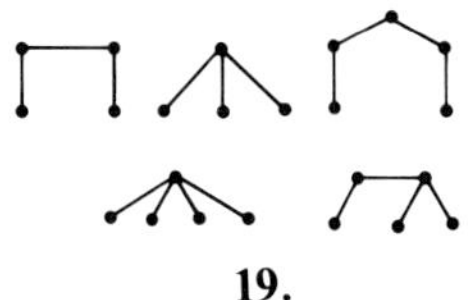

19.

Section 5.2

1a.

A	B	C	D	E	F	H	J	v_1
3	5	5	6	9	10	17	14	17

1b. v_0BDFJv_1.

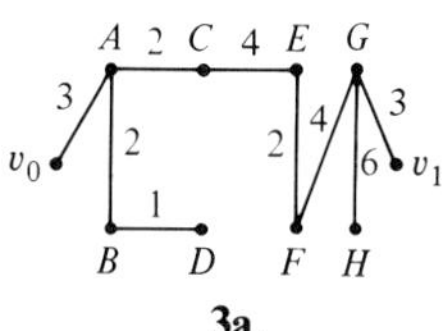

3a.

3b. No. The weight of this tree is 27. The weights of those in Exercises 1 and 2 were 30 and 31 respectively. **3c.** 18. **3d.** No. There is a path of length 18. **3e.** Not necessarily. The shortest path between H and v_1m, for example, is Hv_1, not HGv_1. **5.** The direct route to each city is shortest in every case. **7.** Let $w(AB)=3$, $w(BC)=2$, $w(CD)=2$, $w(DE)=2$, $w(EA)=3$. Then for this weighted graph AB, BC, AE, ED is a shortest A-path tree and it is a Hamiltonian path. But the Hamiltonian path $ABCDE$ is shorter. **9.** Any crossing of edges can be eliminated as in the illustrations. Since the sum of two sides of a triangle is greater than the third side, $AD+CB=AE+ED+CE+EB>AC+BD$. Therefore, substituting AC and BD for AD and CB decreases the total weight of the circuit.

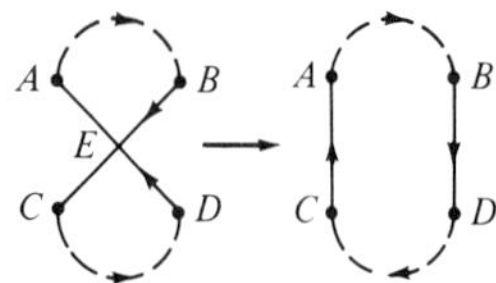

10. If $u_k \notin S$, then $w(u_kv)>0$. Hence the path $u_0v_1 \ldots u_k$ would be shorter, a contradiction. **11.** A shorter path to u_k would give us a shorter path to v, a contradiction. If some $u_i \notin S$, use Exercise (10).
13. $m(m-1)/2$ additions and $m(m-1)$ comparisons.

Section 5.3

1a. The root is 1. **1b.** The leaves are 4, 6, 7, 8, 9, and 10. **1c.** 4 and 5. **1d.** 2. **1e.** 4. **1f.** 4, 5, 9, 10.

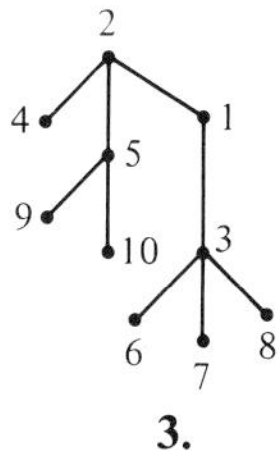

3.

5. No. T_1 has a node with two brothers and two sons on the third level, but T_2 has no such node. **7.** No: 3 has three sons. **9.** 1′, 2′, 4′, 5′, 3′, 6′, 7′, 8′, 9′, 10′.
10. 2.3.5.1, 2.3.5.2, 2.3.5.3, 2.3.5, 2.3.4, 2.3.3, 2.3.2, 2.3.1, 2.3, 2.2, 2.1, 2, 1, 0.

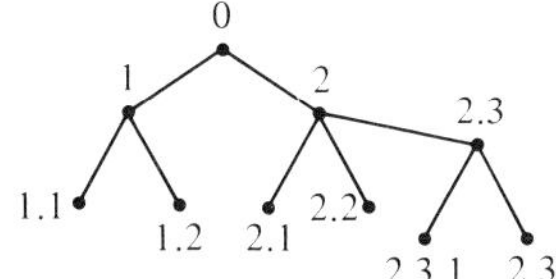

11. $a+[(b-c)\div(d*(e+f))] = +a\div -bc*d+ef.$

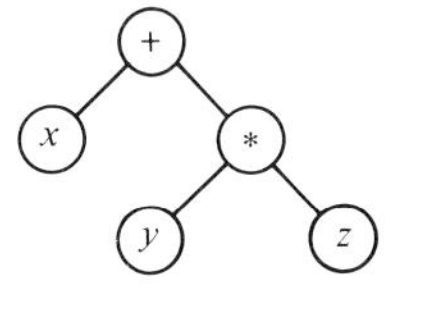

12b.

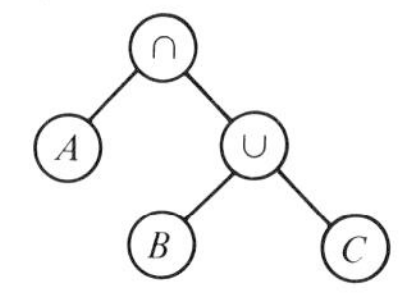

12d.

13b. $+x*yz.$ **13d.** $\cap A\cup BC.$

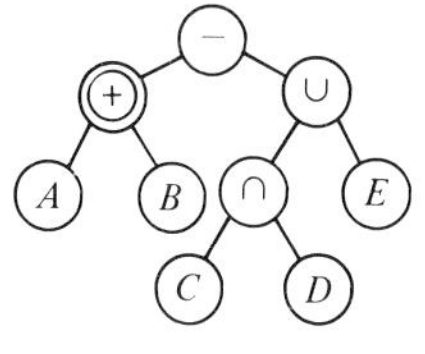

14c. $-\oplus AB\cup\cap CDE.$

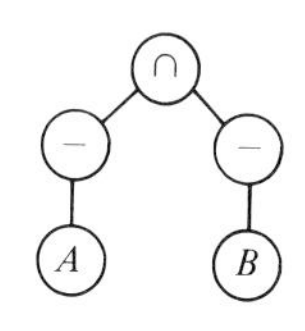

15a.

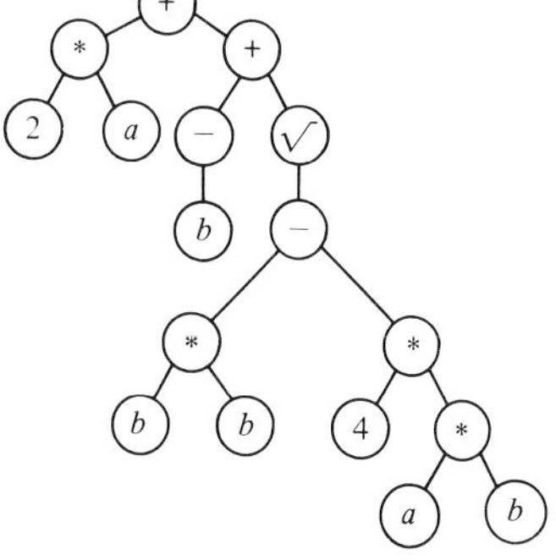

15c.

16c. $A*(B+C).$ **16e.** $(X*Y)+(Z*W).$ **17c.** $ABC+*.$ **17e.** $XY*ZW*+.$
19. 99.

Section 5.4

1a. 4. **1b.** 6. **1c.** 7. **1d.** 10. **3.** The worst case in the sequential search is lack of success after examining 1000 records. In the binary search the worst case is lack of success after examining 10 records.

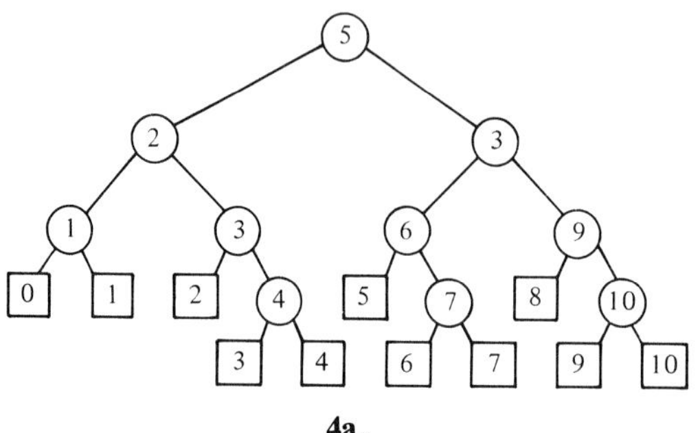

4a.

5a. 4. **5b.** 4. **5d.**

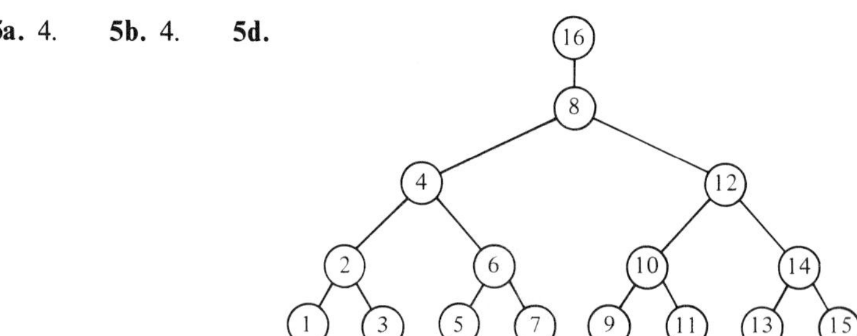

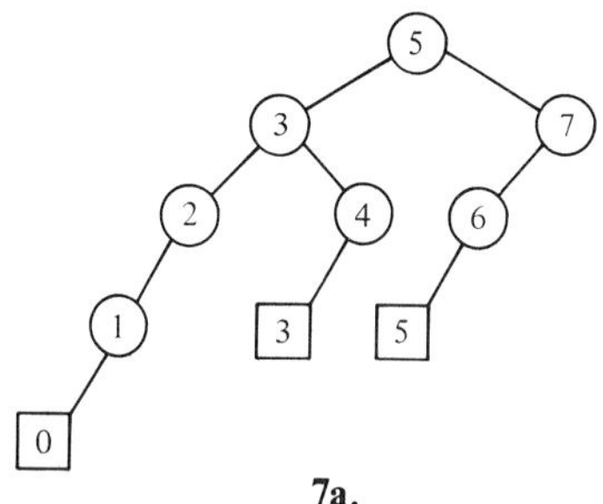

7a.

7b. $F_{r+1} = F_r + F_{r-1}$, $F_r = F_{r+1} - F_{r-1}$, and $F_{r-1} = F_{r+1} - F_r$. **7c.** $a_r = a_r + 1$, $a_2 = 1$. In closed form, $a_r = r - 1$.

Section 6.1

1. $e_1 = (v_1, v_2)$, $e_2 = (v_3, v_2)$, $e_3 = (v_4, v_3)$, $e_4 = (v_1, v_4)$, $e_5 = (v_3, v_1)$, $e_6 = (v_4, v_2)$. **2a.** None. **2b.** $\{e_1, e_5\}$. **3a.** None. **3b.** e_2. **4a.** Semiwalk. **4b.** Neither. **5a.** None. **5b.** None. **6.** Neither has a spanning circuit. **7.** D_1 has the spanning walk $v_1e_4v_4e_3v_3e_2v_2$ and it is, therefore, unilateral. D_2 is also unilateral since it has the spanning walk $v_1e_1v_2e_3v_3$. **8.** D_1 is weakly connected: $v_1e_4v_4e_3v_3e_2v_2$ is a spanning semiwalk. D_2 is weakly connected: $v_1e_1v_2e_3v_3$ is a spanning semiwalk.

9a. $\begin{bmatrix} 0 & 1 & 0 & 1 \\ 0 & 0 & 0 & 0 \\ 1 & 1 & 0 & 1 \\ 0 & 1 & 1 & 0 \end{bmatrix}$. **9b.** $\begin{bmatrix} 0 & 2 & 0 \\ 0 & 1 & 1 \\ 0 & 1 & 1 \end{bmatrix}$. **10a.** $\begin{bmatrix} 1 & 1 & 1 & 1 \\ 0 & 1 & 0 & 0 \\ 1 & 1 & 1 & 1 \\ 1 & 1 & 1 & 1 \end{bmatrix}$. **10b.** $\begin{bmatrix} 1 & 1 & 1 \\ 0 & 1 & 1 \\ 0 & 1 & 1 \end{bmatrix}$.

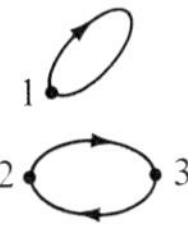

11a.

11b. $\begin{bmatrix} 1 & 0 & 0 \\ 0 & 1 & 1 \\ 0 & 1 & 1 \end{bmatrix}$

11c. None; the 1, 3 entry is zero. **11d.** No; there are zero entries.

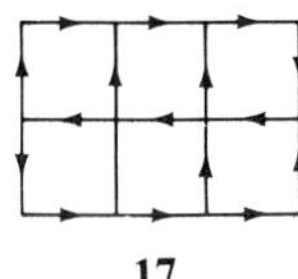

17.

18a. Conners 4, King 4. **18b.** *GCKBES.* **19.** *G* is not the best player in terms of the number of wins and yet *G* leads the walk. **21.** Given *D* with *n* vertices, by induction we may assume there is a walk $v_1 v_2 \ldots v_{n-1}$ on $n-1$ of its vertices not containing vertex *x*. If edges xv_1 or $v_{n-1}x$ are in *G*, add these to the beginning or end of the walk respectively. If not, show that there is an index *i* such that $v_i x$ and xv_{i+1} are in *G*. Substitute these for $v_i v_{i+1}$ in the walk.

Section 6.2

1a. All. **1b.** All. **1c.** None. **1d.** *CH* and *AD*. **2.** $n-1$. Every edge is a bridge since paths between vertices are unique. **3.** None. There are two paths between each pair of vertices. **5-1.** Find any edge *AB* in the graph and orient it from *A* to *B*. **5-2.** Find a cycle containing *AB* and orient it consistent with *AB*. Set $H \leftarrow$ this oriented subgraph of *G*. **5-3.** If $H = G$, output *H* and stop. Otherwise. **5-4.** Do steps 5–8 until $H = G$. **5-5.** Find an edge *CD* in *G* such that *C* is in *H* but *D* is not. **5-6.** Find a path beginning with *CD* that leads back to *H*. (Note: *CD* is on a cycle, by Exercise 4.) **5-7.** Direct this path beginning with *CD* until the first vertex it reaches in *H*. **5-8.** Set $H \leftarrow H \cup \{$all edges in the path we have just directed$\}$. **5-9.** Output *H* and stop. **5-10.** End of algorithm.

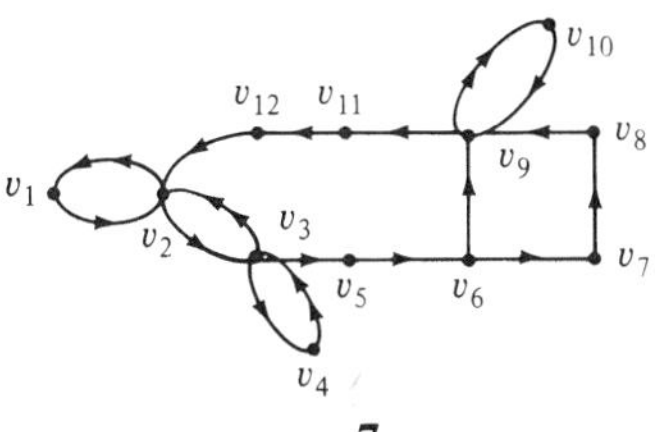

7.

9a. $v_0 = A$, $v_1 = B$, $v_2 = C$, $v_3 = D$, $v_4 = G$, $v_5 = F$, $v_6 = E$, CD is a bridge. **9b.** $v_0 = A$, $v_1 = B$, $v_2 = C$, $v_3 = D$, $v_4 = E$, $v_5 = F$, $v_6 = G$. Graph disconnected. **10.** *KJIFGHDECBA* is one topological sorting. **11.** Suppose every vertex has an outgoing edge. Start at any vertex a_1 and find as long a path as possible $a_1 a_2 \ldots\ldots a_k$. Since a_k has an outgoing edge $a_k b$, b must be one of the vertices in the path. This gives us a cycle. Contradiction! **15.** Let v_0, $v_1, \ldots, v_n$ be the *dfn*'s of the vertices in G. Then Exercises 13 and 14 imply that if we orient the edges in the *dfs* tree from lower to higher *dfn* and those not in the tree from higher to lower *dfn*, then from each vertex v_i $(i \neq 0)$ we can reach a vertex with lower *dfn*, and we can reach all the vertices from v_0. Hence, the resulting digraph is strongly connected.

16. Assume that the algorithm works for n vertices and that we have $n+1$ vertices. After we have found the first w and removed edges into w, what remains is one acyclic digraph on n vertices. Since the algorithm continues by sorting this, and, by induction, sorts it correctly, we are done, since giving w the largest label is consistent with the sorting of the remaining n vertices.

Section 6.3

1a. It has two sources. **1b.** It has two sinks. **1c.** It has a cycle. **1d.** It has a negative weight. **1e.** The same activities must have the same weights. **2a.** D = source, A = sink. **2b.** *DECBFA*. **2c.** *DE, EC, CB, BF, FA*. **2d.** *BA*. Slack equals 6. **3.** Increasing *XE* to 6 makes path *BPFMXE* only 35 units.

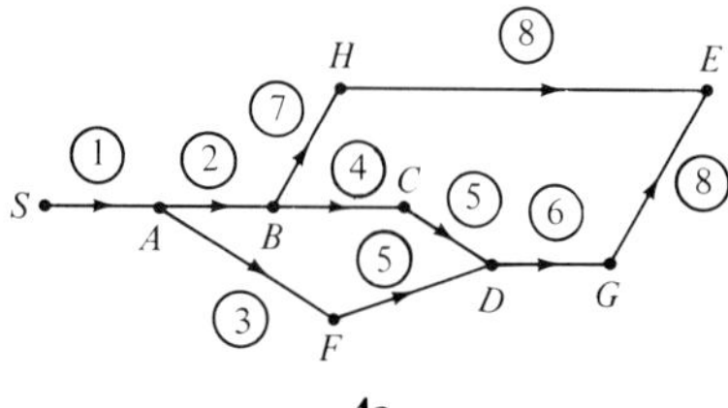

4a.

4b. 41 days: *SAFDGE*. **4c.** *SA, AF, FD, DE, GE*.

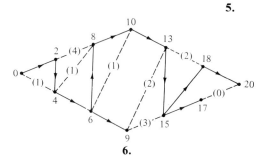

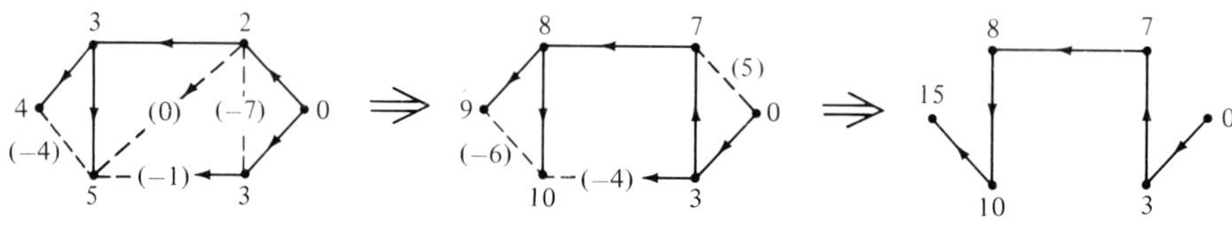

5.

6.

7. A cycle gives infinite length.
13. No. Let $M = 2$ with the following digraph.

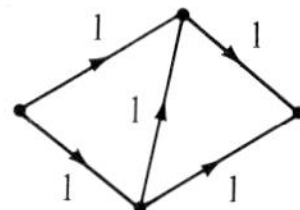

14. Exchanging BC for BD in the following digraph does not put BD on a critical path.

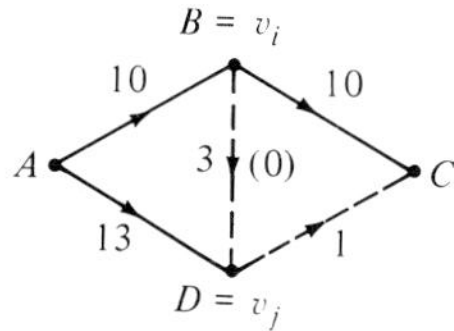

15. In this case the edge v_iv_j becomes part of a critical path, that is, a path having the same length as the old critical path.

Section 6.4

1a. 5. **1b.** 7. **1c.** 7.
2. Max flow = 5.

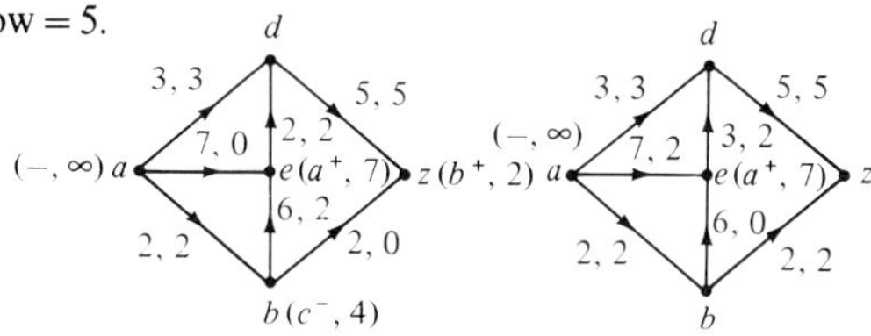

3.

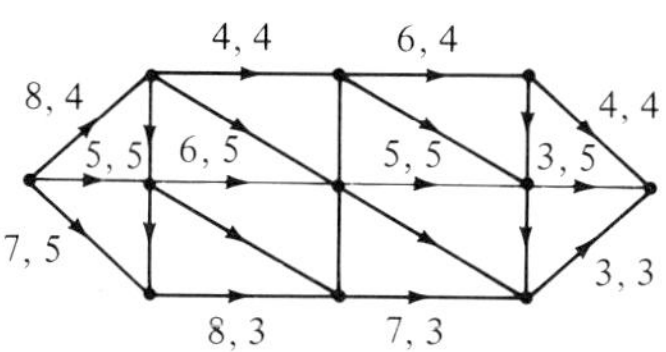

Note: Unlabeled edges have zero flow

4. Yes.

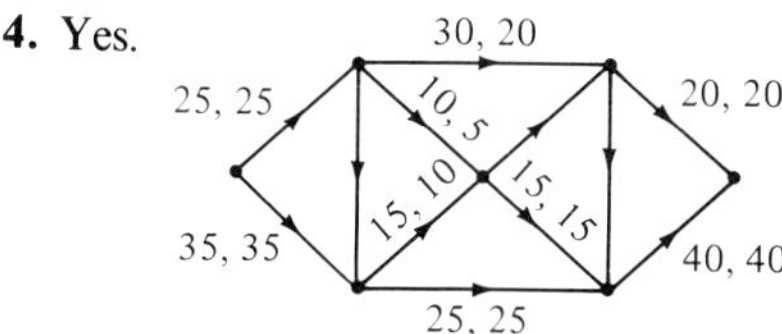

Note: Unlabeled edges have zero flow

5. No: max flow $= 100 < 110$.

6. Yes.

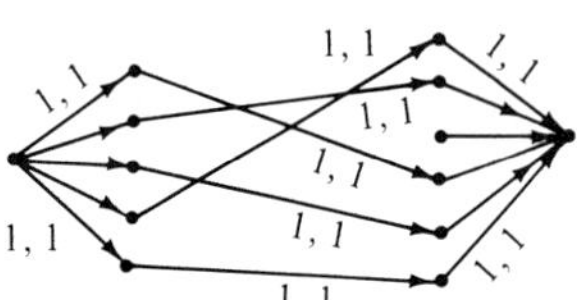

Zero flow in edges between boys and girls not shown.

7. Yes.

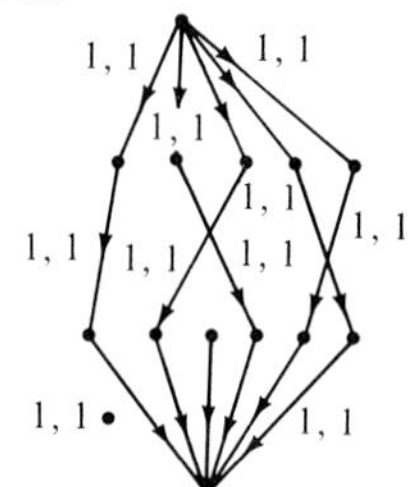

Zero flow in edges between men and jobs not shown.

8a. All edges between A's and T's have capacity $= 1$.

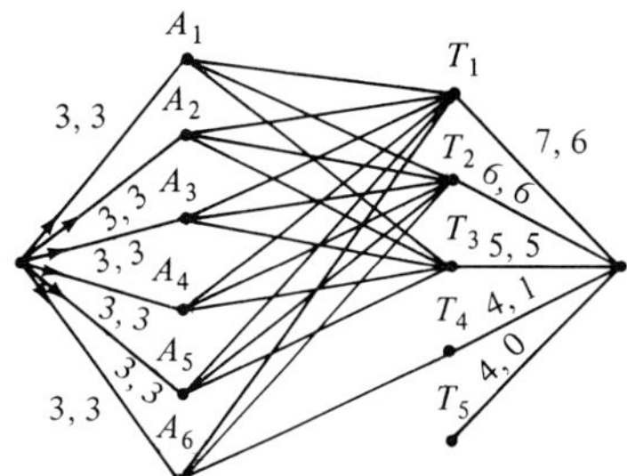

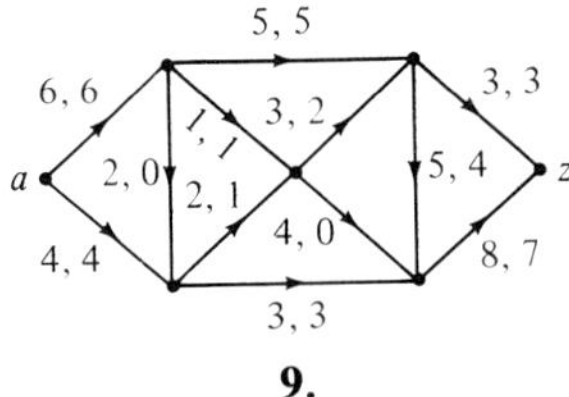

9.

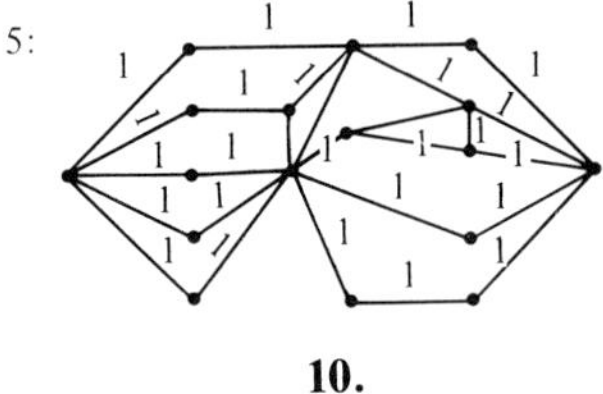

10.

11. Take $P = \{a, b, c\}$.

12. Given network

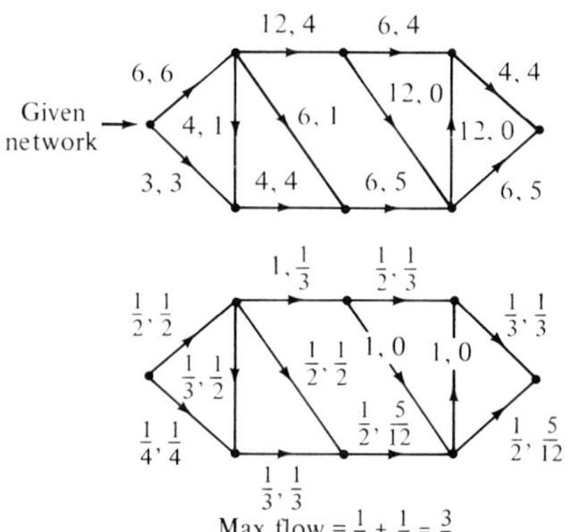

Max flow $= \frac{1}{2} + \frac{1}{4} = \frac{3}{4}$

13. $\sum_{\text{all } j} f(v_j a) = \sum_{\text{all } j} 0 = 0.$ **14.** For $p \neq a$, p in P, $\sum_{\text{all } i} f(p, v_i) - \sum_{\text{all } j} f(v_j, p) = 0$ and so all terms with $p \neq a$ cancel out. **15.** Break each sum in Exercise (14) into two parts. **16.** Both sums run through *all* vertices in P. **17.** The first and third sums cancel each other out. **18.** For the first, note that $\sum_{p \in P, r, \in \bar{P}} f(v_i, p) \geq 0$ in Exercise (17). For the second, $f(p, v_i) \leq c(p, v_i)$ for every p in P, v_i in $\bar{P}$. **19.** The second sum is zero since no edges are directed out of z.

Section 6.5

1. $J(A_1) = \{y_1, y_2, y_3, y_4\}$, $J(A_2) = Y$, and $J(A_3) = \{y_2, y_3, y_4\}$. **2.** $D(A_1) = -1$, $D(A_2) = -1$, and $D(A_3) = 0$. **3.** $D(G_1) = 1$ letting $A = \{x_1, x_3, x_4\}$; $D(G_2) = 0$ since $|A| \leq |J(A)|$ for all $A \subset X$ and $D(\varnothing) = 0$; $D(G_3) = 1$, letting $A = \{x_2, x_3, x_4, x_5\}$. **4.** Only G_2 by Theorem 6.9.
5. For G_1: $\{(x_1, y_4), (x_2, y_1), (x_4, y_2), (x_5, y_5)\}$.
For G_2: $\{(x_1, y_1), (x_2, y_4), (x_3, y_2), (x_4, y_3), (x_5, y_5)\}$.
For G_3: $\{(x_1, y_1), (x_2, y_2), (x_3, y_3), (x_4, y_4), (x_6, y_6)\}$.
6. {(cat, t), (ate, a), (he, e), (ho, o), (co, c)}.
7. The maximum must always be at least as big as $D(\varnothing) = 0$. **8.** Yes. The criteria of Theorem 6.10 are satisfied letting X = either the set of girls or the set of boys. Moreover, this shows that the number of boys equals the number of girls, so either matching is a one-to-one correspondence. **9.** Let $X = \{A_1, A_2, \ldots, A_n\}$ and let $Y = \cup_{1 \leq i \leq n} A_i$. Connect each A_i in X and each a_j in Y with an edge iff a_j is in A_i. Then apply Theorem 6.10. **10.** Yes. Connect each committee with a senator iff the senator is not a member of the committee. Then each committee is connected to eight senators and each senator to eight committees. A complete matching exists by Theorem 6.10.

Section 7.1

1.

{1, 2, 3, 4} = Domain.
{a, b, c} = Range.

$$\begin{array}{c} \\ 1 \\ 2 \\ 3 \\ 4 \end{array} \begin{array}{c} \begin{array}{ccc} a & b & c \end{array} \\ \begin{bmatrix} 1 & 0 & 0 \\ 0 & 1 & 0 \\ 1 & 0 & 1 \\ 0 & 0 & 0 \end{bmatrix} \end{array}.$$

3. 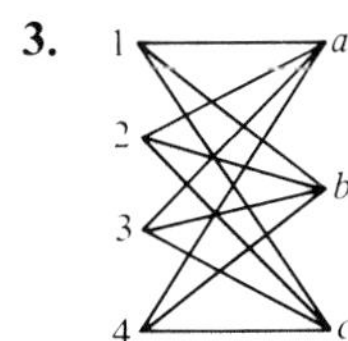

Domain = {1, 2, 3, 4}.
Range = {a, b, c}.

$$\begin{array}{c} \\ 1 \\ 2 \\ 3 \\ 4 \end{array} \begin{array}{c} \begin{array}{ccc} a & b & c \end{array} \\ \begin{bmatrix} 1 & 1 & 1 \\ 1 & 1 & 1 \\ 1 & 1 & 1 \\ 1 & 1 & 1 \end{bmatrix} \end{array}.$$

5. Domain = {a}.
Range = {a, b, c} = B.

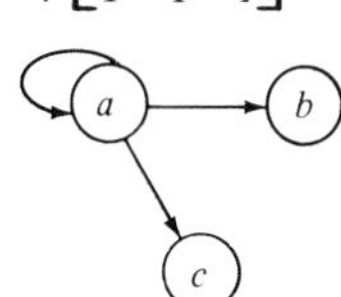

7. $\{(a, 1), (b, 2), (c, 3), (a, 3)\} = R^{-1}$. $\{(a, 2), (b, 2), (c, 2)\} = S^{-1}$.
$\{(a, 1), (a, 2), (a, 3), (a, 4), (b, 1), (b, 2), (b, 3), (b, 4), (c, 1)(c, 2), (c, 3), (c, 4)\} = T^{-1}$.

9.
$$\begin{array}{c} \\ 1 \\ 2 \\ 3 \\ 4 \end{array}\begin{array}{c} \begin{array}{ccc} a & b & c \end{array} \\ \begin{bmatrix} 1 & 0 & 1 \\ 1 & 1 & 0 \\ 0 & 1 & 1 \\ 1 & 1 & 0 \end{bmatrix} \end{array}.$$
$R = \{(1, a), (1, c), (2, a), (2, b), (3, b), (3, c), (4, a), (4, b)\}$. $1Rc$ is true. $2\not{R}c$ is true.

11.

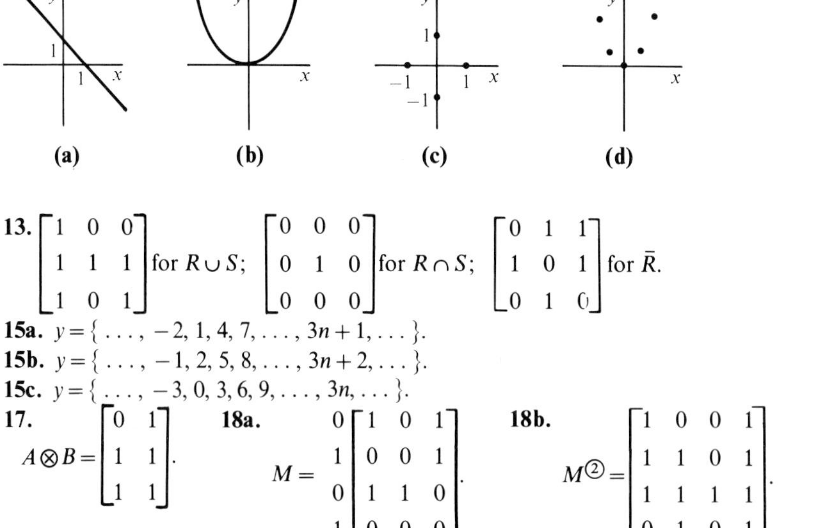

(a) (b) (c) (d)

13. $\begin{bmatrix} 1 & 0 & 0 \\ 1 & 1 & 1 \\ 1 & 0 & 1 \end{bmatrix}$ for $R \cup S$; $\begin{bmatrix} 0 & 0 & 0 \\ 0 & 1 & 0 \\ 0 & 0 & 0 \end{bmatrix}$ for $R \cap S$; $\begin{bmatrix} 0 & 1 & 1 \\ 1 & 0 & 1 \\ 0 & 1 & 0 \end{bmatrix}$ for $\bar{R}$.

15a. $y = \{\ldots, -2, 1, 4, 7, \ldots, 3n + 1, \ldots\}$.
15b. $y = \{\ldots, -1, 2, 5, 8, \ldots, 3n + 2, \ldots\}$.
15c. $y = \{\ldots, -3, 0, 3, 6, 9, \ldots, 3n, \ldots\}$.

17. $A \otimes B = \begin{bmatrix} 0 & 1 \\ 1 & 1 \\ 1 & 1 \end{bmatrix}$.

18a. $M = \begin{array}{c} 0 \\ 1 \\ 0 \\ 1 \end{array}\begin{bmatrix} 1 & 0 & 1 \\ 0 & 0 & 1 \\ 1 & 1 & 0 \\ 0 & 0 & 0 \end{bmatrix}$.

18b. $M^{②} = \begin{bmatrix} 1 & 0 & 0 & 1 \\ 1 & 1 & 0 & 1 \\ 1 & 1 & 1 & 1 \\ 0 & 1 & 0 & 1 \end{bmatrix}$.

19. There is a 1 in the ij position of $I \oplus A \oplus \cdots \oplus A^{(n-1)}$ iff there is a sequence of $n - 1$ or fewer directed edges from v_i to v_j.

Section 7.2

1a. T, U. **1b.** S, U. **1c.** R, T, U. **2a.** Transitive. **2b.** All three. **2f.** Transitive. **3a.** Symmetric and transitive. **3b.** Symmetric. **3c.** All three. **5.** U has only one equivalence class, A. In (2b), each class is determined by three positive numbers, the angles, whose sum is 180 degrees. In (2e), $[n]$ is the class of sets with n elements. And in (3c), the classes are $\{1, 2\}$ and $\{3, 4, 5\}$. **6a.** $[17] = [2] = \{\ldots, -4, -1, 2, 5, \ldots\}$.
7a. $x - x = 0.n$; $x - y = qn$ implies $y - x = (-q)n$; $x - y = qn$ and $y - z = pn$ imply that $x - z = (q + p)n$. **7b.** $[0], [1], [2], \ldots, [n - 1]$. **9.** If $m_{ij} = 1$, then $m_{ji} = 1$. Transitivity implies $m_{ii} = 1$. **10.** Assume $R^n = R^{n-1} \circ R$; use induction on n. **11a.** (a, b) and (b, c) in $\cap R_i$ imply (a, b) and (b, c) in R_i for every i. Transitivity implies that (a, c) is in R_i for all i. So (a, c) is in $\cap R_i$. **11b.** Clearly $R^\infty \supset \cap R_i$, since R^∞ contains each R_i as a subset. Use Exercise (10) to show that R^∞ is a subset of $\cap R_i$. **12a.** $R \cup \{(1, 3)\}$. **12b.** A 4-by-4 matrix with all 1's. **12c.** All cities that are able to communicate directly. **12d.** The ancestor relation.

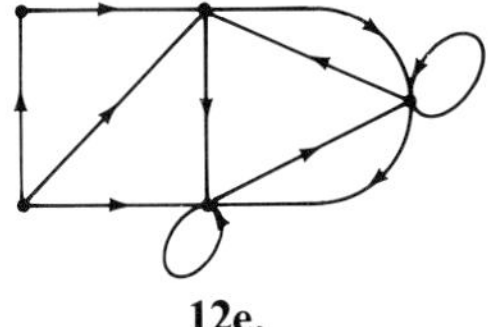

12e.

14a. We will have checked directly that $m_{ik}=1$ and $m_{ij}=1$ imply that $m_{kj}=1$ for $k<j$. In a symmetric matrix $m_{kj}=m_{jk}$. So both must be 1. **14c.** The modified algorithm would output "transitive" for this intransitive relation. **15.** aRb and aRa imply bRa, giving us symmetry. The given condition and symmetry imply transitivity. This gives an equivalence relation. Conversely, in an equivalence relation, aRb and aRc yield bRa and aRc by symmetry. Thus bRc, by transitivity. **16a.** For each of the n^2 entries in the matrix, one can choose either a 1 or a 0.

Section 7.3

1. R_1 and R_2 are antisymmetric; R_3 is a partial ordering.

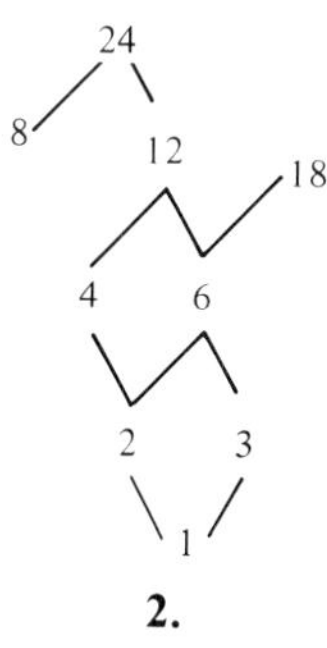

2.

3a. 4, 6. **3b.** 6. **3c.** 8, 12. **4.** 18 and 24 are maximal, 1 is minimal. **5a.** i, h. **5b.** $\{a, e, c\}$. **5c.** $\{c, b, k, j, i\}$. **5d.** $\{a, d, b, e\}$. **5e.** $\{a, k, b, c, d, e, f, g\}$. **6.** 1 is minimal; no element is maximal. **7.** Minimal elements are $\{1\}$, $\{2\}$, $\{3\}$, and $\{4\}$; maximal elements are $\{1, 2, 3\}$, $\{1, 2, 4\}$, $\{1, 3, 4\}$, and $\{2, 3, 4\}$. **8a.** $\{\{1\}, \{2\}, \{3\}, \{4\}\}$.
10. $37=36+1$. Apply Theorem 7.6. **11.** A chain of size $3=2+1$ corresponds to the first case and an antichain of size 3 corresponds to the second; $5=2\cdot 2+1$.
13-1. While $i\leftarrow 1, 2, \ldots, n$. **13-2.** While $j\leftarrow 1, 2, \ldots, n.\ j\neq i$. **13-3.** If $m_{ij}=m_{ji}=1$, output "not antisymmetric" and stop. **13-4.** Otherwise output "antisymmetric" and stop.
15. There is a path from vertex i to itself for each i. So $a_{ii}=1$. No cycles imply antisymmetry. And a path from i to j ($a_{ij}=1$) followed by a path from j to k ($a_{jk}=1$) yields a path from i to k ($a_{ik}=1$). **17.** Given cycle $a \ldots ca$, then aRc by transitivity. But cRa implies $c=a$.

Section 7.4

1a. 36 **1b.** 36. **1c.** 6. **1d.** 1. **2a.** $x\wedge y=\text{lcm}(x, y)$; $x\vee y=gcd(x, y)$. **2d.** 1 is the lower bound; there is no upper bound. **3.** Given $(A, \preccurlyeq)$, let $x\wedge=z_1$ and let $x\wedge y=z_2$. Then $z_1\preccurlyeq z_2$ *and* $z_2\preccurlyeq z_1$. By antisymmetry, $z_1=z_2$. **4a.** $\{1, 2, 3, 4\}$.

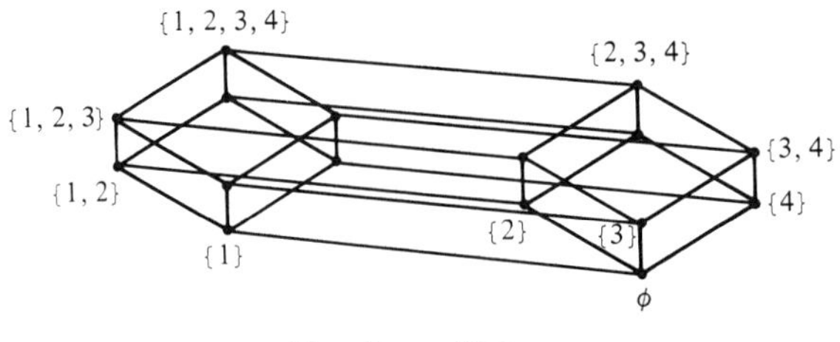

(The other vertices are similarly labeled.)

4c.

5. If O_1 and O_2 are both lower bounds, then $O_1 \preccurlyeq O_2$ and $O_2 \preccurlyeq O_1$ imply $O_1 = O_2$. Similarly for upper bounds. **7.** $a \vee (b \wedge c) = a \vee O = a$, but $(a \vee b) \wedge (a \vee c) = I \wedge I = I$, and $a \neq I$. **9a.** $x \wedge y = \min(x, y)$ and $x \vee y = \max(x, y)$. **9b.** "O" $= 1$ and "I" $= 10$. **9c.** $z \wedge x = 1$ only if $x = 1$. But $z \vee x = 10$ only if $y = 10$. So $x \neq y$. **11a.** $x \wedge I \preccurlyeq x$, and if $z \preccurlyeq x \wedge I$, then $z \preccurlyeq x$ (and $z \preccurlyeq I$). **13.** Suppose that $a \preccurlyeq b \preccurlyeq c$. Then $a \wedge (b \vee c) = \min(a, \max(b, c)) = a = \max(a, a) = \max(\min(a, b), \min(a, c)) = (a \wedge b) \vee (a \wedge c)$. The other cases are similar. **15.** Suppose that $p \preccurlyeq q$. Then, by Exercise (10), $p = p \wedge q$. But $p \to q = \neg p \vee q = ((\neg(p \wedge q)) \vee q = \neg p \vee (\neg q \vee q) = \neg p \vee T = T$. Conversely, if $p \to q = T$, then $p \wedge q = (p \wedge q) \vee F = (p \wedge q) \vee (p \wedge \neg q) = p \wedge (q \vee \neg q) = p \wedge T = p$. Hence $p \preccurlyeq q$.

Section 7.5

1a. Yes: isomorphic to $(P(\{1, 2\}), \subset)$. **1b.** No: $-5 \neq 2^n$. **1c.** No: c and d have no complements. **1d, e.** Yes. see part (a). **3.** The correspondence $\{3\} \leftarrow 3$, $\{5\} \leftarrow 5$, $\{7\} \leftarrow 7$, $\{2\} \leftarrow 2$ gives an isomorphism of the corresponding lattices. **5.** b, d, and e. **6a.** $(x' + y') = (x' + y + z')(y + z')$. **7-1a.** $a = a \cdot 1 = a \cdot (a + a') = (a \cdot a) + (a \cdot a') = (a \cdot a) + 0 = a \cdot a$. **9.** The case $n = 2$ holds by Theorem 7.9. Suppose true for some $k > 1$. Then $((x_1 + \cdots + x_k) + x_{k+1})' = (x_1 + \cdots + x_k)' x'_{k+1} = (x'_1 \cdots x'_k) x'_{k+1}$. The other law is proved similarly. **11.** $i \leftarrow \{i\}$, i a prime; $i \cdot j \leftarrow \{i, j\}$, i and j distinct primes; $i \cdot j \cdot k \leftarrow \{i, j, k\}$, where i, j, and k are distinct primes; $2 \cdot 3 \cdot 5 \cdot 7 \leftarrow \{2, 3, 5, 7\}$; and $1 \leftarrow \varnothing$. **13.** Only e.
14a. $xyz + xyz'$. **14e.** $xy'z' + xyz' + x'yz' + x'yz + x'y'z + xy'z$.
15a. $(A \cap B \cap C) \cup (A \cap B \cap \bar{C}) \cup (A \cap \bar{B} \cap C) \cup (\bar{A} \cap \bar{B} \cap C)$.
15b. $(p \wedge q \wedge r) \vee (p \wedge q \wedge \neg r) \vee (p \wedge \neg q \wedge r) \vee (\neg p \wedge \neg q \wedge r)$.
15c. $(p \wedge \neg q \wedge r) \vee (p \wedge \neg q \wedge \neg r) \vee (p \wedge q \wedge r) \vee (\neg p \wedge q \wedge r)$. **17a.** $\{3\}, \{5\}, \{7\}$.
17b. 3, 5, 7. **17c.** $\{2\}, \{3\}, \{5\}, \{7\}$. **17d.** 2, 3, 5, 7.

Section 7.6

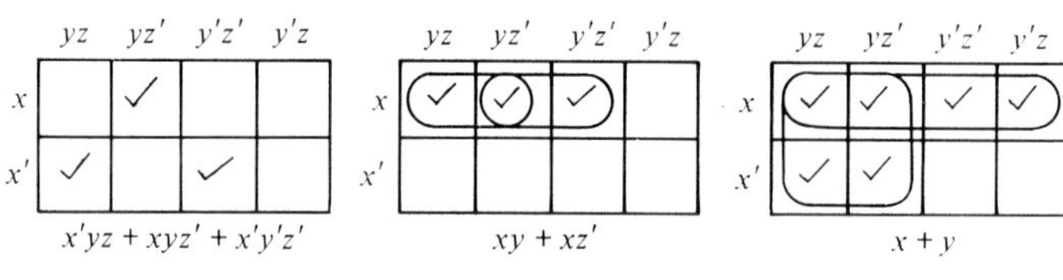

1. **3.** **4.**

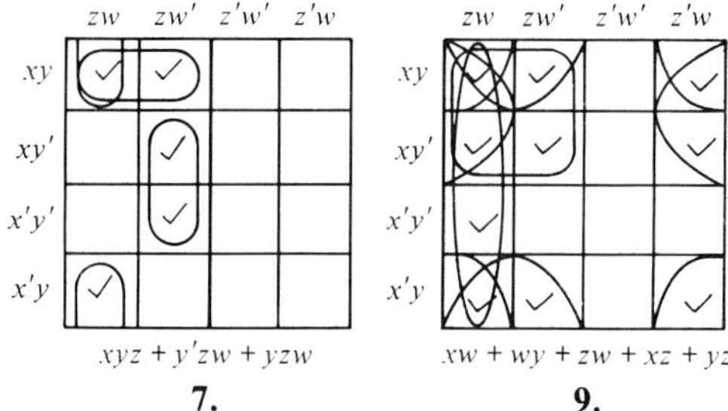

$xyz + y'zw + yzw$

7.

$xw + wy + zw + xz + yz$

9.

11. Minimal as is:

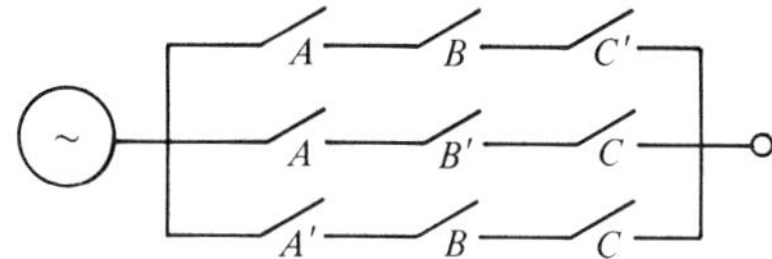

A	B	C	Answer
1	1	1	0
1	1	0	0
1	0	1	0
1	0	0	0
0	1	1	1
0	1	0	0
0	0	1	0
0	0	0	0

13.

Note:
on ↔ 1
off ↔ 0

A	B	C	Answer
1	1	1	0
1	1	0	1
1	0	1	0
1	0	0	0
0	1	1	0
0	1	0	0
0	0	1	1
0	0	0	1

15.

17. No. 13 is $A' \wedge B \wedge C$:

No. 15 is $(A \wedge B \wedge C') \vee (A' \wedge B')$:

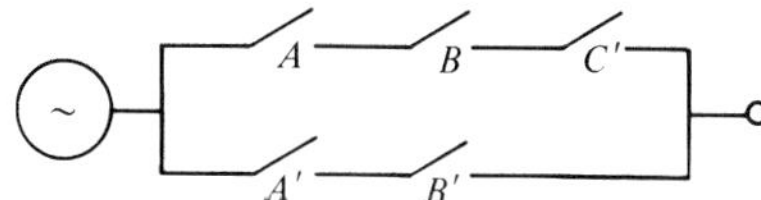

Section 7.7

1. In pairs, 0–0, 1–3, 2–2.

2.

*	0000	1110	0011	1101
0000	0000	1110	0011	1101
1110	1110	0000	1101	0011
0011	0011	1101	0000	1110
1101	1101	0011	1110	0000

3.

$*$	I	R_1	R_2	r_1	r_2	r_3
I	I	R_1	R_2	r_1	r_2	r_3
R_1	R_1	R_2	I	r_2	r_3	r_1
R_2	R_2	I	R_1	r_3	r_1	r_2
r_1	r_1	r_3	r_2	I	R_2	R_1
r_2	r_2	r_1	r_3	R_1	I	R_2
r_3	r_3	r_2	r_1	R_2	R_1	I

5. None of them. **7.** Semigroup and monoid. **9.** None of them. **11.** If G is a group of prime order p, and H is a subgroup of G having n elements, then n divides p. Hence $n = p$ or $n = 1$. **13.** $\{\ldots, -3, 0, 3, 6, \ldots\}$; $\{\ldots, -2, 1, 4, 7, \ldots\}$; $\{\ldots, -1, 2, 5, \ldots\}$. **15.** S_3, $\{I\}$, $\{I, r_1\}$, $\{I, r_2\}$, $\{I, r_3\}$, $\{I, R_1, R_2\}$. **17.** $\{I, r_1\} \cup \{I, r_2\}$ is not a subgroup of S_3. **19.** $f = ff$. Hence $e = ff^{-1} = (ff)f^{-1} = f$.

Section 7.8

1a. $\{1000, 0100, 0010, 0001\}$. **1b.** $\{1100, 1010, 1001, 0110, 0101, 0011\}$. **1c.** $\{1110, 1101, 1011, 0111\}$. **1d.** $\{1111\}$. **3.** $\{1111, 1010, 0000, 0101\}$. **5.** Two, the weight of 0011. **7.** Zero, because one needs a minimum distance of 3 to correct one error. **9a.** $\{0000, 0011, 1101, 1110\}$; 0000. **9b.** $\{1000, 1011, 0110, 0101\}$; 1000. **9c.** $\{1100, 1111, 0001, 0010\}$; 0001 or 0010. **11a.** $0000 * 1101 = 1101$. **11b.** $1000 * 0101 = 1101$. **11c.** $0001 * 1100 = 1101$ or $0010 * 1100 = 1110$.

Section 8.1

1. a and d are linear. **2.** a, b, d, e, and f are linear with constant coefficients; a is homogeneous. **3a.** $x - 1 = 0$; $a_n = A \cdot (1)^n$. **3b.** $x^2 - 3x + 2 = 0$; $a_n = A(1)^n + B(2)^n$. **3c.** $x^4 - 8x^2 + 16 = 0$; $a_n = (A_1 + nA_2)(2)^n + (B_1 + nB_2)(-2)^n$. **4a.** $a_n + 5a_{n-1} + 6a_{n-2} = 0$. **5a.** $a_n = 2$ for all n. **5b.** $a_n = 0(1)^n + 1(2)^n = 2^n$. **5c.** $a_n = ((5/4) - (n/4))2^n + [(-11/20) - (n/20))(-2)^n]$. **7.** $((1 \pm \sqrt{5})/2)^1 \neq 1$. **9a.** -1, -2, and 3. **9b.** $2/3$, $-2/3$. **9c.** $-1/2$, $i/4$, $-i/4$. **11.** $a_n = (1/2)(-1)^n + (1/5)(-2)^n + (3/10)(3)^n$. **13.** $((1 - \sqrt{5})/2)^n$ approaches 0 as n increases without bound.

Section 8.2

1a. $a_n - a_{n-2} = 0$. **1b.** $a_n + 2a_{n-1} - 3a_{n-2} = 0$. **1c.** $a_n - 8a_{n-2} + 16a_{n-4} = 0$. **2a.** $a_n^{(p)} = (1/2)^{n2} + (1/2)n$. **3a.** $a_n = A(1)^n + (1/2)n^2 + (1/2)^n$. **3b.** $a_n = A(1)^n + B(-1)^n + (4/5)2^n$. **3c.** $a_n = (An + B)2^n + (Cn + D)(-2)^n + ((-1/9)n + (16/27))$. **4a.** $a_n = 2 + (1/2)n^2 + (1/2)n$. **5.** $a_n = (P/r)(1 + r)^{n+1} - P/r = [P - P(1 + r)^{n+1}]/[1 - (1 + r)]$. **7.** $a_n = n^2 - n + 2$. **9.** $b_n = 3^n - 2^n$. **11.** $b_n = 2^{n+1}$.

Section 8.3

1. $1 - x + x^2 - x^3 + x^4$. **3.** $1 + 2x + 4x^2 + 8x^3 + 16x^4$. **5.** $1 + x/2 + x^2/4 + x^3/8 + x^4/16$. **7.** $1 + 3x + 9x^2 + 27x^3 + 81x^4$. **9.** $1 - x - 2x^2 - x^3 + x^4$. **11.** $1 + rx + r^2x^2 + r^3x^3 + r^4x^4$. **13.** $1 + 4x + 16x^2 + 64x^3 + 256x^4 + = \sum_{r=0}^{\infty} (4x)^r$. **15.** $3 + 6x + 12x^2 + 24x^3 + 48x^4 + \cdots = \sum_{r=0}^{\infty} (3)(2x)^r$. **17.** $x + x^2 + x^3 + x^4 + \cdots = \sum_{r=1}^{\infty} x^r$. **19.** $\sum_{r=0}^{\infty} ((2)2^r - 1)x^r$.

21. $A(x)=\sum_{r=0}^{\infty}(4x)^r$.; $a_r=4^r$. **23.** $A(x)=[-x-1]/[1-4x^2]=(-3/4)/(1-2x)+(-1/4)/(1+2x)$; $a_r=(-3/4)2^r+(-1/4)(-2)^r$. **25.** $A(x)=(6-5x)/[(x-3)(x-2)]=(1/3)/(1-(x/3))+(-2)/(1-(x/2))$; $a_r=(1/3)(1/3)^r+(-2)(1/2)^r$. **27.** $\sum_{r=0}^{\infty}C(3+r-1,r)x^r=1+3x+6x^2++\cdots$. **29.** $C(13,10)=286$. **31.** $a_r=1+C(r+2,r-1)$. **33.** $A(x)=2x^2/(1-x)^3+2/(1-x)$; $a_n=n(n-1)+2$, $n>0$. **35.** $\sum_{r=0}^{\infty}C(1/2,r)(2x)^r=1+2x/2-(2x)^2/4+(3/8)(2x)^3/3!+\cdots$. **37.** $1+x^2+x^4+x^6+\cdots$.

Section 8.4

1. $(1+x+x^2+x^3+x^4)(1+x+x^2+x^3)(1+x+x^2+x^3+x^4+x^5)$. **2.** $(1+x+x^2+x^3+\cdots)^3=1/(1-x)^3$. **3.** 14. **4.** The coefficient of x^6 in $1/(1-x)^3$ is $C(6+3-1,6)=28$. **5.** $(1+x+x^2+x^3+\cdots)^4=1/(1-x)^4$. **7.** The coefficient of x^5 in $1/(1-x)^3$ is $C(5+3-1,5)=21$. **9.** $(x+x^3+x^5+x^7+\cdots)^4$. **11a.** $(1+x)^9(1+x+x^5+x^{10})$. **11b.** The coefficient of x^{15} in the product is $C(9,5)=126$. **13.** The coefficient of x^{30} in $(1+x)^{50}$, or $C(50,30)$. **15.** Let $m=n$ in Example 7. **17.** $1-x^2/2!+x^4/4!-x^6/6!\pm\cdots=\frac{1}{2}(e^x+e^{-x})$. **19.** $1+1x/1!+x^2/2!+\cdots+x^n/n!$. **21.** $(1+x+x^2/2)(1+x+x^2/2!+x^3/3!)$. **23a.** $(1+x+x^2/2!+x^3/3!+\cdots)(x+x^2/2!+x^3/3!+\cdots)^3=e^x(e^x-1)^3$. 23 **23b.** $r^r-3\cdot3^r+3\cdot2^r-1$. **25a.** $(x+x^2/2!+x^3/3!+\cdots)^n=(e^x-1)^n$. **25c.** $S(n,r)$. **25d.** n^r.

INDEX